Louis Pasteur

LOUIS PASTEUR

Patrice Debré

Translated by
Elborg Forster

THE JOHNS HOPKINS UNIVERSITY PRESS
Baltimore & London

Louis Pasteur, © Flammarion, 1994
English translation © 1998 The Johns Hopkins University Press
Translation sponsored by Fondation Marcel Mérieux, Lyon, France
All rights reserved. Published 1998
Printed in the United States of America on acid-free paper
9 8 7 6 5 4 3 2 1

The Johns Hopkins University Press
2715 North Charles Street
Baltimore, Maryland 21218-4363
The Johns Hopkins Press Ltd., London
www.press.jhu.edu

Library of Congress Cataloging-in-Publication Data will be found at the end of this book.
A catalog record for this book is available from the British Library.

ISBN 0-8018-5808-9

To
Adrien
Matthieu
Timothé

Contents

FOREWORD

S cientific research is an extremely hazardous occupation in as much as nature's secrets are very difficult to elucidate and the results of experiments are often unexpected and hard to interpret. A few highly creative individuals are nevertheless endowed with an uncanny intuition which leads them to ask the right questions, design the effective experiments, and focus on the correct answers. We owe to them the critical observations that have opened new fields of research to generations of scientists and have transformed our daily lives. Louis Pasteur was a giant among these pioneers to whom we owe so much.

Pasteur's discoveries, initially in chemistry and later in bacteriology and medicine, have provided a basis for the disciplines of microbiology and immunology and have established, on solid experimental grounds, the concept of protective immunity, originally introduced by Jenner with smallpox vaccination. To achieve these results Pasteur had to fight courageously and mercilessly against the existing erroneous notions of the spontaneous generation of microorganisms, strongly held by established scientists, as well as struggle to introduce the concepts and practice of antisepsis and asepsis to protect against the dangerous and dirty techniques of reputable surgeons.

He also established the concept of bacterial virulence and of the existence of attenuated strains, derived from the laboratory, which could be used to immunize effectively against the virulent strains. He was wise to take advantage of the opportunities presented by the industrial crisis in the production of silk, caused by the infection of the silk-producing worm, to identify the responsible transmissible agent and to breed uninfected worms. Similarly, to protect the beer producers and improve their technique, he analyzed the process of fermentation responsible for the production of beer and wine and demonstrated the role of microorganisms in fermentation.

While making these important contributions, Pasteur was also a pioneer in the concept of the rapid transfer of scientific information to industry for the benefit of mankind, a process referred to nowadays as "technology transfer," which has become increasingly important in the development of new medicines and therapies. In addition to these outstanding scientific accomplishments Pasteur was a giant also in his ability to inspire and train a generation of brilliant scientists such as Roux, Calmette, and others who continued his work and his traditions in the laboratory.

More important still, with public funding and support, he created in Paris one of the first independent research institutes. The Pasteur Institute, in the spirit and tradition of its founder, has been responsible for many outstanding discoveries, among which are the Bacillus Calmette Guérin, BCG, the effective vaccine against tuberculosis, the first sulfonamide drug, the first antihistamine, and the identification of the HIV virus. Many Nobel laureates trained and did their prize-winning work at the Pasteur Institute, such as Mechnikov, Nicolle, Bovet, Jacob, Lwow, and Monod.

Pasteur's personal life is also fascinating. He was a highly ethical person, a devoted family man, proud of his agrarian ancestry. He would return to the village of his birth and to his family house as often as he could. He strongly felt his dedication to humanity and to society. But, more important, he was endowed with exceptional courage and determination. He could not and would not be defeated. He never would give up. He knew he was right and he was sure that his views and scientific conclusions would eventually prevail. The best evidence of his courage is his determination, after being affected by a stroke at the age of forty-five which left him partially paralyzed, to pursue his work and his struggle for many decades, unfettered and even more determined than before.

Patrice Debré, as a distinguished French immunologist and physician, is eminently qualified to give us an extensive, balanced, and detailed account of the life, struggles, and historical contributions of Louis Pasteur. His biography of Pasteur provides us with an extremely well documented account of his life and family, of his relations with both the French government and the established official organizations of scientists and of physicians.

By now the original scientific documents, such as Pasteur's scientific notebooks, have become available. As a consequence, Debré was able to give us a critical and complete account of Pasteur's discoveries, and of the controversies they raised both with other scientists and occasionally with his closest associates. Since scientific research is an unpredictable, complex, and hazardous activity, it is reasonable to expect that occasionally close associates may not necessarily agree on the most desirable experimental course to follow, as was the case between Pasteur and Roux, as Debré relates. But in the end it is clear that all of Pasteur's group of dedicated scientists, under his enlightened leadership, had but one goal: the discovery of the scientific truth and its practical applications for the benefit of mankind.

Although I was well acquainted with the life and accomplishments of Louis Pasteur, had indeed visited his birthplace in the Jura, and had read many of the earlier biographies of Pasteur, I found Patrice Debré's "Louis Pasteur" highly informative, fascinating, and a pleasure to read. I feel honored to have been asked to write the foreword for the English edition.

Baruj Benacerraf, M.D.
Special Advisor to the President
Dana-Farber Cancer Institute

Preface to the English-Language Edition

The topics of science and health are all around us as we encounter such concepts as biodiversity, the ecosystem, or biotechnology. Science therefore needs to be explained, and scientists are no doubt best qualified to do this. There seems to be no better way to go about it than to retrace the history of scientific discoveries. Science becomes simple, provided we take the time to understand the nature of a problem, the germinating of hypotheses, and the trajectory of theories. Scientific ideas are no different from other ideas; they follow a labyrinthine course, fraught with anxiety and pleasure in the pursuit of a new concept. It is not for nothing that we often speak of a "scientific adventure."

This adventure is naturally part of a scientist's life, which thus becomes a veritable "object lesson." Here one can witness, "live" as it were, the origins of a discovery by observing the scientist at his microscope, seeing the spark of a brilliant intuition. The biography of a scientist thus becomes the main and most effective tool for understanding and explaining science, for asking the right questions and envisaging their answers and their consequences. The history of Pasteur is like science itself: it is worth the effort.

The publication of this biography of Louis Pasteur was intended to mark the centenary of the scientist's death. It was part of an array of ceremonies, seminars, congresses, and articles devoted to the image of Pasteur. In the United States, these discussions were preceded by certain statements and writings of Gerald Geison that the French press came to label "the Affair Pasteur." In the wake of other "affairs," the public was already sensitized to Franco-American controversies over the origins of microbes. Distressed by the accusations leveled at an emblematic French figure, Pasteur's defenders in turn accused Geison of being carried away by virulent criticism and failing to take an objective view of all the pertinent facts and arguments. This book attempts, among other things, to sort out this matter. Readers will find information that does not fit the interpretation of the American historian or, rather, permits them to form their own judgment by observing the experimental procedures of the nineteenth-century scientist.

This book is not only a historical account; it is also relevant to our own time. The scientific thought of another age has a bearing on the preoccupations of today's science. The fact is that science suffers from a serious problem of communication. This is particularly evident in biology, whose implications, above all for medicine, are considerable. In interviews and debates, in the written press and on television, the public is often called upon — indeed pressured — to make major decisions, yet it is not always given the means to fully understand the principles and the results of the research in question. As a result, fear and disillusionment build up because the possibilities are not explained clearly enough, because certain results seem disappointing, and because ethical issues remain fuzzy. Guided by fads or publicity, by the legitimate fear of disease, or by distrust of the commercial interests involved, society is constantly trying to come to grips with scientific matters, yet does not have all the aspects of any given question at its disposal. The root of this problem is the communications gap that even exists between researchers and their employers, be they a government agency or a private foundation.

Retracing the life and the achievements of a scientist in the context of his own time as well as ours is not only a wonderful challenge but also an indispensable means to permit society to make informed choices.

P.D.

PREFACE TO THE
FRENCH EDITION

It was autumn in Touraine. Through the windows of the study where I sat, I saw the reflection of willows in the Cisse, a small river that flows into the Loire at the foot of the hills of Vouvray.

People often go to Madères, the country place of my grandfather, the pediatrician Robert Debré,* when they face important decisions. Filled with memories, this house has seen the comings and goings of many famous physicians, influential politicians, writers, and artists.

That day I had come to collect my thoughts amid the family papers in the spacious wood-paneled library. To be sure, my decision had already been made; I was indeed going to undertake this biography of Pasteur, yet I still had a few qualms: Why had I accepted this challenge, considering that, as a professor of immunology specializing in the field of AIDS, I had never had any dealings outside the world of science and had so far written nothing but highly technical articles (most of them in English!) or administrative reports for the Ministry of Public Health, the university, or the CNRS?

As I was mentally putting together everything I knew about Pasteur, my eyes fell on a stack of brochures sitting at one end of the table. On the top cover I read the title: *L'Immunologie, sa naissance et son développement*, by Professor Robert Debré. It was the inaugural lecture delivered by my grandfather in the grand amphitheater of the Faculty of Medicine when he assumed the chair of bacteriology on 15 March 1933.

Present at the ceremony was Emile Roux, a few weeks before his death. Roux! the student of surgery who had been one of the first to stand by Pasteur in the struggle against his adversaries; Roux! who had essentially

---

*Robert Debré (1882–1978). Famous French physician who founded the French school of pediatrics and reformed French medicine in creating medical centers for treatment, teaching, and research. He was the father of Michel Debré, who drew up the constitution of the Fifth Republic and was prime minister under General de Gaulle, and of Olivier Debré, a renowned abstract painter.

carried out the triumphant experiment of Pouilly-le-Fort; Roux! who was there in the humble laboratory in the rue d'Ulm when it was a small hovel equipped with just one incubator, as well as in the glorious days of the Institute in the street that today bears his name. I imagined that great witness to the Pasteurian epic with his ascetic face, his close-cropped white hair, his pointed beard. I also imagined my grandfather, his neck swathed in a flowing polka-dotted *lavallière*, and I almost felt as if I were hearing his deep voice. His topic that day was to be "Pasteur and Immunology."

"M. Roux is among us this evening," Robert Debré solemnly declared. His address continued with the evocation of "Albert Calmette, a benefactor to humanity; Maurice Nicolle, with whom one never knew whether he would cite Gerard de Nerval's translation of *Faust*, speak of a problem in mathematics, a matter of anthropology, or a scab-forming toxin; Charles Nicolle, who carried the glory of his name and the brilliant light of Pasteurian science to the ends of the earth; Gaston Ramon, a disciple of Roux, the marvelous keeper of the Garches laboratory where Pasteur's life had ended."

The entire lecture resounded to the glory of Pasteur. The speaker meant to show that immunology owed virtually everything to the founder of microbiology. "Guided by Pasteur, French bacteriology turned particular attention to the fundamental problem of immunity. To be sure, it was not surprising that Pasteur should have made this choice. . . . As a laboratory scientist whose revelations were to overthrow existing thinking, he was passionate in his pursuit of practical and immediate results when he studied the fermentation of beer, the manufacture of vinegar, and the diseases of silkworms. He worked in breweries and in silkworm nurseries, among distillers, mulberry planters, and the shepherds of the Beauce. And then he came to the hospital: as a scientist, he felt that the moans of the sick were an appeal to which he must respond. And the result was the birth of immunology!"

It was as if a hand were held out to me across the generations, as if I could approach Pasteur himself. In the light of the fading day I seemed to see vague shadows. I was deeply moved by this Areopagus of figures from a past that was suddenly present.

From that day on, Pasteur has always been with me. He traveled with me on research trips and to conventions, from the laboratories of Harvard to the Faculty of Medicine at Phnom Penh. I often imagined that he was looking over my shoulder in moments of exaltation or disappointment, between failures and promising ideas, when the stakes of our research and their communication to the media are fraught with a heavy weight of responsibility.

Some Sundays I used to take my dreams to the apartments of the Institut Pasteur, rue du Docteur-Roux (formerly rue Dutot), which have now become a museum. I was usually alone there, and when I made the waxed parquet creak under my steps, I once again experienced the gentle and old-fashioned atmosphere of the laboratories of yesteryear.

One day I realized that my office in the Pitié-Salpêtrière Hospital had been burglarized. Forcing open the window, an unwelcome visitor had riffled through all my papers. Not interested in tape recorders and computers, he had stolen only one thing: a book on Pasteur and rabies that Charles Mérieux had given to me to fill a gap in my documentation. Was this the work of a kleptomaniac who could not help himself, an enlightened collector, a well-informed competitor, or merely a groupie of the scientist author? I wondered about this strange thief for a long time.

As a physician, I have always had a definite view of the life sciences. Nothing seems more important to me than scientific research in the pursuit of human health. And, inversely, our work has meaning only if it can alleviate suffering and heal. The patient and the means employed to help him demand a knowledge of the life processes, physiology, and the limits of the environment. Around 1850, at the time when Pasteur was learning his craft, this was not the case. There were many who worked "in the absurd," as the chemist Sainte-Claire Deville put it. Knowledge was sufficient unto itself. Its aim was to push back the darkness. The myth of an eternal beyond was sufficient to foreshadow new territories. There would be no limit to progress.

With Pasteur, everything changed. Pasteur was not a physician, not even a veterinarian. The forcing ground of the Ecole normale had made him into a chemist specializing in crystallography, a field in which French scientists had not particularly distinguished themselves. Nothing prepared him to turn toward medicine, nothing called upon him to forsake the crystal for the hospital. It was within himself that he found the brilliantly designed footbridge that led him from molecular asymmetry to the microbe and the fight against infectious disease.

Understanding a scientist calls for observing him closely as he comes to his crossroads, as he leaves the royal road and takes detours that are his alone. One must find out what motivates him to move from one project to another, to open up new areas of experimentation, and to pass from the known to the unknown.

How did the chemist Pasteur become the man who revolutionized medicine? It seems to me that two moments in his career provide crucial information for answering this question. The first was the cholera epidemic of 1865. Appointed to an investigative scientific commission, Pasteur left the confining atmosphere of his laboratory and was distressed to discover the daily hell of the overcrowded wards of the Lariboisière Hospital. He was convinced that a specific agent was responsible for the disease and thought that he had found it in the heating ducts: above the beds where patients were dying of enteritis, the scientist tried in vain to track down the germ that Robert Koch was to discover in Egypt in 1883. Because of his experiments in connection with so-called spontaneous generation, Pasteur looked for the microbe in the air, whereas in fact it was abundantly present in the soiled sheets. Aside from this error (which, though fraught with serious consequences, was actually secondary), the initial hypothesis was correct, and Pasteur's first foray into medicine can be dated from that day.

However, it was the diseases of the silkworm that eventually led him to put aside his studies on the chemistry of fermentation and put him on the road to biomedical studies. At the outset, this epizootic was of no concern to him: Pasteur was interested neither in the silkworm and its metamorphoses, with which he was almost entirely unfamiliar, nor, indeed, in disease as such. Amid the swan-necked flasks of the laboratory in the rue d'Ulm, he was at this time studying so-called spontaneous generations and fermentation. But he was not only a scientist but also a good courtier intent on furthering his career, and it is quite likely that he did not dare refuse Jean-Baptiste Dumas, and even less Napoleon III, when he was asked to take on a research project which, it was made clear to him, was of vital importance to the industry of France. In fact, it was precisely this aspect of the mission that appealed to Pasteur, for he did not believe that basic research could be totally separate from its practical applications or that there was no connection between them. For him, scientific research was a single entity, especially in a case of emergency, when the task at hand was to save French sericulture. But it is also likely that Pasteur was glad to find a new field of application for the study of germs. As it turned out, the moth of Alès would definitively lead him to medical microbiology.

Pasteur's life work derives its coherence from the microbe, for the infection of silkworm pupae had a connection with the fermentation of grape juice, the efflorescence of pus, and the life cycle of the anthrax boil. From chemistry to immunology, from one stage to the next, changes in Pasteur's preoccupations are only apparent. It was therefore no coincidence that Pasteur became interested in infectious pathology: among the many fields of investigation opened by the concept of biology as introduced almost a half-century earlier by Lamarck, Pasteur clearly saw medical research as the culmination of the life sciences. The fact that he turned away first from the crystal and then from fermentation unequivocally reveals the hierarchy of Pasteur's preoccupations.

Of course it was not going to be easy to obtain recognition for this biomedical revolution. Pasteur had to struggle to convince his peers and to be heard by the physicians. The universities often resembled jousting grounds, where the combatants fought without mercy. It was widely felt that a physician has no need for learning, that he must be a practitioner first and foremost.

Once launched on the study of fermentation, Pasteur used the microbe as his warhorse and the microscope as his weapon of choice. Imagine the astonishment of the surgeons around him when he went to the blackboard to sketch the chains of streptococci responsible for puerperal infections. And yet, ever since the seventeenth century, when Leeuwenhoek described the "marvelous beasties" he had seen, microscopic creatures were no longer novelties. Why, then, were nineteenth-century physicians so reticent? The reason is that Pasteur went beyond observation and asserted that these living beings played a deleterious role. In demonstrating their pathogenic character, he upset existing medical science and showed our world to be filled with danger and lurking invisible enemies. In so doing, Pasteur was not satisfied to open up, thanks to the microscope, new areas of investigation. He called for a rethinking of the relationship between medicine and science, between the world of research and that of treatment. Pasteur not only created medical biology, he also called for new checking and testing, which meant new power: no medical method should be above rigorous experimentation, and the physician must never forget that his work as a practitioner continues on a laboratory bench. Thus hemoculture, a technique that consists of culturing the blood of an infected patient in order to determine the nature of his infection, became symbolic, for now the entire laboratory went to war against the microbe, which could be brought under control by means of the culture medium. The Pasteurian revolution created a close link between theory and practice. It became clear that medicine could no longer do without science and that hospitals must no longer be mere hospices. Scientists and healthcare providers must work together in the laboratory.

The flamboyant style of Pasteur's speeches at the Academy of Sciences, his haughty and at times contemptuous words, show very well that he had to contend with a great deal of rivalry and frustration. Whenever he finished one of these oratorical jousts, there was some approving applause, but from many desks came cluckings of exasperation. Confronted with the crystallographer of the rue d'Ulm, the medical profession closed ranks and draped itself in its legitimacy, assuming attitudes that recall the Diafoiruses and the Purgons of Molière: there was much talk of art—but little of science. Moreover, physicians sometimes felt that since they were responsible for their patients, they also owned their illnesses. At any rate, they were less than pleased that a scientist should come to give them lessons in humility or to ask them to think about what they were doing. Actually, Pasteur was not the only one to encounter this kind of hostility. Before him, Claude Bernard (a physician himself) was scoffed at by some of his colleagues when he abandoned the lint and the lancet of the surgeon to plunge his hands into the bellies of rabbits. Although he had refashioned the science of physiology and created a theory of experimental medicine, he was so weary of the lack of understanding with which his research was received that in his inaugural lecture at the Collège de France he declared with bitterness: "The scientific medicine that I am expected to teach you does not exist!"

Is this situation really different today? Students and professors of the faculties of medicine and science frequently still ignore one another. I myself had studied and practiced medicine for more than ten years when I was told for the first time that I was doing science. And that was in the United States, where my colleagues found it strange that in France I was able to do biology outside the Institut Pasteur. I had indeed learned from my teachers that medicine and science do not necessarily go hand in hand, and I knew that the course of medical education in France conforms to this prejudice: one begins by studying the basic sciences, so that one is sure to have forgotten all about them by the time one becomes a specialist.

It is here that Pasteur's views should set an example. They placed great emphasis on the links that must exist between the patient's bed and the scientist's microscope. From molecular asymmetry to biology, from the bacterium to immunology, from the laboratory to the hospital, the path taken by Pasteur allows us to understand what research should be and compels us to reflect on the role of the research scientist within the medical team. Seen from the outside, this cooperation between medical practitioners and researchers seems natural enough, but in reality there is a profound and ancient reticence about involving anyone who is not a healer in the human suffering of the patient. This explains why for more than three-quarters of a century almost no one understood Pasteur's teaching. In the French universities, double appointments (in medicine and research) were officially permitted only in 1958, and the closed world of the hospital continued to oppose this innovation for even longer. Indeed, sometimes this is still the case, for hospitals are well aware that in our fast-paced society a physician is more likely to become known for effective healing than for the ability to study disease. The idea that the university hospital should be a place for observation and scientific reflection, or even subject to ethical rules, is still difficult to accept for hospital administrators and medical staff.

This is indeed a struggle between two powers. Both sides, the research scientists and the physicians, feel that their freedom will be curtailed. The practitioners hold it against the research scientists that they do not hear the moans of the sick and the cries of the dying; the scientists take a dim view of the medical world, which they criticize for incompetence and at times lack of scientific legitimacy. In order to avoid this dialogue of the deaf, one must constantly try to keep up contacts, and here Pasteur's example is timelier than ever, even if today it is no longer possible to combine the lancet with the microscope. In the era of AIDS, the scientist can no longer be assigned an area of investigation by the physician.

The era of AIDS . . . Today, my own research on that disease has taken me to the frontiers between microbiology and immunology. In the process, I have come upon strange and profound similarities between the specific problems of this disease and those that preoccupied Pasteur. Thus, he found a way to treat rabies, even though he had not seen the virus that causes it and only assumed that it was present in the brains of rabid dogs. Above all, he laid the foundations of immunology when he demonstrated that a virus can change by attenuation and thus become a vaccine, and that, furthermore, this virus/vaccine prevents the disease by exhausting the milieu in which virulence can occur by stimulating the immune system in case of reinfection. Today, the AIDS virus confronts us with the opposite process: we are faced with a virus which mutates in order to become even more virulent and which directly attacks the immune system, thereby protecting itself from antibodies that would neutralize it. But even this new example of the "cleverness" of microbes or viruses remains within the confines of the Pasteurian blueprint.

Rabies and AIDS, then, can be seen as facets of the same struggle. Is this saying too much? Consider the fact that Pasteur tried to use the long incubation period of rabies to stimulate the immune system of those who had been bitten. Does not the prolonged latency of AIDS present comparable, if not similar possibilities of intervention? Moreover, the AIDS epidemic, like rabies in those days, has become more than just a scientific matter; it is a major social issue. From one century to the next, we find a number of common characteristics: the terror of unsuspected contamination, the distrust of medical practices, the sometimes ambiguous role of commissions of experts, the influence of the media, public demonstrations against experimental therapeutic protocols, justified and trumped-up legal proceedings. Yet it is precisely at this level of society that Pasteur's accomplishments are sometimes put into question. There are those who, speaking of a "fissure in the world," claim that Pasteur's work conveys an ideology of security which they hold largely responsible for our blindness in the face of the appearance and the spread of the disease. If AIDS is indeed the sign of a problem in society, it is, scientifically speaking, the society of the viruses, not that of humans, which has a problem.

It would be silly to claim that Pasteur foresaw the AIDS epidemic, even if he sometimes did envisage the invasion of new viruses of African origin; but it would be even more absurd and dangerous to postulate that the epidemic has invalidated the very concepts established by Pasteur. How could AIDS possibly reveal the failure of a system that encourages the formation of a preventive *cordon sanitaire* through hygienic measures and the identification of the agents of transmission? Indeed, one might assert that it was probably because people did not reread and assimilate Pasteur sufficiently that they failed for too long to heat the blood used for transfusions. In other words, the Pasteurian era is not over. For those who are sick today, the example of Pasteur is a source of hope. The solution to AIDS is work and time; there is no need for a scientific revolution. Instead of beating its breast or looking for scapegoats, society would be better advised to look favorably upon the emergence of a new scientist whose work would forge a synthesis of the Pasteurian and the genetic revolutions. If we, the physicians, are lagging behind, it is not because we are led astray by obsolete concepts; it is precisely because Pasteur is no longer with us.

"Any founder in the spiritual realm must endeavor to make himself irresistible," wrote Paul Valéry. Make himself irresistible? It is sometimes said that biographers must above all bring their subjects back to life as if they were the protagonists in a novel, and that they have to captivate the readers by making them share the author's enthusiasm or disapproval, if necessary by covering up certain details that would spoil the picture. When the subject is a man of science, this attitude does not make sense. In this case, what matters most is the history of his discoveries. Large or small, each one of them has its own importance, for they are the only means of demonstrating the validity of the experimental approach used. Pasteur himself believed that one of the steps in the teaching of science was the study of the life of great scientists, that is to say, the history of their scientific work. To experiment is to act, is to be alive.

I have therefore been very careful to retrace this development from one stage to the next without neglecting any one of them. In so doing, I have opted for a methodical presentation of the various research topics (from their premises to their conclusions) rather than for a strict adherence to chronology, for it is often impossible to show the simultaneity of experiments of very different nature or importance. Thus Pasteur moved back and forth between Alès and Paris, between studies on the silkworm and measures to prevent the spoiling of wine, all the while continuing his work on spontaneous generation. A thematic narrative gains in clarity what it may lose in immediacy; above all, it allows every reader to linger over the subject that is of particular interest to him or her, whether it be crystallography or rabies.

In the evening, Pasteur used to dictate to his wife the scientific papers he then sent to the academies. I have read and reread all of these texts. Their content, to be sure, is now out of date, but they still have an expressive vigor that proves that a scientific revolution demands a perfect mastery of language. I also had access to unpublished correspondence and to the large numbers of books of laboratory notes that Pasteur's grandson bequeathed to the Bibliothèque Nationale, where they have recently become available to the public. Descriptions of the set-up of every experiment, temperature readings, the number of bubbles in a drop of beer, the degree of acidity in wine, changes in the condition of the guinea pigs, everything is here. The purple or blue ink has taken on a shadowy hue, but one still seems to hear the scratching of the pen on the paper as the scientist takes down each one of his intuitions. Does this mean that one hears the beating of the man's heart? Not really, for Pasteur reveals little of himself, and no fluttering of the eyelids ever seems to betray his feelings.

That is why I deliberately chose not to develop or to imagine certain psychological situations to dramatize the scientist. This would have had no scientific or historical basis. The man, if we are to look for him, is found only in the progression of his ideas. It is only by following the path of his discoveries that we can hope to decipher the history of the attitudes that motivated Pasteur.

That is also why, for each one of Pasteur's investigations, I have made a particular point of finding out about his motivation. For although he conveyed the impression of working all by himself, he was adept at choosing his collaborators and building on the work of his predecessors. Quite a few of his hypotheses were dictated to him by his time; indeed, it sometimes seems as if Pasteur merely verified the findings described by others and then appropriated them. Yet it was precisely when he picked up demonstrations that had been, as it were, left to lie fallow, that he proved to be most innovative. The salient feature of his genius was his synthetic ability.

Pasteur proceeded like today's research scientists, who begin their work by surveying everything that has been published on a subject. But he was one of the first to do so and to systematize this scientific method, which makes it possible to bring together pre-existing data scattered throughout the literature, thereby often giving them an importance they did not initially possess.

This method is particularly striking in his first studies, which to my mind are the most remarkable of his entire body of work. Having observed the asymmetry of crystals while studying the fermentation of tartar, he turned toward the life sciences, thus establishing contact between two disciplines that had hitherto ignored each other, crystallography and biology. He discovered and proved that the ferment, that is, the microbe, acts as a living being. In doing so, he produced one of the first examples of the biotechnological use of the germ and developed a first model of a culture medium. In the end, chemistry and microbiology became one. Linked in a coherent pattern, Pasteur's intuitions eventually led to applications in industry and medicine.

The specificity of this mode of scientific procedure also illustrates the point that no scientist ever works in isolation and that constant dialogue among peers is a necessity. One can therefore not evoke Pasteur without sketching portraits of the other scientists whose life or work became intertwined with his, from Arago to Marcelin Berthelot, from Littré to Claude Bernard. Bernard, who liked to plunge into the unknown even at the risk of following a false scent, ironically claimed that his illustrious confrere Pasteur had a tendency to answer questions before he had ever started his experiments instead of basing his conclusions on laboratory tests.

In the end, it is from this contest at the heights occupied by the giants of science that we can construct the best portrait of Pasteur: sure of himself, inflexible, even dogmatic, rarely taking a false step while running a virtually flawless race. In a certain sense, I must admit that I was relieved to find in his notes a few unnecessary experiments, the occasional slightly questionable method or evidence of an unattractive attitude. Asymmetry is life, said Pasteur. The biographer might add: Errors are the mark of a man. I have tried, then, to place this account, which aims to be neither hagiography nor debunking, under the sign of lucidity.

THIS BOOK ENDS where it began, at les Madères.

As I worked on its final passage, I took a walk to the neighboring community of Vernou-sur-Brenne. Apart from the nearby vineyards on the slopes of Vouvray, nothing here has any relation to Pasteur and his work: there are two or three streets lined with small houses, a little church, and a few cafés — it is just one of the many French villages that exist everywhere.

I stop for a moment, deep in thought, still absorbed in my day's work. My gaze automatically deciphers the sign indicating the name of the street in which I am standing, and I read: "rue Pasteur." The grateful Republic, then, has honored the great scientist even here. This is something to think about. Is there one community in France that does not have its rue or avenue Pasteur? And how many of them are found throughout the world? Pasteur said that science does not have a fatherland.

My curiosity aroused, I walk around the neighboring streets, noting that the town council has placed Pasteur between Victor Hugo and Aristide Briand. Truly, can one imagine a more splendid academy than the innumerable street corners all across the globe, where the names of the heroes of humanity show the way to those who pass by?

Continuing on my way, I leave Pasteur in this honorable company. At the village gates, the vineyards have already turned purple.

Autumn 1991–Autumn 1994

Acknowledgments

Throughout the writing of this book, I have benefited from the support of the Institut Pasteur. I wish to thank its director, Maxime Schwartz, as well as its secretary general, Marie-Hélène Marchand, and the curator of the Pasteur Museum, Annick Perrot.

Marie-Laure Prévost, curator of the department of manuscripts at the Bibliothèque Nationale de France, kindly guided me through the Pasteur papers.

I also thank Thierry Bodin, Loïc Chotard, Roland Dreyfus, Valérie Guérel, François Haguenau, Charles Mérieux, Philippe Monod-Broca, Daniel Schwartz, and Eduardo Villarreal.

And I salute the memory of Serge Leroy, who was fascinated by the scientific adventure of Louis Pasteur.

P.D.

The translator is most thankful for expert and much needed help with the crystallography chapters provided by Thomas Kistenmacher.

Caroline and Owen Hannaway lent books and made useful suggestions in other areas, and Robert Forster read the entire manuscript aloud in the last checking operation.

E.F.

Louis Pasteur

I

MY FATHER IS A HERO

O nce one is used to working, one can no longer live without it. And of course, everything in the world depends on it; in science, one is happy; in science one rises above all others."[1] Work, science, ambition — the outline of an entire life is already drawn in these words, even though the person who penned them was only a lad of seventeen, a studious young man who regularly wrote to his family: it was Louis Pasteur, a student in his last year at the secondary school of Besançon in 1840.

At the time, this model son was still an anonymous and obscure figure among his classmates. Only a few months earlier he had left his family circle to attend the collège royal of Franche-Comté. *Collège royal*: these two words convey an idea of the heavy atmosphere that weighed down secondary education at the time; the government had wanted to banish the word *lycée*, fearing that it brought back too many memories of the Revolution and the Empire. Yet at Besançon, the buildings of the collège royal were not exactly a credit to the monarchy. Decaying, almost dilapidated, the classrooms were lit by a few dim lamps; all there was in the dormitories were some perpetually sputtering candles. Yet it did not occur to the pupils to complain, let alone revolt. They were there to learn, especially to learn to obey.

Pasteur thus gives the lie to Rimbaud (who was not yet born) and proves that one can indeed be serious at seventeen. Day after day he devoted all his energy to studying, for he was determined to pass his two baccalauréat examinations, so that he could subsequently study for the competitive entrance examinations to the Ecole normale supérieure.

To be sure, this seriousness of purpose was not just a personal trait; it was the attribute of a whole generation, that of Renan and Fromentin, Edmond de Goncourt and César Franck. Nor were there any real exceptions to confirm the rule, for Courbet, Baudelaire, and Flaubert were diligent if not always docile pupils before they became artists who scandalized their contemporaries.

Those who were born in the provinces around 1820 and grew up there would perceive no more than a faint echo of what was happening in Paris. There was a succession of kings — Charles X replaced Louis XVIII, the three Glorious Days placed Louis-Philippe on the throne — but small towns were stuck with the same quarrels among notables, the same questions of money, the same concern for law and order. It is true that there were, especially at Besançon, some opposition newspapers, clubs of intellectuals, and even more or less secret societies, but such things were for the select few, as were the creations of the avant-garde. The plays of the Romantic school, the paintings of Delacroix, the *Symphonie fantastique* of Berlioz reached the provinces only much later. For the children of Besançon who learned to read around 1830, the great man of their town was of course not Victor Hugo, born in 1802; it was not even Charles Nodier, who was Hugo's senior by twenty-two years; it was Joseph Droz, member of the French Academy since 1824 and the immortal author of the *Essai sur l'art d'être heureux*, a work that swept young Pasteur off his feet with enthusiasm: "I have never read anything wiser, more moral and more virtuous. In reading it, one feels one's soul taken over by an irresistible charm and inflamed by the most sublime and the most generous feelings."[2] Here we are very far indeed from the romantic passions and the political debates that rocked public life and people's mentalities throughout Europe. The child Louis Pasteur very likely knew nothing of the battle over *Hernani*[3] or the massacre in the rue Transnonain; but he did read Roman history in Livy or Sallust, and his father did raise him with the cult of the Napoleonic legend. But did anyone tell him about the independence of Belgium or about the Zollverein, which prepared the way for the unification of Germany under the aegis of Prussia? It is most unlikely. The word *adolescence* was not yet used in its modern sense; it was still, as it had been in Chénier or in Bernardin de Saint-Pierre, a synonym of *candor* and *inexperience*. In fact, at the time when a young man's judgment is being formed and his senses come alive, Louis Pasteur did not show a consuming curiosity for the outside world: his horizon was his school work; his universe was his family.

Comtois, Tête de Bois (Franche-Comté Breeds Hard-Headed Men)

The roots convey the sap. Pasteur would have said that the tree's form is reflected in its bark. Throughout his career, he remained faithfully attached to his native Franche-Comté. From time immemorial, the people of the Jura mountains had lived on the European salt trade, which was the mainstay of the area's few industries. In the nineteenth century, Franche-Comté, with its beech and spruce forests, vigorous streams, vineyards, and alpine pastures, was a land that relied upon itself, its craftsmen and its agricultural traditions. In the portraits of Pasteur, one can see on his face traces of this peasant vigor and of his Burgundian antecedents. His temper was exactly the kind that history ascribes to the people of Franche-Comté: proud, tenacious, even stubborn, steeped in the faith of the *charbonnier* but freedom-loving to the point of insubordination.

Pasteur enjoyed citing the saying: "Comtois, tête de bois," which eventually became the inscription on the province's arms in the form of: "Comtois, rends-toi! — Nenny, ma foi" [Comtois, give up! Faith, I never will!] A son of Franche-Comté constantly uses the word *will* as an auxiliary verb in order to appropriate even acts that have nothing to do with his will. Sometimes accused of avarice, the Franche-Comtois is not stingy, he is careful; he spends money when and in such quantity as he has to. When it comes to speaking, he is not a brilliant drawing room conversationalist, but rather a solid and convincing preacher: no Gambetta, but rather a Jules Grévy.

Pasteur was born at Dole on Friday, 27 December 1822, at two o'clock in the morning, into a modest family whose name had never shone in history. The Pasteur family originated in a little village in the Haut-Doubs by the name of Reculfoz, five kilometers west of the ecclesiastical seigneurie of Mouthe. Clinging to the mountainside at an altitude between 1,043 and 1,218 meters, this village is located in the coldest region of France, and life there was harsh. The main occupation was animal husbandry, and the name *Pasteur* indicates that the scientist's ancestors were shepherds. They probably produced the *vachelin* cheese that is today called *comté*.

The first Pasteur we find in the archives is Nicolet Pasteur, who in 1488 signed a sharecropping agreement with the prior of the abbey of Mouthe. The line then passed through Etienne, Jean, another Etienne, to Pierre, called Pierrelon. This Pierrelon had a son, Jean-Etienne, who in turn had eight children, among them Denis, born 5 January 1653. Of these eight children, only two continued farming at Reculfoz. Denis Pasteur, who in 1682 married Jeanne David, lived at Plénisette and at Doye, near Nozeroy before he moved farther north to Lemuy, which is located at the edge of the forest of Joux, the most beautiful pine forest in the Jura.

Denis was the miller for the village seigneur, Claude-François, comte d'Udressier, who belonged to the old Burgundian nobility that had been in the service of Charles V. Denis's son Claude, also a miller, who had eight children, among them Claude-Etienne, Louis Pasteur's great-grandfather. Claude-Etienne was subject to mortmain, that is to say, he had no legal control over his property or his person. When he wished to move from

Lemuy to Salins, a small nearby town that belonged to a different seigneurie, he had to buy his independence from his seigneur. He was granted
his release by an act of 20 March 1763 against payment of the sum of four
louis-d'or or twenty-four livres, to be paid on the spot.

Claude-Etienne Pasteur began working for a tanner at Salins, where
the rapidly flowing waters of the Furieuse River lend themselves to this
activity. He had ten children. The third of these was Louis Pasteur's grandfather, Jean Henri, who also worked in leather. He left Salins for Besançon,
where he opened a shop in the suburb of Les Arênes, on the banks of the
Doubs. He died at the age of twenty-seven. His wife had died at twenty,
after she had given him his only son, Jean-Joseph, in 1791.

Orphaned at a very young age, the child was brought up by his aunts
at Salins. Like his father, he was apprenticed in the leather-currying trade, in
hopes that some day he would return to the family craft of tanning. But his
training was soon interrupted by what was to be the great adventure of his
life, the epic of the Napoleonic Wars.

Conscripted in 1811, Jean-Joseph Pasteur found himself on the battlefields of 1812 and 1813 in the terrible war in Spain, whose horrors were
immortalized by Goya. With his regiment, the Third of the Line, which
distinguished itself in the pursuit of the bands of the famous guerrilla leader
Don Francisco Espoz y Mina, Jean-Joseph was involved in skirmishes at the
mountain passes of the north as well as in ambushes and surprise attacks
under the Castilian sun.

From ordinary soldier, Pasteur was promoted to corporal on 1 July
1812 and to quartermaster sergeant on 26 October 1813. His battalion was
brought back to France the following January, for now the Empire had to
be defended against the Allied coalition. At Bar-sur-Aube, the French were
outnumbered five to one, and Pasteur's battalion fought with such bravery
that it earned the name of honor "bravest of the brave."[4] The Emperor
himself gave out the decorations when Jean-Joseph Pasteur, promoted to
sergeant-major on 10 March 1814, received the cross of a *chevalier de la
Légion d'honneur*. Henceforth he never wore his jacket without that little
red ribbon. However, the imperial epic was coming to an end. After the
disastrous fighting at Arcis-sur-Aube, where Sergeant Pasteur's division
suffered the combined onslaught of Russian, Austrian, Bavarian, and Würt-
temberger divisions, all that was left of the eight thousand men it had
numbered at the beginning of the year were eight officers and 276 men. A
forced march took it back to Fontainebleau. But on 20 April 1814, when
Napoleon reviewed his troops for the last time in the White Horse Court-

yard of that chateau, the soldiers of the Third Regiment of the Line were no longer there to acclaim their emperor. The regiment had been sent to the department of Eure, where it was forced to rally to the restored monarchy and to don the white cockade. On 12 May it was even given the name of Regiment-Dauphin.

Relieved of his functions, Jean-Joseph Pasteur refused to attend the ceremony when the imperial eagles atop the standards were replaced by the fleur-de-lis. He definitively quit the army and its promises of glory and brought back from his campaigns only the mythic memories of the kind of living god that the Emperor had been in his eyes.

Experiencing the restoration of the monarchy and the return of Louis XVIII as a personal humiliation, Jean-Joseph came back to Salins filled with anger and sadness. He had barely heard about Napoleon's return from Elba and the fever of the Hundred Days before Napoleon was gone again, this time to Saint Helena, after the judgment of Waterloo. So the ex-sergeant buried his hopes of revenge once and for all and went back to his work of tanning. Yet when he was asked to hand in his weapons, his pride was wounded: seeing that his saber had been given to a police officer of the new regime, Jean-Joseph angrily tore it out of its new owner's scabbard. This flamboyant gesture on the part of this reserved and taciturn Napoleonic veteran surprised those who witnessed it and earned him the applause of the Bonapartist officers and petty officers who were always looking for ways to express their resistance, even when Salins was still occupied by an Austrian garrison. Ordered by the mayor to punish this outrageous act, the police commander refused to act. Jean-Joseph Pasteur was carried to his home in triumph and hung up his saber between two revered portraits, those of the Emperor and [his son] the King of Rome.

SHORTLY THEREAFTER, while washing his hides in the waters of the Furieuse, Jean-Joseph saw a girl working in a garden on the opposite shore. She was Jeanne-Etiennette Roqui, born into a family that had long been established in the region and had originally come from Marnoz, outside the gates of Salins. A few words exchanged across the raging waters were sufficient to seduce the young woman, whom René Vallery-Radot, Louis Pasteur's son-in-law and his first biographer, delicately and prudently described as "a very imaginative woman who was easily carried away by her enthusiasm."[5] When Jean-Joseph married her on 27 August 1816, she was eight months pregnant.

In a small town like Salins, such a "shotgun" marriage was an unacceptable scandal. Rather than put up with the endless gossip of their neighbors, the young couple left town. There was no need to go very far, and so they settled down at Dole, twenty-five kilometers away. Barely three weeks later, on 18 September, a little boy was born; they named him Jean-Denis, but he died after a few months, as if to make the Pasteurs forget his precipitous beginnings.

Clinging to the sides of a hill dominating the right bank of the Doubs, Dole had lost its position as the capital of Franche-Comté two centuries earlier; it was now a big market town whose houses were clustered around the church of Notre Dame along narrow and twisting streets. At the town's outskirts, where the canal between the Rhone and the Rhine split into two, the tanners had set up shop in order to take advantage of the swift current. The Pasteurs' house was a large stone building with a low-ceilinged cellar; the river bank behind it was cluttered with poles used in the preparation of the hides.

In this austere but work-centered environment, life soon returned to normal. In the spring of 1818 Jeanne-Antoine, called Virginie, was born; she was to marry her cousin Gustave Vichot, who worked with Jean-Joseph before succeeding him as the head of the tannery. Virginie died at Arbois in 1880. Louis was born four years after her, and two sisters arrived later, Joséphine in 1825 and Emilie in 1826. The fate of these two youngest children was very sad, for the first was consumptive and died at twenty-five, and the second, who suffered from epilepsy, died at twenty-six.

Little Louis realized very quickly that he was the only male of the brood; in the letters he wrote as a young man, one can read all kinds of advice and solicitude for his sisters. As the son and heir and also the intellectual of the family, he was quick to chide them for their constant bickering and their jealous tiffs, but deep down he felt only pity when he realized that Emilie was mentally retarded or that Joséphine could not control her passion for the "lottery of gold ingots."[6]

While the first children came into the world at Dole, Emilie, the youngest, was born at Marnoz. The reason is that grandmother Roqui, feeling that she was getting old, distributed her property before her death, dividing everything she owned among her children. Her house at Marnoz fell to Jeanne-Etiennette, and so she moved there with her husband and her children in early 1826, although the location was not very suitable for a tanning operation. On one of the doors of what is today called the "Pasteur House," Jean-Joseph painted the familiar silhouette of the "soldier-worker," that nostalgic Napoleonic veteran who has gone back to the land. His bicorne hat askew on his head and in shirt sleeves, the hero leans on his shovel and dreams of his past glory.

Louis Pasteur's first memories, images of moments forever fixed in his mind, go back to Marnoz, where, for instance, he saw himself running on the road from the village to Aiglepierre. Yet the family was not to stay at Marnoz for very long. Did Jean-Joseph not get along with his in-laws? Had military life made him unbalanced? Whatever the reason, in early 1827 he and his family moved to Arbois, where he was able to rent a tannery on the banks of the Cuisance at the outskirts of the little town. It was a modest establishment. Its front faced the Lyon-Strasbourg road, but the buildings were separated from the road by a courtyard slightly lower than the road. There were two doors, one for the living quarters and one for the upper story. An outside staircase led down to the tannery, which, entirely installed in the basement, had direct access to the river. In the back of the building, a low-lying courtyard was completely taken up by seven round pits lined with oak wood, where the leather was soaked until it became impregnated with tannin. The inside of the house was of a simplicity that bordered on poverty: on the ground floor, a hallway led to two rooms, the pantry and the laundry room, which also contained the bread oven. To the right were three other rooms — two bedrooms, the larger of which was used as a dining room on special occasions, and a large kitchen whose window jutted out over the Cuisance and in which the family took its meals. The upper floor accommodated the curriery and the bedrooms.

This is where Pasteur spent his childhood. He grew up between the river and the vineyard-covered slopes, between the fetid smell of the tannery and the fine aroma of crushed grapes. These first sensations imprinted themselves in his mind: the smell of wool grease and the perfume of fermented juice would stay with him for the rest of his life, cropping up even in his work on yeasts and in his hygienic precepts.

Pasteur's playmates were the sons of vintners, Jules and Altin Vercel, who were two or three years older than he, or the Coulon children. Their games were those of their time: marbles, *jeu de paume*, a kind of racquetball, and capture-the-flag games. In the winter the children went sliding on the ice on wooden shoes or a sled, and in summer they could go swimming and spear fish in the Cuisance. Fishing was a matter of waiting for the fish, while perched very still on a rock, then harpooning it with a fork fixed to the tip of a pole. But the moment of the year most eagerly anticipated by all the children of Arbois was the autumn wine harvest. Louis Pasteur and his playmates took part in the harvest and the attendant festivities. The evenings ended with banquets to which everyone came, to enjoy thick cheese soups and *gaudes*, a dish of cornmeal mush with crème fraîche, which remained one of Pasteur's favorites.

The reputation of the wine of Arbois had spread beyond the confines of the region long ago, which is why the whole town was mobilized for the wine harvest. The streets echoed with the shouts of the vintners and the creaking of carts carrying the grapes. The first task was to make the white Arbois, bottling it early so that it would naturally turn into a sparkling wine, but above all, to take care of the "straw wine," pride of the wines of Franche-Comté, which was pressed only in the spring, after the clusters of grapes had spent the winter in people's attics, lying on beds of straw or hooked onto lattice screens.

A Failed Rendez-Vous with Paris

And then there was school. Little Pasteur seemed to be equally fond of sliding on the snow or fishing from the bridge over the Cuisance and of the lessons at the primary school of Arbois, which was housed in one of the spare rooms of the town hall. In 1831, he entered first grade (*huitième*), and his first teacher, M. Renaud, was a dedicated young man who took a personal interest in his pupils. Later, when he had reached the time of honors and prizes, Pasteur always gratefully invoked the name of the man who had taught him his first lessons. However, until *troisième* (age 14), he was no more than an average student who showed no particular enthusiasm for his schoolwork.

The reason may have been that Louis learned the most basic things at home. Jean-Joseph was deeply involved in his only son's education and acted as his private tutor at every turn. "The father exercised absolute yet wise and reasoned authority over his entire family," reported a neighbor by the name of Vuillame. "Everyone obeyed him, the wife as well as the children; but there was nothing tyrannical about his will, for it was always tempered by love and kindness, and his orders were given only for everyone's good and happiness. He was working extremely hard to provide them not only with the necessities, but even with some little extras, and his example of regular work and persistence instilled in them a love of family virtues."[7] A frequently taciturn man, Jean-Joseph had real conversations only with his son. On Sundays, he would don his old military coat with the ribbon of the Legion d'honneur attached to the wide lapel, and take Louis for long walks on the road which, running through the hilly vineyards, goes from Arbois to Besançon. When the family sat together in the evening, the Napoleonic veteran would go over and over his memories of Spain and the glory of the Emperor. He talked of military honor, of justice, and of duty. Everything he knew and everything he believed came from two bibles, one being the fables of La Fontaine, of which he sententiously declared, "nothing is more apt to make us understand life as it really is," and the other, of course, Las Casa's *Mémorial de Sainte-Hélène*, which had appeared the year after Louis was born.

Raised in the tenets of Bonapartism from the time he discovered the alphabet, the child who became France's most famous scholar would later say about his father: "In teaching me to read, you made sure that I learned about the greatness of France."[8] A father who was as economical with words as with money, an unassuming mother fully occupied with her three daughters—for young children it was a rather cramped way of life, whose austerity was sometimes hard to take for little Virginie and Joséphine. Fifty years later, in 1883, at the inauguration of a commemorative plaque that was placed on the house of his birth, Pasteur would evoke these childhood years with pride and depict the family group in an idealized picture that glossed over the memory of bad times. "Your enthusiasm, my brave mother, you bequeathed to me. If I have always associated the greatness of science with the greatness of the Fatherland, it is because of the sentiments that you inspired in me. And you, my dear father, whose life was as harsh as your harsh craft, you showed me what patience can accomplish when the task is long. It is to you that I owe my tenacity in carrying out the work that needs to be done from day to day. Not only did you have the qualities of perseverance that make for a useful life, you also had admiration for great men and great things."[9]

For the time being, daily life seemed tranquil enough to foreshadow a simple existence. From time to time it was somewhat brightened by the visit of one of the family's rare friends. Jean-Joseph was rather unsociable, and the household had little use for others, except for Louis's classmates and certain elected officials like Dr. Dumont, a former military physician who was practicing at the Arbois hospital, or Emmanuel Bousson de Mairet.

It was thanks to the latter that the Pasteurs moved to Arbois. A man of letters and a philosopher, Bousson de Mairet was the epitome of the local scholar. With the patience of a Benedictine monk, he spent his time reconstructing the history of the people of Franche-Comté in general, and of Arbois in particular. The author of a tragedy about Jeanne d'Arc, he was passionately involved in one of the polemics that rocked the archeological community of the time, namely, the controversy about the exact location of Alesia.

Sitting by the fireside of an evening, this man taught Louis the heroic deeds of his province, insisting on its Burgundian character. For Franche-Comté was one of the last provinces to be taken over by the French crown, having long resisted all outside domination. It was not until the War of Devolution and the Peace of Nijmegen in 1678 that Franche-Comté was taken away from Spain and incorporated into France. After the fall of the Bourbons, the Comtois had to fight the centralization of the Jacobins, who wanted to divide in order to rule: cut up into several departments by the administration of 1791, Franche-Comté nonetheless succeeded in staying united. Nor is it a coincidence that some of the rare conspirators who dared stand up to Napoleon were Comtois: Pichegru was born in Arbois, Malet in Dole.

When Bousson de Mairet brought up these names, Jean-Joseph would frown and send his son to bed. By contrast, he was very glad when his friend reported on more recent events: the rebellion of the people of Arbois who on 4 August 1830 protested against the ordinances of Charles X and were ready to rally to the July Revolution, or the riots of April 1834, at the time of the uprising of the silk weavers of Lyon, when the vintners seized a cache of guns with which to uphold the Republic that had supposedly been proclaimed at Lyon. Bousson de Mairet was not only a brilliant storyteller, he also taught at the collège of Arbois, from which he had to retire prematurely when his hearing failed. However, this infirmity was not too troublesome in one-on-one conversation, and so, starting in the *troisième*, he became the unpaid tutor of Louis, who began to make considerable progress. By the time he was in *seconde* and then in *rhétorique*, the son of Jean-Joseph Pasteur walked away with all the prizes.

But the man whose influence on the destiny of Pasteur was to be decisive was the principal of the collège of Arbois, M. Romanet, who took great pleasure in developing circumspection and enthusiasm in his young pupil. Pasteur admired this official of the ministry of public education, who considered teaching a priestly function. He listened when his principal extolled the advantages of education and described the only place that seemed to him worthy of the abilities he detected in the child: the Ecole normale. Here, for the first time, Pasteur heard about this place of prestige and learning, where the best students of France are made into elite teachers.

Another visitor at the Pasteurs' house was Captain Barbier. An officer in the municipal guard in Paris, this native of Arbois often spent his leave in his hometown. That is how he came to promise Jean-Joseph that he would look after young Louis if it were decided to make him pursue his education in Paris. The father hesitated. Louis had not yet turned sixteen; why should he go so far away? Was it not better to send him just to Besançon once he had finished *rhétorique* in Arbois? And in fact, was it really necessary to send him to Normale? A good position as professor at the collège d'Arbois seemed to him the height of aspirations for his son. And to achieve that, was there really any point in going to Paris, where life was so expensive and so full of dangers for a young and inexperienced provincial? Forgetting that at twenty he himself had roamed all over Europe in the imperial armies without being concerned about danger, Jean-Joseph declared that he could not afford to disburse such sums, even for his son's education.

At this point, Barbier mentioned an institution in the Latin Quarter which, according to the *Almanach du Commerce*, prepared students for "the Schools of Polytechnique, Saint-Cyr, the Navy, and the Forest Service." Located at 3, cul de sac des Feuillantines, very close to the rue du Val-de-Grâce, this institution was directed by a man from Franche-Comté, a native of Pagnoz, a small village near Salins. This Sieur Barbet offered special prices to his compatriots. It also turned out that if his father agreed to send him there, Louis Pasteur would not go to Paris by himself, for his friend Jules Vercel was about to go there to prepare his baccalauréat. Jules Vercel was an easygoing boy with a friendly disposition and no great ambition except to be a faithful friend.

The fact that his son would not be alone in the capital persuaded Jean-Joseph to agree to the sacrifice of a separation. On one of the last days of October 1838, he took the two boys to the post-coach for Paris, which stood waiting, fully harnessed, in the great courtyard of the postal office. It was a bitter cold morning. They huddled under the awning next to the coachman, overwhelmed more by the sadness of the parting than by the excitement of embarking upon adult life. Once they had rounded the sheepfold, the steeple of Arbois disappeared from the horizon. A few kilometers farther, they could no longer see the plateau of the Ermitage, and finally Mathenay, La Ferté, Pupillin, Montigny-les-Arsures as well were left behind. In the distance, the dark outline of the Jura Mountains slowly faded away.

On his arrival in Paris, Louis Pasteur could barely hold back tears, despite the warm welcome extended by Barbet. He was almost sixteen years old, and this was the first time he had left his family. Now he had to face the lack of privacy of dormitories and study halls. Like most of the students at the pension Barbet, Pasteur and Vercel took courses at the collège Saint-Louis, the former collège Harcourt located at the foot of the Montagne Sainte-Geneviève between the Sorbonne and the School of Medicine. But these glorious names meant nothing to the young student from Franche-Comté, who was not used to obeying the roll of a drum or donning a uniform whose military character — tunic with copper buttons and képi — made him bristle. Young Louis became very homesick: "If I could only smell the odor of the tannery," he confided to Jules Vercel, "I am sure I would feel much better."[10]

Barbet did his best to overcome this black nostalgia that turned his protégé into a kind of silent and distressing specter. But he soon ran out of distractions that could make Louis forget his unhappiness, and so he decided to write to his father to tell him that his son could not bear to be separated from his family. In mid-November, Louis Pasteur was summoned from study hall and told that someone was waiting for him at the corner of the rue Saint-Jacques and the cul-de-sac des Feuillantines. Without asking any questions, the boy let himself be led to the office of a wine merchant. There he found his father, who had come to Paris to take him back to Arbois.

No doubt there was regret about having failed to carry through an experiment, shame at having allowed an ambition to be defeated by sentimentality, but also pleasure at returning to the cocoon of the family. Face to face with his father who asked no questions, Pasteur sat in silence in the coach that made him turn back, a rare thing indeed in his life. But a failure of this kind at the gateway of life is not easy to forget; it would take time for this wound to heal.

The Temptation of Art

When Pasteur returned to Arbois at the end of 1838, he needed to come to terms with himself, to regain the serenity that had been badly shaken by his Parisian adventure. He therefore devoted himself to drawing and painting, as if he felt that the face-to-face relation between the artist and his model would permit him to become reaccustomed to the world and to the gaze of others. He was, as it were, carrying out his own psychotherapy through art.

Pasteur had always liked to draw. Since early childhood, he had copied the engravings of his school books in charcoal or pencil. At the school of Arbois his teacher noticed his talent and encouraged it. Later Louis became fascinated with the technique of the pastel portrait. His first models were his friends and the members of his family. A portrait of his mother that he executed at the age of thirteen still exists. It shows her dressed in a plaid dress, her head covered by a white bonnet tied under the chin, her brown curls showing. Her eyes are pale blue, but she does not look straight out at the viewer, and her lips are pinched, so her expression lacks both vitality and gentleness. As for his father, Pasteur depicted him dressed in his coat with the wide lapels, with a prudent expression around the mouth and a wrinkled brow.

Pasteur seems to have been most interested in capturing the actual looks of his subjects, and his portraits form a gallery showing all kinds of physiognomies that are observed with almost clinical patience. Here is a neighbor, old man Gaidot, a barrel maker and a great admirer of Béranger; here are the Roches, father and son, typical provincial notables; the mayor of Arbois, M. Pareau, with his medal of the Legion d'honneur, his tricolor sash, and his hairpiece à la Louis-Philippe; an old nun with her fluted coif, dressed in a kind of white cape against which a cross of wood and ivory stands out; a young boy in a velvet suit; a notary with a big red face. Most of these faces are marked by age and experience, showing frowns and severe expressions, as if the serious effort of the portraitist had rubbed off on them. The only exception to his puritan universe that at times recalls the American gothic may be the picture of a young girl seen from behind, with bare shoulders, in three-quarter profile, who furtively looks behind her as if surprised by the painter's indiscretion. Here the hand of the young artist allowed itself some freedom, and on the lips of the coquette there is a hint of the sensuousness of the Directory.

Yet the outstanding characteristic of these pastels is the extreme classicism of the line, the cold precision of the observation. Pasteur was so obsessed with the representation of reality that he lost all imagination. The effect of these portraits is entirely a matter of extreme precision, of the care given to every detail without recourse to color to create an atmosphere. There is no question that the artist's hand is sure and talented and that his observation is precise and sharp, but there is no inventiveness or poetry in these conventional poses inspired by the official portraits painted by Ingres or David. In a word, Pasteur showed himself to be as conservative in painting as he was to be revolutionary in science.

Meanwhile, at Arbois, the talent of this budding portraitist—who, like a schoolboy, signed his works "Louis Pasteur"—was taken seriously. For a brief moment his family and teachers may have thought about launching him in an artistic career. After all, people in the region were beginning to talk about a young painter from Ornans, a few kilometers from Arbois, who was not yet twenty years old and could do marvels with his brush. Pasteur following in the footsteps of Courbet? Some believed that it would have been possible, for instance the art critic Durand Gréville, who was to write in 1888: "Many of our painters who have won medals at the Salon have never drawn . . . as accurately. No one will regret that Pasteur chose a scientific career. But if he had wanted to, he would have held his own among the painters and — who knows? — become a very great painter."[11] It seems more likely that if Pasteur had taken up painting as a profession, he would have become one of the ambulant artists who went from town to town and tried to earn a few francs by drawing the portrait of the local pharmacist or the police chief. Nor should we forget that in 1839, when young Louis Pasteur was working very hard on his portraits, Arago presented to the Academy of Sciences of Paris an invention that was to have a great future and would soon have put Pasteur out of work, namely, the daguerreotype.

But in any case, Jean-Joseph Pasteur considered painting but a frill; what counted was proper schoolwork. His ambition for his son was a teaching position at a local collège. Nor was he the only one to see things this way. When the prizes were distributed at the end of Louis's year in *rhétorique*, Romanet, the principal, showered young Pasteur with compliments. Knowing how to make this uncommunicative and insecure adolescent feel better about himself, he urged him to go on to the baccalauréat, not to slack off in his efforts, to think about going further, and especially elsewhere. Again, he mentioned the Ecole normale.

Since there was no class of *philosophie* at Arbois and since it was out of the question to go back to Paris, a compromise was made: Louis would go to Besançon. The capital of Franche-Comté was not far away. Jean-Joseph himself went there six times every year to sell the skins of his tannery at the leather fair. At such times he could be seen climbing into the post-coach, carrying the satchel that held his account books under his arm. On the leather flap one could read: "Never think about anything but what you are doing at the time." What a strange maxim, to forbid a man to dream.

A Bachelor's Degree

At the beginning of the school year 1839, Pasteur thus found himself as a student in *philosophie* at the royal collège of Franche-Comté. Initially reserved and distant, he eventually adapted to the courses and to the teachers, but continued to use all his leisure time for his pastels. He now took a course in drawing with Charles-Antoine Flajoulot, who a year earlier had taught Courbet. Pasteur was quite proud of his accomplishments: "A few students here have told me that in Besançon some people are talking about a student at the collège who draws his classmates. That is because, as I told you before, the first portrait I made is on exhibit in the parlor, where lots of people see it when they come to visit a student."[12] This first portrait was that of the headmaster; the next one was of the son of the vice-principal and was hung up in the latter's office. Pasteur was beginning to find out how to get along; he was learning how to make people feel good.

His days were absorbed by schoolwork. In the evening, his eyes were strained by the candlelight. Because of his poor eyesight, he had to wear glasses for several hours a day. He also complained of violent migraines that seized him at exam time. "The final exam has given me a truly fine headache. Actually, this happens whenever we have an exam." Since he devoted most of his free time to painting and drawing, Pasteur did not read much. If he knew *Le Rouge et le noir*, Stendhal's novel published in 1830 in which the notables of Besançon are depicted, he never said so. He was, however, most enthusiastic about the *Méditations* of Lamartine and the edifying tracts of Joseph Droz, which he read on Sundays after mass. Silvio Pellico's story *Mes Prisons* also delighted him, and he recommended it to his sisters: "I would like them to read this interesting work where one breathes on every page a fine scent of religion that elevates and ennobles the soul." Pasteur also drew their attention to *Picciola,* by Saintine, calling it a "very pretty," "very useful," and "very interesting" book.[13] I had been endorsed by the Académie française, which had given it the Montyon prize, he wrote, full of a naive respect for the judgments of the Academy. In his reading, Pasteur showed no originality whatsoever. The books he mentioned were those that were read by all proper and right-thinking young people.

For if young Pasteur was anything, it was right-thinking, perhaps a bit too much so. To realize this, one needs only to read the rather jesuitical account he gave to his parents of a scandal that rocked the small world of the collège of Besançon:

> I am going to give you a piece of news that is still secret, but which, when it comes out, will cause quite a scandal; it is the dismissal (which will take place soon) of our almoner. This man, about whom all the boarding students of the collège have known for a long time, was not known for what he was in town, where his preaching talent and particularly his hypocrisy provided him entry

into the best families of Besançon. Riddled with debts and behaving abomina-
bly within the collège, he was always protected by his gown. You may want to
know just what were this man's crimes, for that is what one must call them,
but unfortunately these are things that one does not dare mention. Later you
will no doubt find out about them, for they will make a lot of noise. . . . It is
unfortunate, truly unfortunate, that religion has such ministers, and it is perhaps even more unfortunate that such a man should have headed a collège for
ten years. Keep this secret, for everything has not yet come out. As for me, you
can be sure that I will not divulge anything. You will be told that he left the
collège because his preaching talent made him wish to become a missionary.[14]

Yet this moral rigidity did not prevent Pasteur from making friends at
the collège of Besançon. Among them was Jules Marcou, of whom he drew
a portrait in his school uniform, which was blue with gold buttons. Born at
Salins in 1824, Marcou was to become a geologist who, as a specialist in the
geology of the United States, participated in the scientific exploration of the
Rocky Mountains and eventually taught at Harvard University.

Pasteur's closest friend, however, was not a scientist but a philosopher. Charles Chappuis was the son of the notary of Saint-Vit. He very soon
became fond of his taciturn classmate. He was able to accommodate to this
rather uncommunicative friend and appreciated his tenacity and his authority when they worked together on their philosophy homework. Chappuis
was to become the lifelong favorite friend and confidant of Pasteur, a man
who respected his choices and guided his ambitions. Along with Jules Vercel, Jules Marcou, and, later, Pierre Bertin, he was one of the few intimate
friends who called the scientist "tu." At Besançon, the two young men were
inseparable, and the apprentice philosopher proved to be the only person
who could bring Pasteur out of his silence and distract him. That is why,
when Chappuis went to Paris, where he was studying for the entrance
examinations for the Ecole normale, he received the saddest letters from the
friend he had left behind at Besançon: "You remember those days last year
when I did not answer when spoken to, when I was as dull as dishwater.
Today I was like that again, and unfortunately this happens to me a lot this
year. The only pleasure that is left to me is to receive letters, those from you
and those from my parents. Therefore, dear friend, write to me often."[15]

Pasteur was a good student and began to gain confidence. He now
started to judge the quality of his teachers. He found that the science courses
left something to be desired, but he was bowled over by his professor of
philosophy. Hippolyte Daunas had entered Normale in 1836, which meant

that he was full of enthusiasm and only ten years older than his students. What captivated them were not only his emotional remembrances of life as a Normalien but also his ability to expose them to the classic problems of metaphysics or aesthetics. "What can be more similar, more equal in every point to my hand or my ear than their images in a mirror? And yet I cannot substitute the hand as it is seen in the mirror for its model; for if it were a right hand, the other one in the mirror is a left hand, and the image of the right ear is a left ear, which also cannot take the place of the other."[16] If at the time Pasteur had occasion to demonstrate this reflection of Kant on the blackboard, it must have made him realize that philosophy can sometimes open up vertiginous scientific perspectives. From the *Critique of Pure Reason* to molecular dissymmetry, the distance is perhaps no more than one step, but the time to take that step was still far in the future.

Pasteur received his degree of Bachelor of Letters at Besançon on 29 August 1840. The minutes of his examination indicate that his answers in geography and history were mediocre, that they were satisfactory in philosophy, Greek, Latin, and rhetoric, and very good in all aspects of the Science program. But, although a degree of Bachelor of Letters was helpful, admission to the Ecole normale required a baccalauréat in science. Pasteur therefore decided to spend a second year at the collège of Besançon working for this indispensable diploma.

When the next school year began in October 1840, Pasteur divided his time between two activities. Not only did he study for his exam by taking courses in advanced mathematics, but, because he had been noticed by the headmaster (the one whose portrait he had done), he was offered a post as substitute teacher. This was a true proof of esteem: in exchange for modest and relatively undemanding work — which consisted of tutoring the pupils of the lower grades — he received room and board, plus a stipend of three hundred francs per year. Three hundred francs was a comfortable sum that enabled Pasteur to pay for his courses and to buy the heating wood, oil, and candles with which he supplemented the conserves and other preserved foods sent by his parents. On Sundays, when he carried out his very thorough examination of conscience during mass, Pasteur concluded that he was a happy student, except that he did not have much company. His life was rather lonely, for aside from a few fellow students at the collège, he did not see many people. His father rarely, and his mother never, came to visit, and he only saw his sisters during vacations, once or twice a year, for it was out of the question to go to Arbois just to spend a Sunday.

So he wrote letters and expected to receive replies. When he did not talk about his own studies, he worried about the education of his sisters. Strongly advocating music for one and drawing for the other, he implored his mother to dispense them from domestic chores that would interfere with their schoolwork and in a censorious tone inundated them with the advice of a model student: "It means a great deal, my dear sisters, to have will power; for deeds and work always follow the will, and work is almost always attended by success. These three things, will power, work, and success, are the mainstays of human existence: will power opens the doors to brilliant and happy careers; work allows us to pass through them, and once we have run the course, success will crown our achievement." For all their rather hackneyed and impersonal character, these letters were the only tie that kept Pasteur connected to his family. The father did not write often, the mother was too busy, and the sisters hesitated to try their hand with this big brother, who was very intractable when it came to spelling, the length of their letters, and their frequency. Pasteur did not hesitate to criticize his parents, to speak in their name and to give lessons in education and morality. Aware that he had already risen above their station in life, he felt he had to make up for lost time. With his savings he bought them the works of Cuvier and a dictionary, that of Napoléon Landais, which he sent them with the rather cruel remark: "I do not need it, as you can imagine, but such a dictionary is always useful to have in the house." This rectitude, even severity, was also a roundabout way of showing his affection and his generosity, for he was altogether incapable of indulging in effusive sentimentality. Feeling guilty about being privileged within his family and considering his years at the collège a debt, he offered to pay for the education of his sister Joséphine, at the time in boarding school at Lons-le-Saunier, by taking on some extra tutoring. And he was always particularly kind to poor Emilie, "who is not hard to please," and full of indulgence when he received long and confused letters from this child who never grew up: "They may not be very expressive or very intelligible . . . and they may not be properly done, but the intention is there, and that means a great deal."[17]

In view of the recommendations he lavished on his family in an authoritative tone, Pasteur was perhaps not quite as exacting when it came to himself. It is true that he was a skillful tutor for others and earned enough money to afford supplementary courses in chemistry and physics, but he still spent a great deal of time on his drawings, took up lithography, and did not totally invest himself in his work. His eyes were on Paris, now that Chappuis had gone there to study for the entrance examination at Normale, together with Pierre Bertin, a young physicist he had introduced to his classmate. Pasteur now wanted to join them and suddenly found Besançon too small. "Every student owes it to himself to go to Paris," he now affirmed. But Jean-Joseph kept his feet on the ground and remembered the disaster of the autumn of 1838.

In August 1841, Louis Pasteur had to go to Dijon to take the examinations for the baccalauréat in science. He was sure of himself. "As for my baccalauréat examination," he wrote in June, "the further I go, the easier I find it, because the subjects seem to shrink as I go along, and those I have studied do not seem to amount to very much."[18] This assessment was clearly wrong. Pasteur did not pass. His hopes for a preparatory course in Paris evaporated, and he had to do a second year of advanced mathematics in Besançon.

This was his second failure, and the blow it dealt him was perhaps even harder than the first, for in 1838 he had felt better about his inability to cope with Paris when he returned to the bosom of his family. But this time he was on his own, and he had to face his family's shaken confidence. His parents began to voice doubts about the son in whom they had placed all their hopes, and there was nothing Louis could do about it: "How often have I cursed that baccalauréat in science that you seem to consider beyond me." The beginning of the school year 1841 found him in a morose state: "One of the study hall teachers is sick at home and I am supposed to replace him, so that I am occupied all day long, and I do not know how long this will go on. You have your troubles, but so do I, and mine are getting bigger from day to day. I can see that I am entering adult life."[19] Failure had brought him down from the ideal sphere of moralizing commonplaces and given him a foothold in reality. About to turn nineteen, he was learning to get over the reflexes of a little boy who looks for praise.

He went back to work, not as proud as before, but more determined. He kept more regular hours ("I go to bed early and get up late" — that is, at half past five in the morning) and he regained confidence in his teachers: "The professor we have this year is very good," he wrote to his parents. "This certainly gives me hope for later. I will accomplish a lot this year, I am sure of it." And indeed, he was ranked second in his class on two occasions, and once even first in his physics class. He did not like mathematics because of its abstraction and was really at ease only with problems that he could visualize: "Nothing dries up the heart like this study of mathematics; it makes one lose all sensitivity. In the end one no longer has before his eyes anything but geometrical figures, letters, calculations, formulas. I, whose soul used to be so expansive, who every evening and every morning (of my year in *philosophie*) prayed to God with such fervor, I have let go of all that."[20] This taste for concrete problems gave Pasteur pause. Being more inclined to pursue concrete achievements than to manipulate abstract concepts, he convinced himself that he should go into engineering rather than into teaching. In other words, he began to study for the entrance examinations at both the Ecole polytechnique and the Ecole normale. Chappuis, to whom he had communicated this decision, warned him, telling him that success at the Ecole polytechnique was not the way to happiness: "You get more glitter there," he wrote to Pasteur, "but on the other side I see that gentle, tranquil life of the professor, which may be a bit monotonous at times, but which would be full of charm for one who likes it."[21] In the end it was another and more serious argument that turned Pasteur away from Polytechnique, and that was its military character. In his letters to Chappuis and to his parents he repeatedly talked about his hatred for army life, and he dreaded the approach of the age that made him liable to conscription. This son of a Napoleonic veteran had no desire to serve under the colors and would do whatever he could, even write to the minister, to obtain an exemption. This he actually achieved in 1843.

Once the thought of Polytechnique had been buried, he had to return to the realities of daily life. First the baccalauréat, then Normale. In August 1842, Pasteur went to Dijon for the second time; he was the only one of his classmates at Besançon to take the examination for the baccalauréat in science. He found the tests "a hundred times more difficult" than those of the previous year and in particular thought it most unfair that the examiners took a malicious pleasure in going outside the program or in exasperating the candidates with questions about minute details. "You are quite right to say that many things in the way exams are administered should be rectified," he wrote to his parents. "What upset me most . . . was their niggling over words," and he added: "But anyway, it's over, and a good thing, too!"[22] He was rather modest about his triumph, but he had reason to be modest. He had indeed succeeded in obtaining the degree of Bachelor of Science, despite a note of "mediocre" in chemistry.

So there was no question of resting on his newly earned laurels. Pasteur did not even go to Arbois to receive the congratulations of his family but returned to Besançon, where he faced the competitive entrance examination for the Ecole normale. This was his main ambition. The written part was very long, lasting from six in the morning until noon. He had to fight fatigue and migraine. The question in philosophy, about which candidates in science worried so much, did not give him much trouble. The subject was "On Evidence," and he handed in three tightly written pages. He felt more comfortable with physics and mathematics, and on 26 August he was notified that he was eligible for the second set of tests. Four years of studying had paid off. This is usually the home stretch of the race, the moment when students forget the time of doubt and discouragement and find the strength for one last push, and when pride and even presumption become legitimate. But what may be true for most students was not true for Pasteur. An extraordinary mixture of self-doubt and pride compelled him to make a dramatic decision: not satisfied with his ranking (he was ranked fifteenth in a group of twenty-two candidates) he preferred not to take the second set of examinations and to reapply the following year.

His refusal to accept mediocrity and uncertainty, his notion that time would improve his chances, his patience at the moment of embarking upon his career, all of this can be seen as Pasteur's first adult decision and indeed — why not? — as the birth certificate of his genius.

From the Pension Barbet to the Ecole Normale

Actually, Pasteur's calculation may have been quite different. His decision to do another year of preparation, in Paris rather than in Besançon, was a matter of stepping back in order to gather momentum for his jump, or rather his two jumps: first to accustom himself to life in Paris near his friend Chappuis, and then to enter Normale.

The big city was a true object of repulsion as well as attraction for this provincial. To be sure, he was most eager to go up to the capital, for it was the only place where he could make something of himself. At the same time he kept rereading in his friend's letters the passages where "he curses Paris, which he constantly calls squalid, sordid, disgusting." But his decision was made. Even though he had dropped the idea of entering Normale right away, Pasteur left for Paris. On the eve of his departure, he executed one last pastel, a portrait of his father, before he left his family, his eyes turned toward the future.

Once again he came to the pension Barbet. This time, however, he was not lodged in the boardinghouse itself but in the adjoining building, No. 1 impasse des Feuillantines. This was a bathing establishment, whose owners, Giroust and Girard, had made an agreement with Barbet. Amid shower stalls and portable bathtubs that served to provide "carry-out" baths, Pasteur shared a room with two classmates, one of whom was from

Salins. He found this situation very convenient, for hygiene was an obsession in the Pasteur family. Fearing that his son would wear himself out, Jean-Joseph gave him positively motherly advice in his letters: "Take some nice baths, for the heavy workload you have to carry right now might heat you up too much, even foot baths would be good. But if you are close to a bathing establishment, take my advice, it will help you sleep, and don't take them too hot."[23] This solicitude was in fact somewhat exaggerated. At almost twenty, Pasteur was no longer the fearful child who had been undone by leaving his home and his family. He was ready to take on Paris, but he still resembled Lucien de Rubempré more than Rastignac. He approached Paris not with the arrogance of the conqueror but with the humility of the beginner. He had come to prepare for the entrance examination to the Ecole normale, and nothing else counted for this student who was accustomed to subordinate his discontent to his ambition.

Although Barbet had again agreed to reduce the price of room and board, Pasteur only had a third of the sum he needed. He therefore had to earn the remainder by tutoring students in mathematics between six and seven in the morning, before going to class himself at the Collège Saint-Louis. This work schedule was not much different from what he had experienced at the collège of Besançon.

Most importantly, in Paris Pasteur was reunited with his friend Chappuis. Admitted to Normale and ranked tenth in the literary section, Chappuis was an assistant in the philosophy course of Jules Simon, who temporarily filled in for Victor Cousin. The encounters between the two friends were practically Pasteur's only distractions. To reassure his parents about his way of life in Paris, he told them that he paid no more attention to Paris and its people than he would if he were not there at all. To this he added: "Whom do I frequent here? Chappuis. And whom else? Chappuis."[24] Life really was not easy. Everything was expensive: heating wood, candles, paper, ink, pens, not to mention the extras, such as a treaty on physics by Benjamin Franklin, a table covering to place over the cracks that interfered with his writing, a hackney cab, or a meal at the Palais-Royal! The parents worried about the extravagance of their son, who had to account for every last sou, not without some impatience, so that he pointed out that he paid for almost all his courses himself. A gulf was opening between the would-be Parisian and the family he had left behind in the country. He constantly had to remind Jean-Joseph that studying for admission to Normale was a necessity, that there was no other way to become a respected professor. But the father still did not understand the difference between a schoolmaster and an *agrégé*.

In bed before ten o'clock every night, Pasteur did not even think of enjoying what Paris had to offer. Not one word in his letters would lead anyone to imagine that he was interested in monuments or museums, nor was there ever a line about the political debates of the day. He opened his eyes only to read, and when he walked in the Luxembourg Gardens with Chappuis, it was only to talk science or philosophy. One of the rare distractions to which he treated himself shortly after his arrival was an evening at the Comédie française, where he wanted to applaud the actress who attracted huge crowds and who kept tongues wagging with her public liaison with the Prince de Joinville, the son of Louis-Philippe: "Yesterday, I saw that woman who causes so much talk, Rachel. She certainly deserves the applause that is so lavishly given to her. I saw her in *Marie Stuart*. At one point the applause lasted for more than ten minutes. People clapped with their hands and their feet. Her features are not beautiful or regular, but because of her energy, she is superb."[25] It was a rare moment of enthusiasm for Pasteur, and all the more remarkable in that the play he saw, an antiquated drama written by Pierre Lebrun and inspired by Schiller's play, was not particularly exciting.

The fact is that Pasteur's really strong feelings were aroused by a different spectacle, namely, by the courses that the chemist Jean-Baptiste Dumas gave at the Sorbonne. Since classes at the Collège Saint-Louis did not start early enough to suit him, he rushed to hear the lectures of "France's first chemist," who was also extremely popular with the students. One of the founders of organic chemistry, and son-in-law of the famous naturalist Alexandre Brongniart, the forty-two-year-old Dumas was both a model for budding scientists and a major figure in intellectual society; his lectures attracted a considerable audience. "The lecture hall is huge, and always filled. One has to be there a half hour early in order to get a good seat, just like in the theater. Here too there is a lot of applause. There are always between six and seven hundred listeners."[26] This first contact with chemistry deeply marked Pasteur. He never forgot how impressed he was when he first experienced the attentive silence of the audience or the passionate exclamations that greeted the lyrical flights of the master, and he was to feel for the rest of his life a veritable veneration for Dumas, "who could set fire to the soul."

Yet physics was still what most attracted Pasteur. Encountering some difficulties in the acoustics course, he decided to take singing lessons. This was an unusual thing to do, and it reveals the profound sense of the concrete that underlay his intelligence: in order to understand a scientific theory, he took a detour by way of an artistic practice. His singing teacher was a young man from Dole, "very good, and he does not have the temperament of an actor at all, except for his love of glory and the artist's enthusiasm."[27] Perhaps one should also see this fanciful step on young Pasteur's part as a certain desire to gain more self-confidence; in learning to sing he also improved his diction and his projection. The example of Jean-Baptiste Dumas had something to do with this, for the young student understood that knowledge alone does not make for a master, that one must also be capable of captivating an audience by the power of one's eloquence. However, Pasteur soon had to give up these lessons because he could not afford them.

All his efforts paid off, and 1843 was the year when everything went right for him. Pasteur participated in the general physics competition, where he received honorable mention, and at Saint-Louis he received the first prize, also in physics. Most importantly, he was admitted to the science section of the Ecole normale, this time with a good ranking. Fourth in his class, he saw no reason not to accept.

To enter Normale in 1846 was truly to become part of a legend. Created by the Convention at the same time as Polytechnique, in 1795, the school had seen a succession of the most prestigious teachers and the most promising students. The first courses were given by such men as Lakanal, Monge, Laplace, Daubenton, Volney, and Bernardin de Saint-Pierre, and among the students who preceded Pasteur one finds such names as Victor Cousin, Augustin Thierry, Louis Hachette, and Victor Duruy. While the collective memory is most familiar with the literary graduates, one should not neglect the scientific ones; their careers may not have been as brilliant, but they produced remarkable work. The mathematician Antoine-Augustin Cournot, for instance, the astronomer Victor Puisieux, or the physicist Paul Desain were there to show that France was still the first nation in the world when it came to scientists. Scientific research was no longer considered a pastime for the amateur or a way to bolster a philosophical argument; it had assumed an existence of its own and even become a factor in material progress. One only has to line up, in no particular order, the names of Laennec, Ampère, Cuvier, Cauchy, Arago, Niepce, Fresnel, Galois, Gay-Lussac, or Pelletier and Caventou to conjure up the extraordinary outpouring of studies and discoveries that revolutionized knowledge in the first half of the nineteenth century.

But intelligence or genius are not always appreciated by the powers that be. For one Baron Cauchy, legitimist and defender of law and order to the point of reaction, how many like Evariste Galois were there — insubordinates, revolutionaries, even men ready to indulge in what we today call terrorism? The state, to be sure, had to train professors, intellectuals, and scholars, but it did not want to deal with opponents, insubordinates, and conspirators. This meant that the Ecole normale was under strict surveillance. Napoleon had wanted to make it into a kind of elite lay seminary (and had even gone so far as to envisage compulsory celibacy for the faculty), and under the Restoration it was attacked by the clergy, which felt that it was bound to develop into a worrisome hotbed of free thinking. The policy of Louis-Philippe's ministers was more crafty: They regained control of the school by promising a physical renovation that was indeed more than overdue. While the first courses had been given at the Muséum, the Normaliens of the years 1820–30 had to make do with temporary quarters scattered around the Montagne Sainte-Geneviève, among them the former Collège du Plessis in the rue Saint-Jacques (actually an annex of the lycée Louis-le-Grand) or the seminary of the Saint-Esprit, rue des Postes (today rue Lhommond). In 1842 the government decreed the construction of an independent complex of buildings in the rue d'Ulm, to be arranged around a kind of cloister and planned for one hundred students. But the students were not to enjoy the new buildings until the beginning of the school year in October 1846.

In October 1843, Pasteur therefore had to put up with humid dormitories and food that he ironically described as "good for young men." Yet unlike many of the students he did not look for a room in town; on the contrary, he asked and was granted permission to move into the school even before the start of classes. And while he continued to work hard, he also began to acquire his own personality. In telling his parents about an official ceremony addressed by the director of the school, Paul-François Dubois, and the Minister of National Education, Abel-François Villemain, he could not resist some remarks about the hollow rhetoric and the wooden delivery of the politician: "M. Dubois gave a speech, and so did M. Villemain; he bellows like an ox. I am sure that outdoors he would make himself be heard by a thousand people. What he said were the usual grand phrases, but it did not mean much."[28] What mattered to Pasteur was his work. He regularly spent twelve hours a day studying, with lectures at the Ecole normale or at the Sorbonne interspersed with practical work in the laboratory, where the students were initiated into the manipulations of chemistry, but he also learned more unexpected skills, such as wood planing and lathing, glass blowing, working with clay, and making keys and locks. Nor was everything strictly focused on the scientific disciplines, for the school, in order to produce scientists capable of communicating their findings, offered free courses in German and English. And in the summer the famous botanist

Adrien de Jussieu, professor at the Muséum, took students on botanizing expeditions in the countryside around Paris. In the woods of Chavilles or Verrières, the learned Normaliens crossed paths with young lovers right out of the *Scènes de la vie de bohème*.

Meanwhile, at Arbois, the whole family was anxiously watching that son on whom it had pinned all its hopes. Since the mother was wholly occupied with her household, the father wrote for two and put down every last piece of advice for everyday life, a collection of disparate and picturesque precepts: rub a few drops of cologne water into your hair to prevent headaches and strengthen your eyesight, put blue lenses into your eyeglasses, do not believe M. Thiers's reports about education, do not get mixed up in demonstrations against the Jesuits. As time passed, the father's news became less relevant to the son's life: "I have bottled some '34 [wine] which I bought especially to drink in honor of the Ecole normale," or, "a couple of days ago we killed a pig," or, "M. Petitjean's son has died." Jean-Joseph also worried about getting his money's worth out of his son's allotment of time. The poor sales of his hides at the Besançon fair made him uneasy, and he saw everything strictly in terms of profit, more attuned to the laws of business than those of science. Fascinated by the young computing prodigy Henri Mondeux, who toured the provinces and made a great deal of money by exhibiting his prowess in mental calculation, he thought that mathematics was the most lucrative and hence the most interesting field. He therefore recommended it to Louis, not without citing a saying of Napoleon. "The Emperor said to a distinguished scholar who kept in the background and lived only for science: 'You must realize, Monsieur, that in this world, one must have money, a great deal of money, or else merit will look weak in the eyes of the world.'"[29]

But Pasteur did what he considered best and pleaded in favor of physics. Not to be discouraged, his father gave him other advice, particularly about health: "For the thousandth time I am telling you that what you need most is good health. You must therefore preserve it by every means at your disposal. . . . I hope that your friend Chappuis will share your leisure activities; you all need them, health above all." The wisdom preached by the old veteran had all the platitude of Béranger's songs. Thus he repeated the foolish assertion that there is more spirit in a hundred liters of good wine than in all the philosophy books in the world, and, in speaking of women, plunged into a diatribe in praise of the housewife: "Women are scatterbrained and flighty, always out to search for happiness where they cannot possibly find it. Outside the family home, there is nothing for women but resentment and sometimes shame. I think I have already told you often enough that the choice of a wife demands a great deal of prudence." Actually, Pasteur was not thinking of getting married, at least not for a long time, but rather envisaged putting one of his sisters, Emilie or Joséphine, in charge of his household, now that he had so closely supervised their education; and they of course could not possibly dream of anything better. The father agreed, and so, apparently, did the sisters. "Neither of them hankers after greater happiness: to serve you, to take care of your health, that is what both of them want to do," Jean-Joseph wrote to his son.[30]

Unlike his classmates, Pasteur did not go in for the quasi-official entertainment of the Normaliens, the usually vehement discussion of all possible and imaginable topics. He preferred spending his hours of leisure alone, in the library. Reserved, almost shy, he lived apart, constantly working. During his three years at Normale, the political or cultural events of the day do not seem to have affected him at all. Neither the capture of Abd el-Kader's retinue by the Duc d'Aumale in 1843, nor the publication of Alexandre Dumas's *Trois Mousquetaires* in 1844, nor the beginning of the restoration of Notre Dame de Paris by Viollet-le-Duc in 1845 had the power to make him deviate by one step from the exclusively academic path he had drawn for himself, making sure that it never crossed the road of current events.

Even his Sunday afternoons were devoted to work, for he spent them in the laboratory of the Sorbonne where Jean-Baptiste Dumas's laboratory assistant gave him private lessons while Chappuis needed all his philosophy to wait patiently, perched on a stool in a corner. When the experiments were finished, he forcibly dragged his friend along on a walk. These were Jean-Joseph's orders, and he did his best to carry them out. But even when Pasteur consented to leave the laboratory to take a few turns in the Luxembourg, the conversation concerned nothing but his course work and plans for future experiments.

Actually, the first experiment carried out by Pasteur made quite a splash. It was a matter of verifying a proposition taught in class. Since his professors had always spoken about the difficulty of obtaining natural phosphorus and put off showing this experiment because it involved time-consuming manipulations, Pasteur felt that this was his chance to make his mark. Did he, perhaps, remember the example of a student at the collège of Arbois, Charles Sauria, who in 1831 obtained a patent for a new technique for producing phosphorus matches? However that may be, he obtained some bones from a neighboring butcher, reduced them to ashes, and added sulfuric acid to them. Having been set aside for twenty-four hours, the mixture produced a precipitate that after filtering yielded a kind of heavy syrup. Then this substance in turn had to be heated to make it give off a vapor of phosphorus that was condensable in water. At the conclusion of these delicate and dangerous manipulations, given the high toxicity of phosphorus, Pasteur obtained a few grams of a transparent, yellow substance. It was his first scientific victory, and he proudly put the residue into a splendid flask with a big blue label to publicize his exploit.

The puerility of the procedure and Pasteur's somewhat naive pleasure might make us smile. And yet this experiment with phosphorus was a defining moment, for it was a true initiation. Until then Pasteur had been a bright, hard-working, demanding, and ambitious student, but he had only dealt with acquired knowledge and was earnestly striving to become a very effective cog in the academic transmission of knowledge. He could have become a conscientious, serious, and meticulous secondary school teacher, eager to become part of the classical culture and to educate the young. In fact, in October 1844, he was flattered when a professor at the lycée of Besançon told him that he would be pleased to have him as his successor and promised to support him; at that time a teaching position at Besançon represented the height of young Pasteur's aspirations.

But when he discovered the thrill of research and discovery, his life took on an entirely new direction. Pasteur suddenly found himself in the grip of that frenzied passion that opens up unknown territory to the greatest among us. To discover something is to break up a wave in order to see the birth of the spray. Having felt the thrill of his first experiments, Pasteur understood that proving a hypothesis can lead to ecstasy.

2

CRYSTALS: A NEW LAW

It was not enough to get into Normale. One also had to get out, that is, to pass the competitive examination called *agrégation*, and then be appointed to a teaching position. Concerning the first stage, everything went flawlessly; in early September 1846, Pasteur learned that he had been ranked third in the physics examination. There were only four openings for fourteen candidates. The reason why Pasteur's ranking was not better was a mediocre performance in the natural history lesson. By contrast, he attracted the examiners' attention in the physics and chemistry orals, essentially for his pedagogical skills. The examiners said that among the newly appointed Normaliens he was the only one who knew how to teach.

As was customary, Pasteur did not have to wait long to find out where he was to teach; within a few days, while he was relaxing with his family in Arbois, he received his appointment as professor of physics at the collège of Tournon in the department of Ardèche. This administrative decision did not please anyone. Jean-Joseph had hoped to see his son appointed at Besançon, closer to the family homestead. Pasteur himself, who now realized that Paris was the only place to start a career, put off his departure as long as he could. He wrote to the assistant director of the Ecole normale, asking him to speak to the minister on his behalf, for he had decided that he would like to stay at the Ecole normale for another year, working as a laboratory assistant in chemistry—and if he really had to teach at the same time, he would be willing to do some tutoring. In fact, with the increasing self-assurance his various successes had given him, Pasteur addressed himself directly to Jean-Baptiste Dumas, who had recently founded the Ecole centrale, asking to be allowed to teach courses there: "Our supervisors' assessment of the oral presentations given at the Ecole in the third year and at the time of the agrégation permit me to tell you that I will be able to teach the material with clarity and energy. I should also like to assure you, Monsieur, that in making this request I am not interested in earning money, nor in the vastly more blameworthy aim of furthering my career by establishing a relationship with a person as highly placed as you are in a science to which I, too, hope to devote my life. My greatest desire, I repeat, is to find a way to perfect my teaching skills."[1]

In a clever apology for his boldness, he then added: "I am too young to have seen you in your early days, but surely, you did not reach the summit you now occupy in one fell swoop, and you too must once have had a great desire to obtain a teaching post that would give you the mastery needed to climb that steep slope whose heights you dominate today."

The Balard Laboratory

It was not, however, Dumas who saved the situation, but rather Balard, a supervisor [*maître de conférences*] at the Ecole normale who had noticed Pasteur. Wishing to keep this promising student close to him, he gave him the desired position. In October, Pasteur could return to Paris, much reassured; his appointment at Tournon had been postponed and he had been hired as a graduate laboratory assistant [*agrégé préparateur*] in the laboratory of Jérôme Balard.

At the age of forty-four, Balard already felt that his career was drawing to a close; he was more interested in his students' work than in his own and allowed them to pursue their studies as they saw fit. This freedom called for a great deal of inventiveness. Balard thought that one should work without instruments. If these became really necessary, the researchers had to construct them themselves. Pasteur accepted these unusual directives and became quite attached to the master.

Balard had discovered bromine in 1826, when he was a twenty-four-year-old chemistry teacher at the collège of Montpellier. As a result of this brilliant accomplishment, he was soon called to Paris to succeed Baron Thenard in the chair of chemistry at the Faculty of Sciences and Jean Darcet in his seat at the Académie des sciences. He eventually occupied the chair of chemistry at the Collège de France as successor of Théophile Pelouze.

At the Ecole normale supérieure, Balard had obtained the use of a few rooms as a research laboratory, which was labeled "collections depot" in order to sidestep an administrative prohibition. The statutes of the institution did not provide for research on the part of a chemistry professor.

So far, Balard had worked only in the back room of a pharmacy or behind its counter. Delighted with the new installation in the freshly inaugurated premises in the rue d'Ulm, he immediately brought a cot into this clandestine laboratory so that he could be there as much as possible. The rest of the time he led a simple life, lodging in a student room that was bare except for a few chairs of fake mahogany. He did not travel much, but when he did, his baggage consisted of one shirt wrapped in an old piece of paper that he stuffed into his pocket. This modest man had never lost his southern temperament and was of a perpetually good humor.

In his effort to win Balard's favor, Pasteur proposed to him a number of different research projects. One of them concerned the influence of atmospheric pressure on crystallization. But Balard did not care for the subject, and so Pasteur did not pursue it. This was just as well because, as he wrote to his faithful friend Chappuis, "a very fortunate circumstance"[2] had just arisen, namely, the arrival at the laboratory of Auguste Laurent, professor of chemistry at the University of Bordeaux. Laurent had taken an unpaid leave to come to Paris to pursue his research. Having found a position at the Mint that allowed him to cover his day-to-day expenses, he came to Balard's laboratory as an unpaid temporary collaborator.

Laurent was already forty years old at the time, and his work in crystallography had earned him a corresponding membership in the Academy of Sciences. A very original mind, he defended some revolutionary ideas, for he was one of the founders of modern atomic theory. His work on the mixing of substances favored the development of the notion of isomorphism and the comparison of crystals: "Since chlorine and bromine are isomorphic with hydrogen, I tried to bring them into various organic substances by substituting them alternatively for hydrogen. In this manner I hope to obtain different substances, which would, however, have the same composition and the same crystalline form."[3] The idea was not without originality and the hypothesis was stimulating. In the presence of Balard, Pasteur listened most attentively to Laurent.

Laurent, for his part, had noticed Pasteur. He wrote to the chemist Charles Gerhardt, "There is a young man at the Ecole normale whom I asked to analyze the chloride of ammonia arsenic." He gave Pasteur his *Précis de cristallographie*, as well as a simple method of analysis by the blowpipe, which he was too modest even to sign. In short, the young Normalien was learning a great deal. Pasteur was also pleased to be working with this calm scientist with his regular but somewhat unhealthy features, of whom he spoke at the time without realizing that the scientist would die of tuberculosis in 1853: "I had started on a project of my own when, a few days ago, M. Laurent kindly asked me to collaborate on one with him," he reported in a letter to Chappuis. "Even if this project should not produce any finding worth publishing, you can imagine that it would be very useful for me to do practical work for several months with such an experienced chemist."[4] However, he added: "During the summer I will try to do some other work by myself." In an effort to avoid jealousies and criticism, incidentally, Pasteur asked Chappuis not to spread it about that he was working with Laurent.

As it happened, Laurent was not to stay at the Ecole normale for very long. In April 1847 he was appointed assistant lecturer to Jean-Baptiste Dumas at the Sorbonne. Although in the end Pasteur worked with Laurent for only a few months, his influence was great, not so much in the area of experimentation as with respect to the formation of hypotheses and theories, as Pasteur pointed out in a note: "One day, it happened that M. Laurent, who was looking, if I recall correctly, at some perfectly crystallized sodium tungstate . . . showed me under the microscope that this salt, though very pure in appearance, was a mixture of three distinct kinds of crystals, which anyone who has some knowledge of the structure of crystals could easily recognize. This example and several others of the same kind made me appreciate how useful a knowledge of the crystalline forms is for chemical analyses."[5] However, Pasteur soon gave up working on Laurent's experiments. A new chemical mixture led him toward another set of manipulations, on which he eventually based his dissertations.

In Pursuit of Underlying Structures

The modesty of Pasteur's first research efforts in Balard's laboratory testifies to his extreme cautiousness. He did not set out to solve a major problem in physics, chemistry, or biology. A conscientious student, he sought instead to advance the work of some of his teachers.

He wanted to be a chemist. Now the principles of the chemistry of his time were based on premises which, established in 1823 by Michel Eugène Chevreul, governed the analysis of organic substances: the individuality of a molecular species resided not only in the nature and the proportion of the elements that composed it, but also in their arrangement. This definition attributed major importance to the structures formed by molecules. However, most of the "classical" chemists—Scheele, Priestley, Lavoisier, and their immediate successors—had been mainly concerned with the nature, the quality, and the atomic proportion of molecules and tended to neglect their organization and their arrangement. It was in this area that Pasteur decided to fill the gaps, for he had a notion that the structures of molecules offered a hitherto poorly understood perspective that might well corroborate the current theories.

In deciding how to conduct his research, Pasteur was faced with two possible experimental approaches: one was the purely chemical one, which consisted of decomposing substances in order to define the pattern in which the molecules are arranged by their path of dissociation; the other sought to reveal certain physical properties, for example, in the case of a crystal, its form. Pasteur chose crystallography both because it pleased him and because he considered it useful. "When I began to pursue specific research, I sought to strengthen my abilities by studying crystals, anticipating that this would provide me with knowledge I could use in the study of chemistry."[6]

Nineteenth-century crystallography amounted to little more than goniometry and polarimetry. The techniques available to today's crystallographer (mass spectrography, the absorption of infrared or ultraviolet light, nuclear magnetic resonance) were of course unknown, indeed unimaginable. The goniometer, the instrument of the mineralogists, measured the axes, angles, and forms of crystals. The polarimeter, that of the physicists, quantified the rotation of light.

Pasteur's turn toward crystallography resulted not only from the needs of his experimental work but also from theoretical considerations. In the seventeenth century, Robert Boyle had shown that there is a relation between the form of the invisible constituent parts of a body and its organization in crystal form. Mendeleyev was to say exactly the same thing when he postulated a connection between atomic composition and the structure of crystals. In one of the books that Pasteur may well have read in the library of the Ecole normale supérieure, Abbé René Just Haüy, the founder of French crystallography (and brother of the creator of the Institute for Blind Children) wrote that substances of the same chemical composition can have different crystalline forms, but that it seemed inconceivable to him that substances of not only the same chemical composition but also the same molecular structure would not have the same crystalline form. This was the first step toward a comprehensive system in which molecular identity was inferred from the properties of the crystal. But this was only done for minerals.

In 1819, the formula "same crystal, same chemistry," of the German chemist Eilhardt Mitscherlich approached the question from a different angle by means of the theory of isomorphism, which was to play a major role in the chemistry of the time and bring fame and glory to the German chemist. Isomorphism postulated that the resemblance of chemical crystals translates to that of their constituent molecules: when Mitscherlich observed that the phosphates and the arsenates of the same metal crystallize in the identical manner, to the point of becoming exchangeable in the structure of the crystal, he concluded that their atomic structures were related.

It was in thinking about these problems that Pasteur conceived the idea of a brilliant shortcut: he decided to assume that the properties of a specific mineral crystal — quartz, whose structure and capacity to deflect light were well known — are also found in the chemical crystal he chose for his study, tartaric acid. The choice of quartz for this argument by analogy was not made haphazardly, for it was the principal model referred to by the crystallographers whose work Pasteur had painstakingly brought together and compared.

IN THE EIGHTEENTH AND EARLY NINETEENTH CENTURIES, mineralogy was one of the principal earth sciences. It was an important field because rocks, most of them in their crystalline state, had been transformed and modified according to laws that did not go beyond the age-old preoccupations of the naturalists.

In his *Essai de cristallographie*, published in 1783, Jean-Baptiste Romé de l'Isle had rationalized the principles that governed the description of crystals and discouraged the fanciful descriptions of the ancient "mineral philosophy." Using the measurement of angles to help him sketch crystals, Romé de l'Isle postulated one original form from which all others were derived when one of its facets was truncated to a greater or lesser extent. But this was still only to describe, rather than to establish, a law. As Pasteur pointed out, this was why the credit for bringing true science to the study of crystals belongs to Romé's successor, Haüy.

Abbé Haüy owed his scientific breakthrough to a mishap, the dropping of a spar crystal. Born in Oise in 1743, René Just Haüy was the son of a modest weaver. He soon attracted the attention of the monks at the abbey in his village by his zeal for religious ceremonies and singing. Because he was gifted in the humanities, they gave him a fellowship to study in Paris. A boarding student at the Collège Cardinal Lemoine, he took courses with the famous grammarian Lhomond, who also taught him to collect plant specimens in the Jardin du Roi, which the Revolution renamed *Muséum d'histoire naturelle*.

Repeating what he had read in Cuvier, Pasteur told the story of the circumstances that led Haüy to his discoveries: Hearing that Daubenton's mineralogy courses were attracting large audiences, Haüy decided to go to one of the lectures. Fascinated by the lecturer and the subject, he gave up botany on the spot and became a physicist. But since he had learned the physical sciences rather late in life, Haüy approached them differently. Thinking of the complicated but constant forms of flowers and fruit produced by each new sprouting, he felt that the simple geometric forms of minerals should also, in a similar if not the same manner, be governed by laws: "How does the same stone, the same salt, assume the form of prisms or of needles without changing its composition by one atom, while the rose always has the same petals, the acorn the same shape, the cedar the same height and the same development?"[7] If one has to know a question in order to answer it, then Haüy did. All he needed now was one of those strokes of luck without which no discovery would ever be made. A careless movement on the part of a friend made him drop a piece of calcareous spar that had crystallized into prisms. Haüy picked up what was left of it and examined its facets, pitch, and angles. To his amazement, they were identical to those found in rhomboid spar crystals and in Iceland spar. This observation was the starting point of his entire theory: a whole new world had just opened to him.

As Pasteur tells the story, Haüy returned to his study, took a piece of spar crystallized into a hexahedral pyramid — it was often called a "pig's tooth" — and proceeded to break it. Again he found the same rhomboid that was characteristic of the Iceland spar, and even the small pieces he picked up were rhomboids as well. Then he broke a third piece of spar, the kind that was called lenticular, and here too found a rhomboid in the center, as well as smaller rhomboids in the fragments. "I get it!" he exclaimed; all the molecules of calcareous spar have one and the same form. By becoming arranged in a different pattern they form crystals.

Haüy did not hesitate to smash his entire collection of crystals, along with those of his friends, to which he also put the hammer. Everywhere he found the same structure governed by the same laws: The elementary crystalline form of a substance depends on its chemical composition and on its structure. Whatever the differences in appearance among garnet, fluorspar, pyrite, and gypsum, their crystals break into lamina parallel to the facets of the nucleus. Their outward appearance is caused by the piling up of the original crystals. Haüy proposed a law that postulated a relationship between symmetry and structure. He imagined that the crystal was produced by the juxtaposition of elementary solids.

This work was rewarded with a seat at the Académie des sciences for Abbé Haüy, whose threadbare and unfashionable cassock stood out among the sumptuous embroidered coats worn at the Institut. But nobody cared, for before him, most crystals had never been subjected to systematic analysis, and their classification had created utter confusion. Nor did these discoveries pertain only to minerals, for in relating the composition of a substance to the form of its elementary crystals, Haüy called for the same kind of analysis for chemical crystals. He not only predicted that differences would turn out to have been overlooked but also foresaw that differences that had apparently been found might not exist after all. But Haüy's hypotheses had their limits, for he identified the crystal not by its chemical molecule but by its structure. Within his system, every mineral species represented a different grouping of identical molecules.

Abbé Haüy left a twofold legacy: on the one hand a collection of crystals that Pasteur could later study at the Muséum, and on the other a *Traité de minéralogie cristalline* which, published in 1801, served as the textbook for the course taught at the Ecole normale supérieure by Pasteur's teacher Gabriel Delafosse. In teaching Haüy's laws of symmetry, Delafosse insisted on their exceptions. Thus he pointed out that in describing the symmetrical forms of crystals Haüy had remarked that some of the quartz crystals were asymmetrical, having one facet more oblique than the others. Haüy, who liked to label every variety of mineral, had called these kinds of minerals hemihedral [*plagièdre*]. He had thus described right and left hemihedrals, depending on the direction in which their facets were slanted. But he had thought that this was a random phenomenon and had failed to recognize that this exception was part of a pattern. The credit for understanding this belongs to Weiss, a German crystallographer who in going over the same ground made a comparison among the hemihedrals and called the phenomenon as a whole "hemihedrism." In speaking about hemihedria to his young students at the Ecole normale, Delafosse pointed out to them that in the past the laws of symmetry had been applied incorrectly. He insisted that molecules could be arranged in different ways in these hemihedral crystals. In the number of facets and the orientation of its facets and its angles, quartz showed hemihedria in both directions. Delafosse concluded that this breach in Haüy's law of crystalline symmetry was caused either by the nature of the crystal or by the pattern of its molecules.

Work in mineralogy and crystallography was closely related to research in optics, which was making great strides in the early nineteenth century. The theories of Etienne Louis Malus on the polarization of light — that is, the modifications light undergoes when reflected by an opaque body or refracted by a crystal — had opened a new chapter in the field of physical optics. He was a fine figure of a man, this revolutionary who made his mark in the army of Sambre-et-Meuse in 1796 and then went on to serve as an officer in the corps of engineers. Having been chosen to participate in the Egyptian expedition, he became, thanks to his friendship with Kléber, commandant of Jaffa, where he formulated his first scientific reflections. In the desert glare of Liesbeh, the African sun led him to think about the existing theories of light. Some of his work reached the Institute of Cairo, but for the most part his discoveries came to grief in the sands along with the baggage of the French army, which evacuated the Near East in 1801. During the next few years, Malus had little time to devote to science. It was only in 1809 that he came to Paris, where he lived in the rue d'Enfer, not far from the Luxembourg Gardens. It was a circumstance to which he owed his principal discoveries in optics.

Legend has it that Malus was in love and that his ardor was unrequited. It is said that in order to observe the woman he loved from afar, and especially without her being aware of it, he invented an apparatus in which two mirrors that reflected each other allowed him to follow the silhouette he desired so much. But he had to do something to adjust the apparatus, for, to the despair of the disappointed lover, the silhouette vanished or reappeared when the reflectors were moved.

Scientific sources describe this event in more prosaic but more concrete terms. Arago claims that Malus was looking through a quartz crystal in the light of the sunset reflected by the windows of the Palais du Luxembourg when, to his astonishment, he noticed two images of different intensity, one of which vanished when he turned the quartz crystal. Arago claimed that Malus was mistaken when he said that he saw one of the images disappear, since the polarization on the windowpanes at the moment of the experiment could only be partial. However that may be, the fact is that in setting forth the properties of reflected light, Malus was indeed the creator of a new branch of physical optics.

To be sure, double refraction had been known for some time. No sunset was involved, but the image of a candle flame seen through a crystal had been talked about before. Before Malus, Descartes, Newton, and Huygens had glimpsed certain properties of refracted light. But what Malus observed for the first time was the deviation that light undergoes when it is reflected. This he called polarization, stating that whenever a ray possesses this remarkable property, it is polarized.

On the very night of his fortuitous discovery, Malus proceeded to establish the laws of polarization in his bathtub. Since the last rays of the sun were gone, he observed the beams of light from a candle directed onto the surface of the water and reflected by a quartz crystal. Noting that only one of the beams traversed the liquid, he established his theories about the modification of light by a mineral crystal.

Malus died prematurely in 1812, at the age of thirty-seven, and thus did not have time to exploit his discovery. But Arago was to reexamine the phenomenon and to show that the polarized rays that traverse a thin sheet of crystal are colored in complementary shades. This led him to the discovery that they are polarized in different directions. However, Arago was not interested so much in the rays themselves as in what modifies them. He analyzed the fading or reappearance of colors in terms of what they signify, namely, differences in absorption. Thus he noted that crystals, from rubies to emeralds, show different hues depending on whether they are examined in fair or overcast weather. Polarized light brings out new attributes. Polarimetry, that is, the measurement of the degree of polarization, thus introduced the analysis of light as a new scientific tool. But Arago did not carry his study any further and abandoned the subject a year later. Eventually it was Pasteur who recovered Malus's legacy and applied it to molecular chemistry, along with Haüy's work on the composition of crystals.

In his first studies, Pasteur also constantly referred to Jean-Baptiste Biot, one of the most venerable figures of French science. Astronomer, mathematician, physicist, and chemist, he became a member of the Académie des sciences in 1803, at the age of twenty-nine. Associated with the greatest minds of the period, Arago and Gay-Lussac, he accompanied the latter on his first balloon ascension. At an altitude of four thousand meters they carried out analyses of the refraction of light through the layers of the atmosphere.

For Pasteur, this remarkable achievement exemplified a career worth imitating. In the course of time, Biot came to consider the young scholar, who was almost fifty years his junior, as his pupil, his protégé, almost his adoptive son. Whenever Pasteur was faced with an important decision, he always received judicious advice from this master, even if he did not always follow it. From scientific critique to more personal counseling, Biot was to play an active role in Pasteur's life from the time when he sponsored his first experimental research in crystallography. It should be noted that a large proportion of Pasteur's papers read to the Academy of Sciences were dedicated to Biot.

Intrigued by the phenomena of the emission of light, Biot built on the work of Malus and Arago on refraction and focused particular interest on the spectrum of colors emitted by polarized light and on the optical properties of quartz. In the course of his studies he invented the polarimeter, which was the first experimental tool used by Pasteur. "Take," said Biot, "a receptacle with two parallel glass facets and fill it with a certain liquid, such as water or alcohol. Then place these liquid shields in the path of a polarized beam before it is refracted by a prism. The image will remain the same, which means that the originally imprinted polarization has not been disturbed." All liquids that allowed light to pass through them unchanged he called molecularly inactive. But unlike water and alcohol, a large variety of other liquids, such as solutions of different kinds of sugars, oils, natural organic products, and oil of turpentine, as well as various liquors, interfered with polarization. The image appeared colored because the plane of polarization had been deflected. Biot said that "these substances are molecularly active."[8] He reported on this notion of active and nonactive molecules and the method of polarimetry that revealed their status in the *Bulletin de la Société philomatique* of December 1815, a few months after the fall of the Napoleonic Empire.

One of the reagents of choice that Biot used and examined in his effort to understand the phenomenon was tartaric acid. He showed that it deviates the polarization of light to the right. He called this acid "dextrorotatory" (right-turning).

In addition to chemical molecules, Biot was interested in minerals, particularly quartz crystals. He made the observation that in a given sample of quartz, some pieces deviated the light in one direction, others in the other. But in order to observe this, one had to work with a crystalline quartz. In its dissolved and noncrystallized state, quartz lost its rotational activity. Biot's observations thus fell in line with Haüy's theory: the effect of organic substances on polarization was of a molecular nature; it was dependent on the chemical individuality of each molecule. In quartz, the phenomenon resulted from the mode in which the crystalline particles were bound together. Deviation from the plane of polarization was thus a function of the nature of the dissolved molecules in liquids, while in minerals it was a function of the physical properties of their crystallized form.

Having come to this conclusion, Biot stopped working on this problem. He never attempted to investigate the chemical molecules on the basis of his observations on minerals. The lessons of the mineral world had yet to be applied to chemistry. As a mineralogist, Haüy had demonstrated the existence of hemihedrals that do not have the same facets. As a physicist, Biot had recognized the existence of two kinds of crystals, dextrorotatory (right-handed) hemihedrals and levorotatory (left-handed) hemihedrals, in quartz.

It remained for the English astronomer Sir John Herschel to make the connection. The son of the discoverer of Uranus and the satellites of Saturn, Herschel was a first-rate astronomer in his own right, as well as a great mathematician and a brilliant physicist. In 1820, at the age of twenty-eight, he submitted to the Royal Society of London an important collection of *Memoirs on Optics and the Rotational Polarization of Rock Crystal*. Bringing together separate observations of crystalline forms and the physical rotational properties of quartz, Herschel wrote: "Quartz samples whose hemihedral facets point in the same direction rotate the polarized light in that direction: a right hemihedral produces right rotation — and vice versa."[9] He thus related the asymmetry of quartz to its rotational activity. But Herschel was not a chemist, and so there was room for another experimental approach, namely, the molecular one.

Mitscherlich's Enigma: A Crystal for a Chemist

While he was gathering data and honing his skills in crystallography as well as chemistry, Pasteur did not lose sight of his career objectives. He was well aware that he was a mere beginner who had yet to prove himself. That is why he wanted to go as fast as possible, even to skip some intermediary steps. While it might take Normaliens in the humanities ten years or more to prepare their doctoral thesis, Pasteur lost no time: by the spring of 1847, his work was well under way. He even considered using the approaching summer vacation for visiting Germany with Chappuis. "You must be, as I am, enormously eager to learn German. Let's go together; we will never be sorry to have made this trip. I must tell you that I really want to do it, and only a very serious obstacle could make me change my mind. I am bound and determined to know German and English a few years from now."[10]

This was an extraordinary perception of the future ways of communicating in a scientific world where Latin no longer played its former role. In the end, however, Pasteur's determination did come up against a "very serious obstacle," namely, shortage of money. He had worked faster than anticipated, so that his dissertations had to be printed that summer. "The expenses for my theses more than ruined me," he was to say with a sigh later.[11]

For his doctorate, Pasteur had to submit two theses, one in physics and one in chemistry. The physics thesis brought together several studies concerning the use of the polarimeter. In chemistry he chose to treat the saturation capacity of arsenious acid. "Studying this question," Pasteur wrote, "would certainly not have occurred to me if MM. Fabvre and Silberman had not spoken about it in terms that contradict the views generally held by chemists."[12] In other words, from the very beginning of his experimental career, Pasteur's principal purpose was to confirm the views of some and refute those of others. The great asset of his methodology was the fact that he approached chemistry by physics, the study of crystals by that of the polarimeter, and the definition of molecules by the rotation of light. Pasteur attached major importance to research in physics. This is proof of great originality in comparison with his fellow Normaliens of the time and even with his teachers of the rue d'Ulm.

Pasteur's first studies showed, or rather confirmed, that two isomorphic substances rotate polarized light to the same degree. When he defended his chemistry thesis in August 1847, the jury gave him an "honorable mention": one white and two red balls. Pasteur was not concerned about these fairly mediocre grades; what counted was that he had his degree of Doctor of Science.

Of this he was quite proud, and it pleased him greatly to receive the congratulations of his father and the rest of the family, as well as of his friends in Arbois. Was not the family's physician, Doctor Bergeret, the best judge of the young scientist's achievement? He cured his patients with ferruginous water after he had bled them, for he knew how to diagnose anemia due to a shortage of iron by the color of the blood he had drawn. Particularly noteworthy among the congratulatory letters Pasteur received was that from the principal of his collège, who thanked him for sending him his thesis. For this functionary of the Ministry of Public Education, language was an integral part of scientific rigor. "Dimorphism is not in the Dictionnaire de l'Académie," he complained.[13] It is there today, thanks no doubt to Pasteur, and designates substances that crystallize into two different forms. Here is one of Pasteur's examples of this phenomenon: in a bulb-shaped flask, saltpeter placed into water crystallizes into rhomboids in the center and into hexagonal prisms at the edges. Substances of the same chemical composition can thus have different crystalline forms. The theories of Abbé Haüy were left behind.

Significantly, this notion interested Pasteur sufficiently to motivate him to follow up his first investigations with a series of investigations of dimorphism. These he carried out with the help of Delafosse and two of his colleagues at the Ecole normale. The first step in this study was to establish a list of all natural and artificial dimorphic substances in which one of the two forms merges into the other. On the basis of his information, Pasteur came to a first conclusion: dimorphism is predictable and must, above all, be looked for whenever there is overlap between two forms, as in sulphur and in naphthalene chloride. This was close to Pasteur's studies of molecular dissymmetry, but this preliminary work was not yet concerned with tartaric acid.

The work he undertook after his two theses kept Pasteur in the laboratory. In the severe winter that was just beginning, he hardly ever set foot outside. For New Year's 1848 he received one of the few letters his mother ever wrote, as far as we know: "My dear child, all the best for the New Year. Take good care of your health, dress nice and warm, I keep thinking that you might get what is called *la grippe*. . . . You can understand why I am bound to worry, not being near you to give you a mother's care. Sometimes I do feel better about your absence when I think how happy it has made me to have had a child who has gone so far, and who enjoys his position so much, as you told us in your letter before last. Whatever happens to you, don't ever fret about it; everything in this life is illusory."[14]

But it was not *la grippe* that was to disrupt Pasteur's work in these early months of 1848. The sound of rioting was heard in the streets, and the first months of the year brought the collapse of the monarchy and the difficult beginnings of a republic torn by factions. A provisional government took over on 24 February, with the poet Alphonse de Lamartine as Minister of Foreign Affairs and the physicist François Arago as Minister of the Navy. While the new regime was being set up in Paris, the provinces, which received news only by optical telegraph, were uneasy: "I learned very late last night that the newspapers predict unrest in Paris and that on this very day there would be trouble stirred up by a reformist banquet. I beg of you, my dear child, do not leave home," Jean-Joseph implored him. He had a point. From April until June, there was constant rioting in the streets, and the repression was terrible: all the talk was of dead and wounded, of summary executions and revolutionary plots. In its fright, the government took a sharp turn to the right and called on General Cavaignac to put down the working-class suburbs. In the face of these threats to law and order, Pasteur decided to come to the defense of the Republic. These days of uncertainty finally brought him out of his laboratory; in this contact with the street he experienced some moments of patriotic fervor: "We learn noble and sublime lessons from the events that take place under our eyes. Why, I am turning into a warrior when I hear all the noise of fighting and rioting, and if it became necessary, I would fight with the greatest courage for the sacred cause of the Republic."[15]

In the end, Pasteur's courage seems to have amounted to little more than a financial sacrifice. Crossing the Place du Panthéon in late April, he noticed a crowd gathered around a patriotic altar where contributions from the citizens were solicited. Pasteur ran home and came back to deposit his salary. Informed of this gesture, Jean-Joseph suggested to his son to publicize this patriotic act by placing a notice in *Le National* or *La Réforme*: "Gift to the Fatherland by the son of a decorated veteran of the Empire. L.P., graduate of the Ecole normale." And the former Napoleonic soldier added, not without clairvoyance: "This will put some water into the wine of our great Republicans here, who would not, I believe, pull one 5-centime piece out of their pocket to save the Republic!"[16]

In the middle of the month of May 1848, Pasteur precipitously left Paris. He was not fleeing the danger, and even less abandoning his studies, but he had received an alarming letter that urgently summoned him to Arbois — his mother was very ill. When he arrived, it was too late; she had just died. On 21 May she had succumbed to a massive stroke.

"Whatever happens to you, don't ever fret about it; everything in this life is illusory."[17] These resigned words from the beginning of the year took on a terrible meaning when Pasteur had to face the loneliness of his father, who needed his support at this trying time.

The father's discouragement communicated itself to the son who, distressed by the situation and by his own role in it, let himself be persuaded that his mother had succumbed to the gnawing anguish of knowing that her son was exposed to the stray bullets of the capital. Once again, Paris became the scapegoat for all his misfortunes: "I know how much my father and my sisters want me to leave Paris," Pasteur wrote to Jean-Baptiste Dumas, who at this moment was asking to be replaced at the Ecole normale. If a university position in the provinces did not become available to him, he would — reluctantly — take a position in a collège. "Obliged to give up the science I loved and to which I was planning to devote my life, I would find it painful to teach a physics course in a collège."[18]

The fact is that Pasteur never really experienced the exaltation of the Parisian street. If he showed some awareness of what was at stake in the riots, he was above all convinced that governing France was not a job for amateurs. Instead of yielding to the excitement that was all around him and becoming embroiled in the heated discussions that resounded in the hallways of the Ecole normale, he always preferred to focus all his energies on his research. Just before he was recalled to Arbois, and at the height of the political chaos, on 15 May 1848, he published his famous memoir, *Re-*

cherches sur les relations qui peuvent exister entre la form cristalline, la composi-tion chimique et le sens de la polarisation rotatoire [Research on the relations that can exist between crystalline form, chemical composition, and the di-rection of rotational activity].

Given the scientific procedures in use in the nineteenth century, few organic substances could produce crystals appropriate to research of the kind undertaken by Pasteur. That is why practically all researchers used tartaric acid, which crystallizes easily and comes in two forms.

The first, ordinary tartaric acid, was known since antiquity and had been analyzed in the eighteenth century. Alchemists in particular had iden-tified it in wine barrels and its presence had been found in many fruits, particularly in the grape and in the serviceberry. Tartaric acid was mainly used in mordanting textiles, a technique that made it possible to stamp a raised pattern onto a tissue. For this reason it was produced industrially in France and elsewhere, and tartaric acid was sold in big milky crystals. The pharmacopoeia listed it as an emetic.

The second form of tartaric acid was discovered in 1819 in the wine vats of an Alsatian industrialist by the name of Kestner. When Gay-Lussac went to Alsace in 1826, he visited Kestner's factory at Thann because he was intrigued by this new acid. He obtained a sample of the product be-cause he wanted to compare it with the traditional tartaric acid and he found that it had identical biochemical properties. He therefore proposed to name it racemic acid, from the Latin *racemus*, cluster of grapes. The great chemist first used this name in his *Principes de chimie*, published in 1828.

A few years later, Berzelius, the Swedish chemist to whom we owe the convention of chemical symbols, carried out a new comparative analysis of tartaric and racemic acid. In its course he realized that they had not only the same atomic weight but also the same molecular composition. He also invented a new name for the racemic variety and proposed to call it *paratar-taric acid*. On this occasion the notion of isomerism made its appearance; in comparing tartaric and paratartaric acid, Berzelius noted that these two substances were composed of an equal number of atoms, but that they were different with respect to their solubility. Isomerism, he postulated, applied to all substances that present the same chemical composition, the same atomic weight, but different properties.

At this point, Eilhardt Mitscherlich was heard from. In order to vali-date his theories of isomorphism, the German chemist, professor at the University of Berlin, repeated Biot's experiments concerning the effect of polarized light, using the two forms of tartaric acid. To his great astonishment he noticed differences in the optical activity of the two isomeric substances. In 1844, he communicated to the Academy of Sciences in Paris a comparative study of their rotational activity: "Tartaric and paratartaric acid, despite their identical chemical composition, their identical crystalline form, and their identical specific weight, differ in their capacity to bring about the deviation of polarized light. Tartaric acid rotates the plane of light, whereas paratartaric acid does not have any effect." "What is so strange about this phenomenon," Mitscherlich added, "is that the nature and the number of the atoms, their arrangement and their distance are exactly the same."[19] Biot, whom Mitscherlich had asked to communicate this note to the academy, was so intrigued that he sought to verify the phenomenon himself. He reproduced the experiment but was unable to come to a satisfactory conclusion.

When this observation came to Pasteur's attention, he too was utterly perplexed: "This note of M. Mitscherlich," he recalled many years later, "preoccupied me a good deal when it was published. At the time I was a student at the Ecole normale. I spent a great deal of time meditating on the fine studies on the molecular constitution of substances and had reached — or so I believed — a thorough understanding of the principles generally accepted by physicists and chemists. This note upset all my ideas. Yet what precision in every detail! Had any two substances ever been better studied, better compared as to their properties? Still, given the state of science at the time, how could one conceive of two substances that resembled each other so much without being identical?"[20]

It is important to understand what was at issue here: if the molecules really had the identical atomic composition yet different properties, then the very notion of a chemical species, as established by Chevreul, had to be revised. How could it possibly be that two substances as close to each other as tartaric acid and paratartaric acid affected light differently?

If Mitscherlich and Biot were unable to elucidate the cause of this phenomenon, it was because they remained strictly within the confines of chemistry. Thanks to his double training in crystallography and chemistry, Pasteur perceived that the only way to reconsider the problem would be to reason by analogy. Following a brilliant intuition, he treated the chemical crystal as if it were a mineral crystal: tartaric acid as if it were quartz. Having read Haüy and taken Delafosse's course, he had an inkling that the similarity between the quartz crystal and the tartaric acid crystal was not simply a matter of words but also one of form, to be precise, of asymmetry.

The Theory of Division

Convinced that his starting hypothesis was reasonable, Pasteur went to work. He set himself the task of carrying out a systematic study of the crystals of tartaric and paratartaric acids and to look for a correlation between the difference in the polarization of light from one crystal to the other and the possibility of asymmetry.

Pasteur began his observation with tartaric acid crystals, which rotated polarized light to the right, examining them very closely with the goniometer after having separated them by hand. He was so convinced of his intuition that he was barely surprised when he noticed that one of the facets of the crystal was longer than the others. This gave them an asymmetrical form comparable to that of the quartz crystal. In their dissolved form, these crystals even preserved the ability to rotate polarized light, since the one, longer, facet pulled the rays to the right. This was the first time anyone had pointed out that the crystals of tartaric acid belong to a class of substances whose images cannot be superposed, just as the images of the right and the left hand reflected in a mirror cannot be superposed.

To Pasteur, this discovery was almost expected. Moreover, it seemed self-evident to him that paratartaric acid, which does not rotate the light, was bound to behave like a mineral in the same situation, for he expected its molecules to have an axis of symmetry. He was therefore extremely surprised when the goniometer showed him that the facets of the paratartaric acid crystals were also asymmetrical.

This, then, was the true enigma: how could one imagine that these two substances, tartaric and paratartaric acid, were so similar that both of them were asymmetrical, yet did not show the same rotational activity, since the first rotated the light and the second did not? More than one experimenter would have stopped right there, feeling that to show the asymmetry of both of these substances was a sufficient achievement. But Pasteur was unwilling to concede defeat. On the contrary, since this problem involved isomorphism, he would attempt to transpose the principles of spatial geometry into the realm of chemistry. To do this he would have to carry his hypotheses even further and to assume the existence of two kinds of crystals, one with left-leaning, the other with right-leaning facets.

This new intuition seemed so plausible to him that he immediately set out to verify it. As he patiently and meticulously examined the crystals of paratartaric acid one by one, he realized that they were composed of two different substances, both of them asymmetrical, but that some of them were right-handed and others left-handed crystals. Pasteur then manually separated the first from the second and prepared solutions of each of the isolated ingredients. The right-handed crystals, which were identical in every respect with those of tartaric acid, also rotated polarized light to the right. Pasteur lost no time carrying out the same experiment with the left-handed crystals, and his hypothesis was confirmed: the light was rotated to the left! And, what was more, this left-handed rotation proved to be as pronounced as the right-handed rotation obtained with tartaric acid.

Pasteur immediately understood the great importance of what he had just established. For all his usual reserve, he could not contain his excitement and his enthusiasm. Rushing out of his laboratory, he ran into a physics instructor in a hallway of the Ecole normale, embraced him, and dragged him out to the Luxembourg in order to explain his discovery to him.

Well, what had he found? First of all, he had cleared up the enigma contained in Mitscherlich's findings. He had, in other words, explained the finding of isomorphism. But above all, he had founded stereochemistry, or spatial chemistry, by establishing the principles of molecular asymmetry: the molecules of tartaric acid, like the quartz crystals, have an asymmetrical form in the spatial dimension.

Molecules have their inverse, called isomers, which act in the identical but opposite manner and rotate the light in the opposite direction. Pasteur furnished experimental proof that a chemical molecule, tartaric acid, can exist in two forms, one right-handed and the other left-handed, which are differentiated only by this inversion. This was the first time that geometry was used to study a molecule in the spatial dimension.

These discoveries did more than carry the work of earlier chemists and crystallographers a step further. Not only did Pasteur establish that the geometric organization of tartaric acid molecules does not conform to the laws of crystallography postulated by Haüy, since they do not have an axis of symmetry; he also, and above all, proved that this asymmetry is responsible for the rotation of light shown by Biot, and that this rotation is governed by the laws of the polarization of light discovered by Malus. It took a prodigious capacity for synthesis to establish the connection among all these phenomena that had previously been studied in isolation.

Pasteur's first impulse was to publish these observations under his own name in a communication to the Academy of Sciences. But Balard advised him against this, not because he wanted to add his name to the publication—he knew perfectly well that he had nothing to do with this discovery—but because custom demanded that new findings be announced by one of the members of the august assembly on the Quai de Conti.

Balard, who immediately grasped the importance of Pasteur's conclusions, spoke about them to Biot, the greatest authority in this field. Through the good offices of his teacher, Pasteur was able to meet with the aged chemist. But the latter, perhaps somewhat skeptical, had made it a principle to repeat all experiments before discussing them. "I will be glad to verify your experiments with you when you have written them up, if you are willing to send them to me confidentially," he wrote to Pasteur, assuring him of his interest "in all young people who work with precision and perseverance."[21]

In the early spring of 1848, Pasteur was thus summoned to a kind of rite of initiation. It was to take place at the Collège de France, where Biot lived. This was not a casual meeting. Biot was an old member of the Academy, Pasteur a young Normalien, and the stakes were high: an unknown scholar was called upon to take his first steps in the world of scientific publications. If Pasteur could convince Biot, he would gain a sponsor of the highest standing.

The circumstances of this first encounter were to remain so vividly imprinted in Pasteur's mind and its consequences were to be so decisive for his career that he was able to recall them precisely more than ten years later.

He [Biot] summoned me to repeat the decisive experiment before his eyes. He gave me the paratartaric acid he had carefully studied himself beforehand and found to be perfectly neutral toward polarized light. I prepared from it, in his presence, the sodium-ammonium double salt which he had offered to furnish me. We left the liquid in one of the slow evaporation cabinets he had in his laboratory, and when it had yielded about 30–40 grams of crystals, he asked me to come to the Collège de France in order to gather them and to separate out the right-handed and the left-handed ones according to their crystallographic character under his eyes. He again asked me if I was really saying that the crystals I would place to his right would rotate to the right and the others to the left. This done, he said he would do the rest. He prepared the carefully weighed solutions in the proper amounts, and when the time came to look at them in the polarization apparatus, he again called me to his laboratory. He first placed into the apparatus the most interesting solution, namely the one that was supposed to rotate the light to the left. Without even taking a measurement, Biot realized from the mere sight of the two images in the polarimeter, one ordinary and one extraordinary, that there was indeed a strong rotation to the left. Then the illustrious old man, visibly moved, took me by the arm and said: "My dear boy, I have loved science so much all my life that this stirs my heart."[22]

Following this memorable experiment, Biot's reaction soon became public: Pasteur's findings were mentioned in the report of the Academy of Sciences as early as May, even before a more extensive report was published in the *Annales de physique et de chimie*.

In the Academy's session of 23 October 1848, Biot again evoked Pasteur's discovery in highly laudatory terms: "The work on which we are about to report not only has the merit of having been directed toward its lofty goal with rare sagacity, these endeavors have also brought the author to a most unexpected discovery, and the procedure he has employed may have the most fruitful applications in the future."[23]

The Law of Biology

It does not diminish Pasteur's merits to point out that a certain amount of luck was involved in his discoveries. In fact, he himself used to say that luck favors those who are prepared. Thus, in most cases, a crystal resulting from the mixing of two isomers is totally different from its original constituent elements. Very few substances remain unmixed and produce crystals of distinct forms, as tartaric and paratartaric acids do. Later, Pasteur was to describe another model, that of aspartic acid, which he obtained from asparagus he grew himself in the basement of his laboratory. But beyond these two examples, almost three-quarters of a century passed before scientists were able to assemble a body of a few dozen of the same observations on molecules that form crystals with asymmetrical facets.

One other circumstance also facilitated Pasteur's observation. Since the crystals of tartaric and paratartaric acids are particularly large, it was possible, with a little patience, to separate the right- and the left-handed ones manually.

Temperature also favored Pasteur. He carried out the experiments on tartaric acid in winter, and in a very poorly heated laboratory. We now know that the separability of racemic acid crystals can only be observed under very special temperature conditions. If Pasteur had carried out these experiments in warm summer weather in the attic rooms of the rue d'Ulm, he probably would not have been able to distinguish right- from left-handed crystals.

If Pasteur was so anxious to publish his description of molecular asymmetry, it may well have been because he realized that counter-examples did exist. In fact, he was to point out subsequently that there were asymmetrical substances that did not show rotational activity and, inversely, forms that rotated the light but were not hemihedral. And finally, it soon became clear that right-handed rotation of the light was not always linked to the right-handed orientation of the crystal's facets. These findings did not upset Pasteur; he simply stated that the form of the crystal is of no more than secondary importance, and that what really counts is its optical activity. Moreover, he was quite willing to admit that his first research would remain valid only in its general aspect, namely, the finding that any right-handed molecule can have a symmetric left-handed counterpart and, reciprocally, that any optically active substance can have its inverse counterpart. What mattered was, to use Pasteur's own terms, that the principle of molecular dissymmetry had been established.

Pasteur built on his first observations on tartaric and paratartaric acids to account for isomerism. These two substances are identical but not alike in the same manner as an eye, an ear, a snail shell, or a spiral staircase are not alike to their counterparts in space; they are molecules of the same composition, but their atomic positions are reversed. Pasteur called this particular configuration "dissymmetry of the non-superposable image." "Looking at material objects, whatever they may be, with respect to their form and the distribution of their identical parts, one is bound to recognize two major patterns, whose characteristics are as follows: some, placed in front of a mirror, will yield an image that can be superposed on their own, while the image of others will not cover their own, even though it does reproduce all the details of the original. A straight staircase, a stem with distichous leaves, a cube, and the human body are bodies of the first category. A spiral staircase, a stem with leaves growing in a spiral pattern, a vise, a hand, or an irregular trait are examples of the second group. These do not have an axis of symmetry."[24]

This discovery enabled Pasteur to make the connection between physical and chemical properties and to conclude that, if the asymmetry of a molecule manifests itself both in its crystalline form and in its rotational activity, this difference does not involve a modification of its chemical properties, at least not in the most minute analysis. In describing this phenomenon, Pasteur defined what thirty-five years later Lord Kelvin was to call *chirality*, from the Greek *kheir*, the hand. And indeed, it would be hard to think of a better symbol than the hand to designate any shape that cannot be superposed on its mirror image.

In order to understand what makes Pasteur's discovery so important, one must of course go beyond the study of crystals and realize that Pasteur not only founded stereochemistry but also gave the initial impulse to developments that allowed biology to achieve greater advances in a few decades than had been made in several centuries.

In fact, Pasteur himself was one of the first to apply his discovery to physiology when he showed that the taste of food has to do with molecular asymmetry. While we are eating, molecules affect the nerve endings of the taste buds differently depending on whether they are right- or left-handed isomers. "When active dissymmetrical substances are involved in producing an impression on the nerves," he wrote in this connection, "their effect is translated by sweet taste in one case and almost no taste in the other."

Thanks to Pasteur, scientists thus began to see that in living beings molecules constitute the functional units of every organism. Forming the cells and the tissues, they mediate all biological events. To understand their organization opens the approach to their specific function. For the effect of every molecule depends on its connection to a receptor, as a key is connected to the lock it is designed to open. If one uses an inverse key, it will not work.

Thus every hormone, vitamin, enzyme, or antibody has an inverse form that imitates it but has no effect. A medication designed to replace it, a synthetic derivative designed to take its place, must necessarily have the same form as the original molecule. This principle governs all biological research and all medical applications. Every active and naturally occurring biological molecule has its definite form, which can be either right-handed or left-handed.

In discovering the principles of molecular asymmetry, Pasteur had done nothing less than to forge a key — and this key has unlocked the door to the whole of modern biology.

3

THE PHILOSOPHER'S STONE

In the summer of 1848, Pasteur's situation resembled that of his country; just as the nation, recovering from the series of shocks it had experienced in February and June, was anxiously awaiting the election of the first president of the Republic by universal manhood suffrage, so Pasteur, shaken by his mother's death, and his head somewhat turned by his academic successes, was unsure about where to turn his ambition. On the one hand, he was thinking about his family, for he had promised himself to support them and knew that they hated to see him return to Paris. On the other hand, he was convinced that his work would be immediately recognized by the academic establishment and bring him a professorship at a university. Not without some naiveté he even conceived the hope that the early retirement of an older colleague would open the way for him.

This was not about to happen. Moreover, Pasteur soon found out that there was not a single vacancy in his field in any university, whether in Paris or in the provinces. So his first regular appointment took him to the lycée of Dijon. Since there was no alternative, Pasteur faced the prospect of devoting all of his time to teaching in a secondary school. He had to contend with overcrowding, sometimes having as many as eighty students in a class, and his daily concerns, which took him very far indeed from scientific research, reflect a teaching situation that unfortunately still exists in many places today. "Don't you think," he confided to Chappuis, "that it is wrong not to limit the number of students to fifty at most? I am having a hard time keeping everybody's attention in the last half hour. I can think of only one thing to try, and I will do it, and that is to have them do a lot of experiments at the end of the class."[1] So this is how Pasteur, in order to keep his high school students interested, invented laboratory practicals.

In this first experience with teaching, the young scientist realized that research counted for absolutely nothing in the educational system of France, and that school administrators were unwilling to make proper use of the real capacities of their best personnel. The hopes and vocations of many foundered in the routine of tedious obligations. Pasteur did love teaching, yes, but not to the point of sacrificing his scientific ambitions to it.

He was exasperated by the energy it took to explain simple notions to pupils who never quite understood. Finding Dijon too small a town for him, he longingly looked from his province toward Paris. Calm had returned to the capital, the time of riots was behind it, and people now expected the presidential elections scheduled for December to restore order once and for all.

Moreover, living in Dijon deprived Pasteur of the chance to pursue his research on tartaric acid with Biot. It exasperated him to think of the aged academician's words: two or three more publications like the one on molecular dissymmetry, and the doors of the Institut would open before him as a matter of course.

Balard, for his part, also tried to help by making the rounds of the ministries lobbying for the creation of an acting professorship at the Ecole normale. Biot supported this effort and appealed to Baron Thenard, the president of the Grand Council of Higher Education. Pasteur was ready to give up his teaching post and to return to Paris as laboratory assistant [*préparateur*] without a teaching assignment. At this point the school administration did make a gesture without really yielding: in the first days of 1849, Pasteur was offered a position as acting professor of chemistry at the University of Strasbourg. He was to replace Jean-François Persoz, who had set his sights on a chair in Paris.

Pasteur was not sorry to leave Dijon and had no qualms about abandoning his fifteen- and sixteen-year-old students. Simply walking out on them in the middle of the school year, he did not even wait for his replacement.

A Grand Marriage

Pasteur arrived in Strasbourg on 22 January 1849 and immediately moved in with Pierre-Augustin Bertin, who lived on the Quai des Pêcheurs, very close to the Faculty of Sciences. Surely, it augured well for the future to be reunited with this old friend, who remembered the times they had spent together at the collège royal at Besançon and at the Ecole normale.

Bertin held the chair of physics at the Faculty of Sciences. A jovial and friendly companion and a man of simple tastes, he was to remain a faithful friend to his impetuous colleague as long as he lived. For the time being, he took it upon himself to introduce Pasteur to proper Alsatian society and to point out to him all the advantages of Strasbourg, a center of academic and commercial activity. It was a "town with many resources for chemistry, given the great variety of Alsatian industries,"[2] as Pasteur emphasized, for he was already interested in finding industrial applications for his research.

Less than a month after his arrival in Strasbourg, Pasteur mailed two letters that were to bring fundamental changes to his life. The first was to the industrialist Kestner at Thann, whom Pasteur asked politely to send him a few kilograms of racemic acid so that he could resume the experiments in crystallography he had been forced to interrupt a few months earlier. Kestner would turn out to be an indispensable supporter of the research carried out by Pasteur, who never failed to acknowledge Kestner in his publications and in the papers he read to the Academy of Sciences.

The second letter was addressed to Charles Laurent — no relation to Auguste Laurent — who had just been appointed rector of the University of Strasbourg. Pasteur had met Laurent only twice, but this was enough to find out that he had two unmarried daughters, one of whom was named Marie. This letter of Pasteur's is an extraordinary document, a self-portrait that speaks more eloquently than all testimonies; it must be quoted in its entirety.

Monsieur,

A request of the highest importance for me and for your family will be communicated to you in a few days, and I feel it my duty to convey to you the following information, which may serve to determine your acceptance or your refusal.

My father is a tanner at Arbois, a small town in the Jura. I have three sisters. When the youngest was three years old, she suffered a brain fever that completely arrested the development of her intelligence. She is a child in intelligence, but a grown-up in body. We are about to place her in a convent where she will no doubt spend the rest of her days. My two other sisters have taken over, as far as the care of the household and the business are concerned, my mother's place at my father's side, for we had the misfortune of losing our mother last May.

My family is in comfortable circumstances, though without fortune. I do not evaluate our assets at more than fifty thousand francs; for myself, I have decided long ago that I shall give my entire share to my sisters. I thus have no fortune whatsoever. All I possess is good health, a kind heart, and my position in the University.

Two years ago, I graduated from the Ecole normale as agrégé in the physical sciences. I received my doctorate eighteen months ago and have

presented to the Academy of Sciences some studies that were very well received, the last one in particular. A very favorable report, which I have the
honor of sending you with this letter, has been written about this study.

This, Monsieur, is the long and short of my present position. As for the
future, all I can say about it is that, barring a complete change in my interests,
I shall devote it to research in chemistry. It is my ambition to return to Paris
after I have gained a certain recognition through my scientific work. M. Biot
has told me repeatedly that I should seriously think about the Institut. In ten
or fifteen years, I may be able to do so if I continue working assiduously. But
even if this dream were to evaporate, it is not what makes me love science for
its own sake.

My father will come to Strasbourg himself to ask you for your daughter's
hand. No one here knows that I have approached you, and I am certain that,
if you should refuse me without withdrawing your esteem, no one would
know about this refusal.

Please be assured, Monsieur, of my profound respect and devotion.

Louis Pasteur

I turned twenty-six on 27 December of last year.[3]

It had not taken Pasteur three weeks to persuade himself that in the
rector's daughter (whose name is not even mentioned in the letter) he had
found his ideal mate. Love at first sight? A cold-hearted career move? In
his personal life as in his experimental strategies, Pasteur was the man of
lightning-quick decisions. We are told that while one of his nephews hesitated to make his marriage proposal, Pasteur spoke to the young lady in his
name and, to everyone's amazement, asked her to respond on the spot.

But Rector Laurent could not be rushed, and the answer was slow in
coming. Actually, aside from studying the indications he had received from
Pasteur, he could do little to find out more about the man who wanted to
become his son-in-law. Bertin, who was glad to serve as go-between, supplied the rector with some additional information about his friend.

But the matter dragged on, and no response was forthcoming. At the
end of March, Pasteur sent two more letters, one to Mme Laurent, the
other directly to Marie. To the former he wrote that there was nothing
about him that would seduce a young woman at first sight, but that if one
came to know him one would surely love him. And to Marie he wrote that
she should not judge him by his cold and timid demeanor. He added that he

would like to spend his life at her side even if this should make him forget all about his crystals, but that for the time being he had to get things done. He therefore insisted that she tell him the exact days and hours when he could call on her, so that he would not interrupt himself for nothing and fall behind in his work. Pasteur believed that feelings should not interfere with research and that social calls and life in the laboratory were two different things.

While waiting for his response he did not, of course, take a break from his work but ardently pursued his work on crystals, regularly corresponding about it with Biot. Biot sent him samples of racemic acid that he had found in his drawers and in exchange Pasteur kept him informed about his latest findings. In order to manage Biot's farsightedness and spare him the trouble of looking through the microscope, he went so far as to carve enlarged cork models of tartaric acid crystals that reproduced all their edges and facets. Biot encouraged him in his work but also cautioned him about certain of his character traits that he considered harmful to his career. In commenting on the form of one of Pasteur's texts, he wrote in a blunt tone that was in fact not unlike that of his disciple: "I have two things to say about the details of your writing. . . . The first concerns the phrase, *And so I shall announce to the Academy a finding that deserves to command the highest degree of attention.* [The paper concerned the hemihedrism of strontium formate.] One must never speak so favorably about one's own work — you should simply say: *which seems to me to be of considerable interest.* . . . My second remark concerns the last seven lines of your paper, beginning with: *I would even go so far as to think* . . . I do not think that it would be useful for you to publish what you think might happen."[4]

The truth is that if Pasteur was a little cocky that day, it was not about molecular dissymmetry. For Biot's letter had reached him the same day as the response of Rector Laurent. It was a laconic and very formal response, but it did accede to Pasteur's request. An exchange of correspondence and a few visits had indeed brought about the desired result: Louis Pasteur married Marie Laurent on 29 May 1849. After a brief honeymoon in Baden, where they barely escaped a shootout between two rival bands of soldiers, the couple took up residence at 3 bis, rue des Veaux, not far from the house where Cagliostro had once lived.

Born 15 January 1826, Marie Laurent was twenty-three years old at the time of her marriage. For the rest of her life she was to devote herself to Pasteur and to create a tranquil family life around him. In the words of Pasteur's son-in-law and hagiographer, René Vallery-Radot, Marie was able "from the very beginning, not only to tolerate, but to approve that the laboratory came before everything else." He adds that it was as if she were fascinated by Pasteur's gaze, "these admirably bright eyes that shone in a grey-blue hue like a certain gem from Ceylon that reflects the changing light."[5]

However that may be, it is true that the laboratory had absolute priority. "I am often scolded by Madame Pasteur, but I make her feel better by telling her that I am leading her into prosperity," Pasteur confided to his friend Chappuis.[6] Another, and later witness, Doctor Roux, confirmed that Marie Pasteur was quite conscious of the role she was supposed to play in her husband's existence: "From the very first days of their shared life, Mme Pasteur intuitively understood what kind of man she had married; she made it her task to keep the difficulties of daily life away from him and to take care of all domestic problems, so that he might keep his mind free to pursue his research. Mme Pasteur loved her husband so much that she learned to understand his work. In the evening she would write down what he dictated and ask him to explain it, for she was really interested in hemihedral facets. . . . Not only an incomparable mate for Pasteur, Mme Pasteur was also his best collaborator."[7]

There is no doubt about it: Marie was awed by her husband's genius. Nothing could reveal the inextricable mixture of common sense and admiration in her attitude better than a brief remark to her father-in-law: "Louis . . . always worries a little too much about his experiments. You know that those he is planning for this year are supposed to give us, if they are successful, a Newton or a Galileo."[8]

Meanwhile, a little girl was born in April 1850 and given the name Jeanne. Eighteen months later a boy, Jean-Baptiste, came into the world, and Biot was his godfather. The least one can say is that the Pasteurs did not show much imagination when it came to naming their children. As was to be expected, Jean-Baptiste, the heir to the family name, received most of the attention. His parents found his baby talk sweet and delightful, and his name soon became "Batitisse" or "Batitim." But Pasteur was also horrified to find out that little children are not always models of hygiene: "Our little Baptiste," Marie wrote to her father-in-law, "is becoming more and more dirty, and this is particularly upsetting to his father, who takes care of him during the night. We have not yet succeeded in breaking him of peeing in his bed, and with the weather we are having, this is something of a calamity, for he gets cold and then he coughs. We certainly do not spare the punishments, but nothing helps."[9]

Even if he changed diapers at night, Pasteur's workday at Strasbourg was punctuated by his research at the University, along with his courses at the Faculty of Sciences and at the Faculty of Pharmacy, where he sought to find a replacement.

Eventually a young Normalien, Adrien Loir, became Pasteur's successor at the Faculty of Pharmacy, a position that gave him the privilege of working with Pasteur in his laboratory. Without doubt, Loir was Pasteur's first disciple and followed in his master's footsteps to the point of marrying Rector Laurent's second daughter, Amélie, shortly after his arrival in Strasbourg. He later became dean of the Faculty of Sciences at Lyon, and his son, Pasteur's nephew, was to become one of Pasteur's most trusted laboratory assistants.

In this peaceful family atmosphere, Pasteur enjoyed the affection and the respect that surrounded him in the laboratory. What he also appreciated at Strasbourg was the calm of provincial life, far from the twists and turns of politics and influence peddling. He knew that if he wanted to realize his ambitions, which were considerable, he had to begin by creating a solid reputation for himself. This could be done best when the country was calm and the government stable. This may account for Pasteur's deeply conservative political attitude. Having wholeheartedly approved of the election of Louis-Napoleon Bonaparte to the presidency of the Republic, he also had no misgivings about the coup d'état of 2 December 1851. "You must be pleased with the political situation," he wrote to his father. "The future looks bright. I do not think we have ever had a government as strong as the present one of Louis-Napoleon."[10]

Yet Pasteur was aware that the government treated the opposition harshly. But he did not protest when his father-in-law was made to retire early, at the age of fifty-five, because he had been less than enthusiastic about the new education law of 1850, the famous Falloux Law, which restored the Church's influence in education. By reasserting its control over the nation's teaching personnel, the government meant to sound a warning to everyone. One of the prefects, Auguste Romieu, head of a task force charged with investigating the state of the French university system, returned from Strasbourg with a report critical of the activities of the rector and suggested that he be replaced by a "reliable man." The position Charles Laurent was offered at Châteauroux would have amounted to a true demotion. He refused to leave Strasbourg and retired to private life.

These events did not disrupt Pasteur's life. "If you come to Strasbourg, you will be a chemist whether you want to or not," he wrote to Chappuis. "All I will talk about with you are crystals."[11] Much later, Pasteur was to say of this period that it was "the time when the spirit of invention flourished."[12] It was indeed a blossoming of experiments, most of them bold, many successful, and all intelligently designed. The Strasbourg period yielded many of Pasteur's crystallographic laws on the properties of racemic acids and on isomerism, some of his comparative reflections on the composition of tartaric and malic acid, as well as new attempts to provoke crystallization by the supersaturation of bases and acids. Biot regularly reported the results of these experiments to the Academy of Sciences, so that they became accessible to an ever expanding public.

The mineralogist Henri de Sénarmont, an old friend of Biot and also a member of the Academy, was among those who had taken an interest in Pasteur's work as early as 1848. In the course of 1852, the physicist Victor Regnault, who had learned of Pasteur's research through Biot and Sénarmont, began to watch the work of the chemistry professor at Strasbourg. Sénarmont nominated Pasteur for a corresponding membership of the Institut. Pasteur was not yet thirty and the proposition was flattering, but he declined, for the position was in physics. This decision was quietly suggested to him by Biot, who felt that he would be better advised to be patient and not to pursue a distinction for which he was not qualified and which would not be within his real competence. In this matter as in others, Biot acted as Pasteur's most trusted confidant. When Pasteur was tempted to leave Strasbourg, Biot advised him to stay so as not to interrupt his research; when he did not know where to publish his work, Biot smoothed the way for him. As a result the Academy of Sciences accepted Pasteur's communications as those of a foreign scholar. The notes and findings he sent to the Institut continued to generate a great deal of interest, and Jean-Baptiste Dumas, who in 1850 had been a minister for three months, even promised him a medal. When Pasteur did not receive this distinction soon enough — for his taste — he finally expressed his disappointment in February of 1852. "I have always found it rather strange when people solicit decorations for themselves. But I find it even more strange that the most famous chemist in Europe, when he has become a minister and solicits a cross for a humble professor of chemistry, is unable to obtain it."[13]

Fortunately, Biot was always there to reassure the young scientist and to teach him how to handle such things: a reward that does not come through immediately, he said, will be even better deserved later. Dumas, who felt quite strongly that Pasteur deserved a reward, considered approaching the President of the Republic directly. "I do hope that M. Dumas will do better by me when it comes to the new position," Pasteur wrote to his father.[14] The decoration was as important to him as a new position, for he was fighting for the recognition of his merits with the same tenacity he brought to his experiments.

The Gentlemen of the Academy

In August 1852, Pasteur went to Paris. He stayed at the Hôtel de l'Empéreur Joseph II, located across the street from the Senate and above the famous restaurant Foyot. Since there was not enough room in the hotel, his wife and the children stayed with Charles Laurent, who was living in the rue Saint-Sébastien. When he had a free moment, Pasteur went to see them, but this was not often, for this stay in Paris was not a summer vacation. As he did every year at this time, he was here to present his work to the Academy.

All the members were present at the session: in the first row sat his faithful supporters, both chemists and physicists: Dumas, Thenard, Regnault, Balard. All in all, the audience was favorably disposed, and Pasteur proudly told his father about the compliments he received. But there was also criticism, for some members were beginning to feel that this young scientist was being a bit too forward. But Pasteur was not overly concerned about encountering some opposition. As Biot had explained to him, "When a colleague at the Institut reads a paper and no one comes afterward to talk to him about it, it means that what he has found is good. But if they want to discuss it, and especially if they are critical, it means that it is very good."[15]

In August 1852 it was hot. Paris was calm. The capital had put riots behind it and France was calmly preparing for next autumn's referendum that would transform the president of the Republic, Louis-Napoleon Bonaparte, into Emperor Napoleon III. Returning to his hotel after a visit to Marie on the evening of 21 August, Pasteur found a note Biot had left for him as he passed by on his daily walk in the Luxembourg: "Please be at my laboratory tomorrow morning at 8 o'clock and bring your samples if possible. M. Mitscherlich and M. Rose will come by to see them at nine."[16]

Mitscherlich had come to Paris to be received as a foreign corresponding member of the Academy of Sciences. With him was the mineralogist Gustav Rose, professor at the University of Berlin. These two men wished to meet Pasteur because they wanted to hear from his own mouth all the details of his discoveries concerning tartaric acid. Pasteur took the crystal samples he had brought with him for his demonstrations at the Institut to the Collège de France, and it was here that on the next day, a Sunday, the three scientists, Pasteur, Mitscherlich, and Rose, met and spent two full hours measuring the facets and angles of the tartaric and racemic acids, discussing the circumstances of the observation, and reflecting on its consequences. Mitscherlich wanted to test for himself the acumen of the man who had been able to take his own findings one step further. Not satisfied with just reading about his conclusions, he wished to interrogate him and push him to the limit. He was not disappointed and did not hide his enthusiasm. Paris for him would always be the city where he had studied French and where he had met Pasteur.

On the following Tuesday, Baron Thenard gave a dinner in honor of the German scientist who had discovered isomorphism. Pasteur was invited. Thenard had heard him speak at the Academy and had at least three reasons for taking an interest in the young investigator: To begin with, Pasteur had exhibited the quality for which Thenard was looking above all others, namely, economy in his experiments, which should be well thought-out, few in number, rigorous, and striking; secondly, he appreciated Pasteur's interest in the industrial application of his research; and finally, the two men shared a taste for the visual arts, for although Thenard did not paint, he was deeply interested in the techniques of painting and had actually developed a cobalt blue paint. Born in the reign of Louis XVI, a baron of the Empire, a peer of France, and a member of the Grand Council of Higher Education, Thenard had received more honors than any other chemist of his time. We owe him the discovery of "oxygenated water" and a classification of the metals, as well as a method for separating sodium and potassium by means of a galvanic current, a recent discovery of Berzelius. A bon vivant and a hearty eater, he enjoyed telling people that his first assignment in the laboratory of the chemist Vauquelin had been to watch the master's stew. But he also looked at foodstuffs with a chemist's eye, ever since he had inadvertently swallowed some mercuric chloride in the course of a lecture and almost died from it. His students were aghast when he cried out, "Gentlemen, I have poisoned myself, send for Dupuytren." This experience had left him with high admiration for every form of experimental research on medical subjects.

On the evening of the dinner, Thenard had brought together a veritable hall of fame of eminent scientists, all members of the Academy, a fact that did not, of course, fail to flatter Pasteur. In addition to Mitscherlich and Rose, Jean-Baptiste Dumas and Victor Regnault were there, but also Eugène Chevreul, director of the Muséum d'histoire naturelle, the pharmacy professor Théophile Pelouze whom Thenard had succeeded at the Collège

de France, Eugène Péligot, professor of chemistry at the Institut agrono-
mique, Louis-Constant Prévost, professor of geology at the Faculty of Sciences, and Antoine Bussy, director of the Ecole de Pharmacie. The conversation was brilliant, and several people alluded to Pasteur's work. They
encouraged him, congratulated him. Better yet, Mitscherlich raised his
glass and drank to the young scientist's future: "I have studied these crystals
long and hard, but I am convinced that if you had not found this remarkable fact when you looked at them anew, my discovery would have been
ignored for a long time to come."[17]

Pasteur modestly made the point that no scientist is ever indispensable, that discoveries are made when they have become possible through
the evolution of knowledge, and that they always come when their time is
ripe. He added that he had benefited from circumstances beyond his control, having had the good fortune of working with experimental models of
exceptional quality: if he had not worked with tartaric and racemic acids, he
surely would not have had the inspiration that had allowed him to build a
bridge to connect mineralogy, crystallography, and chemistry. This striking
shortcut, he said, had occurred to him thanks to the crystals of paratartaric
acid that Kestner allowed him to study.

But he also mentioned his worries. He was afraid that he would have
to stop his experiments, for despite many attempts it had proven impossible
to produce paratartaric acid industrially in the factory at Thann, at a time
when its reserves were being exhausted. Racemic acid, he said, was a kind of
chemical meteor. Mitscherlich was astounded to hear him say this. He knew
an industrial producer of paratartaric acid in Germany, an industrialist in
Saxony who manufactured it from a tartaric acid that he purchased in
Trieste. This statement was a veritable revelation. To understand how paratartaric acid is formed would be to become initiated into the mysterious
workings of nature when it differentiates right- and left-handed molecules.

In characteristic fashion, Pasteur made up his mind on the spot. As
soon as Mitscherlich had assured him that a German method of production
really did exist, he dropped everything and set out to find the elusive molecule. He had been studying crystals for years, hunting them down, and now
he finally had a chance to verify his intuitions. Moreover, he was also aware
that the 1853 prize of the Société de Pharmacy of Paris would be awarded
for the best study establishing whether or not there was a kind of tartaric
acid that contained fully formed racemic acid and whether, consequently, it
was possible to produce racemic acid industrially. In short, Pasteur, eager to
pursue his investigation, spurred by the possibility of winning a prize, and
perhaps also curious to come to grips with a geographical rather than

chemical enigma, was ready to go off to Saxony. The long-cherished project of visiting Germany was finally going to become a reality.

The decision was made instantly, during Baron Thenard's dinner, but Pasteur found out very soon that he did not have the means. Although he was wont to spend a large part of his salary on his research projects, the expenses for this undertaking were more than he could afford. He could not go to Germany without a subsidy. Sure of his support, and convinced that the tour was necessary, Pasteur set out to sweep away all obstacles, as he was able to do whenever he was convinced that he was right. He drafted a letter he planned to send directly to the President of the Republic, asking for his intervention. This would enable him to study on the spot "a problem that France should make it a point of honor to see resolved as soon as possible by one of her children."[18] He may have thought that scientific questions do not know frontiers, but when necessary he knew how to exhibit a fierce nationalism.

However, Biot restrained Pasteur from sending this letter. He assured him that he, Biot, would undertake the necessary steps and even offered to advance him the sum if the subsidy should not arrive in time. Pasteur fretted about these delays and feared that it was too late in the season to study the grapes and collect the tartaric acid. He was also afraid that details of his projects would be spread around through discussions among administrators. In short, he was growing impatient. Fundamentally, of course, he was waiting for only one and obviously essential thing, namely the consent of the Saxon industrialist that Mitscherlich had promised to obtain. For in the end, Pasteur found another way to finance the undertaking: Dumas was ready to have him paid by labeling his tour an official visit to German laboratories. French scientists had long been looking toward German science, which seemed wealthier and better endowed.

Thus Pasteur went off in an official capacity. "My dear Father," Marie wrote to Jean-Joseph Pasteur on 11 September 1852, "Louis started on his trip Thursday morning, but he did not go off without seeking the advice of M. Biot, who approved. Before he left, he had also received a favorable response from the industrialist, who said that he would give him all the desired information about his manufacture of racemic acid."[19]

Between the Rhine and the Danube

Pasteur was embarked upon a veritable initiatory quest. "You know very well," he wrote to his father, "that I am looking for the philosopher's stone, and you too have read about the joys and the disappointments of those alchemists who came before me. They always believed that they were about to grasp it. That is how I feel. But they died in their pursuit. I do hope that my zeal will not falter before I have arrived. My venerable mentor sets the example, and I believe, as he does, that the only thing that can bring joy is work."[20]

Pasteur did not exaggerate. He passed through different landscapes almost without seeing them. Nothing could distract him from his project of discovering new sources of tartaric acid and understanding the causes of the appearance of racemic acid. The letters he wrote to his spouse say more about the comforts of traveling than about the history or the beauty of German towns: "It is no exaggeration to say that in the kingdom of Hanover and in the duchy of Brunswick second class compartments are better than first class ones in France; they are as luxuriously appointed, and the seats are more comfortable."[21]

On 12 September 1852, after a brief stopover in Brussels, Pasteur arrived in Leipzig, from where he rushed off to Zwickau, where he was to see Herr Fikentscher, the industrialist of whom Mitscherlich had spoken to him. There, a cruel surprise was in store for him: the information of the German scientist was inaccurate. When Fikentscher gave Pasteur a tour of his factory, he confirmed that he had indeed obtained paratartaric acid, but that this production had stopped two years ago. The molecule was no longer available in Germany, any more than at Kestner's factory at Thann. However, in the rectangular lead tubs where a thick layer of tartaric acid had accumulated, Pasteur noticed a few of the needle-shaped crystals that are characteristic of racemic acid.

Upon questioning his host about this, Pasteur learned that this deposit had begun to form when the factory had changed suppliers. In the past, Fikentscher had bought his supplies in Trieste, but now he dealt directly with a Neapolitan refiner. As he was leaning over the tubs and listening attentively to the German industrialist, Pasteur's mind was forming one of the bold hypotheses that so often came to him: What if it was the mode of refining, rather than the nature of tartaric acid, that caused the disappearance of the racemic acid?

When he left for Germany, Pasteur had estimated that this investigation might take about ten years. And here he had traveled for only one week, and had already found a scent. Of course it was only a first hypothesis, and the investigation had barely begun. Pasteur spent ten days in Leipzig examining the samples of tartaric acid Fikentscher had shown him, some from Trieste and some from Naples. His German colleagues assisted him in his work, and one of them, Erdmann, most graciously offered him the use of his laboratory facilities.

While he was examining the crystals, Pasteur had the pleasure of receiving Professor Hankel of the University of Leipzig, who had translated all his articles for a German periodical. This meant that Pasteur's work had already crossed the frontiers and that German scientists knew as much about molecular hemihedrism as French ones. Pasteur spent all of his time in his laboratory: "All I know of Leipzig is the street that leads from the hotel Bavaria to the University," he wrote to Marie. "I go back there at night; I have dinner and go to bed." All the letters he wrote to his wife were exclusively concerned with his work or with the minutest details of his daily expenses, which were really quite modest. They usually ended abruptly with the unequivocal words, "My love to you and to science forever."[22]

Pasteur did not take time out to visit any museum, any monument. He barely paid attention to the news, and the death of Marie Lafarge, the famous poisoner, interested him more than that of the duke of Wellington. He spent his days without going anywhere, without hearing anything, devoting himself to his research and to nothing else.

Pasteur's hypothesis slowly proved to be correct, which, as usual, did not surprise him. It turned out that it was the half-refined tartaric acid from Trieste that had yielded racemic acid crystals, but not in sufficient quantity for the development of an industrial production. It was only logical that Pasteur should decide to pursue his research on the shores of the Adriatic. Without hesitating, he left Leipzig and set out for Venice. At the thought of going by himself to the classic honeymoon city, Pasteur yielded for a moment to a kind of romantic yearning: "I shall finally see the sea, and Venice is where I will see it for the first time. Oh! I will write to you from there. I will think of you as I go to see the Palace of the Doges, the portrait gallery, and the black veil that replaces that of Marino Faliero. I will remember the deep impression we received from reading the dramatic story of that doge. I don't know why, really, but it seems to me that in that city one is bound to experience extraordinary emotions."[23]

In order to go to Venice, one had to cross Austria, and to do that one needed a visa. Pasteur therefore had to stop at Dresden. For the first time, and no doubt because there was nothing else to do, he used this stopover to visit the city's famous museum. But this does not mean that he lost his sense of exact measurement. "I spent fully four hours in these galleries, amusing myself with writing down in my notebook the paintings that I liked best.

Those that I found interesting got one cross, and to others I would give two or three, depending on my degree of enthusiasm. I even went up to four!"[24] — no doubt for Raphael's world-famous Sistine Madonna.

After Dresden, Pasteur went to Freiburg, where he was again received magnificently by his fellow scientists. They showed him the university's mineral collections, which absolutely enthralled him. Pasteur was decidedly charmed by the Germanic spirit, but he was also proud to realize that French was the international language of science.

The tour continued, and Pasteur reached Vienna on 25 September. As soon as he arrived, he was given the address of a professor of chemistry, Rettenbacher, who acted as his guide and took him to see the region's chemical plant, which belonged to the industrialist Seybel. This factory used Austrian tartaric acid to produce what was believed to be magnesium sulphate, or Epsom salt. But the descriptions that were given to him convinced Pasteur that the product was in fact racemic acid. A first examination, carried out at the plant itself with a poor-quality and dirty magnifying glass, reinforced his conviction, and then a thorough investigation in the laboratory confirmed his hypothesis: these were without doubt crystals of paratartaric acid. The conclusion was obvious: Unrefined tartar, whether from Naples or from Austria, contains racemic acid, which is destroyed by the refining process.

Having established this, Pasteur could assert that if the French preparations of tartaric acid contained no racemic acid, this was exclusively due to the manufacturing processes. The essential question had thus been resolved. No need to go to Trieste. Pasteur decided to forgo Venice, the Hotel Danieli, and the pleasures of the Lido. "Put off my trip to the sea again. And you won't get your souvenir of Venice. I was planning to buy you either some lace or something made of coral."[25]

Before setting off for Strasbourg, however, he did decide to take a detour through Prague. He had heard rumors to the effect that a Czech industrialist had developed a process for transforming massive amounts of tartaric acid into racemic acid. This seemed so extraordinary to Pasteur that he immediately left Vienna, not, however, without having paid a visit to what he considered Canova's masterpiece, the monument to the archduchess Maria-Christina, an astonishing mixture of cold Neoclassicism and Romantic anguish that he compared, not inappropriately, to Pigalle's monument to Maréchal de Saxe, which he had long admired at Saint-Thomas's church in Strasbourg.

Having arrived in Prague on 1 October, Pasteur rushed off to see Dr. Rassmann, who told him that he could at will obtain as much racemic acid as he wanted. If this was true, the prize of the Société de Pharmacie was in his pocket.

In dealing with this colleague, Pasteur's attitude was ambiguous. He asked for detailed explanations, and carefully repeated them to his informant, not only because he wanted to be sure he understood the process properly, but also because he was afraid that it would run counter to his own conceptions. "You have made one of the greatest discoveries anyone can make in chemistry," he told Rassmann. "Perhaps, unlike myself, you do not fully grasp its importance. But permit me to say that, according to what I know, I consider this discovery impossible. I do not ask you for your secret. But I am looking forward to its publication with the greatest impatience."[26]

There was a certain amount of duplicity in this reaction, for Pasteur feared competition and felt that the question of whose discovery had come first would be very delicate indeed. He therefore mailed a full account of the conversation to Marie, for, as he specifically stated: "I am sure of my own honesty, and I believe in his as well; however, there could come a time when it would be useful to publish what exactly was said between us in order to uphold his rights, those of M. Fikentscher, and mine in this matter."[27] Pasteur's conduct was by no means that of a naive beginner. Whether it was a matter of industrial secrets or of purely theoretical discoveries, he never failed to assert his rights.

In fact, Rassmann's claim was unfounded and he had deluded himself. He had not discovered a method for producing synthetic racemic acid; all he had done — and it was quite a bit — was to draw the correct conclusion from what Pasteur himself had also observed at Zwickau and at Vienna, to wit, that it was possible to isolate the racemic acid that occurs in unrefined tartaric acid. This was reassuring to Pasteur, for it meant that his own conclusions were not in jeopardy. There was nothing to keep him from returning to France.

Yet, on his way to Strasbourg, he made another detour to Thann in order to tell Kestner about his tour. He called on the industrialist as soon as he arrived and explained to him that since his factory lost the racemic acid in the refining process of raw tartaric acid, it would be possible to recapture it by treating the "mother liquors," that is, the effluents of the washing operation. He then and there suggested that Kestner could manufacture and sell racemic acid if he bought the original mother liquors from the Neapolitan manufacturers of tartaric acid. Pasteur was very excited to relate the different stages of his tour and felt compelled to lay out every detail of how his hypotheses were borne out one by one from one phase to the next of his discovery.

In mid-October 1852, upon his return from places whose alleged Germanic barbarism he considered a mere figment of prejudice, Pasteur proudly wrote to his father that he was working hard at "trying to lift a little corner of the veil with which God has covered everything he has made."[28] But his happiness was not complete, for his wife and his children were not with him to share in his triumph. Marie, Jeanne, and Jean-Baptiste were still in Paris with grandfather Laurent. Impatient to see them, Pasteur asked them to come home right away.

> It would be infinitely better if you came home to stay as soon as possible. . . . My clothes (at least some of them) are in bad shape. I need you to fix them. . . . I explicitly order you to take an express train and first-class seats and to leave in the evening. You will place your son on a cushion right next to you. He will sleep perfectly, and so will you. The train stops only at a few stations. You will be in Strasbourg almost before you have time to think about the trip. I order you to do it in this way. I won't pick you up if you come in second class. I mean it. You will receive this letter the day after tomorrow, which is Saturday morning. You must leave by the mail train Saturday night, which only has first-class seating and you will be in Strasbourg Sunday morning. . . . I think you will have enough time to get ready to leave Saturday evening. You cannot put it off any longer. I have no time to lose. Later today, when I get back to the hotel, I will give you the departure time of the train you are to take Saturday evening, and I will be at the station Sunday morning to pick you up. Adieu. See you Sunday. All my love.[29]

What an incredible mixture of tenderness and authority! Pasteur's wishes really were commands. In any case, there was no way to dodge them. Marie did as she was told.

If Pasteur's ideas about his spouse's duties were rigorous, he did not forget his own either. Immediately upon his return, he wrote to Jean-Baptiste Dumas to express his gratitude, for he knew that his tour would have been impossible without the financial support Dumas had provided. He assured Dumas that in order to deserve the trust he had placed in him, he would work "as hard as is humanly possible."[30] Knowing Pasteur's fierce devotion to work, we know that these were not idle words.

In January 1853, Pasteur wrote a brief summary of all the information he had gathered about racemic acid. Paying homage to Kestner and Fikentscher, who had been the true godfathers to his discovery, he emphasized the fact that advances in chemistry had been achieved with the help of industry. Pasteur was also thinking about the prize essay for the Society of Pharmacy; indeed, at this point he was probably overly concerned with the matter of industrial production. He had not yet fully grasped the fact that the theoretical discovery of molecular dissymmetry was much more important than the momentary problem of how to produce racemic acid. But at least this quest for the "philosopher's stone" had given him a chance to discover Germany, to learn firsthand about its rich scientific and industrial development. Forever after, this would be an example for him, a goad that constantly spurred him on.

4

THE WORLD OF THE
LIQUEFIED FORESTS

Strasbourg, 7 December 1853: "I am really afraid that this time I have taken on the impossible. I want to track down the cause of one of nature's greatest mysteries, whose unraveling, it seems to me, would have the most far-reaching consequences."[1] This thought, which he confided to Dr. Godélier, professor at Val-de-Grâce Hospital on his return, makes it clear that Pasteur was aware that his work had taken a new turn. Leaving behind his descriptive and exploratory studies, he had become interested in the specificity of crystalline forms. While he was discovering the importance of determining the spatial dimension of form, Pasteur carried his crystallographical studies to the point of attempting to become a creator himself. For the goal of his quest was no longer tartaric acid, but asymmetry itself. He wanted, in other words, to understand the reasons for the right- and left-handedness of crystals, which would give him the power not only to understand asymmetry, but also to generate it.

Louis Pasteur Vallery-Radot often said that his grandfather's most important discovery was the asymmetry of chemical crystals. And indeed, it is not to detract from his other discoveries to say that here Pasteur touched on the very principles of life. Rarely in the course of his existence was he to feel both so close to a fundamental understanding and so incapable of actually seizing it. He constantly confronted his own limitations: however far the magnifying glass, the goniometer, and the polarimeter might take him, the mystery of the fundamental organization of chemical elements remained unsolved. He therefore had to elucidate their genesis. "Life is dominated by asymmetrical actions. I can even imagine that all living species are primordially, in their structure and in their outward form, functions of cosmic asymmetry."[2]

In his own way, Pasteur was moved by the spirit of Romanticism. When he wanted to transmit to posterity his visionary conceptions of molecular asymmetry, he chose a powerfully poetic symbol, the spiral staircase. It is as legendary in the lore of science as the DNA helix.

A Detour through Quinine

In early 1851, when the Society of Pharmacy of Paris opened its competition on tartaric acid, the task was not only to look for tartaric acids containing fully formed racemic acid but also to set out the conditions under which it would be possible to transform tartaric acid into its left-handed isomer. In other words, to find out whether it would be possible to produce racemic acid artificially.

Upon his return to Strasbourg in mid-autumn 1857, Pasteur could claim to have carried out the first part of the task, but he was still unable to formulate an experimental method for making racemic acid. In fact, he considered it impossible, given the state of knowledge at the time.

In casting about for a different approach, Pasteur was inspired, in particular, by Pelletier and Caventou's studies of quinine. What was the reason for this? *Uncle Tom's Cabin* had just been translated into French and become a best-seller. Perhaps Pasteur, who recommended the book to his father, was struck by the role played by the cinchona bark in the novel. However, it is more likely that he was now systematically studying the optical activity of all substances that he could lay his hands on.

After the discovery of quinine in 1820, other derivatives of the cinchona bark, cinchonia and quinidine, were developed. These three substances were well known for their fever-reducing properties, but chemists were not agreed on their molecular composition. Pasteur was to show that quinine has two isomers, as has cinchonia, and that quinidine is in fact the right-handed isomer of quinine. He observed that through exposure to sunlight, even for a few hours, these substances take on a dark red coloring. This gave him the idea of studying the action of heat on the optical activity of the different isomers. In the course of these experiments, Pasteur found that by modifying the temperature of a substance, one can cause its stability to vary and alter its initial molecular structure.

Never one to neglect the practical applications to be deduced from his experimentation, Pasteur developed a new isomer of quinine, which he obtained by exposing the quinine to very high temperatures; he named it quinicine. This product turned out to be much cheaper to produce than quinine, since it is much more stable. Pasteur therefore envisaged using it against fever and asked several Strasbourg physicians for expert evaluations of the new product. He also asked Biot in Paris to put him in touch with Dr. Rayer, a member of the Institut and professor at the Charité hospital.

In doing so, Pasteur had an ulterior motive, and if he sought to establish a relationship with this influential scholar who had founded the Société de Biologie, it was no doubt in order to gain access to the powers that be. For Pierre-François-Olive Rayer was the personal physician to Louis-Napoleon Bonaparte, who had become Napoleon III a few weeks earlier. Pasteur therefore, summoning his most passionate eloquence, proclaimed that creating a medication is one of the noblest applications of scientific research. To his father, he wrote about the new product, "M. Rayer sent me word yesterday through M. Biot that it does bring down fever, but that before making a definitive statement, he would like to conduct several more tests. . . . It is very likely that on my next visit to Paris I will present my discovery to the Minister of War and that he will then appoint a commission to study my product in the military hospitals of Algeria and in the regions of France where fevers are very intense and sometimes do not even respond to quinine sulfate."[3]

As a medication, quinicine misfired and did not have the success for which Pasteur had hoped. His first contact with the medical milieu ended in a failure that gave the young scholar a first glimpse of the inertia and the indifference of the public health authorities. However, the research he had carried out in developing quinicine led him back to his work on hemihedrism, and he once again turned his attention to tartaric acid.

On the one hand, Pasteur now had at his disposal quinine, cinchonia, and their derivatives, that is, several right-handed and left-handed, active or inactive isomers that were more or less stable when exposed to heat. On the other hand, he was working with tartaric acid, which was very sensitive to heat. Biot had already shown that the optical activity of a tartaric acid solution increases when it is heated. This phenomenon had been described with even greater precision in 1838 by the chemist Edmond Frémy, who studied the changes caused in tartaric and racemic acid by exposure to high temperatures. Pasteur now had the idea of experimenting with compound substances. He therefore prepared cinchonia tartrate and subjected it to heat. The role of cinchonia in this operation was crucial, for it gave stability to the tartaric acid and enabled it to tolerate the heat without deteriorating. As a result the cinchonia disintegrated and lost water, while the tartaric acid was transformed into racemic acid.

On 24 May 1853, in his haste to share the excitement of a great day, Pasteur made use of a brand new invention to inform Biot of his discovery. Thrilled that it would take only one hour (and 12.50 francs) to transmit his message, he sent a telegram: "Transforming tartaric acid into racemic acid–STOP–Please inform Dumas, Sénarmont–STOP–L. Pasteur."[4]

Biot replied the same day in his usual cautious manner: "Dear friend, I am about to go out to communicate your message to the persons you indicated. But you do not tell me whether the product you have obtained actually has the active right-handed and left-handed acids that are found in the natural one, or whether it is just an isomer composed of inactive elements. As long as I do not know this, I shall not make any definitive statement; but please, do send me the means to make up my mind as soon as possible. Today I only have time for these two words: Your friend, J.B. Biot."[5]

But then, on 2 June, he received a letter from Pasteur that dispersed all doubt. The acid he had obtained did indeed possess the same physical and chemical characteristics as racemic acid. The product he had created was in every respect identical to the natural one and possessed right-handed and left-handed isomers that rotated the plane of light in opposite directions. A triumph for Pasteur!

The Interplay of Left- and Right-Handedness

At the same time, Pasteur was also studying another method for obtaining racemic acid from tartaric acid. This method was based on the modifications in the polarization of light that occurs when a substance is brought into contact with another whose optical activity is of the same or the opposite hand.

Pasteur tried first to crystallize active and inactive substances together in order to find out whether the form he obtained presented new physical characteristics. He noticed that in combining an active substance, such as right- or left-handed tartaric acid, with an inactive substance, one does not modify the optical activity of the former. In an imaginative formulation, he wrote that this was tantamount to combining a right hand (which is asymmetrical) with an open book (which is symmetrical): even together they will remain asymmetrical. In short, hemihedrism is preserved in such a case.

However, things went differently when an active substance was combined with another active substance. In this case, a double asymmetry would ensue. The combination of a set of hands and a set of feet potentially allows for four different assemblages. The optical activity of the combined substances therefore becomes stronger if they both go in the same direction and weaker, in proportion to their respective values, if they go in opposite directions. The right foot strengthens the right hand, while the left foot weakens it, and vice versa.

Pasteur used these findings to isolate right-handed from left-handed isomers. When he combined an active substance with a solution of paratartaric acid containing both tartaric acid isomers, the specific crystallization he obtained was determined by the nature of the active substance used. Thus a left-handed base added to the mixture produced a crystallization of the left-handed tartaric acid, while the right-handed acid remained in solution. And vice versa. This method made it easy to separate the two isomers of racemic acid. Pasteur eventually proposed to generalize this method for determining whether a substance does or does not have optical activity, claiming that the asymmetrical character, whether right- or left-handed, of any molecule can be experimentally established by combining it with appropriate active substances.

By means of these different methods, which made use of both physics and chemistry, it thus became possible to transform tartaric into racemic acid and to separate out the different isomers that comprise it. What Dr. Rassmann of Prague thought he had found was finally discovered by Pasteur. And so he wrote to his father: "So here is that famous racemic acid (after which I chased all the way to Vienna), which I have artificially prepared with the help of tartaric acid. For a long time I believed that this transformation is impossible. This discovery will have incalculable consequences."[6]

However that may be, it did have one immediate consequence: thanks to the report he published immediately and to the various papers Biot read on his behalf at the Academy of Sciences, Pasteur garnered the 1,500 franc prize of the Society of Pharmacy of Paris in November 1858. This was a considerable sum, and he earmarked half of it for purchasing equipment for his laboratory in Strasbourg, since the University could not afford to supply him with the materials he needed for his experiments.

Success begets success, and so the prize of the Society of Pharmacy was soon followed by another honor, which, though less lucrative, was more prestigious and perhaps even more ardently desired. The efforts deployed by Jean-Baptiste Dumas had finally paid off, and Pasteur was named Chevalier de la Légion d'honneur. He was barely thirty years old, but Biot almost felt that this honor was overdue: "There was no need," he wrote to Pasteur's father, "to ask that he be given what he had long deserved, but one could boldly make the point that the Institution [of the Order of the Chevaliers de la Légion d'honneur] would have done itself considerable harm if it had waited any longer to admit him to its ranks."[7]

Proudly displaying the red ribbon on his lapel, Pasteur even went so far as to grant himself a few days of vacation at a fashionable seaside resort; in August 1853 he went to Dieppe at the time when the Emperor and the Empress were there. He had no qualms about placing himself where he might be noticed by the sovereigns — no qualms, but no success either. "Due to the uncertain weather, the Emperor and the Empress did not go out while I was there," he told his father. "I have not seen them. Whenever the weather is good, they go out like ordinary mortals; the Emperor often walks. I did what everybody does. I bathed in the sea. This feels much better than I had thought."[8]

The Channel is not the Adriatic, but still, Pasteur had seen the sea, and if he did not mention Marie in the account of his bathing prowess, it was surely because she had not shared it, for she was seven months pregnant. The Pasteurs' third child, Cécile, was born at Strasbourg on 1 October.

These successes and joys were sorely needed to counterbalance the moroseness that came into Pasteur's life because of his father. In the last few months heavy blows had fallen on the house at Arbois: Joséphine had died in 1850, Emilie in 1853. As for Virginie, she was married and occupied with the care of her children. Solitude weighed heavily on Jean-Joseph and he began to feel somewhat irritated by all the honors that came to his son, even though he had at first been flattered. He no longer wanted to read, adopted an unyielding and sometimes aggressive misanthropy, and endlessly harped on the same recriminations. He carried on until Louis felt obliged to justify himself: "I understand you all the better," he finally wrote, "because I was made, or rather because you have made me, in your image. There are days when you are bored, when you get upset by the annoyances that are part of life. Well, I am like that too! . . . You are too skeptical. That is a bad thing. For my age, I am already very skeptical, and I am afraid that when I get to yours, I will surpass my model."[9]

In his bitterness, Jean-Joseph no doubt exaggerated when he claimed that his son was abandoning him; yet to some extent it was true. After all, his research had brought Pasteur to a new stage in life. He who had always docilely followed the directives of his two masters, Biot and Dumas, was now beginning to feel sufficiently sure of himself to conduct his research by himself. Thirty years old and recognized and respected by scientists twice his age, he was susceptible to the pride of the discoverer. The tone of his letters and of his scientific notes became more and more peremptory and he was soon tempted by the demon of generalization. The experiments Pasteur conducted at Strasbourg in 1853 and 1854 clearly show that he was in the throes of a kind of vertigo.

Asymmetry Is Life!

His success with tartaric acid led Pasteur to study the optical activity of a whole series of substances. He meant to prove that his particular findings on the tartrates could appropriately be applied to all substances.

The task to which Pasteur had harnessed himself was both simple and tedious; it consisted of making a study of every molecule and determining its symmetrical or asymmetrical character. To be sure, he was limited to the products he could find by rummaging through the cupboards of the Faculty of Sciences, but at Strasbourg there were enough samples to launch an in-depth study of hemihedral facets. As he meticulously examined these diverse kinds of molecules one by one, Pasteur was intrigued by a strange finding.

Asymmetry, it appeared, has its limits; it stops where the realm of life ends. All the substances that deviate light, presumably because they are composed of asymmetrical molecules, come from plant or animal matter, such as plant extracts and matter secreted by animals or plants. Gums, essential oils, albumin, gelatine, fibrin, and cellulose all deviate light. Conversely, chemical substances that do not modify the plane of light are all of mineral origin: gypsum, pyrite, garnet, chalk, etc. Moreover, these substances are composed of molecules possessed of an axis of symmetry. The inert world of rocks and that of the artificial compounds created by chemical synthesis are composed of superimposable symmetrical molecules. The deviation of the plane of polarized light testifies to the subtle differences between living and nonliving matter. Humans, animals, and plants differ from the mineral world by the spatial geometry of their molecules. The world of the biosphere is oriented in one or the other direction.

This discovery was fundamental in the proper sense of the word. Pasteur did not hesitate to deduce from it that the molecules of life are asymmetrical. Going even further and taking a position that was metaphysical as much as it was scientific, he postulated that life needs asymmetry. Connecting the concept of life with that of procreation, he added that fertilization is asymmetrical in every aspect, from the seed to the egg. Conversely, all molecules that do not come from living matter do not deviate the plane of light; as he was looking at rock crystals or at synthetic products artificially created in the laboratory, he found that they were symmetrical substances evenly organized along an axis.

But what about quartz? Had not Haüy established the asymmetry of the facets of its crystals? Did it not deviate light, as Arago had pointed out? Pasteur demonstrated that this was nothing but a false exception to the rule, for these properties of quartz exist only at the level of the rock crystal. In a solution where the molecules are in suspension, the asymmetry of quartz disappears along with its ability to deviate light. In order to make himself clear, Pasteur once again used an eloquent image: the quartz crystal is analogous to a spiral staircase in which all the steps would be identical wooden cubes. The staircase as a whole is asymmetrical, but when it is broken up, it becomes a set of cubes that do have an axis of symmetry. Carrying this analogy even further, it can be said that in living matter, the steps that make up the staircase are not cubes but parallelepipeds; even when it is disassembled, the steps that are left retain their asymmetry.

Simplifications and symbols led Pasteur to generalize. To him, the difference between the origins of living matter and the mineral world was so obvious, their distinctive characteristics were so clear and evident, and the geometric principles of the two kingdoms so opposite, that he concluded: "Every chemical substance, whether natural or artificial, falls into one of two major categories, according to a spatial characteristic of its form. The distinction is between those substances that have a plane of symmetry and those that do not. The former belong to the mineral, the latter to the living world."[10]

Fascinating as this investigation was, it was soon eclipsed by another set of equally fundamental questions: Why does one not always simultaneously find a right-handed molecule together with a corresponding left-handed one? Why does nature favor one form rather than the other? Without furnishing actual proofs, Pasteur outlined an answer as follows: "At the moment when they come together, the elementary atoms are subjected to an asymmetrical influence and, since all organic molecules that have come into being in analogous circumstances are identical, regardless of their origin and the place of their production, this influence must be universal. It can be assumed that it extends over the entire terrestrial globe, and that it is the cause of the molecular asymmetry of plant matter, products that are also present in almost unaltered form in animals, where they play a mysterious role of which we have as yet no understanding whatsoever."[11]

Thus the world of our liquefied forests is asymmetrical: In a scientist's dream that was both grandiose and excessive, Pasteur imagined that if one reduced the entire plant (or animal) kingdom to chemical solutions, one would without fail observe the deviation of the plane of light passing through these liquids. Having reached this point, he did not pursue his speculation and only pointed out that South African wine contains tartaric acid as right-handed as that of the wine from the vineyards of Arbois and that American tobacco contains the same malic acid as that of Lot-et-Garonne.

Biological knowledge had thus come up against an unknown force that Pasteur called "cosmic." According to him, this force exerted its influence over the seeds of plants as well as over the eggs of animals and humans: "The universe is an asymmetrical entity. I am inclined to believe that life as it is manifested to us must be a function of the asymmetry of the universe or of the consequences of this fact. The universe is asymmetrical; for if one placed the entire set of bodies that compose the solar system, each moving in its own way, before a mirror, the image shown in that mirror would not be superimposable on the reality."[12]

To this asymmetry of the universe, the cause or consequence of a force that is itself asymmetrical, Pasteur indiscriminately ascribed a whole series of phenomena, including the action of a galvanic current on the magnetic needle, magnetic fields, the opposition between the two poles of a magnet or between two electrodes, and indeed all electrical phenomena in general.

In this manner, Pasteur came to wonder about the deep questions that had also preoccupied the greatest of the scientists who had come before him, Faraday, Ampère, Arago. And it must be acknowledged that to this day, despite the great advances in genetics and biology that have led to the development of extremely refined methods for examining the right- and left-handed characteristics of living matter, we are still asking the same question: why does the pine of Aleppo produce a right-handed turpentine while that of Bordeaux is left-handed? And: are the natural or supernatural forces that have presided over the development of our terrestrial environment identical with those that have oriented our molecules in space? If, as Pasteur discovered, life is asymmetrical, why is some of it right-handed and some of it left-handed? How can we attempt to reshape and replicate life if we lack the asymmetry of the natural product?

Ernest Renan has written that right- and left-handed molecules were created like man's two hands so that they can be joined in the gesture of prayer. Pasteur, too, understood that this debate over the first molecule and its asymmetry brought him face to face with the metaphysical problem of origins.

Perspectives and Disappointments

When Pasteur considered asymmetry on a cosmic scale, he went beyond the confines of physics and chemistry to confront the fundamental questions about life that had been asked by previous nineteenth-century scientists. The idea that ecology punctuates or determines the evolution of people and societies was rather pleasing to the biologist who had become familiar with the variations of form that the molecular structure can undergo. Just as explorers, be they geographers or naturalists, stress the role of the environment in human evolution, so the chemist who manipulates molecules soon becomes aware that he can act upon them. Does "to understand life" not imply, at least indirectly, that one can change it?

When, for example, Pasteur crystallized ammonia in the presence of soda, he observed crystals with hemihedral facets that did not form under any other conditions. To create an artificial species of molecules that did not have new physical characteristics was to fall in line with Lamarck's transformism. After all, Pasteur was a contemporary of Darwin, and so he studied the evolution of matter as the English scientist studied the antecedents of man.

> We will not be able to break down the barrier that exists between the mineral and the organic kingdom because of our inability to produce asymmetrical substances in our laboratory reactions, unless we are able to bring influences of an asymmetrical nature to bear on these reactions. Success in this endeavor would give us access to a new world of substances and reactions, and probably to organic transformations as well. In my opinion, it is here that we must place the problem, with respect not only to the transformation of species but also to the creation of new species. . . . The difficulty of resolving these problems must not prevent us from acknowledging their existence. Since it has been possible to find the inverse of right-handed tartaric acid, it will without doubt be possible some day to know all the immediate inverse principles of the existing species. But if we want to go further in the physiological order, and if we want to carry these new immediate principles over into living species by way of nutrition, then the great difficulty will be, I fear, to overcome the development peculiar to each species that is potentially contained in the germ of each one of them, for this germ will always manifest the asymmetry of the existing immediate principles.[13]

The research program outlined here by Pasteur is still relevant, at a time when advances in genetic manipulation make it possible to subject living matter to more and more precise intervention. Could it be that life on the opposite side is the opposite of life? "Who can possibly foresee how living beings would be organized if cellulose, which is now right-handed, became left-handed, and if blood albumin changed from left- to right-handed?"[14] To invert the direction of molecular asymmetry would be to change life!

In his effort to place these dreams of a demiurge on a scientific footing, Pasteur thought up and carried out innumerable experiments designed to modify the spatial aspect of various substances and to obtain a representation of the new forms. He continued until 1870 to develop new ways to subject chemical compounds to all kinds of reactions by using solutions of mineral crystals, experimenting with the influence of magnetic fields, and devising rotational movements brought about by clockwork mechanisms.

Pasteur became a new Cyrano de Bergerac. He attempted, for instance, to germinate and grow a plant under the influence of sun rays that he inverted by means of a heliostat. These were richly imaginative attempts that often caused Pasteur to say that one must get away from the outmoded and impotent methods of the chemists. But results were not forthcoming. When he repeated the experiments he had carried out on racemic acid with all kinds of other substances, Pasteur succeeded only in deepening the chasm that separates the mineral from the living world, the inorganic from the organic. He did not re-create life. "My investigations are going rather badly," he wrote to his father in December 1853. "I am almost afraid that all the tests I have conducted this year will fail and that I will have no important piece of work to show for my efforts by the end of next year. Well, there is still hope. But then, one must be a little mad to take on what I have taken on."[15]

It should be added that many of Pasteur's experiments met with the skepticism or outright disapproval of his peers. Biot in particular tried to discourage his former pupil: "I wish I could turn you away from your idea of trying to study the influence of magnetism on vegetation. M. de Sénarmont and I feel the same way about this. To begin with, you are going to spend a large part of your money, perhaps all of it, on equipment with whose use you are not familiar and which may or may not work. Secondly, this will take you away from the very fruitful path of experimental research you have followed until now, and you have too much to do to run from the certain to the uncertain."[16]

In fact, these experiments did not yield any truly positive results, for Pasteur failed in his attempts to invert a cosmic form that existed only in his imagination or, rather, belonged in the realm of metaphysics rather than science. Yet in the course of carrying out these attempts, Pasteur did realize that he could transform asymmetry, invert it, and modify it, but that he could not create it. Above all, he found that he could easily make it disappear; thus the molecular structure of gum was modified by treatment. And so was that of every other plant substance, as Pasteur showed in a long series of experiments carried out with unflagging persistence. An asymmetrical substance subjected to even moderately energetic chemical reactions loses its asymmetry. In other words, the mystery of life is twofold, for though asymmetry cannot be created, it can be made to disappear.

What the chemists could not bring to bear, Pasteur concluded, was the force of asymmetry itself. When the experimenter in his laboratory "applies cold or heat to elements or their products, he only brings to bear symmetrical forms, so that the resulting artificial products do not show asymmetry." Unable to achieve the breakthrough for which he had hoped, Pasteur gradually returned to the caution and the rigor that characterized his approach and denounced all those who thought they were capable of creating life artificially. "The need to examine the optical activity of all organic products has never been greater," he finally declared. "Henceforth, the identification of a natural product with an artificial product of the same composition can be considered scientifically proven only after careful examination of the latter's crystalline form and its optical activity."[17]

The irreducibility of living matter therefore became one of Pasteur's most firmly held convictions. Even if over the years he came to allow that the difference between the mineral and the physiological world should be considered a matter of fact rather than of principle, he would always oppose such scientists as Marcelin Berthelot, who fervently believed in organic syntheses. As it turned out, more than a century had to pass before the fundamental antagonism between Pasteur and Berthelot was resolved, for it was only in 1971 that H. P. Kagan and his collaborators realized an asymmetrical photosynthesis. And while the yields are very small indeed, the fact that an optically active substance has been obtained in the laboratory is of considerable symbolic value.

5

Fermentation and Life

In the fields of observation, chance favors only those who are prepared."[1] Pasteur was deeply convinced of the truth of this celebrated formula. Yet chance not only intervened "in the fields of observation," but it could also play a role in ordinary life: in early September 1854, Pasteur was promoted. Leaving his position as acting professor at Strasbourg, he was appointed, at the age of thirty-two, professor of chemistry at the University of Lille. This position, enviable as it was in itself, was associated with the highly prestigious appointment as dean of the new Faculty of Sciences. His publications on the chemistry of crystals, along with the influence of Jean-Baptiste Dumas and Biot, had convinced the Minister of Public Education and Religion, Hyppolyte Fortoul, that this young scientist would be an excellent dean for a brand-new faculty of sciences like that of the department of Nord.

This transfer, which may not have been the result of chance — undoubtedly there had been careful lobbying on Pasteur's behalf — came at just the right time. Had he stayed at Strasbourg, where his work on asymmetry was going nowhere, Pasteur might have become stuck in a program of research which, being too focused on fundamentals, faced the risk of becoming sterile. At Lille, by contrast, where he had to divide his attention between his teaching and administrative duties and the specific demands of the agrarian food-industry of the region, Pasteur was bound to open his eyes to increasingly precise practical questions. He gradually — and perhaps grudgingly — turned away from his investigation of crystals. As he was (temporarily, it turned out) bidding farewell to crystallography, he cited Lavoisier: "It is the fate of all those who are engaged in chemical and physical studies to see a new step they must take as soon as they have taken a first one, and they would never publish anything if they waited until they have arrived at the end of the road that gradually appears to them and seems to become longer as they approach in order to follow it."[2]

Dean at Lille

In August 1854, an imperial decree divided the University of the department of Nord into two parts, the Faculty of Letters at Douai and the Faculty of Sciences at Lille. The choice of Lille was not made haphazardly, for it was designed to recognize the importance of a rich industrial region. Nor was the choice of dean fortuitous, for Pasteur's vision of education was in tune with the objectives that justified the opening of this new faculty: he had always emphasized learning and experimentation or, rather, experimentation as a way to learn.

The city of Lille was given special treatment by the government. It was among those Louis Napoleon Bonaparte had chosen to visit in 1852 as he sought to secure his legitimacy and rally a public that still hesitated to accept him. The imperial couple was to return to Lille in 1867 to celebrate the bicentennial of its integration into France.

Lille had replaced Douai as the capital of the department of Nord at the beginning of the century, and by 1850 major public works were undertaken, as they were in Paris, to bring modernity and efficiency to what had become the fifth largest city of France. The suburbs were incorporated and the city wall was replaced by wide, airy avenues bordered by stately homes and small brick houses. Thanks to the development of the railroad, Lille was connected to the capital by the *ligne du Nord*, in which the Rothschilds were the principal shareholders. A new cathedral was also erected. Inspired by early French gothic architecture, it was placed under the patronage of Notre Dame of the Vineyard (de la Treille). The city knew how to take full advantage of all modern inventions and became an important industrial center thanks to cotton and linen spinneries, mechanical construction firms, chemical industries, coal mining, distilleries of beet juice, and breweries. In 1850, manual laborers accounted for one-third of the population of almost one hundred thousand inhabitants.

It was in this booming regional capital that the government created the new university where Pasteur was appointed dean. The site of the Faculty of Sciences, the rue des Fleurs, which was not in the old city center, was chosen above all to accommodate an innovation in university education, namely, the introduction of applied sciences. Following the example of the German universities, the ministry of education wanted to create a new impetus that would allow students to prepare themselves for working in industry. Payment of a modest fee now entitled students to supplement their theoretical courses with practical work in the laboratory. Such work was officially recognized with a scientific diploma, called the *certificat de scolarité de Lille,* which, delivered by the Faculty of Sciences of Lille, competed with the diploma of the Ecole centrale des arts et manufactures of Paris. It is worth noting that prior to this pilot project at the university level, the views and methods of teaching were such that contacts between scientists and industrialists were neither sought by the former nor pursued by the latter. In appointing Pasteur, the Minister of Education, Fortoul, thus chose the path of innovation.

Pasteur held a revolutionary definition of applied science: "There is no such thing as a special category of science called applied science; there is science and there are its applications, which are related to one another as the fruit is related to the tree that has borne it."[3] This position was at odds with the views held by the scientists of Pasteur's time, when science for science's sake was the prevailing ideology. Did not Sainte-Claire Deville, one of the greatest chemists practicing at the time, define himself as working in the absurd? Pasteur was naturally attracted to technical realities by temperament and curiosity. And in this industrial town, he was above all interested in developing processes and research that would be economically useful. Acting upon these ideas, he was equally enthusiastic about the teaching of science and about its industrial applications.

In his inaugural lecture of 7 December 1854, which he delivered at Douai, where the university's administration was still located, Pasteur launched into one of the speeches of public advocacy that soon became his trademark. "Where will you find," he exclaimed, "a young man whose interest and curiosity will not be aroused when you give him a potato and teach him to make it into sugar, and this sugar into alcohol, and this alcohol into ether and vinegar? Who will not be happy to tell his family at the end of the day that he has operated an electric telegraph? And, gentlemen, you can be sure that once such things are learned, they are rarely, perhaps never forgotten!"[4]

When Pasteur spoke in these terms to "the fancy society of Douai," as he put it in a letter to his wife written on the evening of the official ceremony, he was so sure of himself that he was sorry to have left out of his speech a certain "stirring passage" that would have brought even stronger applause. He was quite proud of his peroration, which had "impressed people and made them say the most flattering things about [him]."[5] Little by little the needy and even arrogant young scholar was turning into a brilliant academic who was also a subtle diplomat and a fine speaker.

But of course Pasteur did more than give emphatic speeches; he devoted himself wholeheartedly to his new functions. He was actively involved in teaching, not only in his own chemistry courses but also in setting up new courses which, under his guidance, were being developed by his colleagues at Lille. He personally reviewed, corrected, and commented on proposals for courses in physics, applied mechanics, differential calculus, natural history, and even drawing.

The courses that began in January 1855 in the new premises made available to the Faculty of Sciences by the ministry were neither completely worked out nor fully equipped. Pasteur took charge of the situation and attended to the most urgent problems. If certain courses seemed to aim too high, they should be edited from handwritten notes. If others were too abstract, they should be illustrated with examples, such as: how to work a steam engine, how to induce alcoholic fermentation, how to fatten livestock, and how to develop pesticides. Despite some difficulties, the atmosphere was good. Once the enterprise was launched, success followed. New ways of doing things, brought about in part by ministerial decrees and partly by the dean's program, attracted a large student body: three hundred in the first year. Even workers and women registered for evening courses. Pasteur managed to obtain major funding for construction from the ministry. Soon, teaching was done off-campus. Students learned about the factories in the area and were taken on field trips to nearby distilleries and foundries. These outings took them to Valenciennes, Denain, and Saint-Omer; and there were even plans to go to Belgium to see blast furnaces and metallurgical plants. Six months after the inauguration of what looks like a forerunner of our technical universities, Pasteur wrote to the rector of the University at Douai that practical instruction was becoming the driving force of the Faculty of Sciences and insisted that it was vitally important never to cut the students off from the working world. The educational system, he pointed out, should combat the prejudice that considered educated young people, the future managers of industry, to be totally different from the mass of ordinary employees, for all too often "it is assumed that young people who have studied for too long approach the workers with ideas that they have taken from novels."[6]

Pasteur was happy at Lille, even if he complained about the long winter, the rain, the wind, the cold. Even in mid-June, the laboratory had to be heated. But he was kept so busy by his new activities that he had little time for anything but work. He never left his Faculty of Sciences until August 1855, when he went to Paris. This, however, was not a pleasure trip, but a strictly professional errand; Marie did not go with him ("Right now, I am a widower," he wrote humorously to his father), and his principal purpose was to speak to officials at the ministry and to visit the World's Fair. Pasteur went several times to the Palace of Industry on the Champs-Elysées, where the latest technical innovations were shown: "The more I go to the Fair, the more I realize that it is for me a mine of chemical information the likes of which I will not find again for a long time."[7]

There cannot be the slightest doubt that Pasteur preferred these visits to his university obligations, even if they too constituted a "mine of information": "The Advisory Board is not terribly interesting," he confided to his spouse. "But I do learn about how problems are discussed, and I study the ways of running a meeting. The magistrates of Douai are punctual. So is the archbishop of Cambrai. . . . The rector does very well, and the lawyers talk with easy abundance."[8] Pasteur did not naturally enjoy social events; he declined or forgot certain invitations if they were not useful to his career or if they took time away from his research. But he did not neglect the receptions he had to give at his residence and made sure to invite the prefect, the mayor, the rector or the dean of Douai, serving them straw wine of Arbois or yellow wine of Château-Chalon that he asked his father to send.

Meanwhile, teaching did not overshadow research; on the contrary, Pasteur wished to "set the example of a dean who publishes with the Academy of Sciences." He divided his days between the laboratory and the Faculty. When Marie was not at home, he even forgot to have dinner. Even at Lille, Pasteur was thus a researcher, who could write to Chappuis: "I finally have what I have always been envious of, a laboratory where I can go at any hour, since it is on the first floor of my apartment; and sometimes while I am sleeping, in fact rather often right now, the gas burns all night, so that the operations are not interrupted. . . . Let us all work; it is the only fun there is."[9]

From the Laboratory to Manufacturing

Pasteur does not seem to have done any work on fermentation before he arrived at Lille. It is not certain that he would ever have started doing so if the occasion had not presented itself, particularly in connection with the program of courses that he had to oversee and his approach to teaching; in the heart of a region largely devoted to the distilling of beet juice, it was sensible to teach about alcohol production. On a loose sheet outlining the project for a course, one can decipher these words, which in the most laconic manner announce one of the most important fields of investigation in Pasteur's life work: "What fermentation consists of. Mysterious character of the phenomenon. A word on lactic acid."[10]

At the beginning of the academic year 1856, an industrialist of Lille, M. Bigo, whose son Emile was taking Pasteur's course at the Faculty of Sciences, came to see him. Many manufacturers of beet root alcohol, he said, were having problems with their production; the alcohol was of poor quality and had an acid taste and the fermentation vats gave off fetid odors. Pasteur found it perfectly natural to help these distressed manufacturers. What is more, he developed great enthusiasm for their cause.

What had brought him his first successes in chemistry was the fact that he had found a way to approach it as a crystallographer and to avail himself of polarimetry, a technique heretofore neglected by this science. In studying fermentation, he was to proceed in an equally original, albeit different manner, which lay outside the mainstream of accepted techniques and theories. He did not, to be sure, give up analytical chemistry and its tools, the glass tubes, standing glasses, crystallizing pans. But he broke new ground, or rather went against the customs and habits of the chemists by bringing in the microscope.

To bring a microscope into a biochemical laboratory is a relatively incongruous thing to do, even today. At the time it was a quasi-revolutionary act, the more so since Pasteur did not know what he was looking for, and barely what he was looking at: it was yeast, small round organisms whose presence, it was said, went along with alcoholic fermentation. In the laboratory where he carried out this new research project, Pasteur had, in addition to a simple student microscope, only a primitive coal-heated incubator.

As was his wont, he very soon left the premises of the university to go into the field. Almost every day he went to the sugar factory in the rue d'Esquermes, where he transformed the cellar into a laboratory that would hold his rudimentary equipment. Marie wrote to her father-in-law: "Louis continues to work very hard. He is now up to his neck in beet juice. He spends his days in an alcohol factory. He has probably told you that he is now giving only one class a week, so that he has lots of time, which he uses and abuses, I can assure you."[11]

Just what was he studying? Fermented juice, filtered or unfiltered, which he examined under the microscope, cultivated in his incubator, and analyzed biochemically. One must imagine the different stages of the excitement he experienced: there were times of uncertainty when he explored different approaches, times of disappointment, when he wrote in his notebook: "mistake . . . all wrong . . . no . . ."

But very soon, thanks to two important observations, he was back on the road to success. The first was made with the help of the microscope. Pasteur noticed that the yeast, or the ferment, as it was called at the time, changed its form depending on the state of the fermentation. The organisms were round when the fermentation was normal and took on an elongated rod form when it was defective. The second observation he owed to biochemical analysis. When the fermentation was not going well, Pasteur noted the presence in large quantity of an acid that should not have been there, lactic acid. The more of this product appeared, the more the ferments showed the elongated form, and the less alcohol was produced.

Pasteur could have stopped with this demonstration, or even the rather remarkable one he carried out shortly afterwards, when he showed that distilling fermented juice under the wrong conditions caused hydrogen to attach itself to one of the components of the beet, the nitrites, resulting in the fetid smell. Identifying the presence of lactic acid answered the questions of the industrialist of Lille to the extent that it told him the reason for the bad fermentations. But this did not explain why it had appeared and, consequently, did not furnish the means to avoid this occurrence.

Pasteur was not the man to turn away from a problem before he had expended all of his talent on it. He therefore relentlessly pursued his observations of the different forms of the ferment depending on the state of the distillation. And he kept going until he was ready to state a hypothesis that he alone was capable not only of formulating but of documenting by referring back to the laws he had established earlier in his studies on the chemistry of crystals.

What Is Fermentation?

Most of those who studied fermentation were more interested in the quality of foodstuffs than in their production. Today, when the scientific problem is essentially solved, the phenomenon has lost much of its magical appeal. Yet over the ages it has raised as many questions about its origin — from God to the chemical equation — as it has brought blessings, among them wine and bread. Indeed we are indebted to fermentation for some of the most powerful symbols of our myths, at least in the Western tradition. The ancient Egyptians brewing beer, the ancient Gauls making their bread dough rise with yeast — these images evoke ancestral practices. Yet scientists, including the earliest chemists, from Paracelsus to Robert Boyle, had no convincing explanation to account for the phenomenon. At best, they could build intellectual constructs whose principal shortcoming was their failure to lead to a fruitful theory. The only certain fact was the recognition that alcohol was a product of fermentation.

The eighteenth-century studies of gases, particularly the identification of carbon dioxide, had caused the chemists of Pasteur's time to take a renewed interest in the question. Distinctions were made among several modes of fermentation. Best known was alcoholic fermentation, marked by the transformation of sugar (in grape juice, apple juice, barley, etc.) into alcohol (wine, cider, beer, etc.). This was also called spirituous fermentation, even if Fourcroy had written with high revolutionary fervor that this expression should be abolished, "for the word spirit needs to be banished from science."[12] There was also acetic fermentation, when the resulting product was an acid or vinegar, and lactic fermentation, which was observed in soured milk. In addition to these phenomena that were grouped together in the same category, there was putrid fermentation or putrefaction; it was characterized above all by the fetid smell that emanated from spoiled meat, rotting eggs, and all decomposing organic matter.

Until the end of the eighteenth century, scientists were able to identify these fermentations, at least certain of their characteristics, but they knew virtually nothing about the mechanisms that governed them, even though some observers, particularly certain chemists, claimed that they understood their causes. But their explanations were so murky that one has trouble understanding them. In the end, the questions were as straightforward as they had ever been: Why does grape juice bubble in the vintner's vat? Why does wine sometimes turn into vinegar? How can we account for the fact that in bread making the dough bubbles and rises when yeast is added? Does putrefaction have anything in common with alcoholic or acetic fermentation? Before they became abstract research topics, these problems were of course of practical interest. For the people who ran distilleries, bakeries, or canneries, and for all those involved in agro-industrial enterprises, these were problems of capital importance.

Pasteur was therefore not the first to have tried to find at least a partial answer. Throughout the ages, philosophers and chemists had thought about the matter. The polemics of the eighteenth and nineteenth centuries had kept up debates and quarrels in which metaphysics played no small part. The philosophers were entitled to use the big words! When Descartes contemplated the bubbling in the vat, he talked of forces that could mix and become displaced. He had the explanation: it was mechanics. Malebranche essentially said the same thing when he corrected this concept: fermentation was the communication of a movement caused by invisible bodies, for a body that moves must receive its impulse from another body. Parallel to this futile discussion, certain chemists set forth equally imprecise but more measured hypotheses. In this context, two names immediately come to mind: Lavoisier and Liebig.

Antoine Laurent Lavoisier is credited above all with introducing the notion of balance into chemistry. He was beheaded on the Place de Grève during the revolutionary Terror of 1794, after his accuser had delivered his indictment, which culminated in the famous statement, "The Republic does not need scientists." The tribunal of history, however, has rehabilitated him, even beyond his merits, as one of the founders of a valid theory of fermentation. Lavoisier summarized the phenomenon in a balanced equation and an algebraic formula. He said that if one places sugar on one pan of a scale, it is balanced after fermentation by the sum of the weight of the carbonic acid it has yielded and of the alcohol it has formed. In short, sugar is transformed into an equal weight of two substances, two, and no more. All one needs to know to establish the exact result of the reaction is the formula of sugar, alcohol, and carbonic acid. Lavoisier summarized this reaction in perfectly limpid terms: "The results of the vinous fermentation are reduced, therefore, to separating into two portions the sugar, which is an oxide, oxidizing one at the expense of the other to form carbonic acid, deoxidizing the other at the expense of the first, to form out of it a combustible substance which is the alcohol, so that if it were possible to recombine these two substances, the alcohol and the carbonic acid, we should again obtain sugar."[13]

This is an illustration of the famous formula: "Nothing is lost and nothing is created in the operations of art as in those of nature." Yet if Lavoisier's observations were correct, his conclusions were unsatisfactory. For he neglected the role of yeast. "Since yeast," he wrote, "is reabsorbed in the same manner it has appeared, I cannot take it into consideration."[14] Pasteur attacked Lavoisier's memoir with a certain ferocity: "Excessively deficient in its numerical findings, it is admirable if one looks at it from the point of view of general ideas and philosophy."[15]

Lavoisier's ideas could be confirmed only thanks to a series of considerable errors; Gay-Lussac and Thenard repeated his experiments, having first perfected their methods, but they were only partially successful. Convinced that Lavoisier could not be mistaken, they did not hesitate to modify his results in order to bring them into line with the more apparent than real simplicity of his theory. Inexact experiments and deliberately falsified figures did not, of course, invalidate Lavoisier's idea. All that was left of that, Pasteur said, was the unfounded illusion of a possible equivalency between sugar on one side and alcohol and carbonic acid on the other.

Without in any way taking sides in these debates among chemists, and without praising Pasteur's sagacity too much, one must point out that at this stage of the debate the action of the ferment had been obstinately ignored. Yet Lavoisier had had to add yeast in order to bring about fermentation. Why, then, was this yeast so necessary to the outcome of this experiment? Yeast was the foam deposited at the bottom of the vats of beet juice; it increased during the fermentation of the sugar, and it formed spontaneously if it was not added. Thus, in an experiment to which Pasteur drew attention, Gay-Lussac had shown that in making wine, more was needed than just grape juice, namely, contact with the air. The crushed grapes began to ferment only when they came in contact with the atmosphere. At this point, Gay-Lussac had invoked the theories of Appert who, by heating foodstuffs, had discovered a way to preserve them, postulating that oxygen, whose role in combustion had been elucidated precisely by Lavoisier, made the chemical reaction possible. For, and this was the dogma, everything must be reduced to a chemical formula.

As for Justus von Liebig, he was almost fifty years old when Pasteur became interested in fermentation and set himself the task of establishing its theory. Liebig had studied chemistry in Paris with Gay-Lussac and Jean-Baptiste Dumas and had created in Germany the first European laboratory for scientific studies. A baron from Hesse-Darmstadt, he occupied the chair of chemistry at the University of Munich and was a member of all German academies as well as the Royal Society of London, the principal learned societies of Europe and America, and of course the Academy of Sciences in Paris. The name Liebig may not yet have been a household word, but it already appeared in the advertisement sections of newspapers and in the windows of food stores, for it was he who had invented artificial milk and meat extract — *Liebig's Fleischextract* — as well as the bouillon cubes that were to be so widely used during the Franco-Prussian War of 1870.

For the moment, Liebig was the standard bearer of the official theories on fermentation. Unable to explain the role of the ferment, he limited himself to describing it and attributed to it the property of causing motion. "The yeast of beer and in general all putrefying animal and plant matter transmit the state of decomposition in which they find themselves to other substances; the motion imparted to their own components by the disturbance of their equilibrium is also communicated to the components of elements that come in contact with them."[16]

Lavoisier had assimilated fermentation to algebra; Liebig refined and nuanced this view. He recognized the importance of yeast, only to insist that it was destroyed in the reaction. This destruction, he maintained, created a new equilibrium. Minimized in this manner, the ferment was indispensable only because it disappeared. The German theory, which carried the day in scientific circles, mentioned yeast only in connection with its decomposition. The important thing about the ferment was not its life, but its death. "It is the dead part of the yeast, the part that is no longer alive and undergoing alteration, that acts on the sugar."[17] Besides, Liebig added in support of this theory, had it not been demonstrated that in many fermentations eggs, tainted meat, or old cheese take over the role of the ferment? Soured milk makes milk ferment, sour dough makes bread dough ferment. Vinegar makes wine and cider turn sour. These substances have that effect because they are disintegrating. Their destruction is the basis of a successful fermentation. Thus the putrefaction of the ferment increases fermentation and, conversely, anything that blocks putrefaction inhibits fermentation. Humidity and oxygen favor fermentation, while drying and the creation of a vacuum arrest it. Appert's preserves came on the scene just in time to furnish some additional arguments to the grand old man of chemistry. Chemists had, of course, long known that air and motion favor the fermentation of the grape. The grapes must be crushed, dough must be kneaded, and beets must be mashed in order to induce fermentation by means of movement. The main point to remember in all this is that Liebig did not deny the role of the ferment, but refused to believe that it could act in any other way than by its decomposition, that is, as dead matter.

In addition to these theories, there was one more that went beyond Lavoisier's weighing and Liebig's chemistry of exchange. This different philosophy of chemistry, which acknowledged the role of yeast and denied its destruction, originated with the Swedish scientist Jacob Berzelius and was propagated by Mitscherlich. To explain the role of yeast, Berzelius proposed that it functioned as an ignition mechanism triggered by contact, in other words, that it was a catalyst.

It was, he said, through the particular characteristics of this catalytic force that yeast triggered the reaction. Berzelius believed that its quality as ferment alone enabled it to act through its mere presence, by mere contact with it. Obviously, we are far from strict Cartesian mechanics here. Going even further in this direction, as a matter of semantics more than of experimental precision, Berzelius added: "I shall therefore call this force catalytic force. In the same manner I shall call catalysis the decomposition of this substance by this force." This did not satisfy the chemist Gerhardt, an associate of Liebig, who wrote: "Calling the phenomenon catalytic does not explain it, it only replaces a name of ordinary language by a Greek word. This is not particularly helpful to science."[18] Yet this chemistry which spoke of the activity of the ferment without outlining its precise contours found favor in the eyes of Mitscherlich, even though he did not agree with its vocabulary. For him, the chemical process involved was "a decomposition or a combination by contact."[19]

With a certain partiality and, as it turned out, wrongly, Pasteur rejected the theories of Berzelius along with those of Liebig. Summarily dismissing the champions of a Cartesian chemistry, he was equally firm in rejecting the ideas of Liebig and Berzelius. Yet the theories Pasteur did put forth and defend were not his alone. Although he misunderstood or ignored them, others had also looked into the microscope and recognized the biological role of yeast. Ever since the turn of the sixteenth and seventeenth centuries, when a certain optician of Zealand had found a special way to combine a set of lenses, the microscope had revealed the infinitely small; and by the end of the seventeenth century the Dutch naturalist Leeuwenhoek had used this instrument to describe red corpuscles and spermatozoa and to identify brewer's yeast for the first time. Barely visible under the microscope, yeast raised questions as to its nature even then. In the early nineteenth century, Thenard, seeing that it released ammonia, spoke of its animal nature, but it never occurred to him to envisage its role in fermentation other than in the form of a chemical compound.

Returning to Leeuwenhoek's observations after more than a century, Baron Cagniard de La Tour went much further. As a physicist, he had discovered the principle of the siren in 1819; as a chemist, he brought to light a phenomenon of prime importance, the budding of yeast, that is to say, its capacity to reproduce itself. Convinced that this budding played a role in fermentation, he wrote in 1835: "It is very probably through the effect of this growth that the yeast globules take carbonic acid from the sugary solution and convert it into an alcoholic solution."[20]

In Germany, at the same time, Theodor Schwann had done more than observe and had begun to conduct experiments on the role of yeast. Questioning Gay-Lussac's experiment on the possible effect of oxygen in the fermentation of grape juice, he showed that when the air was heated, its oxygen did not produce any effect; in that instance no fermentation took place. From this he concluded that oxygen was not responsible for the reaction and that heating destroyed a plant germ; better yet, he showed that this germ was sensitive to arsenic, like a plant, and insensitive to the strychnine of nux vomica, like animal matter. Thus his point was proven: living microorganisms were needed to bring about the alcoholic fermentation of grape juice.

One does wonder why these discoveries, which seem so obvious today, did not carry the day at the time. It is true that Schwann did not adequately report his observations, for he published only a brief and laconic memoir announcing a more detailed future publication, which never appeared. The fact is that Schwann and Cagniard de La Tour encountered the opposition of men more powerful than themselves. Berzelius received their reports with disdain. He wrote that microscopic observations were useless, since yeast was no more alive than some precipitate of aluminum. For his part, Liebig forcefully contradicted Schwann's conclusions. We have already spoken of his theories: yeast was nothing to be concerned about, since it was unstable and eventually died. Moreover, he said, why would the explanation be any clearer if it involved a living being? The ferment can perfectly well be replaced by dead matter; after all, fermentations other than the alcoholic or the lactic ones — putrefaction, for example — take place without yeast.

Yet one fact should have given pause to these critical minds and made them more cautious, at least with respect to alcoholic fermentation: informed of the observations of Cagniard de La Tour, the Paris Academy of Sciences, upon the recommendation of Thenard, commissioned a study of the role of egg white in alcoholic fermentation. It turned out that egg white was a good substitute for brewer's yeast. Analyzed by Turpin, the deposit formed in the course of the fermentation was teeming with new yeast. This meant that it was not necessary to add more germs; the decomposing egg white alone was sufficient to make them proliferate. But these experiments were soon forgotten. In science, nothing is more difficult to destroy than a dogma; especially in this case, where Liebig himself summarily dismissed Turpin's findings. Translating them for a German publication, he added his own commentaries that contradicted Turpin's.

If the theories of Lavoisier, Liebig, or Berzelius were written up in every textbook and accepted as true, it was because they were in tune with the mindset of most contemporary scientists. The chemists of the nineteenth century gloried in the thought that they were capable of establishing a clear-cut frontier between living and inorganic matter and of attributing specific characteristics to each of them. Today, such an approach to a complex subject elicits from us a mixture of amusement, puzzlement, and disdain. Yet the frontiers of life are almost as hazy to us as they were a century ago.

In the eyes of the nineteenth-century chemists, their discipline was superior to the physiology of a Claude Bernard, for chemistry had equations to deal with atoms. No wonder, then, that they saw the refutation of the official theories of fermentation as a step backward for science. They considered any attempt to find the effect of living microorganisms in fermentation retrograde. To put fermentation back into the realm of biology was to retreat.

Admittedly, yeast was a nuisance for them, not because it was necessary, but because it was alive. This extra dimension contributed by biology further complicated a matter that was difficult enough to simplify as it was. What fermentation needed was not a chemist, but a biologist. And this biologist would have to know the language and the processes that would enable him to reduce the problem to the molecular level without overlooking the necessarily complex intervention of a living organism.

From the Crystal to Life

Pasteur came to study fermentation as he was thinking about the nature of life. Looking at his first discoveries, one can have no doubt that he had long been haunted by what his contemporaries called the life forces and what he himself named the life processes [*le vivant*]. His work on yeasts gave him the first opportunity to experiment with a biological phenomenon. Called upon to state what he dimly perceived, he not only had to contradict Liebig in order to define what fermentation owes to the life processes but also to criticize Lavoisier by pointing out that the life processes do not conform to chemical formulas but use them. In order to make this double assertion, he had to make sure of the role played by yeast and establish the correlation between its function and its biological state.

In relative agreement with his contemporaries, Pasteur thought that a biological phenomenon cannot easily be considered systematized or even amenable to systematization. Referring to the complexity of fermentation in order to carry his argument even further and stress the role of a living organism in the process, he wrote: "Fermentation is not a simple and unique phenomenon, but a very complex one, as is often the case with phenomena closely correlated with life, which give rise to a variety of products, all of which are necessary."[21]

That Pasteur should have turned to biology when he studied the yeasts or described the characteristics and the multiplicity of derivatives of fermentation he had encountered in his research initially struck his peers as more odd than convincing. Nonetheless, the reflections ensuing from this work were close to the mainstream of the nineteenth century's major philosophical and scientific investigations concerning the role of life. Pasteur, however, was far ahead of his contemporaries because he constantly subjected his findings to the laws, results, and interpretations of rigorous experimentation.

The findings Pasteur presented to the Academy of Sciences of Lille, and subsequently that of Paris, seemed very different from the studies he had undertaken previously. He was known as a specialist on crystals, and now he had become a theoretician of fermentation. Ranging from polarized planes of light to culture media, his reagents had little in common. Yet the preoccupations that guided Pasteur's thinking at that period were not really different from those that had haunted him for a long time: he wanted to understand the relationship between life and molecular asymmetry. Although his first papers on fermentation are dated from his time at Lille, they were based on reflections going back to 1849.

At the end of that year Biot had drawn Pasteur's attention to amyl alcohol, which had the ability to deviate the plane of light. Having been unable to obtain an alcohol of impeccable quality after six months of study, the young scientist had put aside this work, stating that he would like to return to it when he was better prepared, for he was still baffled by what he had observed. "Amyl alcohol is a complex milieu composed of two isomers; one, which affects light, rotates the plane of light under the polarimeter; the other, which is inactive, has no optical activity. Yet these two have exactly the same physico-chemical properties: smell, solubility, boiling point, specific weight. What is more, they form the same hemihedral facets, the same crystals with a comparable asymmetry."[22]

This last finding, which was troubling for the man who had tied optical activity so closely to asymmetrical crystallization, could not easily be seen as referring to two different substances. What was the reason for their appearance? And why did one of them rotate the light? In one of the pirouettes that he often performed when he was sure of his ground, Pasteur placed this problematic matter into a general context. He declared the two isomers of amyl alcohol to be the first exception to the correlation between crystalline hemihedrism and optical activity. What characterized the asymmetry of crystals in the last analysis, he concluded, was not the form of a substance, its facets leaning to one side or the other, but their ability to deviate the plane of light. The crystal is the unstable expression of asymmetry; the asymmetry of the facets is neither necessary nor sufficient. With this conclusion, Pasteur disavowed his original assertions.

But he did want to understand these atypical crystals and therefore felt it important to reexamine the conditions that favored the appearance and the crystallization of the two isomers, only one of which rotated the light like a biological product. The beet root, which was available in Lille in large quantities, gave him the opportunity to do so, since the raw oil pressed from its roots contained a mixture of the two amyl alcohols, the active and the inactive kind. This alcohol appeared, of course, in the course of fermentation.

To explain this synthesis, Justus von Liebig had given the official interpretation: the amyl alcohol proceeded from the sugar, not from the fermentation. But this did not account for the appearance of the optically active amyl alcohol. At this point, Pasteur remarked that if Liebig was correct, then the optical activity of the amyl alcohol should be identical to that of the sugar. However, this hypothesis was not borne out by observation: amyl alcohol examined by the polarimeter differed too much from the image of sugar to have inherited its asymmetry. Pasteur also found that each time he tried to trace the rotational characteristics of a substance that underwent a chemical reaction, these characteristics either disappeared or remained the same, for, as we have seen, asymmetry was an unstable characteristic. Hence, the appearance of amyl alcohol could not be due to the simple breakdown of the sugar. An additional intervention must have taken place. Pasteur concluded that the amyl alcohol that rotated the plane of light must come from the ferment rather than from the sugar.

But then Pasteur believed that life alone is responsible for asymmetry. If one does not want to give up this ironclad assertion — and Pasteur emphatically did not — one has no choice but to recognize that the ferment has a particular quality, that of being alive. To acknowledge the role of the ferment in the chemical process as described above was to acknowledge that the ferment is life.

Fermentation thus had to be considered as a natural biological occurrence, not as an artificial chemical synthesis. Once he had arrived at this point, Pasteur began to fear that his research would take him too far afield from crystallography, which had brought him his first rewards. He therefore initiated a series of experiments on the fermentation of tartaric acid. In the course of these he established not only that the action of a ferment is a way to isolate the left isomer of paratartaric acid, but also that this method is the most effective means of separating the two isomers. This, incidentally, proved to Pasteur that fermentation is responsible for the absence of racemic acid in the polluted mother liquors of manufacturing plants and finally provided a valid explanation for its absence in wine vats. Something of a showman in his own way, Pasteur enjoyed making an *a posteriori* connection between his first reflections on amyl alcohol and his later thoughts on the appearance of an asymmetry in the course of certain fermentations.

As we have seen, in these two models of fermentation (beet root and tartaric acid), the ferment acted as a living organism, for it was possible to make a connection between fermentation and molecular asymmetry. This fact allowed Pasteur to link his second field of investigation, fermentation, to his first one, crystallography. This filiation became especially compelling when his research once again called for the use of the polarimeter, the instrument of his great discoveries about tartaric acid.

Still, he had not yet proven that the ferment really was a germ of life and directly responsible for the production of alcohol. The hypothesis was bold, for it went against tradition and questioned the passive role of the ferment. It was imaginative, for it postulated a role for biological matter, the "life effect" [*effet vital*]. It was also daring, for he based it on apparently limited observations, the appearance of an optically active substance, amyl alcohol. However, Pasteur had yet to carry out the measurements and the interpretations called for by the new phenomenon. For the moment, we must retain his starting hypothesis: the ferment is life.

To say that the ferment acts on the beet juice because it is alive and that fermentation is a biological and not a chemical occurrence was — and this point cannot be stressed enough — to combat the active majority of the mandarins of chemistry and to buttress the discredited hypotheses of a few visionaries who had been unable to assert themselves. Pasteur had the reputation of a serious scientist and an excellent teacher, but the tone he now had to adopt was that of a crusader for a cause. Anyone who dared challenge the sacred shadow of a Lavoisier or the authority of a Liebig at the pinnacle of his glory not only had to be very sure of his ground — and Pasteur was used to that — but also had to know how to convince others. It should be noted that if fermentation no longer particularly interested philosophers and chemists, since they believed that they had elucidated its mystery, it continued to arouse the passionate interest of students, readers of scientific reviews, and particularly industrialists, to whom Pasteur primarily addressed his arguments.

Pasteur's notebooks indicate that he seems to have planned back-to-back studies of lactic and alcoholic fermentation. But he soon concentrated on lactic fermentation, which he eventually used as a case study to establish his theories. His first deductions came from this study.

The choice of this experimental model may seem surprising. The formation of lactic acid was not an industrial priority, and lactic fermentation was of little interest to the distillers of Lille, except in cases of unsuccessful alcoholic fermentation. Yet at that time Pasteur always made sure that his studies yielded some side benefits for the industry of the department of Nord.

One must therefore assume that there were several reasons for this choice on the part of the scientist. First of all, lactic acid was produced in great abundance in defective fermentations of beet juice; it was logical to try to find the reasons for this. Secondly, lactic acid appeared as a by-product of amyl alcohol, an optically active substance. And finally, working on lactic fermentation held an additional twofold attraction for Pasteur. The first was that it provided him with a much less overworked subject than alcoholic fermentation, which had already given rise to multiple studies. This meant that Pasteur did not have to deal with the many published findings on this subject and with their critiques. The second advantage was that lactic fermentation led directly into the arena of the controversies over the role of microorganisms since no scientific description of lactic yeast existed. Better yet, Liebig thought that lactic fermentation did not need a yeast and even used it to generalize and minimize the role of yeasts in all modes of fermentation.

What was known about lactic acid? It had been isolated in 1780 by Carl Wilhelm Scheele, a Swedish chemist who is also credited with the discovery of oxygen in 1773. Inaccurate research had obscured its properties so much that in 1813 Henri Braconnot, director of the Botanical Garden of Nancy, described lactic acid anew and named it acid of Nancy. He reported its presence in fermented rice, turned beet juice, and moistened baker's yeast. Subsequently a number of chemists, most notably Gay-Lussac, became interested in this substance and precisely described the conditions under which lactic acid is produced when sugar ferments in the presence of a variety of protein substances ranging from casein to albumin. In this wide range of ferments, no one had observed the slightest trace of yeast. It can be assumed that if no yeasts were seen by the investigators, it was because they are difficult to identify in preparations of milk or casein. One had to know what to look for, and, above all, an original instrument, the microscope, and a special process, culturing, were needed. For if the yeast was reluctant to show itself, the scientist had to find the means to recognize it by providing for its multiplication in a favorable milieu.

Meanwhile, Pasteur's attention was attracted by a detail he found in a report published by the Academy of Sciences in 1843: "As the production of lactic acid proceeds, the action of the ferment becomes inhibited. This can be prevented by adding chalk to sugar water to maintain its neutrality."[23] Why chalk? Because it was well known that chalk is an alkaline substance that would saturate the acid as it was forming. This was the experiment that Pasteur reproduced as a way to verify his hypotheses and to understand what a ferment was by very rigorous observation.

As he was carefully looking at a lactic fermentation, Pasteur noticed some gray spots that were forming above the chalk and the ferment, sometimes stretching all the way to the surface of the deposit, and clinging to the walls of the glass. Microscopic examination of this grayish matter revealed nothing that could be distinguished from casein or fibrin. But here again, Pasteur's originality was that he did not stop with this first analysis. Picking up this deposit, he proceeded to sow it into an extremely simplified medium consisting of chalk to neutralize the acid he hoped to produce, sugar he sought to bring to fermentation, and a nutrient liquid.

Once again, Pasteur's talent as an experimenter was manifest in his choice of this nutrient liquid. It had to be clear so that a possible formation of yeast could be seen. But he also thought that in order to stimulate the production of this microorganism, he had to provide it with food appropriate for its multiplication. But since he had not yet proven the existence of lactic yeasts—he only supposed it—he was certainly not in a position to imagine a milieu that would lead to their reproduction. He therefore reasoned by analogy: if it was a yeast, it would have to be similar to brewer's yeast. Imagining that a soluble extract of brewer's yeast in full growth contains the elements—one might also say the food—needed for their multiplication, Pasteur decided that the milieu into which he would sow the gray matter taken from the glass vessels containing lactic fermentation—a milieu that must be clear so that the contents can be seen and enriched so that they can multiply—would be a decoction of brewer's yeast, a lysate of yeasts which after filtration would contain nothing of these organisms but their own extract.

The combination of all these conditions spelled success. Pasteur observed that the gray matter made the sugar ferment and allowed him to isolate a new class of yeast, lactic yeast, which differed in form and size from brewer's yeast and the yeast involved in alcoholic fermentation. The fermentation of sugar by means of lactic yeast cultivated in this manner was better than that carried out with the usual lactic ferment, soured milk. "It is not necessary," wrote Pasteur, "to have lactic yeast on hand to prepare it. It will 'spontaneously' come into being. I am using this word to describe a fact, stating explicitly that this has nothing to do with spontaneous generation."[24]

Clearly, Pasteur was already thinking of what was to become one of his most important studies. For the time being, he had understood that the casein, fibrin, and albumin used in the experiments of his predecessors, or the decoction of yeast in his own, only played the role of nutrients. The nutrient medium was like a fertilizer that favored the growth of the yeast. Pasteur showed that ferments can be cultivated like plants. A major new step had been taken.

Pasteur gathered these first discoveries together in a *Mémoire sur la fermentation appelée lactique* dated April 1857 and published in August of the same year. It showed not only that he had identified a new class of yeast and linked its presence to fermentation, but also that he knew how to grow it, and even to select it by an appropriate nutrition. Anticipating that fermentation would be subject to all kinds of variations, he even conducted one experiment that can be called a case of asepsis when he established that onion juice, which is naturally acid, inhibits the action of yeast (so that the onion cannot ferment) and also that heating removes this inhibition.

Upon reading this first memoir, one concludes that Pasteur had already found almost everything. In reality, he had written almost everything, but he had not discovered everything, for, as he himself acknowledged, a large part of what he said was still hypothetical. "If anyone were to tell me that in these conclusions I am going beyond the facts, I would reply that this is true, in the sense that I am unequivocally dealing with an order of ideas which, strictly speaking, cannot be irrefutably proven. But this is how I see this matter." But then he immediately added: "In the present state of my knowledge of the subject, anyone who impartially examines the results of this work and the findings I shall publish in the near future will agree with me that fermentation is a correlative of life and of the production of globules, rather than of their death or putrefaction, and that it cannot be considered a phenomenon of contact in which the transformation of sugar takes place in the presence of ferments that add nothing to it and take nothing from it."[25]

This tranquil self-assurance was one of Pasteur's strengths: his scientific rigor compelled him to admit a slight lack of rigor. Yet it must also be noted that this proceeding was useful in itself, for it was based on a serious hypothesis inspired by other models and other systems, among them the chemistry of optical activity, and on observations that fit into one another like a set of Russian dolls.

Failure at the Academy

Meanwhile, that is to say, in the intervals between the experiments he carried out at Lille, Pasteur repeatedly traveled to Paris, for he had been proposed for membership in the Academy of Sciences, where a seat in the section of mineralogy and geology had become vacant. Pasteur realized from the outset that this election would have its problems, even though his work, as he blithely wrote to his father on 6 November 1856, placed him "far above the four other candidates for this seat." Two circumstances in particular were detrimental to his candidacy. First of all, Pasteur did not live in Paris, but in Lille, and this was against the regulations of the Institut, which made exceptions only for very eminent personages, like Monseigneur Dupanloup, the archbishop of Orleans and a member of the Académie française. Secondly, questions were raised about Pasteur's field, for he was a crystallographer and definitely not a geologist. Yet the campaign for the academy was important to him. "I am very anxious to get this affair over with. I am becoming more and more preoccupied with it; not that I would become terribly upset if the outcome were unfavorable to me, but I think that this is a milestone in the career of a scientist and that it is far from unimportant whether one becomes a member of the Academy at one time or at another." At the end of this same letter to his father he added, thinking of the year 1857 that was about to begin: "God grant that by then I may sign my letters 'membre de l'Institut.'"[26]

As a result, Pasteur set out to mobilize his assets. In the antechambers of the ministries and in scientific laboratories, he did his best to foster his goal: he wanted to become a member of the Academy. Since his residency in Picardy was an obstacle, he sent a conditional letter of resignation to the Minister of Education: if he were elected, he would leave Lille. In other words, he was ready to give up his administrative responsibilities in the provinces, his innovations in teaching, and his contacts with industrial circles for the honor and the glory which he felt his work legitimately deserved.

Careerism? No question about it. Yet that is not the whole explanation: Pasteur was truly convinced that he was entitled to this distinction, and that it was simply a matter of fairness to bestow it upon him. Besides, had he not just received the Rumford Medal of the Royal Society of London? This was a rather significant prize endowed with 1,750 francs, with which he hoped to purchase a garden along the Cuisance River near his father's house. Now that England had recognized his merits, surely France should do the same.

Pasteur made all the necessary calls. Moreover, he did not lack support: Sénarmont, Regnault, Chevreul, and of course Biot were campaigning for him. Biot was even taken to task by the Parisian daily paper *La Vérité*, which criticized him for misusing his influence — to which he disdainfully responded, "I am not acting as anybody's *godfather*; I do not have, nor would I wish to have, the influence you attribute to me. I have great respect for M. Pasteur's talent, but I do not *protect* him."[27] For the committee of the Academy of Sciences, Sénarmont produced a report summarizing Pasteur's work in crystallography: "He has been able," the report concluded, "to rise continually and with equal success from the theoretical conception that imagines a fact to the experiment that demonstrates it, and from that demonstration to new speculative views, so that logical induction and material observation mutually reinforce one another by turns, and in a continual sequence of corollaries and verification." This praise was flattering, to be sure, but perhaps ill-advised. For it only stirred up the hostility of those who opposed Pasteur's candidacy, people for whom he, in turn, had little indulgence. "Everyone knows," he calmly wrote to Marie, "that I am the valid candidate. They all know that they should vote for me. But they are afraid (at least many of them are) of chemistry. They are saying that chemistry wants to take over everything. This is why I have all the naturalists against me, especially the most ignorant ones. You know what it's like in the section of botany, which has been filled with stop-gaps and mediocrities in the turnover of the last three years. They are solidly against me!"[28]

The voting held in mid-March 1857 confirmed this diagnosis: with sixteen votes when he needed thirty, Pasteur was not elected. He had to yield to the mineralogist Gabriel Delafosse, his former teacher at the Ecole normale. For this respectable but not brilliant scholar, who in the 1840s had been the first to notice the correlation between the optical activity of crystals and the hemihedrism of their facets, this election rewarded a long career; but it must also have been a dreadful humiliation to have as his rival a former student who was his junior by thirty years.

Despite his failure, Pasteur asserted that he was not unhappy; to be proposed had given him visibility and made his work more widely known. Yet he could not altogether stifle his resentment when, on his return to Lille, where Marie was waiting for him, he exclaimed: "I will work off the rage in my heart. How happy I'll be when I go back to read them a fine paper, crying out to myself: 'Idiots that you are, go ahead and try to do this yourselves!' "[29]

Yeast and Alcohol: The Decisive Experiments

And indeed, upon his return to Lille, Pasteur threw himself headlong into further work on fermentation. He needed to refine his first conceptions, and above all to confirm them by incontestable experiments. Having opened up the terrain by means of a model which, since hardly anyone had studied it, was unlikely to be attacked, namely lactic fermentation, he felt ready to tackle the most famous of them all, alcoholic fermentation. In principle, this was where he could expect to encounter the opposition of chemists who were wedded to simple equations.

Liberated from the constraints of his campaign for the Academy of Sciences, Pasteur was no longer pressured by the need to make the rounds of potential supporters and even less by the desire for multiple publications. He was taking his time. After several partial reports starting in 1857, he published his *Mémoire sur la fermentation alcoolique* in 1860. Its tone was resolute, at times ironical. It was that of a scientist who had a cause.

It started with a flourish of trumpets: "I call alcoholic fermentation the fermentation that is undergone by sugar under the influence of the ferment that bears the name of brewer's yeast." Further on, a note of clarification: "M. Berthelot has applied the label alcoholic fermentation to phenomena which in my opinion are in every case, without exception, of the same nature as lactic fermentation. For there can be no hesitation, yeast brings about fermentation, and not the other way around."[30] There was always an element of polemic in Pasteur's self-confidence, whereas in his reports the precision of the chemist is often visible behind the passion of the biologist: "Having made this point, I shall now give the detailed analysis of a fermentation." Here follows the meticulous description of what had been observed in the laboratory on 10 December 1858 at noon and at 4:00 P.M. on 11 December, including the precise spot on the bottom of the vial where the little fragment of yeast had fallen and where one saw "the continual rising of extremely fine bubbles." Other precise notations concerned the readings of that same night at seven o'clock, and then on 13, 14, and 15 December and on until the following January.[31]

What mattered most, of course, was the statement of the facts that had been established: the quality of the yeast, its size, its form, its composition, and the rate of its multiplication determine the specificity of the reaction and of the product that has been formed, alcohol in some cases and lactic acid in others.

What did this memoir add to the first deductions Pasteur had drawn from lactic fermentation? He now asserted that he was dealing with a living microorganism. That was his most original contribution. Having been unable to create life or to change it by means of molecular asymmetry, he had set out to study its reproduction. In doing so, he had come to a revolutionary definition: Fermentation is the act of reproduction of the living germs that constitute yeast.

Showing a masterful understanding of the phenomenon, Pasteur precisely enumerated the ingredients necessary for the growth of alcoholic yeasts. As a result, he was able to define the relationship between yeast as ferment and inert ferments such as casein, albumin, and all the other substitute products on which Liebig, who considered them products of organic decomposition, and Berzelius, who explained them by contact, had based their theories of fermentation.

Once he was launched in this direction, he also postulated that the elements necessary for fermentation, sugar on the one hand and the organic product of the ferment on the other, were nothing more than food that allowed the yeast to reproduce in the manner he had brought to light. The most difficult part of this was to imagine that the living yeast was making use of the components of the dying yeast for its own benefit. In the case of the production of alcohol by means of brewer's yeast, Thenard had been the first to raise this possibility. He had shown that after fermentation there was less yeast than there had been when it started. And yet Thenard had not found, as he had expected, the nitrogen that should result from the destruction of the yeasts. Pasteur seized upon this idea and demonstrated that the nitrogen produced by the dead yeasts is used by the living yeasts for their own synthesis. It does not appear in the form of a gas but as an organic compound. And whether one artificially adds a simple nitrogenous substance, such as ammonium chloride, or a complex nitrogenous compound such as gluten or the casein of whey, the lactic yeast will use the nitrogen for its own growth. The inert ferments, or proteins, thus serve the living ferments. It is thanks to them that yeast cells are able to survive and, depending on their quality and quantity, to reproduce by budding. As for the sugar, it behaves in the same manner; its carbon is used by the new yeasts. It now remained to identify what other compounds might be necessary for this synthesis. Pasteur's evaluation established that mineral salts, and particularly phosphates, were needed, among them magnesium phosphate.

This was a masterful way to define a synthetic culture medium and to bring to light the elements that yeast needs to stay alive, particularly the basic elements of the equation of fermentation, sugar and ferment. Thus he could conclude: "I have attempted here to prove that brewer's yeast placed into sugar water lives at the expense of the sugar and its nitrogenous and mineral matter, which is soluble or can become so because of the mutations that take place among its constituent principles during fermentation."[32]

In the end, the yeast that Pasteur saw, depicted in drawings, and discussed became so much of a living being that it was easy to understand that it should behave differently depending on its age, independently even of the modifications that can be caused by the nature of its environment and external conditions.

Convinced that Lavoisier's simplification was false and that in fact the production of alcohol was a phenomenon as complex as a biological act, Pasteur defined fermentation as an act of reproduction of the yeasts. At the same time he drew up an inventory of the components formed from sugar in the course of this chemical act. He established that in the course of this fermentation glycerine and succinic acid are formed. He even showed that sugar contributes to the formation of cellulose and fatty acids, thereby confirming an intuition of his teacher Jean-Baptiste Dumas, who had indicated that fatty acids could form out of beehive sugar. These products, which no one had looked for or found, came from the sugar and not from the yeast. Pasteur went so far as to quantify them and to compare their presence in the wines of Burgundy, Bordeaux, and the Jura.

What is more, he showed that in principle an alcoholic fermentation by means of brewer's yeast does not produce either acetic acid or lactic acid. It is only through an accidental contamination by lactic yeast that this phenomenon can occur.

A few years after the request of the industrialist Bigo, Pasteur had thus established beyond a doubt that the lactic acid in the vats in the rue d'Esquermes came from an unfortunate contamination with this yeast. He even suggested the means to get rid of this contamination, namely heating the beet juice to destroy the lactic yeast and reseeding it to reproduce the alcoholic yeast. Then, leaving behind alcoholic fermentation, he returned to the subject of lactic fermentation because he wanted to determine the proper conditions for its cultivation.

In the end, what Pasteur considered necessary for any kind of fermentation, whatever the yeast involved, was what he felt was needed by every living cell. Speaking of the lactic microorganism, for example, he said that its biological function was the transformation of sugar, much as the function of the cells of the mammary glands is to use the nutrient elements of blood to produce milk. Comparing a yeast to a mammary gland—what a symbolic image to use for lactic fermentation! An immense future lay ahead for the yeasts, for this was the time when great efforts were under way to make an industry out of all that human beings had learned from nature and reproduced empirically, without understanding it, for many centuries.

A First Setback for Panspermatism

In his studies on fermentation, undertaken before he had given any thought to spontaneous generation, Pasteur intuitively brought some partial answers to that problem when he opposed panspermatism, the theory that saw germs everywhere. Fermentation, like the growth of germs, does not take place indiscriminately. Only precise cultivating conditions allow for the selection and growth of the different types of yeast. This is how fermentation takes place in nature: the juice of ripe grapes, whose natural acidity prevents the development of certain microbes, exclusively permits that of alcoholic yeast. But what about those famous clear and shiny crystals of tartaric acid that have formed there? If one adds these crystals to sweet pear juice, which naturally contains a flora of polymorphous yeasts, only the alcoholic yeast continues to multiply.

Moreover, the nature of the alcoholic yeasts differs with the barrels used and with the time of the harvest or the fermentation. The microorganisms involved are not identical with those of brewer's yeast or the microorganisms that produce vinegar. In his studies on lactic acid, Pasteur also insisted that its appearance was due exclusively to a contamination by air. "Here we once again encounter what looks like spontaneous generation. If one prevents any contact with ordinary air, or if one brings the mixture of sugar, ammonium chloride, phosphate, and chalk to the boil, allowing only calcined air to enter, neither lactic yeast, nor infusoria, nor ferments of any kind will be formed."[33]

Even before he launched his crusade against the theory of spontane-ous generation, Pasteur already knew from what he had observed in yeasts that he could grow specific kinds of yeast, which would be a blow to the defenders and the mandarins of undifferentiated panspermatism. But since his discoveries were not yet sufficiently coherent to constitute a full-fledged counter-theory, he simply recorded his observations with a view to assem-bling them in a bundle of proofs at a later date.

Thus, in the course of his manipulations of butyric acid, which is responsible for the bad smell of rancid butter, Pasteur was surprised one day to see a motile microorganism. The first yeasts he had perceived under his microscope, both brewer's yeast and lactic yeast, had not moved, and so, like a good naturalist and despite his reservations about the classifications of the botanists, he had placed them into the plant kingdom. The new organ-ism he observed in the butyric fermentation, however, moved about and multiplied by dividing. Pasteur therefore ascribed to these organisms char-acteristics close to those found in infusions and placed them into the cate-gory of infusoria. He also attributed animal life to them and therefore called them animalcules.

The analogy Pasteur established between microorganisms and mam-mals was so close that he initially feared that the animalcules might eat the plant ferments to which he — mistakenly — attributed the production of butyric acid. Soon thereafter, when he noticed the simultaneous appearance of motile microorganisms and acids, he realized his error and accurately ascribed the fermentation to this new organism, to which he gave the name "butyric vibrios."

At a time when mobility was considered a characteristic of the animal kingdom, the discovery of microorganisms that could move led Pasteur, in the lineage of Buffon and the other naturalists of his era, to classify them as a new kind of fauna. What mattered most at this point, however, was not so much this act of classification as that here Pasteur entered into a new field of investigation. He opened the door to a new world vastly more active and extensive than that of the yeasts, the world of bacteria.

This is the context of a puzzling observation that Pasteur exploited with magnificent insight. Since he left nothing to chance, all microorgan-isms were taken from the incubator to the microscope. Pasteur examined them batch by batch, the butyric vibrios more than the others, for they moved and especially multiplied, so that one had the impression of witness-ing an active phenomenon and participating in it.

Smearing a drop containing butyric vibrios onto a glass slide, Pasteur noticed that the microorganisms at the edges of the smear were immobile, while those at the center were motile. This detail, which many observers would have ignored, intrigued Pasteur. Several years earlier, in the course of his experiments with crystals, he had already noticed variations in the forms of crystals depending on whether they were located at the center or at the periphery of the preparation. In the drop of water in which the butyric vibrios were swimming, the concentration of oxygen was not everywhere the same. High at the periphery, it was lower in the center. On the strength of his observation, Pasteur thought that perhaps the quantity of oxygen had a bearing on the motility and the growth of microorganisms. Passing from the hypothesis to experimentation, he directed a stream of oxygen onto a preparation of vibrios. And there he saw the fermentation slowing down and finally stop. The oxygen was harmful to the life of the animalcules. Oxygen, which was believed to be absolutely indispensable to life, now marked the frontier for a new kind of microorganisms, those that live without air.

At the same time that he observed them, Pasteur also gave a name to these microorganisms, calling them *anaerobes*, as opposed to the microscopic creatures that need air to live, the *aerobes*. He showed that the frontier between aerobes and anaerobes is not always well defined and that some germs can live both with and without air. Finally, he also showed that these hitherto unknown organisms that live without air produce carbon dioxide. All these observations eventually turned out to be of considerable importance for the progress of medicine, especially when it came to understanding the mechanisms of infection.

The perpetual battle between the microorganisms and the higher organisms might easily have remained outside the precise and methodical preoccupations of Pasteur as he divided his attention between his microscope and his incubator. Who among the scientists of his time would have bent over a test tube, however finely calibrated it might be, seeking to contemplate the infinitely small and at the same time to understand its role in the universe? Who else would have known enough not to confine the microorganisms to the process of fermentation and to envisage their ecological role as guardians and grave diggers of our remains? Was there anyone else who would have thought of bringing the cycles of nature together in this manner?

One can imagine that Pasteur did ask himself the truly fundamental question as to the ecological role of fermentation and putrefaction. In connection with life without air, Pasteur sought to understand the complex links between microorganisms and the decomposition they cause, between the conditions under which germs develop and the consequences they produce. The fact that the butyric vibrio lives without air yet is also a ferment suggested to Pasteur that there might be a connection between these two characteristics. Brewer's yeast, which can live both as an aerobe and as an anaerobe, was chosen for his experiment. In a medium rich in oxygen, it multiplied without fermenting; in an anaerobic medium, it played its role as a ferment. Pasteur concluded that the yeast, forced to live on the medium that surrounds it, takes oxygen, as well as carbon and nitrogen, from the nutrients within its reach — hence fermentation. The yeast can survive only by breaking down its prey. Fermentation is brought about by life without air.

Described as an act of slow combustion, fermentation was akin, at least according to the tenets of philosophical and scientific thinking, to putrefaction. Devoting himself to the study of this phenomenon without giving much thought to the danger and the unpleasantness it involved, Pasteur showed that putrefaction is brought about by an anaerobic vibrio. Death, he said, was really a kind of physical and chemical life. Gangrene sustains the life of putrid microorganisms. Thus buried animal and plant matter disappears, so that fermentation and putrefaction "come together to accomplish the great destruction of organized matter, which is the necessary condition for the perpetuation of life on the surface of the globe. Life directs the work of death at every stage. The perpetual return to the atmospheric air and to the mineral kingdom of the principles that plants have taken from them is correlated to the development and the multiplication of organic activity." Putrefaction restores to the atmosphere the water, the carbon dioxide, hydrogen, and ammonia without which life cannot exist. Extracting the oxygen and rejecting the carbon dioxide that will be taken up by the plants, the anaerobes are necessary to the cycle of life, for "the continual breakdown of dead organic matter is one of the necessities for the perpetuation of life."[34]

In 1862, Pasteur summarized his convictions in a letter to the Minister of Public Education: "After death, life reappears in a different form and with different laws. It is inscribed in the laws of the permanence of life on the surface of the earth that everything that has been part of a plant and an animal will be destroyed and transformed into a gaseous, volatile, and mineral substance."[35]

Pasteur had traveled a long road from the vats of beet juice at Lille and the first titration of lactic acid. He had now discovered the microorganism and found out what it gives to us and what it takes from us, both as a parasite and as a helper that unites the two kingdoms, the mineral and the living world, through the ebb and flow of its own respiration.

Pasteur and Liebig: Two Lions in the Arena

To be sure, Pasteur did not leave those whom he had so vigorously contested speechless. The first, Marcelin Berthelot, lodged a complaint with the Academy of Sciences and contested the priority of certain of Pasteur's observations on the formation of alcohol. He also criticized Pasteur's new theories of the role of the yeasts: "As for the vitalist opinions adopted by M. Pasteur on the real causes of the chemical changes accomplished in alcoholic fermentation, I do not think that the time has come to discuss them as fully as they deserve to be discussed." Pasteur responded indirectly: "Our writings and our personal conversations had made it sufficiently clear how much we differed in the interpretation of the facts, and even if he (Marcelin Berthelot) believed that the time had come to debate the views I hold, he knows very well that he would not convince me. Nor would I presume to believe that I could make him change his mind. Both of us should therefore preserve the independence of our views and, while waiting for the time to discuss them, follow Buffon's principle: let us gather the facts that will give us ideas."[36]

Other, though less authoritative, voices accused Pasteur of *lèse-majesté* toward Lavoisier's doctrines. To the physician Jacques-Antoine Béchamp, for example, Pasteur replied disdainfully that his criticism did not interest him because it was lacking in rigor, "not the absolute rigor for which we always strive without ever attaining it, but that relative rigor which is demanded and indicated by the state of knowledge in our field of study."[37]

In point of fact, the majority of scientists, at least in France, assessed the value of Pasteur's work correctly. In 1859, even before he published his major monograph on alcoholic fermentation, the Academy of Sciences awarded him its prize for experimental physiology. The word *physiology* is important here, for it means that the jury, headed by Claude Bernard, recognized fermentation as a biological phenomenon. However, the conclusion of the committee remained cautious; it stated that the general definition of the word *fermentation* still presented serious difficulties. What is more, it blandly stated that the ferment always comes from a living or formerly living being. This remarkably cautious formulation was the work of Claude Bernard, who was not completely convinced that fermentation was a corollary of microbial life. Yet, in the presence of a microorganism, he did have to acknowledge a biological intervention. "The physiologist," he wrote, "is bound to recognize that this research has brought to light true chemical agents that play a physiological role."[38]

But these subtly shaded critiques amounted to nothing compared to the awakening of the old lion, the high priest of fermentation seen as a chemical rather than biological process: Justus von Liebig. Liebig took ten years to reply to Pasteur, and he did so in a long memoir read before the Royal Academy of Sciences of Munich in 1869. Liebig, whose argument was bolstered by a great international reputation, was not convinced. What is more, he was certain that Pasteur was mistaken; he refused to admit that fermentation was a biological phenomenon and claimed that his view was proven by the fact that it was impossible to reproduce certain of Pasteur's experiments concerning the life of yeasts in a synthetic medium. Reducing the ferment to a microorganism, the organic milieu to a feeding ground, and the sugar to a substratum destined to be broken down by the yeast for its own survival — all of this, he felt, was unreasonable, wrong, and unfounded.

At the time of this response, Pasteur had already gone far beyond Liebig's arguments. By now he had penetrated deeply into the domain of the fermentations and was about to combine his studies on molecular asymmetry with those on fermentation for a general study. Thus he was able to write that in the phenomena of fermentation sugar was, like tartaric acid, endowed with optical activity. Going from the polarimeter to the incubator, he imagined how stereochemistry and the development of microorganisms might be related: "If yeast is naturally constituted of asymmetrical materials, it will not accommodate to the same degree to an element that is itself asymmetrical depending on whether it is in the same or in the opposite direction."[39] He then proceeded to work out which mineral culture media and organic components — right- or left-handed substances — will promote biological multiplication. He advocated a synthetic chemical conditioning which is used to this day, with slight modifications, in all microbiological laboratories in carrying out precise and reproducible experiments. A visionary, Pasteur went beyond fermentation to enter the realm of cellular physiology.

But this did not keep him from replying to Liebig. Expressing his outrage, he published a memoir in which he demanded that the scientific community decide between the two opposing theories, French biology and Liebig's German chemistry. He said that he was willing to submit to such a judgment, and indeed demanded it. He asked for the appointment of a jury of objective commissioners, members of the Academy, which would judge the reproducibility of his experiments and their veracity. The experiments would be repeated under the eyes of the judges as many times as they deemed necessary, on condition that Liebig indicate the type of fermentation he wanted to have tested. But Pasteur added, rather craftily, that Liebig would have to assume all the expenses occasioned by these experiments; since he considered himself the offended party, he felt entitled to choose the weapons in this duel of scientists. This long memorandum, which sounded like a plea in court, fell flat. Liebig did not even take the trouble to reply to it. Nonetheless, Pasteur went to Munich, insisting on meeting with his opponent in order to defend his cause and debate the arguments face to face. While Liebig did not close his door to him, he received his visitor standing, clad in a dark morning coat, and did not even ask him to be seated. He refused any discussion, claiming that he was not well. Pasteur was disappointed when he left Liebig, for he did not know whether the old chemist was sure that he was right or too proud to admit that he was wrong.

The controversy over fermentation was thus far from over. The fundamental opposition between chemists and biologists was to be compounded by the Franco-Prussian War to such a point that skirmishes on this subject marked the entire last quarter of the nineteenth century. Even though by then Pasteur had many other things to think about, he continued to defend his positions; for years he took on all comers, always fighting with great spirit. Thus he curtly replied to the obscure Trécul, who contested the distinction between aerobes and anaerobes, "My classification is what it is, you can take it or leave it, that is your business; I happen to think that it is excellent."[40]

One can say that when he finally gave up the debate on fermentation, he did not turn his back on it. For unlike Liebig, he never failed to reply to his opponents. He indefatigably sought to convince those who contradicted him not only that they were wrong because he was right, but indeed that he was right and they were wrong, period. Yet it would not be fair to say Pasteur was right at all times and on every subject.

If he was responsible for prodigious scientific advances in the field of biology, it is also true that he paid very little attention to the purely chemical interpretation of fermentation. In particular, he failed to take into account the discovery of soluble ferments that have absolutely no living ferments, namely the diastases, which are today called enzymes. When they were brought to light, notably by Marcelin Berthelot, they gave a boost to Berzelius's theories of a catalytic force. However, it should be acknowledged that it was only in 1897, two years after Pasteur's death, that the German Eduard Buchner definitively elucidated the complex mechanism of fermentation and showed that in the absence of any living yeast it can be induced by an enzyme.

Yet the final victory of the enzymes should not be permitted to overshadow the discovery of the microorganisms. In the pursuit of questions related to their origin and their effect, science was led to the study of spontaneous generations and microbial diseases. Pasteur's research on fermentation created microbiology, which became the field of his next investigation.

6

RUE D'ULM REVISITED

If Pasteur learned anything from his failure to be elected to the Académie des sciences in March 1857, it was that in order to succeed, one had to be a Parisian, that all important matters were decided in Paris, and that it was essential to be near the center of power. The young dean had every reason to be proud of his work at the Faculty of Sciences at Lille; still, he was just a provincial. This being the case, he felt that he had no choice, and if he had to leave a whole educational program in the planning stage, so be it; he would seize the first opportunity to return to Paris.

As usual, Pasteur did not waste any time. Less than a year after his failure at the Academy, he received his new assignment, which was also a further promotion. On 22 October 1857, he was appointed administrator and director of scientific studies of the Ecole normale supérieure.

The twofold task that awaited him in Paris was not simple, for the school was going through a difficult time. Pasteur, who had always kept up with his school, noted that it was "but the shadow of its former self,"[1] but then he was not one to shy away from even the most difficult challenges. In his desire to prove his gratitude to the school that had formed him and, above all, his impatience to return to Paris, he gladly accepted the flattering offer extended by Désiré Nisard.

Nisard, the newly appointed director of the Ecole normale, had been given a completely free hand in choosing his staff and carrying out reforms. A formidable academic and an influential critic, Nisard had been a fierce adversary of the Romantic movement. As a professor of Latin eloquence at the Sorbonne, he had published a harsh and much-noted study, in which he implicitly compared Victor Hugo and his followers to the Latin poets of the period of decline. At one time or another, he had served in every important institution, from the Council of State to the Ministry of Public Education, from the National Assembly to the Collège de France; and in 1850 he had become a member of the Académie française. According to the historian Ernest Lavisse, who was a student at the Ecole normale at the time, he liked to wear a black morning coat and vest, light gray trousers, and narrow patent-leather ankle-boots. Elegant in his manners as in his attire, Nisard felt that the only occupation worthy of a "gentleman" was "the rational administration of masterpieces." Distinguishing the morality of the state from that of the citizen, a stance for which he was criticized by the Republicans, he set out to direct the school from above, which meant that he had little contact with the students and placed much responsibility on his two lieutenants, Pasteur and, for literary studies, Paul Jacquinet.

The arrival of the Pasteur family at the beginning of the academic year 1857 was not celebrated with great fanfare, for there was not much money for housing the new administrator. Marie, of course, followed her husband to Paris with the youngest of the children, Cécile, but the two older ones stayed in the Jura: Jeanne (age seven) and Jean-Baptiste (six) were sent to a boarding school. But even though it was dispersed, the family continued to grow, and on 19 July 1758 the fourth child of Louis and Marie Pasteur was born. It was a little girl, whose first name sealed her parent's union: she was named Marie-Louise.

Pasteur considered the return to the rue d'Ulm and the birth of Marie-Louise good omens. Barely thirty-five years old, he had already produced a sizable body of work, and a bright future lay ahead of him.

Rue d'Ulm! The magic of this name was more powerful than Arbois, Strasbourg, or Lille. Located midway between his family origins and his teaching positions in the provinces, it was the true place of his birth, his birth to science. It was also an attic where he had to fight to create a true laboratory; and finally, it was a prestigious institution to which he returned as a teacher, having been a lowly student before. He was to leave it only when his great reputation and the proliferation of his projects propelled him to the institute that he was to found.

Grandeur and Servitude of Administration

It was only normal that Pasteur should be put in charge of scientific studies at the Ecole normale. But it is harder to understand that he should also be given administrative duties, namely the restructuring of the school's economic, sanitary, and disciplinary regime, as well as its liaisons with the students' families and with other scientific and literary institutions. Jean-Baptiste Biot, who had of course given his opinion on his protégé's new position, interpreted this as a maneuver on the part of the young scientist who was hoping to obtain a sinecure. "They have appointed him administrator," he sneered, "let's let them believe that he will administer."[2] For once, Biot was wrong.

Pasteur actually took his new responsibilities very seriously. In fact, one can read in his pocket notebooks some remarks that testify to preoccupations that took him as far as possible from yeasts or alcoholic components: "Food service: Find out from the Ecole polytechnique how many grams of meat they are giving per student." There are also reminders: "Have sand spread in courtyard, air out classroom X, have door of dining hall fixed."[3] The administrator was thus involved at every turn in the life of the students, nor did he neglect to take note of chipped paint, loose floorboards, dilapidated doors. At that time the rue d'Ulm, which fifty years earlier had been built starting from the transept of the Panthéon, had not yet been carried much beyond the intersection with the present rue Louis-Thuillier. It ended in a cul-de-sac closed off by a gate attached to the porter's lodge, so that it was a particularly calm and tranquil spot. The neighborhood still had the atmosphere of the convents that had once flourished there.

In 1858 the rue d'Ulm was extended. At that time a pavilion symmetrical to the porter's lodge was also built. In October of that year, Pasteur put in a request for the planting of trees to fill the space between the new gate and the facade of the building. In 1861, he complained that there was no longer enough privacy in the courtyard and requested a tight-meshed screen to train ivy over the main gate.

With thoroughly scientific precision, Pasteur also informed parents about their sons' work, mentioning every minute detail of school life: "I include a table indicating how often every first-year student has been called on in the different lectures in the first three months of the school year 1859–60."[4] When he studied the students' schedules, he found that first- and second-year students had twenty minutes to dress, another twenty minutes to walk from the school to the university, as much to walk back, and so forth. All in all, they wasted three and a half hours of their working day, he protested.

Unskilled in personal contact, Pasteur communicated by notes and decrees. Sure of himself, he issued formal orders outlining the functions and responsibilities of everyone on the staff, from the director to the housekeeper. When, for instance, the director went to the country, Pasteur considered it a breach of hierarchy that the housekeeper should open his mail; on the strength of his authority, he therefore decreed that "in the absence of the inspector general, the administrator carries out the function of director and assumes the title of interim director-administrator for the purpose of official acts." Pasteur defined his own competencies. Hence, "article 2: the administrator of the School is solely responsible for economic and sanitary matters. . . . He alone is in charge of enforcing the regulations concerning internal and disciplinary matters. No deviation from these regulations can occur without his consent."[5]

His relations with the students were even more difficult than those with his colleagues. In the Second Empire, the atmosphere of the school had changed from what it had been under the July Monarchy or the Second Republic. Among the students in arts and letters, in particular, the love of debate had turned into a libertarian spirit; the Normaliens had a tradition of resistance, and the government would have loved to channel these unfocused energies by which it felt threatened. In his *Grand dictionnaire universel du XIXᵉ siècle*, Pierre Larousse has left us an astounding description of what went on during recess in the rue d'Ulm:

> After the lectures, study hall was interspersed with long visits to the . . . how shall I put this . . . let's speak English . . . to the *water closets*. They were very big, these water closets, and in front of them was an antechamber where the students held their club meetings during the long study halls of the morning. Everyone can't belong to the Jockey Club! After two hours of speaking the language of Bossuet, the language of Cicero and that of Demosthenes, it was nice to use a bit of the argot of the boulevards, and the place to do it was the above-mentioned antechamber. And here the talk, naturally, was of the fashionable *cocottes* and, not quite as naturally, of the renowned actors and actresses; here also falsely acquired literary reputations were raked over the coals. . . . Sometimes just a word thrown into the discussion by one of the students would trigger a tightly argued, insistent, and tumultuous discussion; everyone became heated and fired up; no one stood aside without becoming involved. . . . It was a time for endless discussions, and whenever politics or religion were brought up, it felt as if one sat right in the Legislative Body. Speaker after speaker climbed onto the big stove of the room; people applauded, interrupted, grumbled, demanded or yielded the floor; they looked for all the world like budding legislators.[6]

Needless to say, Pasteur was not pleased to see these improvised sessions, which easily turned into political rallies and gave the school the atmosphere of a Voltairean club. This teacher had no sense of humor, and the least one can say is that he did not bring the slightest touch of whimsy to his job. When the students asked to be allowed to wear regular clothes, he became quite indignant and reminded them that the wearing of the uniform was mandatory: "Are the students at Saint-Cyr or the Ecole polytechnique allowed to go out in a straw hat or in sailing pants?"[7] If they

wanted to be seen in civilian clothes outside the school, the students therefore had to hide gray trousers under the regulation black ones or, if they
could afford it, rent a room in town.

These administrative functions, which demand a certain diplomacy,
or indeed a little dose of demagoguery, were totally unsuitable to Pasteur's
inflexible character. One is not surprised to see him make unpopular decisions that brought him the worst troubles.

Pasteur professed a quasi-military sense of order and hierarchy that
demanded the same respect for rules as it did for the law. The school, he felt,
must be governed by regulations as nature is governed by determinism.
Detention rained on such offenses as talking in class, wasting time, or
dressing improperly. Schedules had to be kept to the minute, as is attested
by this note, one among many: "Any student who is not present at the
School on 2 November at 10:00 P.M. will be considered to have resigned."[8]

The year 1864 saw the first noisy demonstrations about religious
matters. Alarmed by the alleged conversion to Protestantism of one student, Pasteur exclaimed: "It is absolutely unthinkable for me to permit a
Catholic student to cease taking part in Catholic ceremonies on his simple
assertion that he is no longer a Catholic. If he is really a Protestant, he must
prove it by an attestation of a minister of that faith; and if he is led to assert
that he does not belong to any religion recognized by the State, he must be
dismissed from the school."[9] He justified his position by pointing out that
the Minister of Public Education was also Minister of Religion.

According to Sainte-Beuve, Pasteur was not liked because of his frequently authoritarian stance and his incredible rigidity. Not that he cared.
Nisard was too far removed from day-to-day affairs, and Jacquinet, the director of literary studies, was too mild-mannered to counterbalance Pasteur's influence in disciplinary matters. The science students, who had
closer contacts with him, more or less accepted his difficult temper, but
those in the arts and letters did not take kindly to his tutelage. They protested against it with frequent noisy demonstrations, which Pasteur of
course repressed and condemned. He was to remember this resistance for a
long time to come. In his parlance the label "literary" became a criticism
designating a poor scientist; it was a manifestation of the resentment Pasteur was to harbor against the humanities graduates of the rue d'Ulm.

One of the episodes in the permanent conflict between Pasteur and
the students became known by the name of "the bean revolt." One day, two
tables in the dining hall refused to eat an allegedly wretched mutton stew.
Pasteur was furious, did not listen to any of the complaints, and announced

that the same stew would be served the following Monday. A week later, almost all of the tables refused to eat the stew. Pasteur, beside himself with anger, came to the dining room and announced that this dish would stay on the menu for every Monday. Only those students who had shown a spirit of order and submission would go unpunished. The mutton stew divided the classes and separated the reactionaries from the revolutionaries. Resistance seemed to have gained the upper hand when, a few days later, Pasteur gave another speech in the dining hall after he had caught some students smoking. He warned the students that anyone caught smoking or returning from a smoke would be instantly dismissed from the school: "There is no point asking whether the punishment is proportionate to the offense. That decision is up to me. And the offense will not be the smoking but the failure to obey the injunction you are receiving at this moment."[10] Seventy-three of the eighty students resigned from the school as a result of this decree. It took the skill of Nisard and the intervention of the minister to make them change their minds.

In the end, however, and despite these more picturesque than serious incidents, these elite students had to buckle under to this iron discipline. No doubt it was their forced submission that made a young journalist by the name of Emile Zola write about the Normaliens: "Their brains all have the same insipid and musty smell. They recognize each other by their need for the rod, by their smothered and impotent desires of old bachelors who have not made it with women. . . . Silly little schoolmasters, all of them, every one of them!" In fact, all that Pasteur was doing was to carry out literally the wish of Napoleon I, who wanted the Ecole normale to be run "in the spirit of a convent combined with that of a barracks."

For all of his repeated blunders, Pasteur played a very important role in the school's history. There is no question that he was much more talented as a director of scientific studies than as an administrator. As soon as he was appointed, he began sending notes concerning scientific teaching at the Ecole normale to the minister. He firmly insisted on the need for two distinct scientific *agrégations*, one in mathematics and the other in physics / chemistry, for he was convinced that it was a most harmful practice to entrust all the teaching to one mathematics professor. Another proposition was to administer a competitive agrégation examination without a quorum at the end of the third year.

For the first and second year, Pasteur proposed courses to be given at the school itself—which would thus no longer be dependent on the Sorbonne. The rue d'Ulm, he felt, should become self-sufficient. The *licence* was obtained after two years of study; Pasteur asked that the Normaliens be exempted from this rule. To be sure, these reforms would cost money. Additional chairs in mathematics, physics, and chemistry would have to be filled, and their salaries would have to be brought in line with those paid by the Ecole polytechnique, the Sorbonne, or the Collège de France. Unless he was listened to, he said, the rue d'Ulm would continue to recruit "very estimable but totally undistinguished professors whom the scientific community places into the second or third rank."[11] This was harsh criticism, but it happened to be realistic.

Pasteur also became involved in the debates about the modes of the agrégation and its examinations, written as well as oral. He suggested that the number of students should be increased and added that the extra expenses this would entail should not be exclusively made up by saving on food. And of course he intervened in the programs. One of his innovations consisted of inculcating in the students a sense of the history of science, or rather the history of experimental thinking.

> I know that most scientific discoveries can be stated in a few words and that their demonstration requires only a small number of decisive experiments. But if one tries to find out how they came about, and if one carefully follows their development, one realizes how slowly these discoveries have come into being. That is why one can teach about them by two different methods. One consists of stating the law and immediately proving it . . . ; the other is more historical and traces the individual efforts of the principal inventors. . . . This second method appeals to the student's intelligence. If this faculty is expanded and cultivated, it can learn to produce on its own and to develop the ways of the inventor. Historical examples show that nothing of durable value can be accomplished without hard work. Teaching the mind the habits of modesty, they ask young people to have respect for authority and tradition.[12]

Two of his administrative actions brought him considerable prestige: the hiring of graduate *préparateurs* and the founding of the *Annales scientifiques de l'Ecole normale*. What top administrator would not be pleased to have succeeded in these two domains, expansion of the staff and new means of communication?

The positions of préparateurs had previously been held by functionaries unconnected with the institution. Pasteur succeeded in increasing their number from three to five, and starting in 1858 these positions were given exclusively to graduates of the school. Pasteur had enough vision to understand that he could train excellent teachers only if he offered them the opportunity to do scientific work and, conversely, that scientific research needs the technical proficiency, the enthusiasm, and the strenuous work of these young doctors destined for higher education. Henceforth, experimental science had found a home in the teaching institution. An entering student who was willing to work hard could assert his interest in science, develop it, and maintain it without deviating from the rules. This new dispensation, which combines academic teaching with scientific research, is followed to this day. Pasteur was a pioneer of this twofold commitment.

These new dispositions lengthened the most successful *agrégé*'s stay at the Ecole normale by two years. Teaching positions in physics, chemistry, mathematics, zoology, botany, geology, and mineralogy were filled, along with "a position in the scientific library." These paid temporary posts were the only way for young graduates to conduct scientific research. The functions of the préparateurs, who were expected to supervise the practical work in the laboratory, varied with the expectations of their director, but their role seems to have been midway between today's technician and investigator. Several of these agrégés benefited from this apprenticeship, which Pasteur granted only to the candidate whom he considered the most meritorious. Did those who were not fortunate — or perhaps not capable — enough to be included among the elect feel great regret? This is not an idle question, for Pasteur took advantage of his administrative prerogatives to select his own candidates, young men whose careers he meant to take in hand and who thus had to submit to his tremendous authority.

Pasteur's other major contribution was his proposal, submitted as early as 1859, to publish a periodical that would keep up the contact among the school, the former préparateurs, the teachers, and the students. Some academic institutions, such as the Muséum d'histoire naturelle, the Ecole des mines, the Ecole polytechnique, or the Conservatoire des arts et métiers, already published their own reviews, but France as a whole was lagging in this domain. In Germany, places where young scholars could publish their discoveries and advance new hypotheses had long been available. At that time, Ernest Renan wrote to the director of the *Revue germanique*, "In France, we do not allow ourselves to place our work before the public before it is completely finished. In Germany, they bring it out in a provisional state, not as a pronouncement from on high, but as an instigation to further thinking and a ferment to the minds of others."[13]

"I propose," Pasteur wrote in this connection, "to call this review *Annales scientifiques de l'Ecole normale* published under the direction of Monsieur Pasteur, director."[14] In principle, the minister favored the project. But it was still necessary, on the one hand, to overcome Nisard's scruples concerning the parallel publication of a literary revue and, on the other, to find a publisher. Pasteur insisted that the review be published free of charge. Gauthier-Villars, the principal scientific publisher of the period, agreed to assume all the risks and received no subsidies from the ministry or from the Ecole normale. The first issue appeared in July 1864 and contained an article by Désiré Gernez, one of Pasteur's first students, on the optical activity of certain liquids and their vapor. Under the iron rule of its director, the review developed rapidly. "May the *Annales*," he wrote, "enhance the honor and the strength of an institution whose well-being is inseparable from that of public education in our country."[15] Pasteur was to remain on the editorial board until 1871.

Private Tears and Public Honors

In the cramped apartments of the rue d'Ulm, the Pasteur family led a daily life of extreme austerity. They rarely went out. Except when he was teaching or lecturing, Pasteur lived in his laboratory, which was adjacent to his living quarters at the school. It was thus very easy for him to return in the evening to continue his experiments; his assistants also lived close by. Working days began at eight in the morning and ended at about six. In the evening, Pasteur read, wrote, or pondered. Marie copied letters and reports to be sent to the ministry or to the Academy. No manuscript ever went to the publisher without having been meticulously reread, checked, annotated. Evenings spent at social events were exceptional, and vacations were rare, amounting at most to a few days at the seashore at Saint-Georges-de-Didonne near Royan.

But these years were not exclusively devoted to teaching and research. Pasteur lived through all kinds of private and public ordeals and experienced the inescapable pangs of mourning as well as the solace of public honors and official recognition.

The little cemetery at Arbois, sheltered behind a gate decorated with revolutionary emblems, stretches between the vineyards and the Cuisance River, which encircles the village with its torrential waters. The church of Saint-Just dominates the tombs. The colors of the vineyards brighten their inscriptions. Four times Pasteur had to come to Arbois as chief mourner.

In August 1859, while Pasteur had to stay in Paris for the agrégation examinations, his oldest daughter, Jeanne, was stricken with typhoid at Arbois. Fortunately, Marie was with her and wrote almost every day. The famous physicians Trousseau and Andral were consulted. By the end of the month, the news was reassuring and Pasteur, feeling that he should relax a little, took his son Jean-Baptiste to the theater for the first time. They went to the Comédie-Française, which was playing *Les Enfants d'Edouard* by Casimir Delavigne, a somber tragedy in which innocent children are murdered. It was a sad omen, for while everyone expected a slow recovery, Jeanne suffered a sudden relapse. Pasteur rushed off to Arbois but arrived too late: the child died on 10 September at seven o'clock at night.

On her grave, between those of the Meunier and Thevenot families, one can read the simple words: "Pray for Her." Pasteur did not speak, walled into his silence. "I heard the sound of the coffin and of the cords that took it down to the bottom of her grave, and the sound of the earth falling on that wood, both so empty and so full. . . . What are those letters from strangers, from friends, even family? Vain talk — drops of water taken from the fury of the ocean." And to Marie, who stayed at Arbois, he wrote: "Yes, when you come back, I wish that between us, with us, there will be nothing but our love, our children, their upbringing, their future, along with my dreams of science. For you, for them, life will be made beautiful by my work, by the success of new discoveries, and by generous feelings."[16]

Fortunately, Pasteur did not count on his work alone to bring solace to his spouse. Four years after this terrible moment, in July 1863, Marie gave birth to a fifth child, another girl, Camille. But this newfound happiness was ephemeral, for Camille was to live for only two years. Stricken by a liver tumor, she wasted away in a few weeks under the eyes of her father who, though "stunned by pain," as he wrote, also spoke with cold lucidity of the impotence of medicine.

Camille died six years, almost to the day, after Jeanne, on 11 September 1865. Pasteur was distraught. That same evening he wrote to Jean-Baptiste Dumas: "My poor child died this morning; she was so lucid to the very end that when her little hands were getting cold, she constantly asked to place them into mine, which she had never done throughout her long illness." And because work was his only recourse, he added a few lines farther down a word on his research: "Since I had written to M. Lachadenède that I would send him the galleys of the reports this week, I would like to ask you the favor of handing in my work and ask to have it put on the agenda of the session."[17] Pasteur took his child's coffin to Arbois, where he led the funeral procession with great dignity.

Three months earlier, he had also entered the small town's cemetery in a formal mourning ceremony for his father, Jean-Joseph Pasteur, who had died on 15 June 1865. The old Napoleonic veteran had passed away at sixty-four, immersed in his memories of the imperial era and toward the end probably more interested in the echoes of the past than in the career of his son, whose work he could no longer understand. A loner like his son, and often equally irascible, he had sometimes frightened his grandchildren. Louis and Marie had regularly written to the "grandfather of Arbois," who had often asked Marie for details of their day-to-day life rather than for exact information about the experiments, the theories, or the academic achievements of his son.

Upon learning that his father was in extremis, Pasteur immediately left Alès, where he was conducting a study for the government. But when he arrived at the tannery in Arbois, he saw that his cousins from Salins were already dressed in black. The coffin was closed. On the night of the burial he wrote a long letter to his wife, who had remained in Paris. In this letter he reverently gathered together his memories and attempted to exorcise a vague feeling of guilt by hinting that his father had not really understood him. "The touching thing about his affection for me is that ambition was never part of it. Remember how he said that he would be pleased to see me as headmaster of the collège of Arbois. He said these things because he saw behind every advancement the work that would bring it about, and behind the work my health, which might suffer. And yet, such as he was, and such as I see him better today, some of the successes of my career in science must have given him great pride and filled him with joy. This was his son, his name. It was the child to whom he had given guidance and counsel. Oh my poor father! It makes me very happy to think that I have been able to give you a few satisfactions."[18]

In the presence of the grave, Pasteur always wanted to justify himself, to respond to the subconscious reproach he leveled against himself, for he suspected that his passion for science was more important to him than the affection of his family.

Nor was this the end of his trials. In May 1866, Pasteur lost a third daughter, Cécile, who also was carried off by typhoid fever at the age of twelve and a half. Detained once again at Alès by his work, Pasteur was not present when the illness broke out, since Marie and her two daughters were spending the summer at the home of Marie's brother-in-law, Charles Zevort, who was superintendent of schools at Chambéry. Torn between the urgency of the experiments he had started and the alarming news he received from the department of Isère, Pasteur did not immediately rejoin his family. When he finally arrived — Marie's letters had become more and more desperate — he found his daughter smiling and fully lucid. The illness had suddenly abated, so that Pasteur felt he could shorten his stay. He therefore spent only three days near his daughter, the reason being that despite the strictest instructions he had left with his co-workers in the department of Gard, he did not trust them to carry out the manipulations correctly.

And then what had happened with Jeanne seven years earlier happened again: Cécile's recovery was but an illusion, a few days' respite. The young girl died on 23 May, and her father never saw her again. Her body was taken to Arbois and interred beside those of her two sisters and her grandfather. The words "Pray for Her" that can be read on Jeanne's tomb are not found on the otherwise identical gravestones of Jean-Joseph, Camille, and Cécile. It appears that for the first time in his life Pasteur gave in to despair: "My dearest Marie, so they will all die one by one, our dear children, you my poor Cécile whom I loved so much, and you two others who are already gone and who call her to be with you. I too long to join you, my dear children."[19]

Upon his return to Paris, only religion brought him some consolation; thus he gave a rosary in memory of his child to the daughter of the minister Victor Duruy, Hélène, who had been Cécile's best friend. Of Louis and Marie's five children, only Jean-Baptiste and Marie-Louise survived. Although one cannot doubt Pasteur's faith, it should be said that he was not given to mystical exaltation or flights into the irrational. He sought comfort above all in his work and in the public recognition of his achievements. The excitement of discovery and the certainty of being right justified every sacrifice and eventually healed all wounds.

Meanwhile the accumulation of misfortunes increased Pasteur's needs: a brilliant career in the institutions of the state was no longer enough; more than the esteem of his peers he now sought that of the public. The outlines of a new direction in Pasteur's work were gradually emerging. No doubt because his discoveries in crystallography and his studies of fermentation had already allowed him to bring about a fundamental revolution that was all his own and to overthrow the existing theories of the life sciences, he was now intent on answering more concrete questions whose symbolic and material stakes were accessible to a larger public. Pasteur wanted to be a scientist for town and gown.

For that he needed a forum. The press was a perfect vehicle; in 1862 the editor of the prestigious *Revue des deux mondes* began to ask him for articles for a nonscientific public. Pasteur refused on the grounds that he was not competent to do this. In fact, he had set his sights even higher, for he eventually published two much-discussed articles — one on Lavoisier and one on Claude Bernard — in the *Moniteur universel*, the official organ of the Empire. At the same time he also gave a series of lectures to the *Société philomatique*. But Pasteur's dream was elsewhere, and his deepest ambition had to do with revenge, with making up for a failure, and indeed with an obsession. He wanted a seat at the Academy of Sciences.

In 1861, he had already declared his candidacy for a vacancy in — of all things — the botanical section. Biot, who faithfully campaigned for him, claimed that his work on microscopic mushrooms amply qualified him, but the regular botanists were not sure. As the most respected among them, the naturalist Horace Moquin-Tandon replied to Balard, who vaunted his candidate's merits: "Well then! Let's go to Pasteur's house, and if we find in his library a single volume on botany, I'll put him on the list." Such a witticism can kill, and it is to be assumed that Pasteur's library was not well stocked in this field: the winner was the obscure Duchartre.

It was only in 1862 that Pasteur started a serious campaign. Biot had died on 8 February, and Sénarmont on 29 June. Pasteur could legitimately present himself as crown prince to both of them; but he chose to pursue the succession of the latter in the mineralogy section. Yet his election was not a shoo-in.

Pasteur's adversaries were finding new grounds for opposing him. They argued, for instance, that he had long ago given up his work in crystallography, which to the mineralogists was his principal claim to membership, and that the field had progressed without him. A German mineralogist, Rammelsberg, had developed a theory which, according to him, was diametrically opposed to that of Pasteur. He asserted that right-handed crystals deviate the light to the left, and vice-versa. What Pasteur had written in his studies on tartaric acid was that right hemihedrism deviates light to the right. This was obviously a false problem, a case of true "Germanic hair-splitting." It was simply a matter of different conventions in defining orientation, which did not diminish Pasteur's brilliant discovery in any way. But his detractors seized upon it and tried to use it against him. Whispering and insinuations circulated in the ranks of the members of the Academy. The campaign soon turned into a cabal.

Pasteur was irritated and angry. He ordered a cubic meter of pine wood and, taking up a hatchet, a plane, and a file, made an enlarged model of right- and left-handed tartaric acid, sculpting the facets to the proper inclinations and covering them with green paper. Armed with his models, he called for a special session and delivered a definitive lecture on molecular asymmetry, which he concluded with a haughty put-down of his adversaries: "If you understand the question, where is your conscience? And if you don't, why do you meddle?"[20] Pasteur was never more eloquent than when he spoke in cold anger. He emerged victorious from this struggle and was elected to the Academy of Sciences on 8 December 1862 by thirty-six out of sixty votes.

Shortly after this election, on 19 March 1863, Pasteur was presented to the head of state as was customary for recently elected members of the Institut. Jean-Baptiste Dumas introduced the forty-year-old scientist as he entered the gilded salons of the Tuileries Palace. Ten years after the summer when he had tried to catch the eye of Napoleon III as he was walking along the beach at Dieppe, Pasteur finally met the Emperor. The day after this meeting, he marveled at the sagacity of his interlocutor's reflections: "I assured the Emperor that my whole ambition was to arrive at an understanding of the causes of putrid and infectious diseases. The Emperor approved of the direction of my work and added that he believed that some animalcules might well play a role in the development of these diseases. He spoke of the malaria of the Roman countryside and in that connection cited some unusual facts he had observed."[21]

Always receptive to flattery from high places, Pasteur temporarily succumbed to the charms of social events at the imperial court. He could sometimes be seen in its salons, especially at the residence of Princess Mathilde, daughter of Jérôme Bonaparte. There he rubbed shoulders with Sainte-Beuve, who solicited his vote for certain candidates for the Institut, among others Théophile Gautier, who was dying to belong to it. But whenever he was asked for his support, Pasteur invariably cited Jean-Baptiste Dumas's words: "When it comes to election to the Academy, I am not interested in what the election does for the candidate, but in what the candidate's election does for the Academy." But like everyone else, Pasteur went to the princess's to plead his own causes, for he never stopped talking about the role of yeasts in wine making or the need for new ways of vinegar making and lamenting the poor public image of laboratories and the inertia of government agencies. Pasteur's presence at these soirées amounted to a kind of command performance. Being seen at court was quasi-mandatory for the recipients of academic honors.

However that may be, these endeavors soon bore fruit, at least as far as Pasteur's personal career was concerned. On 18 November 1863, a chair of geology, physics, and applied chemistry was created for him at the Ecole des Beaux-Arts. It is true that Pasteur had executed a few nice-looking pastels in his youth, but he had done nothing of the kind since, or even shown any particular interest in the arts.

A great revolutionary in the sciences, Pasteur was more than conservative in his artistic taste. He would always ignore innovative artists, among others the Impressionists, and favor some second-rate painters. Pasteur liked the canvases of Thomas Couture and Jean-Jacques Henner, the statues of Paul Dubois or Jean-Joseph Perraud. He paid no attention to Courbet, who lived and worked a few kilometers from Arbois. Contemptuous of Manet, he felt that Monet, Sisley, or Pissarro did not even come close to any pupil of Meissonier.

At the Ecole des Beaux-Arts, his classes for architects were essentially concerned with public health issues. Pasteur showed the effects of poor ventilation and inveighed against the polluted air of prisons. He told the students that in 1815 the wounded had died less rapidly in the crowded abattoirs than in the hospitals. He considered the isolation wings of the Lariboisière hospital that were just then under construction in Paris a model of medical architecture.

His classes for painters also addressed technical matters, such as the handling of oils and varnishes and the proper use of gum, glue, egg white, quick-drying oils, and linseed oil: "In the art of painting, the smallest things can be important. Do you realize that if the oxygen of the air did not combine with the oil of linseed, carnation, hazelnut, and hemp and thereby harden it, oil painting would not exist, could never have been invented? For neither olive oil nor almond oil is capable of drying."[22] Did these young art students really listen when Pasteur spoke to them in this manner? One may doubt it. However, even today, many an artist would benefit from looking into this basic chemistry of painting, which for the most part is still valid. There is a great deal of talk about technique, and there are the obligatory complaints that it is too often neglected, but in fact Pasteur's lessons do not seem to command any more attention today than they did a hundred years ago.

There is little point in wondering who was the first to become tired of the other, the students or the teacher. At any rate, Pasteur did not continue this pedagogical experiment, for which he was hardly suited. He resigned from the position in November 1867.

The Affaire Sainte-Beuve

This resignation, in the autumn of 1867, was surely not a major incident in Pasteur's life, but it did coincide with an event, a veritable earthquake in the microcosm of the Ecole normale, that brought about an appreciable change of direction in the scientist's career and took him further and further away from teaching and definitively turned him from a professor into a research scientist. This was the so-called "affaire Sainte-Beuve."

By a decree of 22 October 1857, Sainte-Beuve had been appointed lecturer in French literature at the Ecole normale. However, the famous critic did not get along with Désiré Nisard, his old foe in the literary quarrels of 1830, whom he moreover accused of conspiring with Guizot to deprive him of a teaching position to which he was entitled as an *agrégé*. On the other hand, he was on good terms with Pasteur, who occasionally came to his classes. Sainte-Beuve was not at ease with the students: his elocution was poor, and as a man of the written word he did not know how to improvise, so that he was persuasive only when he read his lectures. Moreover, in this school where republican sentiments were prevalent, he was considered a pillar of the Empire. Realizing that only the strict discipline practiced at the school saved him from hostile demonstrations, Sainte-Beuve preferred to resign in 1861. This gave him the leisure to deliver to the *Constitutionnel* a new series of *Causeries du lundi*, in which he could fully express his subtle judgments without having to worry about the reaction of an audience of obstreperous young men. In 1865, thanks to the friendship of Princess Mathilde and Prince Napoleon, he was given a lifetime appointment as senator of the Empire.

In the Senate, Sainte-Beuve, aware of his oratorical deficiencies, kept silent for a long time. But in March 1867, when Baron Chapuis-Montlaville criticized Ernest Renan in barely veiled terms and accused the author of the *Life of Jesus* of "spreading doctrines of atheism and irreligion among the masses," Sainte-Beuve rose from his bench: in the name of freedom of thought, he was going to defend a man of whose friendship he was proud and who, he said, "upholds honorable and respectable philosophical opinions."

"This is not what you are here for!" Senate President Troplong interrupted him curtly, whereupon the matter was declared closed. Three months later, Sainte-Beuve had the opportunity to explain himself further. Some one hundred inhabitants of Saint-Etienne had sent to the Senate a petition in which they protested the municipality's choice of certain books that would constitute the basic holdings of two municipal libraries. In the Senate, the relevant committee chairman proposed to send the petition to the ministry of public education, which was competent to deal with the matter. The procedure would no doubt have followed its normal course had not Sainte-Beuve asked on 21 June 1867, for the first time, to speak from the rostrum. His intervention, he said, did not have the ambition to defend works that had no place in a people's library, but he nonetheless wanted to make sure that "the occasion of what might be a controversial or blameworthy fact is not used to institute in our free France a kind of index of condemned books as is done by Rome. And besides, just what books are these? Voltaire's *Zadig* and his *Candide*, Jean-Jacques Rousseau's *Confessions*, Michelet, Balzac, Renan, and George Sand." Senator Lacaze, a reactionary Bonapartist, took the floor and tried to justify the attitude of the petitioners of Saint-Etienne. But Sainte-Beuve, without missing a beat, continued his speech: "I will point out one thing that is missing. Your list, long as it is, is incomplete, gentlemen, it does not include Molière, it does not include *Tartuffe*." Lacaze (rightly) felt that this was directed against him and asked that the Senate be called to order. The ensuing uproar was such that no discussion was possible and the voting took place without it.[23]

Outside the Senate chamber, this session created an enormous stir. The *Moniteur*, which had to, and the opposition newspapers, which enjoyed it, printed Sainte-Beuve's speech and his plea for freedom of thought. Delegations of students and workers acclaimed it. The author received several hundred letters of support from Paris and the provinces; and the agitation, far from dying down, grew stronger. Lacaze, offended by Sainte-Beuve's rejoinder, challenged him to a duel. To this Sainte-Beuve replied, "I do not accept as easily as people seem to assume this summary jurisprudence, which consists of strangling a question and doing away with a man within forty-eight hours." He added that among his devoted friends there was no man who knew anything about handling weapons. "My friends," he wrote, "usually know about the things of the mind, of writing and speaking, which does not mean that they are therefore less firm or men of less honor."[24] This was truly a slap in the face for those who wanted an iron-fisted Empire . . . Other letters were exchanged, and in the end Sainte-

Beuve handed the entire file of correspondence to the press. *L'Avenir national*, *Le Journal de Paris*, *Le Courrier français*, and *La Patrie* eagerly seized upon it.

But no one was happier than the students of the Ecole normale about such an act of temerity on the part of their former professor, whose past awkwardness was instantly forgotten and who now became so popular that he was referred to as "Uncle Beuve."

It was early summer. Every year at this time, the students organized a lottery to benefit the neighborhood poor. It was the custom to ask for donations from the most famous men of letters and scientists. François Lallier, a second-year student in the literary section, was given the task of soliciting Sainte-Beuve. "We have thanked you once before for defending our beleaguered and persecuted freedom of thought; today, when you have again pleaded its cause, we hope you will again receive our gratitude. . . . It takes courage to speak in favor of the independence and the rights of the mind in the Senate. But as the task becomes more difficult, it also becomes more glorious. At this moment, statements are being sent to you from every side, and we hope that you will forgive the students of the Ecole normale supérieure for following the general example and sending their own statement to Monsieur Sainte-Beuve."[25]

"Statement" [*adresse*] — the ensuing debate centered upon this term. Although Lallier probably used it as an ironic manner of expressing himself, he did give his letter a collective character and public significance. The petition was, in fact, signed by eighty of the school's students. Twenty-one of them had refused to associate themselves with it, among them Edouard Branly, the future inventor of the wireless telegraph. Lallier's move was bold, for it was a troubled time: a few weeks earlier, when Czar Alexander of Russia and King Wilhelm of Prussia had come to visit the World's Fair, a young Pole had fired on the imperial carriage. This assassination attempt had touched off a show of loyalty on the part of the country's constituted bodies. The "statements" the Emperor received in this context were so many acts of submission, whose political intent was obvious. In this context the very terms used in Lallier's letter clearly expressed a spirit of insubordination.

Wily as he could be, Sainte-Beuve did not react to this allusion; he simply thanked the students for their support and delivered a passing dig at the direction of the school by taking note of its prudent silence.

This would no doubt have been the end of the matter if two of Lallier's classmates, Henri Marion and Gaston Maspéro, had not been in close touch with one of the most passionate pamphleteers of the opposition, Etienne Arago. Arago obtained copies of the exchange of letters and published Lallier's statement in the *Avenir national* of 2 July 1867. The administrators in the Ministry of Public Education were upset. The minister, Victor Duruy, was on an inspection tour, so that it fell to his chief of staff (a former Normalien by the name of Danton) to order an investigation. The next day, 3 July, the students of the rue d'Ulm learned that François Lallier had been suspended until further notice.

Since Pasteur was in charge of disciplinary matters, the suspension order went out over his signature. The students therefore naturally turned to him to plead for indulgence for their classmate. Not only did Pasteur not want to hear anything of the kind, slipping away by the back door of the school's kitchen in order to avoid meeting the students, he also asked them to inform against those who had given the letter to Arago. The students dodged the issue and tried to meet with Nisard, who was perpetually "out." Faced with what they considered a dereliction of authority, they decided to walk out of the school in a body. Since a recent decree had abolished monitors, they did not encounter much resistance and simply left without being stopped. This symbolic gesture was accompanied by a written declaration stating once again that the Normaliens supported freedom of expression and therefore demanded the reinstatement of their classmate.

The families of the young Normaliens were not pleased with the events in the rue d'Ulm; many of them wrote to Pasteur to solicit his indulgence and ask him to forgive the students. He even received epistles in verse, some of which recalled that "the Good Shepherd [*Bon Pasteur*] of the Gospel liked to bring the lost sheep back into the fold."[26] All of these letters asked him to reconsider. The ministry too was concerned about the speed with which the punishment had been meted out and put pressure on Pasteur to show some flexibility. But the administrator replied: "If vigorous repression were not to be the consequence of Lallier's wrongdoing and that of the two students who took this ridiculous and unpardonable statement to a newspaper without even consulting their fellow students, I would not hesitate to offer my resignation to His Excellency."[27] He added that in his opinion the minimum sanctions should be temporary suspension for Lallier, permanent suspension for Marion and Maspéro, and respectful compliance for all who wanted to come back to the Ecole normale.

Sainte-Beuve was stunned. Even though his opinions, both in politics and in religion, were the opposite of Pasteur's, he respected the scientist, who reciprocated these sentiments. The critic therefore sent to the administrator an urgent letter asking for indulgence and pardon. But Pasteur would not budge. Etienne Arago therefore decided to plead guilty and to clear up the matter by assuming full responsibility for everything that had occurred because it was he who had made the letter public. But it was too late.

On 7 July 1867, Victor Duruy, the Minister of Public Education, examined the matter and sided with Pasteur. Considering that discipline in the rue d'Ulm had been most scandalously flouted by three separate acts—the statement, its publication in the press, and the collective walkout—he ordered the closing of the school.

In the press, the minister's decision exploded like a bombshell. Comparisons were made with the Polignac ordinances, which in 1830 had triggered the July Revolution. Pasteur was the first one to be blamed: the *Courrier français* accused him of having acted out of fanatic Catholic opposition to the libertarian attitudes of the Normaliens. Everything in the administrator's conduct was erratic and inconsistent, one read in the *Journal de Paris*: "It is inconceivable that a single student should be suspended for wrongdoing unless M. Pasteur is relieved of his functions on the grounds of negligence in permitting this wrongdoing to occur."[28] Nisard's discretion thus bore fruit: the blame was not directed at him, but at his executive officer. However, the minister himself was not spared either. *Le Temps*, *La Presse*, *Le Journal de Paris*, and *L'Avenir national* attacked the government, while *Le Figaro*, *L'Epoque*, and *La Situation nationale* tried to defend Duruy by claiming that he had agonized over his decision.

But whatever his personal merits, Duruy was the Emperor's minister and as such a rather more interesting target than the administration of the rue d'Ulm. Hence, the criticism was deflected. In the *Journal de Paris*, Duruy was criticized for having helped the careers of his two sons, Anatole and Albert, one of whom was his father's chief of staff and the other secretary of the prefecture. Anatole and Albert Duruy went to the paper's editorial office to challenge the author of the article to a duel. When the journalist declined, they slapped his face in the presence of witnesses.

Now the entire liberal press was in an uproar and the Ecole normale became a symbol: its closing amounted to a major purge for the system of higher education. The task now was to restore the authority and the dignity of the teaching profession, the freedom of thought, and the dignity of literature, following the fine example Sainte-Beuve had provided single-handedly at the Luxembourg Palace. Pell-mell, everything became grist for the mill of the liberal press: the petition of Saint-Etienne, Sainte-Beuve's speech, the "statement" of the Normaliens, the slap administered by Duruy's sons. The affaire Sainte-Beuve served the purposes of the opposition. Thiers and Jules Simon considered an official parliamentary challenge to the government.

The affair calmed down under the influence of a provision for conciliation worked out by Ernest Bersot, the future director of the Ecole normale. Moreover, reassuring news came from Duruy, who permitted the students to retake the agrégation and prepared for the reopening of the school. Lallier was appointed professor in the upper grades of the lycée of Sens, which was an excellent position. Marion went back to the Ecole normale in November. Maspéro, disappointed by the system, left for South America. He eventually became a brilliant Egyptologist and was elected professor at the Collège de France, having first served as one of the directors of the Ecole des hautes études [School of Advanced Studies]. Rebellion can lead far.

Meanwhile, the authority of Nisard, and especially that of Pasteur, his right-hand man in charge of discipline, had been so badly damaged that they could not continue in their positions. The government thus had no choice but to sacrifice two of its most faithful supporters. On 24 October 1867, Nisard was replaced by the philosopher Francisque Bouillier, while Pasteur was moved to a rather more desirable position, succeeding Balard in his chair at the Sorbonne. By an irony of fate or a clever manipulation, his old classmate Pierre-Augustin Bertin was appointed to his position as head of scientific studies at the Ecole normale.

A few weeks later, a personal intervention of the Emperor once and for all repaired the damage that had been done; a laboratory of physiological chemistry was created at the Ecole normale especially for Pasteur. The erstwhile administrator thus did not have to leave the rue d'Ulm, but he no longer had to deal with the "literary types," those troublemakers who had made his life so difficult. All things considered, this was a satisfactory epilogue, for it would have been absurd if, as a result of an overblown disciplinary matter, a scientist of this caliber had been made to molder in a closet.

The Laboratory That Could Not Be Found

The casual walker who enters the rue d'Ulm from the Panthéon can see on the right side, almost at the end, a modest little two-story house. This was Pasteur's laboratory. On its façade, a bronze medallion and a marker list the principal discoveries that were made here. The municipal council of

Paris placed them there while Pasteur was still living, at the time of the hundredth anniversary of the Ecole normale, in 1895. Since then, the outward appearance of the house has been modified by the addition of an attic story, but the windows are the original ones, with the first four giving light to the laboratory and the fifth one to a hallway. One is inclined to think that this is a more than modest installation, and unworthy of Pasteur. This is no doubt true, especially if one thinks of what can be done in today's laboratories, but at the time such a facility represented a true and hard-won privilege.

Ever since his return to Normale in 1857, Pasteur's principal scientific goal had been to find a way to continue his investigation of fermentation. Knowing himself, he was aware that he was a teacher more than an administrator and indeed more suited to be a research scientist than to fulfill the role of a professor. At the Ecole normale, as in Lille or in Strasbourg, he saw himself above all as a man of the laboratory: "Laboratory and discovery are correlated terms. . . . Outside their laboratories, physicists and chemists are like unarmed soldiers on the battlefield."[29] But at the time laboratories did not exist as such. Research was done by teachers in higher education who, devoting to it whatever spare time they could muster, worked with variable and frequently very modest means. University budgets provided only minimal resources, most of which were used up by course preparations. This left almost nothing for real research. The only facilities available to the professors were their classrooms, from which they had to remove all their apparatus on days when the students came in. Officially without a locale, without equipment, an operating subsidy, and a full-time director, the research laboratory existed only on what it was able to steal from teaching.

In the provinces, the situation varied from place to place. When he was dean at Lille, Pasteur had had the use of a few rooms that he had wrested from the administration, even if he had to push his apparatus to one side for the students' laboratory sessions. Teachers and researchers in Paris were hardly better off. Higher education, particularly in the sciences, was neglected. The Sorbonne had not undergone major modifications since the time of Richelieu. The first stone for a new building had duly been placed in 1855, but the second one had still not been added in 1867. At the Muséum [d'histoire naturelle], the galleries were cluttered with miscellaneous collections of stuffed animals and minerals, looking like theater sets. At the Collège de France, Claude Bernard worked in a cramped laboratory where the humidity made the walls sweat, a situation that eventually made him ill. Working conditions were such that people often derisively depicted laboratories as the scientists' graves!

While he was teaching at the Ecole normale, Balard had, with some difficulty, secured two somber rooms on the first floor, which he used to prepare for his courses. And even this was a special favor, which was given only to the holders of a chair. In October 1857, Pasteur could not ask for anything comparable, so that he had neither a place nor the money to continue his work.

The only laboratory at the rue d'Ulm was occupied by the biochemist Henri Sainte-Claire Deville, who in 1851 had replaced Balard upon his appointment to the Collège de France. The rooms were dark, and with an endowment of 2,800 francs per year, Sainte-Claire Deville had been able to buy only a few instruments. Yet his research on aluminum, which he conducted thanks to original methods of preparation carried out in the village of Javel (at the time located at the gates of Paris), had permitted him to produce metal ingots of such high quality that they were sent to the World's Fair of 1855 and earned their creator the Légion d'honneur. But then it was less costly to place a red ribbon into a man's buttonhole than to endow a laboratory worthy of the name.

Sainte-Claire Deville was an agreeable colleague. Brought up in the Antilles, he was of perpetual good humor and much beloved by the students, with whom he liked to be on friendly terms, sharing their dining hall as well as their library. Always bustling with activity, he stimulated their minds. What a contrast to the calm demeanor of Pasteur, his uncommunicative, cold, authoritarian, and often cantankerous attitude! Despite this difference in temperament, Pasteur was genuinely fond of Sainte-Claire Deville—not, however, to the point of sharing a laboratory with him.

More than anything, Pasteur wanted to be independent. This meant that he had to make do with the last few square meters still available in the rue d'Ulm, namely, a small nook in the attic that nobody had claimed since it was considered too uncomfortable, particularly because of the summer heat that made work there impossible. Moreover, these rooms were infested with rats, which first had to be eradicated. Pasteur had no subsidy for installing a laboratory. He wrote to Gustave Rouland, Minister of Public Education at the time: "I believe, Monsieur le Ministre, that I am following a part of your instructions by devoting all of my leisure-time to the advancement of science." This was a roundabout way of asking for funds to purchase some makeshift equipment. He was informed that funds were to be "entirely devoted to the maintenance of the buildings and not to construction projects requested for the convenience of persons lodged in these buildings."[30] But Pasteur refused to give up and finally did obtain some small subsidies. But the installation was necessarily provisional.

The equipment that Pasteur needed at that point was modest. His research on fermentations required only an incubator, a microscope, some chemicals, and glassware. But even though the ministry had seen fit to give him a little money to convert the attic, there could be no question of providing for equipment or operating expenses. "There is no rubric in our budget under which we could allocate fifty centimes to you for the cost of your experiments."[31] Consequently, Pasteur had to pay for the equipment and the running expenses of his laboratory out of his own pocket. This installation cost him more than two thousand francs, a considerable sum for that time.

In fact, Pasteur was to face this kind of struggle throughout his life. The first thing a researcher must find are subsidies. To a journalist who criticized him for his constant pursuit of scientific prizes, Pasteur replied in exasperation: "In my laboratory, there is not one object, however small its value, that is not my personal property." He always drew a parallel between the honor or the profit France stood to gain from his discoveries and the insignificant sums he needed for his laboratory and was regularly denied. He strongly felt that a researcher was entitled to a laboratory as much as a professor was entitled to a salary. At the same time, he stated what is today considered a truism, namely, that the authority and the usefulness of a research director are directly related to the success of his own work. "There is in the life of every man engaged in the pursuit of experimental science a period when time is of inestimable value. This is the fleeting period when the spirit of invention flourishes, and when every year must be marked by a major advance. Would it be too bold of me to add that I feel myself to be in this period of life and that I beg your Excellency not to let me remain under the pressure of obstacles arising from the insufficiency of resources assigned to the progress of science in our country?" With this kind of argument, Pasteur often scored his point.

More seriously yet, Pasteur was not even entitled to the help of a laboratory assistant [*préparateur*] when he first arrived at the rue d'Ulm. For such an assistant, as one of Pasteur's close associates of these heroic times was to put it, was the "dog that keeps the cutler's back warm": one could get along without it, but then one had to keep a fire going.

To be sure, laboratory assistants were included in the budget of the Ecole normale (after all, Pasteur had been Balard's), but they were assigned with a draconian parsimony that made little sense since they were not even agrégés. One can imagine the astonishment and the suspicion of the bureau chief in the ministry when Pasteur requested not only a laboratory assistant to work with him on his research but specified that he must be an agrégé.

After endless applications submitted with unshakable tenacity, he finally won his case. But the decree that officially created the position of *préparateur agrégé* stipulated that the person in question would remain at the disposal of the school administration and could, if necessary, be temporarily employed in a provincial lycée. Otherwise, he would receive room and board in the rue d'Ulm and Pasteur would be entirely responsible for his work.

These dispositions, which were not auspicious for research and would be unthinkable today, were in fact the logical consequence of the administrative practices of the time. After all, the Ecole normale was supposed to prepare teachers for secondary schools, and students who did not eventually hold a position in a lycée were looked at askance.

When the first of these laboratory assistants, Jules Raulin, who had entered Normale in 1857 and written a remarkable study on the microscopic fungus *Aspergillus niger*, came to work for Pasteur, the latter had already found a way to leave his attic and move to the pavilion facing onto the rue d'Ulm. Permission to use this little building, hitherto occupied by the department of architecture, had been a conspicuous favor: "Your fermentations are costing us a great deal of money," the minister had written Pasteur, "but I am hoping for fine results."

The place was certainly inconvenient: five tiny rooms spread out over two floors. Since the lack of space made it difficult to find room for an incubator, Pasteur had it installed in a closet built by sacrificing part of the staircase; accessible only by a ladder, it provided kneeling room only. Yet Pasteur spent many hours there. "This little hovel, which one would nowadays hesitate to use as a rabbit hutch," Raulin's successor Emile Duclaux was to testify later, "was the starting point of the movement that revolutionized every aspect of the science of physics."[32]

In 1862, the laboratory was enlarged, though not by much. A newly built room, lit by one window looking out onto the courtyard and the garden, was added to the first floor. This room was attached to the existing pavilion, creating a single space ten meters long and four meters wide; the department of architecture pushed its generosity to the point of decorating it with a bust of Lavoisier. One must imagine this place: in the center a polished oaken table; on a shelf, orderly rows of clean glassware. There was no cleaning staff; it was part of the laboratory assistant's work to clean the floor and the utensils. Nor was there an easy-chair, for Pasteur did not see the need for comfortable seating in a laboratory where he was always standing.

Pasteur was a solitary worker. He hated to be disturbed. "Pasteur was so absorbed by the concentration of his thoughts," reported his nephew

Adrien Loir, who was also his laboratory assistant, "that he did not notice anyone around him. Duclaux sometimes had to wait quite a while before he was asked what had brought him there."[33] When the time came to reflect about his working hypotheses, Pasteur would immerse himself in his laboratory notes and isolate himself from everything and everyone around him without ever looking up from his notes. In the throes of these solitary meditations, he might spend hours pacing up and down the room without saying a word, and it was well-nigh impossible to gain his attention. Quite often, these silences continued at home. Marie and the children were used to it.

"Pasteur does not want anyone next to him," Loir also testified. "He trusts only himself. What he wants, what he needs, is the most complete concentration on his own powers. . . . Pasteur would only think of one thing at a time. He was completely absorbed by the subject he had chosen to study. He would dissect it and experiment with it in all kinds of ways in silence and without allowing anyone to interfere." And he added: "Pasteur never got angry, and if something did not work out as he wished, he would say: 'Oh Lord! . . . Oh Lord!,' and he would pace about like a caged lion. That was all."[34]

The discipline in the laboratory was severe, almost military. The assistants had to work in silence without disturbing the master, except when absolutely necessary. Conversely, when they were doing important work, Pasteur kept all intruders away so that they would not be distracted. Tobacco was prohibited in the laboratory, and the assistants could only smoke when the boss was not there. The working hours were strictly regulated, and leisure was frowned upon. No experiment was to be undertaken without thorough reflection, and the initiative was strictly reserved for the master.

Every manipulation was designed with meticulous precision. The work was started as soon as the protocol had been established. Not much time passed between conception and execution, and this was undoubtedly one of the reasons for Pasteur's great scientific productivity. In case of failure, he would repeat the same experiment, growling, "the important thing is not to let go of the subject."

Pasteur was not interested in laboratory techniques as such, only in what they could accomplish for him. He wanted clear and unambiguous answers to his questions. In fact, his experiments often turned out to be easy to carry out, although they demanded extreme care in setting up the details as well as minute observation. Pasteur spent hours in silent observation, his eye glued to the microscope. "Nothing escaped his myopic eye," Roux was to say later, "and we used to say jokingly that he saw the microbes grow in the broth." Pasteur did not entrust the keeping of the records of the experiments carried out in his laboratory to anyone; he himself wrote down the information his assistants gave him. Nothing was recorded unless he had seen it. Once it was recorded, a fact became a truth, and the notebook a Bible.

More than solitary, Pasteur was secretive, as well as authoritarian and demanding. Long accustomed to needing no one but himself, he tended to treat his assistants more as menials than as fellow scientists. Duclaux, who knew of what he spoke, did not hesitate to ask the essential question: "Could Pasteur really have collaborators? The term collaborator implies working together, continual exchanges, constant intellectual communication in pursuit of a shared goal, and mutual help. Pasteur never told his entourage anything about what he was doing."[35]

In fact, his successive assistants did participate in Pasteur's experiments, occasionally even co-signed a publication, but they executed rather than directed this work and were almost never allowed to discuss strategies. It should be added that Pasteur had trouble recruiting assistants, for the career of a young agrégé who came to work in his laboratory became problematical, since it deviated from the normal path of teaching and from the usual progression, considering that the position of laboratory assistant was not yet part of the university hierarchy. Moreover, Pasteur was doing research in new fields (first microbiology, then veterinary and human medicine) and his projects, despite the public recognition they received, often seemed too risky to the young Normaliens. And finally, Pasteur's personality was not very attractive: he trusted only himself, checked out everything, and did not talk much. Pasteur assigned everyone his task and then forgot all about him. He demanded a great deal but gave very little of himself.

"Budgeting Science"

Under these circumstances, the outcome of the "affaire Sainte-Beuve" in 1867 can actually be considered a godsend; as director of a new laboratory whose specialty, physiological chemistry, had been defined to fit the direction of his own research, Pasteur could now become even more demanding than he had been before.

In order to obtain what he considered the minimum, he addressed himself directly to Napoleon III in a letter he submitted to him in September 1867 through General Favé, his aide-de-camp. "Sire, my work on fermentation and the role of microscopic organisms has opened new paths in physiological chemistry, which is now beginning to benefit agricultural industries and medical science. But the field that lies before us is enormous. My greatest desire would be to explore it with renewed ardor without being at the mercy of insufficient material resources. I should wish to find in the premises of a fairly spacious laboratory a place where experiments could be performed conveniently and without danger to the public health. . . . These investigations and a thousand others, which in my thinking relate to the great act of transformation of organic matter after death and to the inevitable return of everything that has lived to the soil and to the atmosphere, can only occur if a large and richly appointed laboratory is set up. The time has come to free the experimental sciences from the fetters of poverty."[36] On the very next day, Napoleon III asked Victor Duruy to contract for the construction of a new laboratory within the premises of the rue d'Ulm.

However, for all the authoritarianism of the regime, the wishes of the Emperor alone were insufficient. In December 1867 an obstacle arose that almost stopped the creation of the laboratory at a time when Pasteur was dreaming over the building plans: "I think I told you," he wrote to Raulin, "that the laboratory would be built between my present laboratory and the private house in the rue des Feuillantines that stands at a right angle to the garden of the Ecole normale. It would abut that house and be connected to my present laboratory by a glassed-in gallery. I hope it will have a basement and an upper floor. . . . I hope that my laboratory will have a sufficient endowment for me to create a first position paying 4–5000 francs, to be augmented in time and depending on a variety of circumstances."[37]

In reality, the imperial coffers were empty, and the additional moneys needed for the construction were not allocated by the administration of the Beaux-Arts. Money was found to build Charles Garnier's new opera house, but none was forthcoming for scientific research. Pasteur was furious. In order to mobilize public opinion, he wrote an article for the *Moniteur*: "The boldest concepts and the most legitimate speculations cannot assume body and soul except when they are consecrated by observation and experiment. Laboratory and discovery are related terms. Do away with laboratories, and the physical sciences will become the image of the sterility of death. They will be reduced to something that is taught, something limited and impotent, and cease to be sciences of progress and the future. Return them to their laboratories, and they will bring back life with all of its fecundity and power."[38] As always, he was able to call upon current events: Pasteur pointed out what the electric telegraph, the daguerreotype, anesthesia, and other equally admirable discoveries owed to laboratories. As far as Germany was concerned, he said, it had invested in spacious and rich laboratories long ago. New laboratories were being created every day throughout the world. Centers of chemistry had been set up in Berlin and Bonn; in Saint Petersburg an institute of physiology had been founded; England, America, Austria, and Bavaria were making great sacrifices. But in France, whether in Paris or in the provinces, scientists lived in cellars and produced their results amid rats in basements that were more favorable to humidity than to creation.

Pasteur's conclusion was not without power: "Who would believe me if I said that there is not in the budget of Public Education a single penny set aside for laboratories for the physical sciences; and that it is thanks to a fiction and to administrative tolerance that these scientists, placed into the category of professors, can obtain from the public treasury some funds to cover the expense of their personal research, though to the detriment of the allocations for their teaching duties?"[39]

When Pasteur handed this hot potato, soberly entitled "The Budget of Science," to the *Moniteur*, where he had already published his academic articles on Lavoisier and Claude Bernard, and which normally avoided all controversial subjects, the managing editor was uneasy. He consulted the director, Dalloz, who submitted the proofs to the private secretary of Napoleon III. Pasteur was politely but firmly informed that the *Moniteur universel*, the official organ of the Empire, was surely not the place to publish these pages.

However, Napoleon III was a smart man and knew how to profit from criticism when he felt it was justified. He spoke about his misgivings to Duruy: was it really true that the administrative buildings of some sub-prefectures had been funded before laboratories? Duruy was only too happy to confirm this. Sensitive to Pasteur's arguments, he did what he could to bolster them. Pasteur's article was published in the *Revue des cours scientifiques* of 1 February 1868.

A few weeks later, on 16 March, the birthday of the Crown Prince, the Emperor held a meeting with Rouher, Marshal Vaillant, and Duruy, all members of the government, and an array of famous scientists: Milne-Edwards, Claude Bernard, Sainte-Claire Deville, and of course Pasteur. Napoleon III asked each one to express himself. Rouher stated that science applied to industry was gradually replacing fundamental science. Napoleon III exclaimed: "But what if the sources of these applications have dried up?"[40] Pasteur, reinforcing this argument, pointed out that neither the

Ecole polytechnique nor the Muséum d'histoire naturelle was in the vanguard of research any longer, and that an increasingly prosperous industry
attracted the best graduates of Polytechnique. This was a veritable brain
drain; already there were no candidates willing to work as professors, tu-
tors, or examiners.

Here Pasteur brought up the creation of the post of graduate laboratory assistant. The ministry, he said, must be urged to keep young graduates
in Paris, so that the best students are initiated into research; moreover,
Polytechnique should be made to adopt what he (Pasteur) had already
instituted at the Ecole normale: teaching by, for, and with research. The
example of Germany was convincing, for that country was building laboratories and supporting the recruitment of research scientists. The best minds
were attracted by the quantity and the quality of the equipment and by the
financing. Look at the German scientists, Pasteur urged his listeners, they
live next to their laboratories. Let us follow their example!

Going even further, he felt that the French cities themselves should
have a stake in the work and the reputation of their scientists. "When we
speak of the University of Paris, those of Lyon, Strasbourg, Lille, Montpellier, Bordeaux, and Toulouse — which, taken together, form the French
University System — we should establish between these cities and their institutions of higher education some of the ties that exist between the German universities and the places whose prestige they enhance." Napoleon III
listened closely as Pasteur concluded: "To safeguard the future, the greatest
task that must be accomplished is to ensure France's scientific superiority."[41]

If Pasteur dared speak in these terms in the presence of the Emperor,
it was because he felt sure of the almost unconditional support of his minister. Victor Duruy was surely one of the most remarkable politicians of his
time. Not only did he take hold of the Ministry of Public Education firmly
enough to get it out of the deep ruts that had been dug over the last half-
century by the likes of Fontanes, Guizot, and Falloux, he had also gained
the esteem and the trust of thoughtful people by the careful hearing he gave
them and by his awareness of the importance of research in teaching. Pasteur had understood very quickly that he was dealing with an interlocutor
of a new kind, a man who felt that the imperial regime owed it to itself to
practice a kind of institutional patronage. The scientist therefore entered
into a truly personal relationship with his minister, a relationship he was
not to have with any of his successors, not even Jules Ferry. It should be said
that Duruy was a graduate of the Ecole normale and that this fact probably
did a great deal to strengthen the bond of real friendship between the two

men. "You seem to have altogether forgotten me," Duruy would sometimes wail, "and yet you know how very interested I am in your work. Where are you? And what are you up to? Surely getting close to something."[42]

Victor Duruy was already a respected historian when Napoleon III called on him for a special and rather unusual mission. He was to research the life of Julius Caesar in view of a book to be signed by the Emperor. In other words, Duruy had been tapped as an official ghost writer. In 1863, while he was on an inspection tour of secondary education that took him all over France, an imperial decree appointed him Minister of Public Education. At the time, communications were so poor that the news did not reach him for several days, for the telegram followed him from town to town without catching up with him. But Duruy was to make the point that the sluggishness of the mails was nothing compared to that of the educational system.

One of his first measures was to introduce contemporary history in the lycées. He felt that it made no sense to teach the history of Pyrrhus to students who knew nothing about the reign of Napoleon I. Under his influence, the principle of free primary education made its way, although Duruy was unable to establish that it should also be compulsory. Occupational secondary schools were created and series of free public lectures were offered: sponsored by the local authorities, state-certified professors gave lectures on literary or scientific topics; this was a direct blow against the clergy in the provinces. In late 1867, Duruy went further still, instituting lectures exclusively for young ladies, a step that at the instigation of the famous Dupanloup, the bishop of Orleans who had the ear of Pius IX, caused bishops all over France to rattle their croziers.

In addition to these academic reforms, Duruy's tenure at the Ministry of Public Education was also marked by the attention and the help given to scientists in general, not only to his friend Pasteur. Duruy's door was always open to scientists, particularly if they were Normaliens. It was said that Sainte-Claire Deville sometimes left the rather noisy dining hall in the rue d'Ulm to eat at the minister's table and ask for a subsidy. When he saw him coming, Duruy would meet him head-on: "Well, how much? I'd just as soon know right away."

One of Duruy's decisions was particularly welcome to Pasteur, namely the decree providing for the creation of research laboratories in all the universities. One of the drafts of the text stipulating the applicability of the decree clearly shows that Pasteur had been very carefully listened to. "The research laboratory will be useful not only to the professor, but even more to the students, and will therefore ensure the future process of science. Then we will see students who have extensive theoretical knowledge, who have learned in these research laboratories the basic techniques of instrument use, the elementary manipulations, and what I would call the classic exercises. These students will then form a small group around an eminent master, take their inspiration from him, and under his eyes become proficient in the art of observation and the methods of experimentation. Involved in his studies, they will carefully preserve all of his thoughts, help him pursue his discoveries to their very limits, and perhaps begin to share in them. . . . Institutions of this kind have allowed Germany to accomplish the widespread development of the experimental sciences that we are studying with uneasy sympathy."[43] No question about it, Duruy had paid close attention to the master's lessons.

In the end, this shared conviction of Duruy and Pasteur that science constitutes an essential part of the national patrimony, along with the personal support of Napoleon III, did triumph over the unwieldy administrative apparatus: the ground breaking for Pasteur's laboratory of physiological chemistry took place in the summer of 1868. The Ministry of Public Education obtained an appropriation of 30 million francs and the Emperor's household contributed an equal sum.

These were considerable sums, but Pasteur continued to feel cramped in his laboratory. This was no longer a matter of equipment or lack of space due to the growing number of students; it was the number and the complexity of his experiments that caused a constant shrinking of the space where, intensely absorbed in his manipulations, by turns tormented and elated, the scientist paced like a caged animal. These basic and truly unlimited needs made Pasteur's temper increasingly difficult, but he never showed any ingratitude toward Duruy. From the controversies over spontaneous generation to the final victory over rabies, the laboratory of the rue d'Ulm was to be the birthplace of the most prodigious discoveries. This was Pasteur's way of thanking a minister who, not satisfied with giving him the means for his research, also did what he could to find symbolic rewards for him. Duruy thus personally saw to Pasteur's advancement in the Légion d'honneur and even persuaded the government to award him a special Grand Prix for his work on vinous fermentation at the World's Fair of 1867. The official ceremony took place on 1 July, two days before the dismissal of Lallier triggered the "affaire Sainte-Beuve."

It is a splendid summer day. From the Tuileries to the Place de l'Etoile, masses of people are filling the streets. Surging crowds try to catch a glimpse of the carriages rolling down the Champs Elysées, for the ceremony will be held at the Palais de l'Industrie in the presence of Napoleon III. The Empress, wearing a sparkling tiara, is at his side, accompanied by the Prince Imperial, a child of twelve in gala uniform. A whole string of foreign princes further enhances the splendor of the occasion: here are the Sultan Abdul-Aziz, the Prince of Wales, the Crown Prince of Prussia, the Grand Duchess of Russia, the Duke of Aosta.

Pasteur shares the honor of being a prize-winner with sixty-three other leading lights of the Empire, among them the painters Gérome and Meissonnier. Also here is Ferdinand de Lesseps, who receives a tremendous ovation for his project of a canal connecting the Red Sea and the Mediterranean. And now it is Pasteur's turn to receive his prize. He must walk through the crowd, mount the steps of the podium, bow before the throne. There is something chilling about the slow and solemn walk of the scientist. A secretary of Duruy's who watches him even finds that he has the look of an undefinable sadness, like a vague resignation. No one is particularly curious, let alone enthusiastic, about this little man whose studies seem rather complicated. He has not yet become a legend.

7

THE SO-CALLED
SPONTANEOUS GENERATIONS

Throughout Europe, the 1860s were marked by an extraordinary effervescence and upheavals that transformed the world. Krupp built his first canons, Maxwell discovered the electromagnetic waves, Mendel described the laws of heredity, Broca located the language center in the brain, and Charcot studied multiple sclerosis. Meanwhile, Hugo published *Les Misérables*, Dostoyevsky, *Crime and Punishment*, Tolstoy, *War and Peace*, Karl Marx disseminated *Das Kapital*, and Richard Wagner composed *Tristan und Isolde*.

During this time, Pasteur hardly ever left his laboratory, where he stored up experiments and observations. The research he carried out in solitude was as revolutionary as the most spectacular discoveries or the most innovative works of art. His work on ferments led him to the study of spontaneous generation in a logical yet bold progression, for he was aware that he was taking on a problem whose scientific and social implications would be incalculable. "Among the questions raised by my research on the ferments in the narrow sense, none are more worthy of attention than those relating to the origin of the ferments. Where do they come from, these mysterious agents, so feeble in appearance yet so powerful in reality, which, with minimal weight and insignificant external chemical characteristics, possess exceptional energy? This is the problem that has led me to study the so-called spontaneous generations."[1]

It was not the first time that Pasteur had attempted to explore the origin of life. Both his work on molecular asymmetry and his discoveries concerning the microorganisms responsible for fermentation had brought him face to face with the need to define the very nature of life and, hence, the conditions of its appearance. The Strasbourg experiments exploring the possibility of producing a synthetic asymmetry also prove that Pasteur did not dodge one of the most fascinating questions of science: is it possible to create life out of nothing? Inevitably compelled to enlarge the scope of his questions, he who usually worked by himself in highly specialized fields was thrust into a general debate that predated his own work. As he became involved, he had to confront scholars and philosophers, both his contemporaries and earlier thinkers, whose long-established theoretical presuppositions were largely based on poorly constructed, though frequently inventive experiments.

The theory of spontaneous generation, which constituted a traditional alternative to the concept of procreation, had been so thoroughly described, represented, and praised that large numbers of supporters had a stake in its preservation. Many of them had gone from hypothesis to theory with little or no rigorous proof. The polemic was aggravated by the fact that spontaneous generation had partisans among serious scientists, while others neither dared nor wished to experiment in this domain. Concerning these arguments, which were often presented with considerable oratorical skill, Pasteur roundly declared: "One must not assume that an understanding of science is present in those who borrow its language."

However that may be, the question was indeed in the air; the entire century was working on the notion of the species. If Darwin went off to the Galápagos, Pasteur was riveted to his microscope. From the rivers of Peru to the microorganisms grown in the swan-necked flasks of the rue d'Ulm, scientists tried to identify life. Darwin theorized on the birth of a new world and on the evolution of the species; Pasteur did the same for the origin of germs and their ecology. This comparison surely is not unwarranted, for Pasteur's preoccupations concerned the interaction between the germ and organic matter, that is to say, the question of whether one can evolve from the other, and thus touched on the origin of life, perhaps its very purpose.

A Letter to Pouchet

"Having reached this point in my studies on fermentation," Pasteur wrote in the early 1860s, "I had to form an opinion on the question of spontaneous generation. Perhaps it would furnish me with a powerful weapon in favor of my ideas on fermentation in the narrow sense. The research on which I am now reporting has therefore been a mere digression, although a necessary one, from my work on fermentation. This is how I came to work on a subject which has hitherto only engaged the sagacity of the naturalists."[2]

Several of Pasteur's experiments on fermentation, in particular the effect of heating and of oxygen on the production of alcoholic and lactic yeast, were so many indications speaking against the spontaneous generation of microbes. But Pasteur knew that he needed to do experiments designed specifically for this purpose. He wrote to Chappuis that he was hoping to "take a decisive step by resolving the famous question beyond the shadow of a doubt."[3]

At that point, however, he was not yet deeply involved in this debate. One can date his entry into the arena by a letter he sent to the Norman scientist Pouchet in early 1859 in response to a question he had been asked early in his research on fermentation. At the time, Félix-Archimède Pouchet, a naturalist and physician, was the director of the Muséum d'histoire naturelle of Rouen. Born in that city in 1800, a Protestant and the son of a manufacturer, he had studied at the Rouen hospital under the guidance of Dr. Achille-Cléophas Flaubert, Gustave Flaubert's father. A brilliant professor of natural history at the Faculty of Medicine, Pouchet had acquired a certain reputation in scientific circles by his diverse studies in animal and plant biology on such topics as the digestive tract of insects, ovulation in mammals, the characteristics of plants of the Solanaceae family, or the respiration of turtles. He had published numerous well-researched books on Aristotle's natural philosophy and the advances of experimental science in the Middle Ages. His status as a provincial notable and the diversity of the studies he had undertaken had earned him the honor of being appointed a corresponding member of the Institut, a position that allowed him to have his papers read to the Academy of Sciences, particularly those on spontaneous generation, to which he was passionately committed.

Pouchet therefore took offense when Pasteur, in his *Mémoire pour servir à l'histoire de l'acide lactique*, stated that leavening has its origin in the air of the atmosphere, for if one cuts off all contact with air and brings the reagents to a boil, fermentation does not occur. Pressing his fellow scientist to explain himself, Pouchet repeated what he had said in his papers communicated to the Institut.

Pasteur first replied to Pouchet in a letter dated 28 February 1859, whose extremely polite tone does not hide a certain disdainful haughtiness: "I therefore think, Monsieur, that you are mistaken, not so much to believe in spontaneous generation, for in such a question it is difficult to avoid preconceived ideas, but to assert that spontaneous generation exists. In the experimental sciences it is always wrong not to have doubts as long as the effects produced do not make an assertion necessary; however, I hasten to add that if your opponents, on the basis of the experiments I have just indicated, claim that the air contains the germs of the organized productions found in the infusions, they are going beyond the findings of the experiments and should simply say that there is in ordinary air something that is a precondition for life, which is to say that they should use a vague word that does not prejudge the most delicate aspect of the question."[4]

The matter, then, is clear: Pasteur was opposed to spontaneous generation because the experiments he had conducted—and which he urged Pouchet to repeat—did not necessitate recourse to that concept. But as yet he did not have the means to prove his intuition and to offer a rigorous characterization of the role played by air (or the atmosphere) in the phenomenon he observed, namely the appearance of fermentation in his laboratory flasks. Constrained by the logic of these first findings, guided by his questions about the nature and the origin of life, and spurred on, as he often was, by the wish to rectify erroneous opinions—or perhaps the desire not to let a theory he considered erroneous develop any further—Pasteur now set out to deepen his understanding of this matter by an extensive series of experiments. When Biot, Dumas, and Sénarmont learned that their favorite disciple was planning to embark on this new study, they tried to dissuade him. They were afraid that the young scientist would abandon the true scientific problem, which was fermentation, in order to pursue a chimerical question. Biot fretted that he was about to make the same mistake he had made at Strasbourg seven years earlier.

The old academicians' distaste for such a fundamental question must no doubt be laid to their age and their experience. In the minds of all these scientists born into the science of the first years of the nineteenth century, the debate over spontaneous generation was inevitably tied to the famous quarrel that had pitted Cuvier and Geoffroy Saint-Hilaire against each other in 1830, a quarrel that Goethe had considered vastly more important than the fall of Charles X! Those who espoused transformism, Lamarck and Geoffroy Saint-Hilaire, had to conceive of the possibility of the passage from one species to another, a concept that requires a theory that would explain evolution leading from primitive organisms to advanced species. Cuvier, on the other hand, a defender of "fixism," thought that the divisions defined by the naturalists must be upheld, since they were confirmed by the discontinuities found in the fossil record. For him, unity of composition in the creation did not exist. We now know that Cuvier's vision prevailed, in large part because of the philosophical implications of his position, since transformism led to a materialism that was at the time almost unanimously rejected by the scientific community. This accounts for the cautious attitude which the question of spontaneous generation encountered at the Institut in 1860, at a time when Darwin reopened the debate on evolution in England. The problem was further compounded by the fact that the experiences on spontaneous generation conducted over the centuries proved to be so lacking in rigor that they lost credibility. Was it because the subject was so closely identified with obscurantism that Pasteur's best advisers counseled him against becoming involved in it?

Yet it is clear that, at this stage of his career, Pasteur was of sufficient stature to take it up. He combined theoretical rigor with experimental precision, and his assertion that nothing less than "the clarity of a mathematical reasoning [was needed] to convince the adversaries of one's conclusions" showed that he knew where his strength lay.

Facing one of humanity's oldest problems, he entered the battle armed with two technical weapons: calcined air and the swan-necked retort. When it was over he had demonstrated that life does not spring from nothing and discovered that "the study of acts occurring under the influence of plant or animal life, even in their most complicated manifestations, is in the last analysis identical with the discovery of phenomena connected with the cell."[5]

Mice, Maggots, and Other Animalcules

Theories about spontaneous generation had not sprung up yesterday and had not traveled through the centuries without many a debate — given the absence of valid experiments. When Pasteur broached the problem, he began by taking a critical look at the many definitions and controversies that had gone before.

Ever since antiquity, naturalists, whether botanists or philosophers, and such thinkers as Aristotle, Theophrastus, Diodorus of Sicily, or Pliny, had stated as an undisputed truth that certain creatures are capable of not coming from animals of the same species. Worms, bugs, or slugs exist without having been procreated and are spontaneously generated by the soil or the sun. Aristotle, for example, had asserted that every dry body becomes humid and that every humid body that dries engenders animals.

By the seventeenth century little progress had been made on the road to truth and reason. Thus the Flemish scholar Jan Baptista van Helmont said that frogs came out of the miasmas of marshes and even described an experiment he claimed to have carried out: "If one presses a dirty shirt into the opening of a vessel containing grains of wheat, the ferment from the dirty shirt does not modify the smell of the grain but gives rise to the transmutation of the wheat into mice after about twenty-one days."[6] Helmont even specified that these were adult mice. Elsewhere he proposed a strange recipe for creating scorpions, which was done by carving a hole into a brick, filling it with dried basil, and placing it in the full sun.

The theory reached a first turning point toward the end of the century, when a famous German scientist, the Jesuit Athanasius Kircher, felt confident enough to assert that these experiments could only rarely be reproduced and that in any case they required a fertilizing power coming from a living organism. Of a whole organized body, one part would be able to continue living after death and to conserve an independent existence. Under this assumption, the new recipe for obtaining living scorpions would henceforth call for a corpse, although it would still have to be steeped in the same basil leaves exposed to the sun!

Notwithstanding the dogmatic difficulties this position raised at the time with respect to God's role in the creation, and however outdated it may have appeared to Pasteur's contemporaries, it is intriguing to note that in our own day certain biological and therapeutic acts — organ transplants or the multiplication *ex vivo* of blood cells — represent a contemporary extension of the frequently baroque intuitions of the first naturalists of the early modern period; we now know that cells from a living being can be cloned and made to reproduce outside the organism that has given birth to them. What does separate our modern vision from that of the adherents to spontaneous generation is that they always assigned a function to decaying bodies and their putrefaction, whereas we work directly with living matter.

To these observations on the decomposition of flesh we also owe the first experiments that made a dent in the theory. In the second half of the eighteenth century the Italian Francesco Redi showed that flies are not born from fermenting meat but from the eggs deposited by other flies. A piece of gauze was enough to prevent the birth of fly larvae. This suggested the idea that the worms found in apples do not come from the rotting of the fruit but that they facilitate it and in fact owe their generation to insects that deposit their larvae.

In insisting on describing these observations — many of which were picturesque and even ridiculous, as he himself pointed out — before turning to the presentation of his own work, Pasteur adhered to one of the principles of sound scientific procedure: he investigated the history of the subject. However, he visibly enjoyed dwelling on the experimental aberrations that had occurred earlier, especially when they discredited the idea of spontaneous generation in the reader's mind.

At the same time, he did not gratuitously lay them out in every detail, for they permitted him to seize, among other things, the new problems that appeared when the microscope burst upon the scene in the seventeenth century. For in addition to identifying the complexity of the organisms of the smallest insect, the microscope also opened the door to an unknown world, that of the lower one-celled organisms observed in stagnant waters, the so-called animalcules.

Unable to give a precise accounting of the conditions under which they appear, scientific opinion had assumed until the middle of the eighteenth century that animalcules are not born spontaneously but that they are engendered by microorganisms from the atmosphere. However, some scientists who believed that they were bringing greater rigor to their demonstration soon began to doubt that microorganisms could be transmitted by the air and for that reason proposed the hypothesis that these rudimentary organisms come into being by spontaneous generation. Almost a hundred years before Pasteur, as he himself pointed out, all it took to launch the debate between the partisans and the opponents of bacterial reproduction was a closed vessel, some heat, and some atmospheric air. In the course of the eighteenth century, the two sides of the controversy, which was fairly confused even then, were led by two churchmen, the Englishman Needham and the Italian Spallanzani.

A skillful observer and a Catholic priest (a circumstance that, Pasteur remarked, amounted to "a guarantee for the sincerity of his views"), John Needham had published in 1745 a book in which he described the following experiment, which he considered fundamental for the thesis of spontaneous generation: a closed vessel, or one he assumed to be closed, hermetically sealed and filled with previously heated meat juice, was buried for several minutes in hot ashes, whereupon, a few days later, it presented profusion of microbial growth.

According to Needham, the animalcules could only come from the infusion, in which they were generated spontaneously. Observers were aware that the extremely simple organization of these microorganisms excluded any possibility of sexual generation. Furthermore, they were so plentiful, had such strange shapes, and so closely resembled those that could be seen in the course of putrefaction that the observers, Pasteur said, came to formulate a specious theory, that of the organic molecules. This idea was further developed by Buffon, who, in the "brilliant, vivid, and highly authoritative style" to which Pasteur paid tribute, made Needham's conclusions his own: "The substance of living beings preserves a remnant of vitality after death. Life resides essentially in the last molecules of bodies. These molecules are arranged as if in a mold. There are as many molds as there are beings, and when death puts an end to the organization of a body, that is, to the power of this mold, the decomposition of the body ensues, and since the organic molecules, all of which survive, now float freely in the dissolution and the putrefaction of that body, they will pass into other bodies as soon as they are sucked in by the power of a different mold."[7] The animal thus disappears as an individual of the species, but its molecules reappear in the form of an inferior organism.

When Needham came to Paris, he collaborated with Buffon in the dissemination of their theories and experiments. The discussion continued on a speculative level for twenty years, until Abbé Lazzaro Spallanzani became involved.

Spallanzani approached the subject without preconceived ideas and began by repeating Needham's experiments, which he in fact initially confirmed. Postulating, however, that the heating time might not be long enough to have an effect, he brought some modifications to the model, namely a larger receptacle, prolonged heating (45 minutes), and sealing the plug after stretching the neck of the receptacle. Under these conditions, microbes no longer appeared. In a series of further experiments, Spallanzani showed that the number of animalcules was directly related to contamination by air.

The matter caused a great stir. Voltaire took up the cudgels for Spallanzani and against Needham and Buffon, declaring that "it has been demonstrated to the eyes of reason that there is no plant or animal without a germ." Needham, a good loser and a clever operator, arranged to have his rival's experiments translated into French and published, but the criticisms he added seemed so pertinent that Spallanzani felt unsure of himself and repeated a number of his experiments.

In the end, there was no winner in this controversy. Needham declared that Spallanzani had "hermetically sealed nineteen vessels filled with different plant substances and boiled them, closed in this manner, for one hour. But, given the manner in which he treated and tortured these nineteen plant infusions, it is obvious that he has not only greatly weakened or perhaps totally destroyed the vegetative power of the infused substances, but also that the exhalations and the heat of the fire have entirely corrupted the small amount of air that remained in the empty part of these vials. It is therefore not surprising that these infusions, having been treated in this manner, did not give any sign of life; it could not be otherwise."[8]

On a strictly scientific level, one realizes that the problem was essentially how to interpret the effect of air and heat in this context. This question was broached a few years later by the industrialist Nicolas Appert, a *confiseur* who made his fortune when he discovered and commercialized the principle of food canning. Without entering into theoretical discussions, Appert applied Spallanzani's techniques to his culinary preparations, in particular to peas, which he was thereby able to keep from becoming moldy. In so doing, Appert brought experimental support to Spallanzani's theses.

But Needham's reservations were justified anew by the discoveries of Gay-Lussac. In 1810, Gay-Lussac, in an effort to understand the new conservation process, analyzed the air of Appert's preserves and found that it contained no oxygen. From this he deduced that the absence of this gas must be a condition for the conservation of animal and plant substances. These experiments were in turn contradicted by the German Theodor Schwann, whose research Pasteur had already used in his work on fermentation. In 1837, Schwann had invented a way to bring heated and then cooled air into sterilized meat juice. In this case, oxygen was indeed present, yet no rapid microbial multiplication occurred. Nor were these the only discrepancies, not to say differences. No sooner had it been verified for putrefaction, than the same technique led to the opposite result in alcoholic fermentation: in sugar, the liquid did ferment. Undaunted, Schwann made the following deduction: "In alcoholic fermentation as in putrefaction, it is not the oxygen, at least not the oxygen from the atmosphere alone, that causes them, but rather a principle contained in ordinary air that can be destroyed by heat."[9]

The simple fact that heat has a deleterious effect should have eliminated or at least greatly weakened as complex a hypothesis as the appearance of life *ex nihilo*: if heating is sufficient to impede generation, it cannot be spontaneous. Pasteur therefore criticized all these experiments which canceled each other out without arriving at a convincing conclusion. However, he knew how to recognize and use the often judicious experimental procedures deployed in the process, whether it be purification by sulphuric acid or the filtering of air through cotton. This last procedure is used to avoid bacterial contamination in laboratories to this day, which goes to show that practices are often more durable than theories.

Heterogenesis According to Pouchet

It is evident that by the time Pasteur entered the debate, the matter had gone far beyond Van Helmont's mice or scorpions, and if we are to understand what was at stake in the debate, we must take care not to reduce the hypotheses of the supporters of spontaneous generation *a priori* to a handful of chimerical notions inherited from the past. At this point the theory no longer concerned anything but a small group of inferior creatures, for which science had not discovered the eggs that would account for their appearance. Actually, Pouchet avoided the pitfalls of ordinary language; instead of organizing his thinking around the traditional but highly charged expression "spontaneous generation," he preferred to develop the concept of "heterogenesis." In late 1859 he published with Baillière et fils, publisher to the Academy of Medicine, a thick volume entitled *Hétérogénie ou Traité de la génération spontanée*. In its preface, one could read: "When I had thought about the matter long enough to become convinced that spontaneous generation is still one of the means that nature uses to reproduce its creatures, I set out to discover which procedures would allow us to bring these phenomena to light."[10]

This preliminary statement amounted to a metaphysical postulate: if the phenomenon exists, it is because God has wanted to use it. It also elaborated a definition: spontaneous generation is the production of a new organized being, all of whose primordial elements have been taken from living matter; it was a conceptual hypothesis: animals are being produced out of a "plastic manifestation" that tends to group molecules together and to impose on them a specific mode of vitality that eventually results in a new being; and it is also a credo: organisms are not extracted from raw matter in the narrow sense but from the organic particles of earlier generations of animals and plants. On this basis, Pouchet could arrive at a limpid formulation of his theory: "Spontaneous generation is the production of a new organized being that lacks parents and all of whose primordial elements have been drawn from ambient matter."[11]

According to this doctrine, mineral molecules are not being organized; rather, organic particles are called to new life, reanimated by a "molding force" [*force plastique*], which, according to Pouchet, is the expression of vitalism. He thus contested the existence of microorganisms, and as far as the effect of heat was concerned, he felt that one did not have to believe the interpretations that were being proposed: simply stating that heat blocked fermentation, Pouchet did not modify the data of the problem. To be sure, the passing of time, new discoveries — among others those of Pasteur — are bound to make us speak derisively of such imaginative concepts, but one has to admit that they had a certain poetry: nature "exhaled" rather than "breathed," and successive generations arose as a result of the innate wisdom of matter, as a manifestation of heterogenesis safeguarded by divine providence. Nor should we overlook the fact that Pouchet could seek shelter behind a brilliant company, ranging from Aristotle to Lamarck by way of Lavoisier and of course Needham and Buffon.

Alongside, or rather in advance of his *Traité*, Pouchet published an experiment that gave Pasteur a chance to show his mettle as an acerbic observer and an ironic critic: on 20 December 1858 Pouchet communicated to the Academy of Sciences a "Note on animal and plant protoorganisms born spontaneously in artificial air and oxygen gas." In this note Pouchet, who had repeated Schwann's experiments, claimed that in following the same procedures, and even in varying them and giving them a higher degree of precision, he had consistently obtained positive results, that is, a proliferation of microbes. By substituting artificial for atmospheric air, he had furthermore demonstrated that airborne contamination had nothing to do with the phenomenon.

Here is what Pouchet had done: He filled a flask with boiling water, closed it hermetically, and plunged it into a mercury trough. Once it had cooled, he added oxygen and a small bundle of calcined hay. By this procedure he obtained a maceration apt to ferment, and in fact fungi consistently appeared—and as a result Pouchet felt justified in stating that atmospheric air is not and cannot be responsible for this proliferation. He concluded that this was a case of spontaneous generation.

In this struggle, Pouchet was not alone. Two other scientists, Nicolas Joly and Charles Musset, supported him. Joly was professor of physiology at the Faculty of Sciences of Toulouse; Musset, his student, was working on a thesis on spontaneous generation. This work was respectfully dedicated to Archimède Pouchet: "Coming after you, we can only glean; in truth, you have said everything." "Said everything" was of course an exaggeration, even though Pouchet was without contest the head of this little team and its principal driving force, at least in his writings.

At the Academy of Sciences, by contrast, there was a great deal of reticence. Henri Milne-Edwards, Jean-Baptiste Dumas, and Claude Bernard in particular sought to criticize certain points and objected to the assertions made by Pouchet, Joly, and Musset. However, no one had the evidence to show that their experiments were flawed. In view of the renewed interest in this question aroused by Pouchet's *Traité* and Pasteur's studies on fermentation, the Academy of Sciences therefore proposed the following topic for the 1862 Alhumbert prize in natural science: "To attempt by means of well-designed experiments to cast new light on the question of the so-called spontaneous generations."

This prize, endowed with a 2,500 franc stipend, was to be awarded following the report of a committee composed of Milne-Edwards, Flourens, Brongniart, Serres, and Geoffroy Saint-Hilaire *fils* — a committee whose majority was opposed to Pouchet's theory. Such a prize was clearly a helping hand stretched out to Pasteur. And, of course, Pasteur did not hesitate to seize it, pleased that he had succeeded in convincing the academicians that it was both possible and necessary to reopen the traditional debate on spontaneous generation without having to resume the ideological quarrels of the early years of the century.

So Pasteur set out to respond to Van Helmont, Buffon, and above all to Pouchet and his key experiment: "What objection would you raise against M. Pouchet? Would you tell him: 'The oxygen you used may have contained germs'? 'It did not,' he would reply, 'for I took it out of a chemical combination.' And so he did, and consequently it could not contain any germs. Perhaps you would tell him, 'the water you used contained germs.' But he would answer, 'this water, which had been exposed to contact with the air, might have received some germs, but I made sure to place it into the vessel when it was boiling, and at that temperature, if germs had been present, they would have lost their fecundity.' Then you would say: 'it was the hay.' It was not, for the hay came out of an oven heated to 100 degrees. But this objection was raised, for there are some unusual organisms which do not perish when heated to 100 degrees; but he responded, 'I can easily rectify that!' and he heated the hay to 200 and 300 degrees. He even said, I believe, that he went so far as to carbonize it. Well, I admit it, the experiment conducted in this manner is irreproachable, but only with respect to the points that engaged the attention of the author. I shall demonstrate that there was one source of error that M. Pouchet did not notice, that never occurred to him, that had never occurred to anyone before him, and that this source of error makes his experiment completely useless, and as bad as that of Van Helmont's pot of dirty linen. I shall show you where the mice came in. I shall demonstrate that in every experiment of the kind that concerns us here, one must absolutely rule out the use of the mercury trough."[12]

Here Pasteur launched into a demonstration of the role of the different kinds of dust, recalling that one cannot always see them "for the same reason that by day we do not see the stars of the celestial vault." When they are lit, these small particles of dust that float in suspension in the atmosphere become as visible as the stars of the night. Pasteur as poet? Perhaps, for not content to project a utilitarian beam of light on the visible world, he was an astronomer of the microscope who observed the starry vault of the infinitely small.

Experimental Strategies

However, it was not going to be that simple. In order to show Pouchet that he was theoretically rather than factually mistaken and to assert that spontaneous generation definitively does not exist, Pasteur had to begin by developing a new experimental strategy, which was to remain exemplary for its precision, its chronology, and its inventiveness. These are the experiments Pasteur described in the notes of 7 May, 3 September, and 5 November 1861 and then published in July 1863 in the form of a *Mémoire sur les corpuscules organisés qui existent dans l'atmosphère*. They contain the principal proof of his scientific objections.

In the first phase of this series, Pasteur demonstrated that microorganisms are in the air. To do this, he invented an apparatus for bringing in outside air that would prove remarkably useful. It was a kind of vacuum tube with a hydraulic mechanism that pumped in air from the rue d'Ulm or the garden of the Ecole normale. This air was then filtered through a wad of gun-cotton. If one wanted to examine the dust caught in the cotton, one could wash it in a watch glass; the particles were then examined under the microscope and sketched. In order to eliminate one possible criticism — cotton is an organic matter that might have produced the germs — Pasteur soon replaced it with a wad of asbestos and obtained the same result.

Germs, then, were in the atmosphere, but were they capable of reproducing themselves? In order to answer this question, Pasteur endeavored to reproduce Pouchet's experiment with the mercury trough and, like his predecessor, obtained a proliferation of microbes. But he parted ways with Pouchet when he affirmed that one cannot conclude from this observation that spontaneous generation has taken place but that one must try to find out how the germs have entered. A more skillful and above all more precise experimenter, Pasteur discovered that the source of contamination was in the mercury. He was thus able to show that dust particles from the atmosphere proliferate in liquids that serve as their nutrient milieu, regardless of whether they consist of mercury, sugar water, albumin, urine, or milk.

In the incubator, common and diverse organized productions proliferated, and some of them had the prettiest names: mucoraceae, torulaceae, molds, myceliums, vibrios, penicilliums, species of the genus *Ascophora*. And everything had to be sketched, cultivated, and compared. "All I have done," said Pasteur, "was to provide solid proofs to the adversaries of heterogenesis. Heat destroys airborne germs which can proliferate abundantly if given the opportunity. In mercury or other compounds which receive microbial spores, all it takes to propagate this possible source of contamination are the nonsterile hands of a careless experimenter."[13]

But Pasteur was not yet satisfied with his demonstration. It was not enough to say that germs deposit themselves and proliferate in preparations; it also had to be demonstrated that the fewer germs deposit themselves, the more the liquids will remain sterile, untouched by any microbial contamination.

Pasteur had already shown considerable inventiveness when he installed the vacuum pump to capture the air of the neighborhood, but at Balard's prompting he now adopted another experimental model: vessels that communicated with the outside air by a neck that could be given different forms and lengths. These vessels were all large and their necks were long, thin, inclined, bent back, rounded, or swanlike. Serving to keep the germs away from the receptacles, these long necks were perfectly suited to the conclusions Pasteur wanted to formulate. If one places a fermentable liquid into these vessels and then boils the liquid, and if one then deposits the vessel in a place where the air is still, the liquid will remain clear for months.

Since the receptacle does communicate with the atmosphere, one is forced to conclude that the sharp bends in its neck have prevented the germs from entering into the liquid and reseeding the previously sterilized milieu. The swan-neck keeps out the dust particles; the germs lose their way in the twists and turns of the glassware and tumble down its slopes.

It was an illuminating experiment. The observer had a choice between only two hypotheses: placing the origin of germs either in solid particles (fragments of wool or cotton, starches) that float in the atmosphere, or in the spores of molds or the eggs of infusoria. Pasteur said: "I prefer to think that life comes from life rather than from dust."

These experiments were exactly what the Academy of Sciences had had in mind. In 1862 they earned Pasteur the Alhumbert Prize attached to the question of spontaneous generation. In fact, Pasteur was the only contestant to present his work to the Academy, for Pouchet had withdrawn, feeling (not without reason) that the jury had made up its mind in advance. As a result, Pasteur pocketed the 2,500 francs, but he had yet to convince his adversaries in the heterogenicist camp.

Throughout the years 1860–61, the latter doggedly persisted in presenting ever more elaborate, sometimes ingenious but usually puerile experiments. Thus Musset came out with the idea that in order to experiment in rigorous conditions, one must gather air inside a pumpkin or in the air bladder of a fish. He also noted that this air always made calcined hay germinate and brought out spores, sometimes of unknown species, one of which was pompously given the name *Aspergillus pouchetti*. But the fact is that mercury was always present to vitiate the experiment.

The main problem was that Pouchet did not believe in the presence of germs in the air, at least not to the extent envisaged by Pasteur. He examined dust from the deepest past, which he gathered in the most diverse regions of the earth: in his own laboratory at the Muséum of Rouen of course, but also at the abbey of Fécamp (where the dust was five or six centuries old) and even at the Egyptian sites of the Karnak temple at Thebes or the pyramid of Giza. All he found there was debris of grains and minerals, fragments of silica, insect skeletons, butterfly scales, plant down, specks of pollen; and this led him to conclude that germs were very rare and represented an exception which, according to him, was not sufficient to account for the constant multiplication of microbes in his vessels.

He went even further. If, as Pasteur believed, air was the vehicle for germs, what was one to think of the action of hurricanes, which surely would churn up astronomical quantities of germs? He therefore subjected his infusions to a huge draft, which he produced by inventing the aeroscope, a set of ventilators turning at five hundred revolutions per minute and capable of filling the entire nave of Notre Dame of Paris. Pouchet observed nothing that he had not found in vessels left in a quiet place. Moreover, his findings did not change when he sprinkled various kinds of dust, fragments of cork, or sand on his infusions. Pasteur's theories, he said, were thus disproved by experiment.

In fact, it would be difficult to find a more tenacious and less rigorous experimenter. Nor was he the only one; Musset, for his part, was working in Toulouse, where he sought to deny the very existence of germs. Under his microscope he was unable to see any in nasal mucus, in bronchial phlegm, or in saliva, even if it was taken early in the morning! Nor was he able to find germs in snow, whether it came from Rouen or from Toulouse.

It should be said, however, that these deficiencies had more to do with the limits of the experimental equipment available at the time than with bad faith on the part of the experimenter. The supporters of heterogenesis were not just anybody, and outside the Academy included the names of respectable scholars, such as the Italian Mantegazza, who had been publishing notes on spontaneous generation since 1852 and claimed that he had observed the birth of a microorganism by keeping his eye glued to the microscope for sixteen hours. Lauras in England and Wigman in America were also supporters of heterogenesis.

Actually, Pasteur's experiments could also be criticized, not only because one could always carry out other and contradictory ones that would argue in favor of heterogenesis, but even in themselves, for one could draw from them a theoretical conclusion that would be difficult to defend. Pouchet was well aware of this, and Pasteur found himself embarrassed by what he admitted to be a sensible and judicious criticism: "If the proto-organisms that we see proliferating everywhere and in everything," Pouchet wrote, "had their germs hidden in the atmosphere in the proportion mathematically required to this effect, they would totally darken the air, for they would have to be much more tightly packed than the globules of water that form our thick clouds. Every cubic millimeter of air would have to contain infinitely more eggs than there are inhabitants on this earth. . . . In that case, the air in which we live would almost have the density of iron."[14]

In the face of this argument, Pasteur had to abandon the thesis of "integral panspermatism," which saw germs everywhere. He therefore developed the notion of "semispermatism," which supposed that the generative capacity of the atmosphere is neither constant nor immutable and that it must vary with the place and the season. This meant that the debate over spontaneous generation gradually turned into a debate on the analysis of the atmosphere.

Pasteur himself had to turn from an experimenter glued to his little laboratory in the rue d'Ulm into an explorer and go off in search of germs, just as he had once gone to track down crystals all over Europe. From the low ground of Paris to high mountaintops, he always dragged behind him the contradictory experiments of his determined adversaries.

"What will be the outcome of this battle of the giants?" a journalist of the *Moniteur scientifique* grandiloquently asked himself in April of 1860.

Scientists on Mountaintops

The proliferation of experiments on both sides had a simple purpose; Pasteur wanted to show that if a fermentable liquid is exposed to the purest possible air, no generation will take place — a finding that would not only confirm his theory of ambient germs but also combat the principle of spontaneous generation. Conversely, the supporters of heterogenesis sought to establish that proliferating growth is found in every case, regardless of the quality of the air used and the precautions taken — a finding that allowed them to uphold their convictions. In order to test the quality of air and its degree of microbial fecundity, Pasteur began to use small round flasks filled with an alterable liquid. He then boiled these flasks, in which the evaporation of water created a vacuum. During this operation the neck of the flask was pulled out into a slender point and sealed by means of an enameler's lamp. This meant that the flasks would remain sterile and could be opened anywhere. And indeed, when the neck was broken open, the air rushed in with a whistling sound, carrying germs and dust particles that became suspended in the fermentable liquid. Thereupon the flasks were immediately resealed as cleanly as possible by means of a torch and transported to an incubator kept at a temperature favorable to the development of microbes.

In the course of these experiments with collecting air, Pasteur soon noticed that the contents of certain flasks underwent alteration, while in others they remained intact. This led him to conduct a series of experiments essentially for the purpose of showing that these flasks do not become infected just anywhere and for no reason. No need for statistics; just the facts. Early in 1860, air was collected at the Paris Observatory. Of the ten flasks opened in its vaults, only one underwent alteration, while the twelve that were filled in the courtyard all showed a proliferation of microbes. But Pasteur did even better. As the summer vacation approached, he decided to take his experiment to Arbois. Sorry to see that his old friend Chappuis could not accompany him on his hike in the Jura, he went off by himself.

Early one morning, Pasteur left Arbois with seventy-three sterile flasks he had brought from his laboratory in the rue d'Ulm. On his way he met some vintners, baskets strapped to their backs, who were intrigued to see him. In his black suit and tie, a pince-nez on his nose, and carrying his retorts in a willow basket, Pasteur was indeed a strange sight. Here he was, with a whole assortment of tubes, flasks, workbooks, and notes, by turns crouching down and standing up, opening one after the other of twenty of his flasks. He broke off the ends of the necks, subsequently sterilizing them with burnt alcohol. He collected air at different points of his walk, not far from his father's tannery, on the road to Dole, and on a path leading to the Mont de la Bergère. The result: eight of the flasks showed a proliferation of microbes.

Pasteur then proceeded to Salins, to the source of the brackish waters that feed the brine pit of Arc-et-Senans. Then he climbed Mont Poupet. Of the twenty flasks opened at 580 meters altitude, five showed alteration.

But even that was not enough. Since he could not go up in a hot-air balloon, Pasteur set out for the Alps. Arriving at Chamonix on 20 September 1860, he looked for a guide who would take him to the summit of the

Montanvert glacier near the Mer de Glace. A mule carried a basket containing thirty-three flasks. Pasteur, anxious to preserve the precious cargo, walked along the precipice holding on to the crate to keep it from falling. He reached the first firns, but nature, or rather the ice, thwarted the experimenter: "For the purpose of resealing the tips of the flasks after having collected the air sample, I had brought with me a blow torch fueled by alcohol. The sunlit icefield was so bright that it was impossible for me to see the jet of burning alcohol vapor, and since this flaming jet was somewhat agitated by the wind, it never stayed on the broken glass long enough to melt the point and hermetically reseal the flask. Whatever means I might have had at my disposal to make the flame visible and then controllable would have inevitably created sources of error by spreading extraneous dust particles through the air. I was therefore obliged to carry the flasks I had opened on the glacier back to the little inn of Montanvert without resealing them."[15]

Thirteen other flasks were opened that same evening at the inn, and all of them became infected; of the seven others tested the next day on the Mer de Glace only one underwent alteration.

On 5 November 1860, Pasteur reported on his excursion before the Academy. He had the honor, he said, to place on the desk of the Institut sixty-three flasks, adding this bold and futuristic thought: "However, I have not finished with all of these studies. The most desirable course would be to carry them far enough to prepare the way for a full-scale study of the origin of various diseases."[16]

But for the time being, the question to be settled was still that of spontaneous generation. For Pasteur's opponents immediately responded to his experiments. Since he had gone to the Observatory and to the Mer de Glace, Pouchet and his friends also departed. Italy and Sicily were first, and air was collected on the very slope of Mount Etna. Then they went to the Pyrenees to study the atmosphere at Rencluse and Maladetta, a glacier at a higher elevation than that in the Alps. They were also careful to keep the guides at a distance lest they contaminate their flasks filled with infusions of hay. It is eight o'clock one evening. As night falls, the three scientists take off their stove-pipe hats, solemnly lift the flasks above their heads, and uncork them. Suddenly, Pouchet cries out: "We are at an altitude of 2,083 meters, 83 meters higher than the Montanvert glacier. That is 83 meters better than Pasteur, but it is not good enough! We have to go higher."

So they decide to spend the night under an overhanging rock. In the early morning chill, having made their way through a jumble of rocks, they arrive, dead-tired, at one of the highest glaciers at the foot of the Maladetta range. They are at an altitude of 3,000 meters. A deep and narrow crevasse seems suitable for their experiment. Keeping their flowing ties around their necks but pushing up their sleeves, the three scientists open their flasks and then close them again, taking precautions that Pouchet finds excessive. The return was even more picturesque, though hazardous. Joly took a false step near a precipice and survived only because of the iron grip of a guide "whose arm was as strong as his shin." Finally the three heroes arrived back at Luchon. What heterogenesis had gained in this climb was an air sample from one of the highest elevations ever used: thanks to the infusions of hay as well as to the atmosphere, eight flasks filled up with microbes. The experimenters concluded: "Everywhere, absolutely everywhere, air is constantly fecund."[17] Life is born spontaneously, even in the purest of atmospheres.

"Not so!" Pasteur responded. Pouchet, Musset, and Joly's work is seriously flawed. They have opened only four flasks at each point. How can such a small sample be considered conclusive, and how could anyone think that it amounts to a survey? Moreover, they have broken their flasks with short files rather than long-handled pincers. The fact is that while Pasteur had indeed taken this indispensable technical precaution, he had not particularly concerned himself about the statistical implications of his own experiment; but then it is always easier to criticize one's adversary than to set an example.

The academician Milne-Edwards, for his part, added that all one could conclude from Pouchet's experiments was that there were more germs in the air in the place and at the time when he opened his eight flasks than there were in the Jura when Pasteur opened his.

Pouchet, Musset, and Joly resumed their experiments, the former at Rouen, the two latter at Toulouse. Twenty-two flasks were fecund even though, Joly insisted, they had used pincers with gigantic handles. Pouchet concluded: "At whatever place on earth I will gather up a cubic decimeter of air, as soon as I bring it in contact with a putrescible liquid sealed into hermetically closed glass tubes, they will without fail become filled with living organisms." And, in a dig meant for Pasteur, he added: "When Frederick criticized Voltaire for worrying too much about Fréron, the immortal philosopher replied: 'I worry about him only enough to spit on him.'"[18]

The debate was becoming nasty. In the preface to a work entitled *Nouvelles expériences sur la génération spontanée et la résistance vitale*, Pouchet wrote: "We have repeatedly seen M. Pasteur present his flasks as the ultimate in science, destined to dominate the world as a result of his findings." "This assertion is false," Pasteur responded. "I protest that I have never spoken or written these ridiculous words, and I expect M. Pouchet to be fair enough to provide a public rectification."[19]

Pasteur did not obtain a public rectification, but he did even better when he compelled his adversaries to submit to an official investigation: the Academy, exasperated by this discord that had repercussions even in its own ranks, and also yielding to pressure brought by the heterogenicists, decided to set up an investigating committee. Appointed on 4 January 1864, it included among others Dumas, Balard, Flourens, and Milne-Edwards. Pasteur asked that their expert evaluation be carried out soon, whereupon it was set for early March, even though the heterogenicists, fearing the cool temperatures, asked that it be put off until the summer. How can we be sure, they asked, that it will not freeze in Paris? All you have to do, Pasteur replied ironically, is put the flasks into an incubator, underlining the point that he, for his part, was ready to repeat his experiments in any season.

In the end, the Academy waited until June 1864 to convoke the committee and designate the place, namely, Chevreul's laboratory at the Muséum d'histoire naturelle. Each of the experimenters was asked to operate with his own series of flasks and his own culture medium; Pasteur used yeast water, Pouchet hay infusions.

Upon his arrival on 22 June, Pasteur presented three of the flasks containing air gathered in the Jura, which had not become infected and had been kept sealed in the cellars of the rue d'Ulm. The committee's first decision was to examine the quantity of oxygen in the neck of one of the flasks. Once it was broken and filled with the air of Chevreul's laboratory, it was easy to show that it could become infected. Pouchet and his acolytes expressed their astonishment that Pasteur had only three flasks. Pasteur retorted that he could have added many more, having brought seventy-six back from the Jura.

Thereupon the two teams set out to operate under the eyes of the committee, using new sets of both flasks and reagents. But as Pasteur was filling his sixty flasks one by one with the liquids he had heated to the boiling point and sealing the necks of the flasks, Pouchet argued about the temperatures. The last flasks cannot be compared with the first, he said, in which the reagent has rested and has had time to cool.

The jury, tired of such quibbling, dismissed this criticism and asked Pouchet to go on with his experiments. At this point Pouchet, Musset, and Joly changed their minds about their own experimental program. Complaining that they were reduced to repeating Pasteur's manipulations, they demanded that the decisive experiment, the opening of the flask, be preceded by a microscopic analysis of the air in the amphitheater of Chevreul's laboratory, followed by that of a liter of beer. Dumas and Milne-Edwards felt that all these complaints were nothing but useless caviling that wasted their time. The committee refused. Pouchet and his friends protested, accused the members of the jury of bad faith, and formally refused to limit themselves to a single type of experiment. They walked out on the debate and gave up the comparative test without a fight.

Pasteur was now the only one to carry out the examination of his sixty flasks under the eyes of the jury. He successively opened them in the amphitheater of the Muséum and then in various locations in the area around Paris where the committee went with him. The observations, which were easily confirmed by the jury, turned out to be identical with those he had published: some of his flasks became infected, others did not.

However, in order to be scrupulously fair to the heterogenicists, Pasteur and the jury agreed to say that it would be necessary to carry out experiments with hay infusions. Even in the absence of Pouchet, Musset, and Joly, it must be possible to carry out the test according to their own protocol. But all of this took a great deal of time, the summer was drawing to an end, and it was rightly pointed out that the bundles of hay no longer had the required quality. The experiments were therefore put off until the following spring and eventually abandoned altogether. There was to be no second report on the question.

Pouchet and his collaborators actually made a serious mistake when they abandoned the contest as they did and allowed their about-face to discredit them, for it is highly probable that their macerations of hay would indeed have produced a proliferation of microbes, despite all the precautions preached by Pasteur. A few years later, it was found that these substances possess heat-resistant spores that are fully capable of reproducing themselves.

Pasteur at the Sorbonne

Leaving aside the conclusions reached and even the scientific truth, it must be recognized that in this debate the attitude of the Institut was not altogether impartial. There is not the slightest doubt that the academicians supported Pasteur, whom they had just elected, though not without some difficulty, to join their ranks. But in the end the question of spontaneous generation was too important to remain at the level of personal quarrels because it implied a profoundly ideological debate.

One of Pouchet's main shortcomings was that he failed to recognize this fundamental dimension and to have ensconced himself in a personal terrain from which he launched a series of ad hominem attacks against Pasteur. After his walkout at the Muséum, he tried to alert public opinion. Having obtained Victor Duruy's permission to expose the matter to the Faculty of Medicine, he turned his speech into a plea designed to explain away his defeat. He embroidered his polemic by spreading the rumor that Pasteur constantly dodged him and that his own supporters were not given the opportunity to express themselves.

For his part, Pasteur operated in a politically more subtle manner and was perfectly willing to draw the obvious conclusions. He had entered this fight knowing exactly what was at stake, and whose side he had taken: it was that of official science. Significantly, he defended his position at the Sorbonne, the temple of the imperial University system, where on 7 April 1864 he gave a lecture of whose symbolic importance he was fully aware.

A large crowd had come to hear him. In the hallways one recognized Alexandre Dumas, George Sand, Princess Mathilde. Pasteur would speak to the cream of Parisian society. He had harsh words for Pouchet, but also for the writers who had taken part in the debate. Was not Michelet, for instance, one of the greatest defenders of heterogenesis in his writing when he celebrated a primitive creation in a drop of water? "This drop, what will come of it? Will it be a plant fiber, the light and silky down that one would not take for a living being, but which already is nothing less than the first-born hair of a young goddess, the sensitive and loving hair that is so well named Venus' hair fern? This has nothing to do with fables, this is natural history. This hair of two natures (plant and animal) into which the drop of water thickens, it truly is the eldest child of life."[20]

To these ravings Pasteur retorted: "As I show you this liquid, I too could tell you, 'I took my drop of water from the immensity of creation, and I took it filled with that fecund jelly, that is, to use the language of science, full of the elements needed for the development of lower creatures. And then I waited, and I observed, and I asked questions of it, and I asked it to repeat the original act of creation for me; what a sight that would be! But it is silent! It has been silent for several years, ever since I began these experiments. Yes! and it is because I have kept away from it, and am keeping away from it to this moment, the only thing that it has not been given to man to produce, I have kept away from it the germs that are floating in the air, I have kept away from it life, for life is the germ, and the germ is life."[21]

Applause. The audience gives an ovation to Pasteur, or at least a large majority does, for the speaker has not convinced everyone. The science journalist Louis Figuier left the hall saying, "I came here without having an opinion on spontaneous generations, but I leave convinced that M. Pasteur is wrong, and I shall say so."[22] In fact, this was the second time that he said so, for in 1860 he had already written about the experiments on spontaneous generation: "Since Pasteur is a chemist, the insufficiency of the physiological work he is doing can only harm the reputation he has acquired by his other work. But if he is to continue the studies on which he has now embarked, he will have to engage the services of a microscopist and a logician."

To this Pasteur responded in a letter to the editor in one of the next scientific columns of the *Opinion nationale*, saying that this was an affront to the dignity of science, and that mendacious insinuations were a shabby way to slander people. This exchange shows that the debate had gone beyond the strictly scientific realm and reached the public at large. Soon Pasteur was accused of being in the service of a doctrine, and the hostilities fanned by the press continued to pull the opposing parties out of their laboratories, away from their incubators and their microscopes.

The questions now raised became metaphysical, and they mobilized journalists and philosophers, in short, the group that was not yet called "the intellectuals." Abbé Moigno, a Jesuit who studied the diffusion of scientific knowledge and had long been the editor in chief of the periodical *Cosmos*, saluted Pasteur's theses and was happy to see that a scientist was able to convince skeptics and atheists by his experimental thoroughness; and he was particularly pleased that a "good Academy" had finally taken steps to correct the materialist tendencies that had taken hold in the Institut under Napoleon I.

Conversely, the anticlerical Edmond About took up the cudgels for spontaneous generation: "M. Pasteur has preached at the Sorbonne, amidst a concert of applause that pleased the angels." And, bringing up the origin of the species, he added, "If a small creature no bigger than the tenth of a pinhead could be born spontaneously, then there is no reason why nature should have been unable at other times and in different circumstances to form out of its own strength whales, elephants, lions, and even humans." This type of argument makes us understand what the defenders of spontaneous generation had in mind: Pouchet's theory did away with the need for a creator and thus allowed them to distance themselves from religion and dogmas. This is why France, as well as other countries, experienced an explosion of heterogenicist publications. Schopenhauer, for instance, spoke of the "will to life" that impelled chemical elements to re-form into an organized living being. He believed that something else than direct shock and countershock is hidden behind nature; and he criticized as extravagant the assertion that "everywhere and always millions of germs of all possible fungi float around in the air, along with millions of eggs of all possible infusoria, until one or the other of them happens to find the milieu that is suitable to its development."[23]

Newspapers, literary reviews, and public opinion seized on the scientists' swan-necked vessels. The Pasteur-Pouchet debate fascinated them as much as Ernest Renan's *Vie de Jésus* or the dissemination of Darwin's ideas. Inevitably, this publicity falsified the facts, and it is not easy to pick one's way through the confusion that took over, for the debate was really no longer about science. What came to the fore once again was the perpetual opposition between the spiritualists and the materialists.

"Man," wrote Guizot in his *Méditations*, "did not come into the world by spontaneous generation, that is, through a creative and organizing force inherent in matter." We owe a debt of gratitude to Pasteur, he added, for having "shed the light of his painstaking critique on this question."[24] But this argument was flawed by the fact that Pouchet never claimed to be a materialist, that he refuted the ideas of Darwin with whom his supporters tried to associate him, and that he even cited the Church Fathers in support of his experiments.

This type of lumping together was no doubt bound to occur, particularly when the support it received from the political regime forced those who opposed it to look for ammunition everywhere. For there was also a political interpretation of spontaneous generation. The dissident paleontologist Boucher de Perthes, who in 1861 published a pamphlet entitled *De la génération spontanée: avons nous eu père et mère?* [On Spontaneous Generation: Have We Had Fathers and Mothers?] showed just how far this question could be taken, for he ended by questioning the notion of the family, the inheritance of imperial power, and even the authority of the pope. The important part of spontaneous generation was the spontaneity, that is to say, liberty!

These, of course, were not the questions that preoccupied Pouchet; he was sorry to see that his experiments were discussed not so much for what they were but for what they implied. When he left the Muséum, declining to submit to the verdict of the Academy, he sent the Institut a letter of protest that was to become the standard of the heterogenicists: "In the face of the completely unexpected obstacles that were placed in my way, my conscience told me that I have only one choice, namely, to protest in the name of science and to safeguard the right of the future."[25]

What did he mean by this sentence, which substituted the word *con-*

science for *science*? Simply that the critique he directed, or rather reserved the right to direct, against Pasteur would target the protocols of his experiments. In every case, Pasteur had claimed that after having collected the air, he had either heated it or filtered it through cotton; however, in doing this, he might well have eliminated not only germs but also other organic matter conducive to generation.

In order to disarm this objection, Pasteur had to demonstrate that the atmosphere in the flasks, once they had been sterilized, neither killed nor infected the contents. To this effect, a way had to be found to place into the flasks, under sterile conditions, a biological product "taken from living matter" and therefore capable, according to Pouchet, of producing spontaneous generation if left in the open air.

These experiments were carried out by Pasteur with the help of Claude Bernard. He had the idea of placing a blood sample taken as aseptically as possible from a dog into a sterile flask. The blood gathered in this manner did not undergo any alteration. There was neither fermentation nor putrefaction. Better yet, there was not even a change in the oxygen within the flask. The same observations were made on urine.

This finding about living matter that decomposed and did not generate life was more than Pasteur had hoped for. It even gave rise to one observation that recalled the scientific experiences of his early days, for he noted that under such conditions "blood crystals form with remarkable ease." Crystals again, crystals forever! In this case, Pasteur marveled at finding the presence of fibrin crystals in the clotted blood. At a time when he was embroiled in a fierce polemic, encountering these crystalline forms with which he had so brilliantly launched his scientific career was a good omen indeed and made him feel that he was going in the right direction.

Lingering Controversies

Pasteur really won this contest by default. The Academy declared his experiments conclusive, considering that Pouchet and his collaborators refused to adopt their protocols, showing a lack of sang-froid of which their adversary was quick to take advantage. The polemic thus died down, but it was not over.

Meanwhile, Pasteur had learned a great deal from Pouchet's technical errors, in particular because they had made him understand the role of mercury as a vehicle of germs. This had taught him that contamination can be found in highly unexpected places. As if in a Trojan horse, microbes are carried in the most unlikely liquids. These first models therefore firmly implanted in Pasteur's mind the notion that one must not neglect anything in the search for contamination. In this spirit he was to continue to fight a whole series of opponents, who continued until the end of the century to step up their objections in order to keep Pouchet's theses alive. At Pouchet's death in 1872, Pasteur saluted him in a funeral oration that was both ironic and condescending: "This conscientious scientist deserves the gratitude of all of us for the good and useful work he has done, and even his errors are worthy of every respect."

The scientists, some of them shrewd, others clumsy, who set out to do battle with Pasteur were named Meunier, Trécul, Donné, Frémy. All of them claimed to have observed spontaneous microbial proliferations while manipulating with the greatest care. Exasperated by these rear-guard actions, Pasteur showed no indulgence and often used a biting tone to finish off those who contradicted him: "There are two parts to your note," he explained to Edmond Frémy. "The first has nothing scientific about it, and in the second you do nothing but formulate gratuitous hypotheses."[26] When the same Frémy contested that milk can be preserved in swan-necked flasks, Pasteur forced him to drink some of it to verify that it had not altered. And when Frémy, his back to the wall, claimed that Pasteur's experiments were not valid because small quantities of grapes do not ferment, Pasteur brought him a series of miniature flasks into which he had placed one drop of crushed raisin; breaking them open before the eyes of his fellow academician, he forced him to note the fermentation. Thereupon, Frémy kept quiet.

Often ironic, Pasteur could at times be positively harsh and, treating his adversary's arguments as mere logomachy, exclaim impatiently: "M. Frémy, trying to look profound, makes a radical but altogether gratuitous distinction between alcoholic yeasts and molds. That is not the issue. Whether the yeast comes from heaven or earth, from this or that, does not make any difference; the point is that it comes from the outside."[27]

What is astonishing to see in this period, which in the end lasted longer than the original debate with Pouchet, is the meticulousness, the passion, and the forcefulness of tone and conviction that Pasteur brought to the fight against each of his opponents. We must imagine him as an ardent and eloquent advocate and even preacher: "When all is said and done, what do you want to accomplish, all you declared partisans of heterogenesis or accommodating or unthinking supporters of that doctrine? Combat my assertions? Then go after my experiments. Prove that they are imprecise instead of constantly doing new ones that are nothing but variants of mine, although you do bring in errors that afterwards I have to point out to you."[28]

At times he became so discouraged that he thought the heterogenicists, luckier than the inventors of the perpetual motion machine, would enjoy the attention of the scientific community for a long time to come. But he also had to respond to serious challenges, for example when he faced the opposition of Charlton Bastian, with whom he exchanged some of the most acerbic and polemical barbs, even though this controversy proved most fruitful for science.

Like Frémy, Bastian entered into this discussion without being properly prepared. He had tenacity, he had experimental know-how, and he knew it. Professor of pathological anatomy at University College, London, he had published a series of studies whose principal experiment consisted of boiling urine in order to destroy all germs — so that it remained clear — and then add to it a solution of potash before placing it into a heated incubator. After a few hours the preparation would swarm with microbes. Bastian concluded that spontaneous generation had taken place.

The physicist John Tyndall, who tried to invalidate this experiment, called on Pasteur for assistance. Bastian himself also referred to the French scientist, asking him to replicate his experiments and, if possible, confirm his findings. Pasteur, skeptical as usual, and exasperated by these demands, nonetheless promised to cooperate.

He repeated the experiments and corroborated the facts described by Bastian. However, in this case as in that of the mercury in Pouchet's troughs, he thought that the profusion of microbes came from the outside. First he incriminated the water, showing that it contained microbes; Bastian boiled it without changing the results. Pasteur now blamed the solution of potash; Bastian increased the heating time, also pointing out that it seemed impossible to him that germs could thrive in such a caustic medium. The discussion continued, experiments were sent back and forth, but there was no solution: Bastian continued to see the phenomenon; Pasteur continued to deny it. The Frenchman's replies were sometimes embarrassed, often critical, and at times unfair.

This episode serves to reveal one of the limitations of Pasteur's scientific method: he was usually correct in what he asserted and seldom at fault for what he overlooked or failed to assert. When he repeated these experiments with a large volume of the solution of potash and found negative results, he did not convince Bastian, who thought that the excessive amount of potash, its heating, and its toxicity impeded or should impede spontaneous generation. The dialectic was confusing for everyone. Exasperation mounted until Pasteur finally challenged Bastian to come and repeat his experiments in the presence of a competent jury. Bastian accepted with pleasure. He said that he would be very happy to spend three days in Paris, where a laboratory was placed at his disposal at the Ecole normale, but he wished to conduct only one experiment, which he would design as he saw fit. The jury, more demanding than it had been in Pouchet's day, asked for all kinds of verifications. After several days of fruitless discussions, Bastian returned to London without having reproduced his experiments, and convinced as before that he was right.

Actually, the interpretation of the facts was a much more delicate matter than Pasteur initially claimed. Neither the potash nor the air alone were responsible. Germs, in the form of spores, can withstand very high temperatures, and the glass receptacle can also be contaminated. As Pasteur was to show later (1877) in collaboration with Joubert, germs can withstand washing and attach themselves to glass surfaces. In a half-full flask, the dry portions of the glass are not sterilized at the same temperature as the wet parts.

In the final analysis, the conflict with Bastian turned out to have been useful, since it sharpened Pasteur's techniques of sterilizing receptacles by flaming. This led to the autoclave, which was to become the basis of surgical sterilization. In 1882, Tyndall was to use this information in developing a sterilizing procedure using intermittent moist heat.

What science and the history of science remember from these successive controversies is essentially Pasteur's victory. He savored his triumph, understandably so. For if there was one topic that in the course of his career wrought a profound change in his personal way of doing things and at the same time forced the scientific community to react, it surely was spontaneous generation. Whenever one broaches the question of the origin of life, symbolic values are at stake. It is no coincidence that Pasteur's notoriety among the public at large dates from this period.

Pasteur's position was therefore not without ambiguity, even if he refused to venture into a terrain that lay outside of science: "This has nothing to do with religion, or with philosophy, or with systems of any kind. Assertions and a priori views do not count; we are dealing with facts."[29] Pasteur prudently asserted that he did not mean to prove that spontaneous generation can never occur; in matters of this kind, all one can prove are the negatives. Was this a manifestation of his scientific rigor and his sincerity, or did he keep within himself a shady area, the secret garden that every scientist shields from the bright light he shines on the facts he examines?

"Spontaneous generation," Pasteur was to protest toward the end of his life, "is something I have been looking for without finding it for twenty years. No, I do not consider it impossible. But on what grounds do you think you can say that it was the origin of life? . . . Who tells you that the steady advancement of science will not oblige scientists living a hundred years, a thousand, ten thousand years from now . . . to affirm that life has existed from all eternity, but not matter? . . . Who can assure me that ten thousand years hence people will not consider that it is impossible not to think that life does not cross over into matter."[30]

Life, to him, was the living breath for which he had been looking from the very beginning, whose influence or presence he had sensed from the time of the crystals to that of the germs. Yet in the last analysis, it had always remained a silent question for him, the legitimate unknown quantity.

Does this mean that Pasteur was bound to a religious ideal? His attitude was that of a believer, not of a sectarian. One of his most brilliant disciples, Elie Metchnikoff, was to attest that he spoke of religion only in general terms. In fact, Pasteur evaded the question by claiming quite simply that religion has no more place in science than science has in religion.

But one must also recognize his honesty, which forced him to speak out against those who wanted him to state the impossibility of spontaneous generation simply because he could see nothing that would warrant such an affirmation. Pasteur knew the limits of human rationality and probably hoped, consciously or unconsciously, that those who would come after him would solve the questions that remained in his mind.

The primordial soup, the identification of the first fragments of nucleic matter in selected chromosomes, the force, whether it be divine or chemical, that has brought about or will bring about their assemblage . . . all of these hypotheses were unknown to Pasteur. He could only tiptoe around the origins of life and think intently about the set of problems it raised. He was a wise man who stopped when he reached the limits of his observations, yet he foresaw and accepted in advance that some day science would go beyond them. A biologist more than a chemist, a spiritual more than a religious man, Pasteur was held back only by the lack of more powerful technical means and therefore had to limit himself to identifying germs and explaining their generation.

8

THE MICROSCOPE AND THE SILKWORM

At the age of forty, Pasteur had become the most prominent among the young scientists of the Second Empire, owing both to his position as director of scientific studies at the Ecole normale supérieure and to his stand in the debate over spontaneous generation in the Academy of Sciences. Nor were his capacities strictly limited to the scientific domain, for in addition to showing exceptional rigorousness of experimentation and outstanding synthetic abilities, he also knew how to deal with institutions and administrations and did not hesitate to become involved in fierce polemics when he felt that important scientific or political issues were at stake. In short, for the ruling powers Louis Pasteur was a reliable resource.

How could such a man be made to serve the regime? The answer was not long in coming; he would be called upon to use his abilities to solve problems affecting the state. As a result, Pasteur found himself entrusted with official assignments that forced him to venture into new areas of experimentation and kept him from becoming locked into purely theoretical speculations. Step by step, the scientist was led to investigate concrete problems in commerce and industry, so that he had to delve into areas of which he knew little or nothing: silkworms, wine making, beer brewing.

At this point, Pasteur's career does seem to have been guided by the special providence that watches over a genius. At times he seemed to disperse himself, to be impelled by duty or obedience to the powers that be to accept tasks that were far from his own preoccupations, to forsake his laboratory to rescue the wealth of great merchants or the authority of prefects. Yet in reality he steadfastly pursued his life's work, not obsessively, but with the suppleness and the intuition of the researcher who knows how to react to the realities from one detour to the next. Whether at this stage he knew it or not, he was now embarked on the course that would lead him to the elucidation of infectious disease.

In the spring of 1865, Pasteur was approached about investigating the diseases of silkworms. The request came from Jean-Baptiste Dumas, who was charged with reporting to the Senate on this scourge that was devastating France's silk industry. The old chemist turned to Pasteur, who had never heard of these problems, and asked for his help: we must save our silk!

Pasteur hesitated. Dumas pressed harder, urged him to accept, and offered to have him sent on an "extensible" assignment—meaning that it would initially not be altogether official—in the department of Gard. "This would allow you to speak on my behalf, or on anybody else's for that matter, to the best educated silkworm breeders. . . . You can tell them that you are very sorry but that you know nothing about this matter, and that the government could have chosen a better commissioner, but that you happened to be the one, strictly on the basis of your good will."[1]

"I really do not know what to do," Pasteur replied. "Your proposition is very flattering, its purpose is noble, but I am very worried and embarrassed about it. Please keep in mind that I have never even touched a silkworm." But then he added: "Remembering your many kindnesses to me, I realize that I would feel very badly if I refused your initiative. I therefore place myself into your hands; do as you see fit."[2]

Pasteur had shown on several previous occasions that he loved scientific adventures and was willing to travel in their pursuit. The departments of Gard and Vaucluse were not really so far from Paris, were they? No doubt this undertaking was attractive to the scientist, not only as a way to show his gratitude to Dumas, but also because the scientific challenge had major industrial implications. He expected his success to bring him popularity and publicity beyond the benches of the Institut.

Between 1865 and 1871, Pasteur therefore devoted six years of his life to the techniques of silkworm raising in the countryside of the Gard. Since he was not a veterinarian, let alone a physician, he approached these diseases with the enthusiasm of the newcomer and the advantage of inexperience. Knowing how difficult it is to pull a scientist away from a current research project, one is bound to wonder whether Pasteur was not impelled toward this new adventure by a deeper motivation at a time when he had just discovered the world of microbes and surely had not yet fully explored it in his test tubes. The new undertaking did not carry a salary, and even less a special subsidy; Pasteur and his assistants were to be paid the per diem of an agricultural inspector on assignment. Under these circumstances, it does not seem far-fetched to think that the scientist accepted because he perceived Dumas's request as logical: more than the idea of defeating a disease and rescuing an industry, it was the prospect of investigating the possible impact of the microbe on living matter that passionately interested him.

From fermentation to spontaneous generation, his work had taught him what chemistry, and then what biology owe to the microbe. The sponsors of the project explicitly stated that in view of what he had already learned about microorganisms and their biological mechanisms, Pasteur would be more likely than anyone else to find a solution to the problems of silk production. Thus he went south prepared to deal with the microbe, having already described its fermentative powers and dimly perceived its pathogenic role. From the ferment to the animal, it was indeed the germ that would form the connecting link between these two investigations, and if one was started before the other was finished, the microbe, the microscope, and infection were to remain their common denominators.

It is clear, then, that this undertaking marked a new turn in Pasteur's life. Success, when it came, was of course most beneficial to the sponsors of the project and to the silk industry — after all, Pasteur had well and truly saved the "golden tree" that resounded to the insistent gnawing of the silkworms as in Gounod's opera — yet it also went beyond the needs of industry, for the diseases of the silkworm gave Pasteur an understanding of the causes of epidemics. In the end, the caterpillar led him to man.

Disaster Strikes the Silkworm

"Misery is greater here than anything one can imagine,"[3] Jean-Baptiste Dumas wrote in May 1865 in an effort to sensitize Pasteur to the distress of the silkworm breeders. Born in one of the most sordid back streets of Alès (at the time it was spelled Alais), the chemist never forgot his southern background, despite the outstanding achievements of his scientific and political career. Founder of the Ecole centrale des arts et manufactures, professor at the Faculty of Sciences of Paris, at the Faculty of Medicine, and at the Collège de France, he developed equations to express chemical reactions and, in collaboration with Auguste Laurent, somewhat reluctantly laid the foundations of molecular atomism.

Minister of Agriculture and Commerce for three months in 1850, he became a senator in 1865, but he was too modest and too cautious to speak up in political debates and dutifully voted with the government, seeing this attitude as a means to wield a certain amount of influence. By contrast, he became quite verbal when the debate concerned industrial, commercial, and scientific affairs — and positively unstoppable when it came to silk!

In the first half of the nineteenth century, silk had been a flourishing industry in France. Long a monopoly of Asia (in China the export of silkworm eggs, which were jealously kept within the Great Wall, was punishable by death), silk had entered France in the thirteenth century through Provence, whence it had been brought from the comtat Venaissin, a territory belonging to the pope. By the sixteenth century, the peasants had developed the cultivation of the mulberry tree, the golden tree whose leaves are the main food of the silkworms. These food-producing trees were soon planted around the royal châteaux: Louis XI brought the worm to Touraine and cultivated it at Plessis-lèz-Tours; Catherine de Medicis acclimatized the mulberry trees in the region of Orleans, and Henri IV planted some in the park of Fontainebleau and in the Tuileries Gardens.

Between Plessis-lèz-Tours and the silkworm nurseries of the Cévennes, it took the efforts of Olivier de Serres, and later Colbert, to develop a true silk industry. Pioneering enterprises sprang up in the future departments of Gard, Hérault, Ardèche, and Lozère. In these desolate regions, chestnut trees were cut down and aqueducts were built to irrigate the mulberry trees. This industry prospered until the Revolution, which not only caused heads to roll, but also tore the finery of the rich to shreds — or at least considered it suspect. Wool took the place of silk.

The ostentatious ways of the Directory and the Consulate gave a new start to commerce, and beginning in 1800 the invention of the Jacquard loom and then Chaptal's policies provided new stimulation to sericulture. Lyon became the European silk capital, and the Cévennes Mountains were turned into a mulberry plantation. In these parts of the Midi, all business was suspended when it was time to harvest the leaves of the golden tree.

By 1853, the prosperity of the southern provinces was largely based on sericulture, and Napoleon III's government could take pride in the fact that France accounted for a tenth of the world's silk production, with 26,000 tons of cocoons per year. The cultivation of mulberry trees had more than doubled in twenty years and represented a considerable share of the national budget. It was a sector that operated on many levels, for all the revenues did not go to the rich, and many *magnaneries*, or silkworm nurseries, were run by small proprietors, who employed day laborers paid in mulberry leaves, silkworm eggs, or hundredweights of cocoons. In this manner, the "caretaker" (that is, the person who raised the silkworm) benefited as much as the owner before becoming one in his or her turn. Women and children were often in charge of watching the worms, for it was a simple task that required great vigilance only while the worm was spinning its cocoon; otherwise, labor was needed mostly for harvesting the mulberry leaves.

The silk is spun by a worm, more precisely a caterpillar, that of the silkworm moth, which is a little larger than the caterpillar that feeds on the leaves of fruit trees. Its body shows six rings and eight pairs of legs, a small head with articulated jaws that can grind the mulberry leaves. An enormous sack that functions as a stomach performs the digestive function and serves to store the raw material for the silk, which, starting out as a liquid, subsequently passes through a capillary tube, where its consistency thickens before it is extruded through an orifice (the silk gland), dries in the air, and hardens into a thread of silk.

The caterpillar develops out of eggs, which were called "seeds" because of their resemblance to plant seeds. As soon as it is formed, it starts to feed and undergoes four successive metamorphoses. The fourth of these is followed by a period of extreme voraciousness, in which the worm greatly increases in volume until it reaches its maximum size; this was called the feeding frenzy [*la grande frèze*]. Once this period is over, the caterpillar stops feeding. It becomes agitated and looks for a sheltered place. The silkworm producer uses heather and branches to build a cradle into which the worm can climb. Once it has found a suitable place, it grasps it with its ten hind legs and begins to spin silk in the form of a yellow or white cocoon. Once it has started, the worm does not stop spinning until it has pushed out a thread about a kilometer long; this takes seven or eight days.

Once it has emptied out its silk, the caterpillar becomes engorged and undergoes a sixth metamorphosis. It turns into a pupa from which the moth can emerge. Male and female moths mate, whereupon they die within two weeks. But first the female lays a considerable quantity of eggs, which will lie dormant and not hatch before the following year. Two months have passed between one crop of "seed" to the next; this is a short cycle favorable to observation and experimentation.

If one wants to obtain eggs to multiply and reproduce worms, one must make the moth "give seed" after it emerges from the cocoon, but if one wants the silk, one smothers the pupa before its last metamorphosis. On the average, twenty-seven kilograms of mulberry leaves are needed to produce one gram of silkworm eggs; and with one hundred grams of eggs one can obtain 100 *quintaux* (hundredweights) of cocoons.

France's sericulture was rapidly expanding, and mulberry plantations were multiplying, when it was struck by a terrible blight that seriously threatened the industry at each harvest. In the past, it had not been too difficult to deal with *muscardine*, an episodic disorder caused by a fungus. Certain disastrous harvests had also been attributed to bad hibernating conditions for the eggs, as well as to the inexperience or carelessness of the caretakers. But around 1850 it was obvious that silk producers were faced with a new disease that decimated the breeding chambers and which, because it was transmitted through the eggs, seemed to be hereditary.

The first sign had appeared in 1849. In the beginning, the harvests remained good, for the damage could be repaired by importing new eggs from abroad. But it soon became difficult to find healthy eggs; the disease was spreading. It reached northern Italy, then Spain and Portugal. One had to go further and further afield, to Greece or Turkey. Then these countries succumbed in their turn. Supplies were imported from the Near East, then from the Far East. Soon even China was affected.

In the early 1860s a breeder from the department of Drôme went to Japan and brought back eggs of excellent quality. By the time Pasteur attacked the problem in 1865, healthy eggs could be obtained only in Japan.

Although the disease was spreading by degrees in the manner of an epidemic, the severity of its effects varied from case to case. Generally a considerable mortality of the worms was observed right after hatching, or sometimes after the first metamorphosis, which did not go well. The worms failed to grow, and the breeding chambers on the shelves of the silkworm nurseries were sparsely populated. In some places the production went well until the fourth metamorphosis, but then the worms took on a rusty color, crawled away from the leaves, showed blackish spots, and did not spin. In yet other cases, the worms went through the first four metamorphoses without apparent difficulties, but once the last one had taken place, the caterpillars became sluggish and did not fall on the leaves with their normal voracity. Gone was the grinding noise of their jaws, which was usually so loud that in the breeding chambers it resembled the sound of rain striking the leaves during a thunderstorm. The thin and feeble worms lay motionless, and the tables were littered with dead bodies.

Everyone in the Midi, from the merchants and silk producers to the simple peasants, was well aware of these disasters, but the public at large in the rest of France knew almost nothing about them. The government, to be sure, was concerned. The Academy of Sciences, always on the lookout for critical situations, was alarmed at the destruction of the silkworm. But several inspection tours by eminent scientists, and a number of reports directed by men of science who had a particular interest in this matter, such as Quatrefages or Dumas, did nothing to solve the problem. At best, these reports noted that the vegetation of the mulberry tree seemed to be normal, so that silkworms' food could not be incriminated. What, then, could be recommended, aside from recourse to foreign eggs? The government put the Imperial Society for Zoological Acclimatization in charge of distributing eggs of the "yamamai" variety, which the tycoon of Japan had sent the Emperor to help the industry. But this was to hide the problem, not to cure it, even if the harvests of 1863 and 1864 seemed reassuring: from a production of 26,000 tons for the year 1853, yields had plummeted to 5,800 tons in 1861 and then risen to 6,500 in 1863. But the results of 1865 (only 4,000 tons) dashed all hopes. The distressed silkworm breeders sent a petition to the Senate. Signed by the municipal councils of the principal landowners in the departments of Ardèche, Gard, Hérault, and Lozère, it did not lead to action, not out of lack of goodwill, but out of helplessness.

Joseph Favre, deputy from Gard, exclaimed in the Chamber: "Agriculture, gentlemen, and our local industries, spinning and milling, are not the only ones to be compromised by this epidemic; the industry of Lyon, that is to say, the most prosperous and the wealthiest of our industries, is also affected. . . . There are entire cantons where the land no longer provides its owner with the means to pay his taxes. . . . In the face of this terrible situation, what has the government done? What steps has it taken to stop this disease before it had a chance to spread? And what has it done since to alleviate the ensuing misery?"[4]

What did the government do? It sent Pasteur.

An a Priori Idea

Pasteur did not embark upon this investigation without experiencing a moment of bewilderment. To be sure, his previous studies had prepared him to envisage the existence of possible agents of infection, namely, the microbes, which might make the diseases of the silkworms similar to a fermentation. But for the moment this possible link could be no more than a working hypothesis. Actually, if Pasteur envisaged any kind of relation between the disease of the worm and what he knew about microbes, he was more inclined to believe that the disease favors the development of germs rather than that it is their consequence, for he considered the diseased host as a culture medium that nourishes the germs. He tenaciously clung to this idea that the germ benefits more than it kills, just like a ferment.

Pasteur came from crystals. Owing to his scant knowledge of animal biology, he was somewhat apprehensive about experiments on animals. As soon as he had accepted Dumas's assignment, he therefore went, along with

his assistant Emile Duclaux, to the physiology course taught by Claude
Bernard at the Sorbonne. There he took notes and humbly relived his years
of training in the halls of the university. But he found it difficult to learn a
whole new field; and indeed, since he had neither the time nor the patience
to do this, he soon preferred to form his own ideas on the problem at hand.
This was not so easy, for caterpillars and mulberry trees were not available
in Paris. Yet Pasteur set out in all haste to familiarize himself with the world
of insects, enlisting the help of an assistant in natural science who dissected
larvae in his presence. He filled in his knowledge by attending a few sessions
of the Imperial Commission on Sericulture, where he gained a better understanding of the scope of the economic disaster, although he came away
more discouraged than enlightened.

There was no time to lose, for the research was linked to the life cycle
of the silkworm, and in Provence the breeding period was drawing to a
close. In early June, Pasteur decided to leave Paris and went to the department of Gard. Alès at the time was a small town located in a deep valley facing the Cévennes Mountains, from which it was separated by the Gardon
River. Several bridges crossed the river, one of them leading to the suburb
of Rochebelle. The tall houses pierced by windows wedged atop dark ar-
cades testified to an austere Protestantism. In the *garrigues* with their poor
soil where vegetables could not grow, mulberry trees were found everywhere. And every farm had a piece of land set aside for raising silkworms.

Pasteur was welcomed by the town's mayor, Dr. Pagès, dressed in his
morning coat and stovepipe hat. As soon as he was settled in his lodging,
Pasteur went to work at a small silkworm operation in the suburbs. Morning and evening the citizens saw him walking through town looking pen-
sive and purposeful. With his short beard and his shortsighted eyes, this
frowning scientist seemed out of place.

Before going any further, he sought to understand. The entomologist
Jean Henri Fabre, the Homer of insects, and a man just one year younger
than Pasteur, held a position at Avignon at that time. The Parisian scholar
therefore went to call on his Provençal colleague, who later told the story:

> A few words were exchanged on the ravages of the disease and then, without
> preamble, my visitor said: "I would like to see some cocoons; I have never
> seen any and all I know about them is their name. Would you be able to find
> me some?" — "Nothing could be easier. My landlord sells cocoons, and we are
> right next door." I rushed over to my neighbor and stuffed my pockets with
> cocoons. Back at my house, I show them to the scientist; he takes one and

rolls it around between his fingers. Full of curiosity, he examines it as one would examine a strange object brought from the far end of the world. He shakes it in front of his ear. "It makes a sound," he says, surprised, "is there something inside?" "Of course there is." "Well, what is it?" "The pupa . . ." Then, without further ado, the cocoons disappeared in the pocket of the scientist, who was going to study that great novelty, the pupa, at his leisure. I was struck by his magnificent self-assurance. Knowing nothing about caterpillars, cocoons, pupae, or metamorphoses, Pasteur had come to regenerate the silkworm. The gymnasts of antiquity came to the wrestling match naked. A brilliant fighter against the scourge of the silkworm nursery, he too came to the battle naked, that is, without even the simplest notion concerning the insect he was to rescue. I was dumbfounded; better than that, I was enchanted.

Later, Fabre added: "Impressed by the magnificent example of Pasteur listening in amazement to the sound of cocoons held up to his ears, I made it my rule to adopt the method of ignorance in my research on insects. . . . Ah yes! Pasteur did well not to know about the pupa!"[5]

In certain respects, particularly because of its slow but steady advance toward eastern Europe and then Asia, the disease of the silkworms had the characteristics of an epidemic and contagious disease. Yet in other respects, it turned out to be unusual. Early in his investigation, Pasteur observed the evolution of a mixture of eggs of two different colors and noted the following fact: Even though they had all been raised together, only certain worms hatching from the eggs of one of the two colors became sick. This meant that the spread of infection among worms raised together was not the whole story.

The signs of the disease, incidentally, were rather difficult to define. The academician Quatrefages, who had studied this matter on several occasions, had seen on the skin of the worms tiny blackish spots, which he had likened to scattered peppercorns. He therefore proposed to call the disease *pébrine*, a word derived from the langedocian dialect term *pébré* [pepper]. But this symptom was inconsistent. The animals could present this pébrine without otherwise being sick, and, conversely, some of them could produce diseased eggs without themselves showing the black spots. As Pasteur pointed out rather ironically, Quatrefages had raised his description of a cutaneous condition to the level of a scientific study and a theory. This, of course, was not enough, and Quatrefages had to admit that there were two aspects to the scourge: one, the pébrine, can be observed; the other, the predisposition to the disease, can be imagined, and is very probably aggravated by the intervention of factors related to the season, the place where the eggs are harvested, and the weather. Quatrefages's publications, though amounting to little more than compilations of rather superficial observations, nonetheless were the first to give a name to one of the forms of the disease.

While this work was done in France on the basis of rather confused observations, other studies were being pursued in Italy, where certain naturalists, examining the tissues of diseased worms under the microscope, noted the presence of shiny, oval, and relatively well delineated corpuscles. These corpuscles, which were often associated with the disease, were thought to testify to its presence. But this sign too was inconsistent. Philippi, a zoologist at Turin, reported the existence of corpuscles in healthy moths, while another naturalist, the Frenchman Félix Guérin-Méneville, suggested (mistakenly) that the corpuscles were related to another disease of the silkworm, muscardine, even though its episodic manifestations were clearly different from those of pébrine. In fact, there was no agreement on the significance of these famous corpuscles, any more than on their origin. To complicate matters further, the corpuscles were also found in another insect, the cochineal. Since no one knew what to do with them, they were provisionally classified as plants.

Yet in 1856, Emilio Cornalia, a naturalist of Milan, had already published a series of detailed observations on the corpuscles. He noted their abundant presence in the bodies of diseased moths and indicated their existence in the eggs, a phenomenon that his predecessors had failed to notice. Cornalia showed that the proportion of corpuscular eggs was higher in the egg masses of moths infected with pébrine. One of his colleagues, Orcino, built on these observations and completed them. He considered the presence of corpuscles a useful indicator for separating the good eggs from the bad and proposed examining all eggs before incubating them and destroying those that appeared affected. However, this procedure did not seem effective enough, so that two years later Orcino suggested examining the pupae as well and making them suffer the same fate.

This judicious idea could have led to the solution at which Pasteur arrived several years later, but because it was purely speculative, it was simply written up and never used for serious experimentation. In view of this shortcoming, Pasteur once again stated one of the articles of his scientific credo: "In the experimental sciences, truth cannot be distinguished from error as long as firm principles have not been established through the rigorous observation of facts."[6]

In the absence of solid bases, in France and elsewhere, experimenters looked high and low, but in vain, for a remedy, which caused one observer to write: "Today the pharmacopoeia of the silkworms is as complicated as that of humans."[7] Everything, from chlorine, sulfuric acid, rum, sugar, and sulphur to wine, absinth, and vinegar, was tried for treating the worms as well as the mulberry leaves. Experiments with electricity, which science had not yet mastered, and which was therefore considered a universal modern procedure, were also carried out; but electroshock was no cure for the disease of the silkworm. New therapeutic trials were conducted every year. Silkworm breeders ruined themselves buying medicines and trying out marginal and untested procedures that became popular fashions or crazes. Whether regulated or uncontrolled, these attempts to develop bold and imaginative new treatments invariably failed.

When Pasteur warned Dumas that he knew nothing about silkworms, the latter had replied, "all the better, for then all your ideas will come from your own observations."[8] However, Pasteur did leave Paris with a preconceived idea. When he arrived in Alès, he was struck by the confusion of symptoms and remedies. Rather than plunging into the copious literature on improvised tests and unsystematic descriptions that he could find in the countless local books and newspaper articles on the subject, he chose to concentrate on his observations in order to substantiate an intuition he had; he wanted to find out all about the corpuscles about which he had read in Quatrefages's publications. That author placed great emphasis on the corpuscles described by the Italian authors and considered them the equivalent of abscesses, that is to say, as symptoms and not causes of the disease. Pasteur was inclined to share this view.

In the field, he found it easy to corroborate this opinion, for the silkworm nurseries were filled with diseased moths and worms, most of them corpuscular, while, conversely, the few noncorpuscular specimens seemed healthy. But things were not as simple as that, and this approach was to lead Pasteur into a labyrinth of contradictory observations.

At the silkworm nursery where he set up his work station, he began by following two broods of worms from two breeding chambers. The first one flourished; the worms hatched at the fourth metamorphosis, moved about vigorously, and, once they had fiercely devoured their mulberry leaves, produced abundant silk for their cocoons. The second brood looked poor from the start. The worms were sluggish, did not feed, did not grow, did not spin a cocoon. In examining the pupae and the moths of the first brood, Pasteur found corpuscles in all of them, while they were rare in those of the second brood. This troubling and paradoxical observation was not unique.

Pasteur soon encountered many similar examples. These findings could have led him to the conclusion that the corpuscles do not invariably represent a symptom of the disease. But as a rigorous scientist, he was not given to hasty claims and therefore did not give up but followed the further development of the bad brood. At the nursery where he examined the worms day after day, he now found that the corpuscles, though almost or completely invisible in the young worms, appeared and then multiplied in number as the worm aged. Moreover, even though the worms seemed healthy, the pupae were frequently, and the moths always, corpuscular. From this Pasteur concluded that the corpuscle is a late-appearing sign that is always present at the final metamorphosis of the diseased animal.

The question thus seemed to become a little clearer: there was indeed a disease that threatened to ruin the silk industry; and there were also the corpuscles, symptoms of an advanced stage of the disease. Being pressed for time, Pasteur used these observations as the basis for a first series of conclusions, which proved to be both incorrect and useful. He stated that it was a mistake to look for the symptoms of the disease exclusively in the worms or in the eggs, since these, even though diseased, may not present the corpuscles. Considering that these were frequently found in the pupae, and always in the moths, further study should concentrate on these two stages. And since the most important objective was, after all, the production of high-quality eggs, Pasteur felt that all that needed to be done was to use only noncorpuscular moths at the time of reproduction. Pasteur was thus led to propose a method of sorting the eggs that was virtually identical with that recommended by Orcino several years earlier. If that method had failed, Pasteur asserted, it was because its author lacked self-confidence; and this, of course, was not true in his case.

A few weeks after his arrival in Alès, Pasteur was thus already in a position to communicate his conclusions to the *Courrier agricole*. In September, he explained his findings at greater length to the Academy of Sciences, where he outlined the conditions for raising a good brood of silkworms: "The means to achieve this consists of isolating each couple of moths at the time of mating. Once the mating is accomplished, the isolated female will lay her eggs, whereupon she, as well as the male, will be opened to look for corpuscles. If they are absent, this batch of eggs can be considered healthy; it will be given a number and raised with particular care the next year."[9] According to Pasteur, this method made it possible to establish the prognosis of the disease before it was fully apparent. But he was criticized even before the first practical tests had taken place. It was true, of course, that the technique had already been described by Italian naturalists, and the silk producers found it unfortunate that the French government, instead of calling on a specialist in silk or an experienced veterinarian, had seen fit to entrust the task of elucidating so mysterious a disease to a chemist. As is often the case, the scientific polemic was sidetracked by political, economic, or diplomatic interests.

A more serious flaw, however — and this was not pointed out at the time — resulted from Pasteur's belief that the stigma of the corpuscle, if it was present, was so diffuse that in studying it under the microscope it was sufficient to use a piece of skin removed by scissors. Because this method did not examine the whole animal, it was to lead Pasteur into faulty interpretations and onto false tracks concerning the nature of the corpuscles, at least during the entire following year.

The Microscope of Pont-Gisquet

On his return to Paris in the late summer of 1865, Pasteur brought back to the rue d'Ulm the healthiest possible moths and pupae, which had not been easy to obtain since the whole region was infested and most of the healthy worms were made to spin. Two breeders in the small town of Anduze who had heard him expound his ideas had brought him five moths fed on mulberry leaves still on their branches. This mode of rearing them was no better than other remedies, but in this particular case it at least had the merit of yielding four healthy female moths whose eggs Pasteur could study. It was a disappointing experience, for some of these select moths eventually produced corpuscular worms.

Despite these first contradictory observations, Pasteur's campaign strategy remained unchanged: he would continue to watch the broods of worms that had hatched from the eggs of noncorpuscular moths. In order to carry out this plan on a large scale, however, he had to be present. Pasteur therefore returned to Alès in early February 1866 and stayed until late June. This time he took along two former students who had become his assistants, Désiré Gernez and Eugène Maillot. Emile Duclaux joined them later. Jules Raulin had to finish his thesis and could not go along; he was to catch up with the subject later. Pasteur thus not only committed many months of his own time to this study, he also involved his best students, who seemed to have followed the master without ulterior motives, wishing to serve him, rather than to further their own careers — unless they developed a passion for a subject that Pasteur himself had initially approached with a certain reluctance.

Pasteur, Gernez, and Maillot first stayed at a hotel in Alès, but they soon began looking for a house that they could transform into a laboratory. That laboratory would have to be in the country. In the suburb of Roche-belle they rented a house belonging to a sieur Combalusier; it was a low and modest structure whose main room and attic they transformed into a silkworm nursery. The place was dark and humid. Even during the day, the breeding chamber had to be lit by candles. The landlord came home every night and was liable to disturb the ongoing experiments. "For several weeks," Gernez tells us, "this was where Pasteur passed all of his days. Looking through the microscope next to a window, he only left his place to go into the attic, a veritable dark incubator where by the light of a candle he followed the movements of the worms he was testing."[10] But he had to return to the hotel for lunch and dinner, which he considered a waste of time. Also, the house was located at the intersection of two roads, where peasant carts, heavy drays, mules and horses passed by in a steady stream, disturbing him with their noise.

Maillot was soon put in charge of finding another locale. Chance led him to Pont-Gisquet, beyond the outskirts of the town. There was a house leaning against the slope of Hermitage Mountain; it was rather isolated, and located 1,500 meters from the town limit. Nothing was left of the old silkworm nursery but a few mulberry trees interspersed within a grove of narrow-leaved olive trees. This is where the three scientists settled down for good: the orangery was turned into a laboratory, and the house had room for everyone. "After much searching," Pasteur wrote to his sister Virginie, "we were able to rent a house which, situated more in the countryside than in town, is very clean, well furnished, and very well located above a ravine between two mountains. . . . We have hired a very competent cook, who returns to Alès in the evening and comes in in the morning bringing sup-plies for the day. We give her 45 francs a month, and we have agreed to pay 600 francs in rent for the house for the duration of our six or seven months' campaign."[11] The house at Pont-Gisquet was like a haven for the scientists accustomed to the close quarters in the rue d'Ulm. "In this laboratory we were very far from Paris,"[12] Duclaux was to remember nostalgically.

Pasteur also kept fond memories of the days he spent in the Gard. Yet it was a time of hard work. One had to rise at 4:30 in the morning to check on the worms, for Pasteur was extremely particular about the care of his silkworm operation, even more meticulous and rigorous than he was in his laboratory in the rue d'Ulm. The hatching chamber was kept scrupulously clean: the floors were scrubbed with water, then sprinkled with copper sulfate, the walls were whitewashed. The worms were reared in baskets. In order to avoid raising clouds of dust, the premises were never swept but cleaned with a wet rag. Strict hygiene was the order of the day.

"Now began a period of intense work," writes Gernez. "Pasteur undertook a large number of tests, which he carried out himself with attention to the smallest details; we were only called upon to do similar operations, which served as controls for his. . . . The second task was to clear the terrain by deciding which of the countless assertions made in France and abroad seemed to have some validity. And finally, a number of presumably infallible remedies that had been proposed had to be checked out very carefully before anything could be said about their effectiveness. Hence, the experiments were piling up, as were the requests for information that came in from all sides concerning the most diverse and unexpected points."[13] Duclaux, for his part, adds: "We were not, needless to say, those who thought up strategies, for at that time M. Pasteur was still a secretive worker who kept his ideas and his thoughts to himself. But we guessed what they were, or thought we did, and that was enough to make the thousands of experiments that took up our days intriguing and interesting."[14] The rhythm of daily life was thus established by the broods of silkworms. The scientists had to keep track of their experiments, keep out the mice that wanted to nibble at the silkworms, and frantically pick mulberry leaves when rain threatened. There were practically no boundaries between work and private life; the group formed one household in that large, clean, and comfortable dwelling. The students had no families. Their family was Pasteur, their passion, the silkworms.

On the rare occasions when Pasteur was not with his students, particularly during his daughter's illness, he left them peremptory instructions that left them very little leeway: "Meanwhile, and during my possibly rather prolonged absence, you are to follow the tests on the medicated and unmedicated worms. Add new ones, using the indications to be found particularly on the first page of a ruled piece of paper kept with my notebooks. A new notebook has been started. Feed them once a day when you give the medications. Be sure you watch the metamorphoses. Do not think you know from experience."[15]

After the death of Cécile, on 23 May 1866, Marie and little Marie-Louise came to join Pasteur at Alès. But they had not come for a vacation, and everybody participated in the work: microscopic experiments, watching the broods, picking leaves. Jean Baptiste had stayed in Paris by himself because he had to go to school. Marie wrote to him in June 1866, two weeks after his sister's death: "Your father asks me to tell you how sorry he is that he cannot write to you. He is so busy and has so much to do when the moths are hatching, because he must immediately examine them so that he cannot take a moment away from his work." And she added: "Your father goes up to the mountain himself to pick the mulberry leaves with which he feeds his silkworms. . . . Our Zizi [Marie-Louise] really enjoys going to the mountains with the silkworm-people, but she does not like to touch the worms, any more than I do, because I find them quite disgusting. All I do is peel the cocoons, count them, sort them, and place them in nice, clean baskets in order to facilitate the hatching of the moths as well as the scientific experiments."[16]

At Pont-Gisquet, Pasteur gradually improved his methods of obtaining eggs. On the basis of his observations he was able to carry out selections and to study the offspring of diseased moths over several generations. In doing so, he obtained strains of eggs that were superior in yield and quality to those produced in Japan.

This selection process was made possible by the microscope. In order to decide whether cocoons should be smothered and sent to the spinning mill or preserved for reproduction, he had the idea of raising the temperature to hasten the hatching of some hundred moths, which were then examined under the microscope. When the moths had hatched from their cocoons and mated, they were separated. Each female was placed on a small square of cloth, where it laid its eggs. It was then pinned to that same cloth, where it gradually dried up. Later, in autumn or winter, the dried moth was dissolved in water, ground up, and the resulting suspension examined under the microscope. If the slightest trace of corpuscles was found, everything was burnt, including the eggs that would have perpetuated the disease.

Pasteur was tenacious in his effort to instruct the sericulturists of Alès in the use of the microscope. The method was so easy, he claimed, that his little Marie-Louise was capable of using it. To his detractors he replied: "A woman, even a child can take care of it. Suppose the eggs have been laid at the nursery of a peasant who does not have the facilities to examine them right away. Instead of discarding the moths after they have mated and laid the eggs, he can place a large number of these moths into a bottle half-filled with spirits and send them to a testing station or an experienced person."[17] As for the different stages of the operation he recommended, Pasteur did not stint on the details. To begin with, he said, one must make sure that the objects placed on the table are perfectly clean; the lighting must be meticulously regulated, the best solution being to place the preparation next to a window whose half-closed shutters permit only a small amount of light to filter in. The microscope must be to the left, the reserve of water to the right. Once the moth has been ground up and the preparation placed between the two slides, the corpuscle can be recognized by applying slight pressure on the slide; it is taken up by the movement and will roll about.

Reading these instructions, one can imagine the amazement of the silkworm breeders of the Gard. Some of them were converted and came to learn their lessons from the master or his students; duly trained under Pasteur's severe supervision, they scrupulously followed his directives. But there were also skeptics who continued to criticize the man and his method and felt that new remedies would bring a solution.

Pasteur was always convincing, even when he was not totally correct. Keeping absolutely silent as he experimented with his silkworm nursery, he became positively wordy when it came to reporting the slightest advance. Since he was working for himself, but also on behalf of many breeders, he owed them explanations and predictions. He distributed eggs and asked the sericulturists of the department to carry out these experiments on their own. On the other hand, breeders also brought him their eggs so that he could diagnose whether they were good or bad for spinning or for reproduction. In *Madame Bovary*, Flaubert ridiculed the agricultural associations of the time in the pastiche of a grandiloquent speech filled with every cliché of provincial eloquence. However, life is not a novel; and the very skill that can produce an undeniable literary achievement can also perpetrate a historic injustice. The fact is that the agricultural associations and the research of the agronomists of the Second Empire contributed some extremely important information to the debate on the improvement of society and progress.

Bénigne Jeanjean, the mayor of Saint-Hyppolyte-du-Fort and secretary of the agricultural association of Vigan, became the champion of Pasteur's methods. "The early start of this year's tests has had the additional benefit of allowing us to make important observations on the merit of M. Pasteur's procedure. And while it is only proper that we reserve to this worthy scholar the honor of reporting in person the findings of his interesting experiments, it is appropriate to note that the findings of the early experiments corroborate the facts established through the microscopical examination of the moths."[18]

Many others also testified in his favor. Lachadenède, for instance, the president of the agricultural association of Alès, asserted his high respect for Pasteur's work, to which he attached the greatest practical importance; in the *Revue universelle de sériculture*, Ducrot, assistant at the imperial school of agriculture in Aix, testified in his favor when he described the worms he had raised by Pasteur's method. Was not, he asked, the number of healthy silk-worms he had obtained magnificent and apt to produce even greater yields in the future? The journals of practical agriculture, such as the *Moniteur des soies*, the *Revue universelle de sériculture*, the *Messager agricole du Midi*, all featured reports, counter-reports, expert opinions, and testimonies, most of them favorable to Pasteur.

Later, in June 1868, Dr. Pierruges, the mayor of Callas, was to publish in the newspaper *Le Var* the following letter to Pasteur, who had been asked to evaluate the quality of two lots of eggs:

> As a service to the sericulture of our department, you agreed to publish the practical conclusions drawn from your examination in a letter which, according to your wishes, was published in the journal *Le Var* on 30 April and subsequently reprinted in a Toulon paper. As you may well imagine, your communication aroused strong feelings among the breeders who were working with one or both of these two kinds of eggs. At first, no one knew what to do. Should one toss the eggs you had condemned into the fire? In other words, should one trust the predictions of science? Well, I hate to admit it, the hesitation did not last long. . . . People told themselves that after all the microscope is not infallible, that the judgments of science are sometimes reversed on appeal, and so forth. The breeders continued to raise the worms about which you had warned them as if the problem did not exist. . . . Well, the facts have spoken. The broods raised from these two kinds of eggs have completely failed. Thus, Monsieur and most distinguished Master, the evaluation of the pathogenic qualities of the eggs in broods no. 1 and 2 that was made by you on the basis of your microscopic examination of April have been rigorously confirmed by the facts in this community.[19]

Criticized by some — mainly those who refused to apply it — and verified by others, Pasteur's method was in fact nothing more than what it claimed to be, namely, a procedure for obtaining better eggs. Indeed, when he proposed this empirical but effective stop-gap measure, Pasteur was aware that he had yet to discover the cause of the disease and to analyze all of its aspects.

Meanwhile, there was one fact that never ceased to intrigue Pasteur in the course of his experiments and observations, and that was the absence of a strict correlation between the corpuscles and the disease; he even found that it was not impossible to obtain healthy worms (in small numbers, to be sure) from eggs laid by diseased moths. He hesitated between two interpretations, thinking that both might be equally valid. Perhaps, he thought, some eggs were not very sick and produced worms that cured themselves as they grew up. Or else some eggs simply escaped contamination.

By 1866, doubts had arisen about the hitherto generally accepted hypothesis which, entirely based on Pasteur's first interpretation, postulated that the corpuscle was a symptom testifying to the disease. Perhaps, it was now felt, another plausible explanation should be envisaged, namely, that the corpuscle was in fact the infectious agent responsible for the propagation of the disease.

When Pasteur was wrong, he deployed as much vigor as when he was right. In this case, he refused with a kind of obstinacy to conceive of the possibility that this disease, which definitely seemed to be contagious, might be due to the corpuscle itself. He passionately rejected this possibility for two reasons, which he laid out in a note presented to the Imperial Society of Sericulture in January 1867. In the first place, he said, the disease is constitutional, for it precedes the appearance of the corpuscle. Thereupon he argued on the basis of experiments in which he had attempted to infect the worm through mulberry leaves covered with a dust of corpuscles. But the worms that had eagerly devoured this corpuscle-laden meal had died within an extremely short time without presenting any corpuscles. In conducting two series of these observations and using them to combat the hypothesis of the corpuscle as agent of infection, Pasteur had been dealt an unlucky hand. As he was to learn later, pébrine was actually compounded by another endemic disease, called *morts-flats* or *flacherie*, which created overlapping symptoms that confused the picture. In this particular case the worms that had fed on the corpuscle-laden meal had not died of pébrine. But at the time Pasteur did not know this, and he not only believed less and less that the corpuscle caused the disease but also hesitated to consider it a completely reliable symptom.

A third factor also led Pasteur astray. His studies on fermentation and spontaneous generation, and more specifically on germs in general, had familiarized him with their development and the reproduction of yeasts and bacteria: budding, followed by scissiparity. Had he not spent long hours drawing, inventorying, and describing microbes with extraordinary conscientiousness? Each one of their movements had been shown. The world of germs and that of microbes and fungi was familiar to Pasteur, but the corpuscle he observed in the viewer of his microscope was a different kind of microorganism, a parasite. This was a new kind of microbe, completely unknown to him. What he discerned was a structure surrounded by an amorphous membrane [*gangue*]. Instead of multiplying, this mass invaded the tissues, a kind of protoplasm with barely visible contours. The substance that formed the corpuscle was homogeneous and translucent in the beginning, but soon developed rows of more and more distinct granulations, which eventually broke away and became as many distinct and clearly delineated corpuscles. This appearance seemed to him to be extremely different from that of bacteria and fungi, especially since the corpuscles "were not free to come and go" because, unlike the animalcules he had studied previously, they were immobile.

Pasteur may have been at a loss for an interpretation, but he did know how to observe the phenomenon. He sought to describe the corpuscle with extreme precision and even went so far as to "have its portrait drawn" by a professional. Yet at the same time he continued to define it in a manner that led nowhere, calling it necrosed tissue, a kind of late-developing abscess, an inconstant sign of pébrine. And, since he always had to reach a conclusion, he wrote to Jean-Baptiste Dumas: "I am quite convinced that these are neither animals nor plants."[20]

While Pasteur himself did not envisage the possibility that he might be dealing with a parasite in corpuscular form, his assistants were convinced that this was the case. However, since the master said very little, indeed nothing at all, in part no doubt because he did not want to be influenced by other opinions, and in part because he kept his communications for scientific meetings, his collaborators did not know exactly what he was thinking. To such a point, wrote Duclaux, "that while his mind moved in this direction and did not want to be diverted from it, his assistants, to whom he said nothing about his thinking, were persuaded that he was firmly attached to the idea of the corpuscle as cause. They therefore could not understand why he was not doing the requisite experiment, namely, feeding the healthy worms corpuscular food in an attempt to give them, not the rapidly evolving disease . . . , but the other disease with its slow evolution and the concomitant development of parasites."[21]

Confident that his interpretation of his first experiment was correct, Pasteur felt no urgency about recommending a second one. And so it was not in the field, at Pont-Gisquet, but in the rue d'Ulm that he repeated the experiment which consisted of making healthy worms ingest tainted mulberry leaves. Despite the difficulty of obtaining silkworms in Paris, the plan of the experiment called for feeding several lots of worms mulberry leaves, some of which were regular, while others had been dipped into an infusion of corpuscles. Some of the worms were fed these leaves at the third meta-morphosis, the others after the fourth. Pasteur noticed that those who had ingested a meal of corpuscular food became sick and spun an inferior co-coon, and that this happened sooner according to how early they had been contaminated.

"Here we have," Duclaux comments, "a rare occurrence in Pasteur's life, an experiment that did not immediately assume its full and entire significance."[22] For the results were clear: the healthy worms selected by examining the eggs under the microscope stayed healthy unless they were infested, and the effect of the ingestion of contaminated corpuscles de-pended on the moment of the feeding. And yet, to the amazement of Gernez and the other assistants, Pasteur was unable to draw the obvious conclusion from these observations. In particular, he did not see their most important implication, the demonstration that the corpuscle is the agent of infection.

Consequently, in the two successive reports he submitted in Novem-ber 1866 and in January 1867, Pasteur was still wondering whether the disease was parasitic and wrote: "I do not at this time believe that the corpuscles are parasites. If I assimilate them to organelles, blood corpuscles, pus corpuscles, etc., it is because I have never seen them reproduce, any more than one sees the corpuscles of blood or pus, spermatozoa, globules of starch etc. engender offspring. As long as the mode of generation of the corpuscles is not demonstrated, the idea that they are parasites will be unfounded. . . . I am quite prepared to accept the opinion of any scientist who will be able to demonstrate that he has gone further than I have done in his investigation of the generation of the corpuscles, for which I have looked as a possible sign of parasitism without being able to discover it."[23]

Pasteur was mistaken. He was not to change his mind until sometime in 1867, and when he did, he pursued the course he had hitherto rejected with equal vigor and self-assurance. For the contagion caused by the cor-puscular food could not escape Pasteur forever. In the course of his research campaigns of 1865 and 1866, he had selected eggs that he now cultivated in his experimental silkworm nursery. He now repeated the experiment for the third time. He obviously realized that nothing was easier than to communi-cate the corpuscle disease to healthy worms. Another incident put him on the right track. Not far from Pont-Gisquet lived a family of silkworm breed-ers, the Cardinals; they had obtained two batches of healthy Japanese eggs which, to their great surprise, produced corpuscular worms. This was reported to Pasteur, who immediately imagined what had happened: on the shelf placed above the Japanese eggs was a breeding chamber full of diseased worms, whose droppings had fallen onto the lower beds. This situation reproduced the experimental conditions of contamination.

The fact that Pasteur was slow to come to a correct interpretation of what he had seen was probably related to the specificity of animal biology as compared to the manipulations of the laboratory: the need to take incubation times into consideration may explain the errors made by an experimenter accustomed to instantaneous results. Once he had meticulously repeated his experiments and realized that he had gone astray, Pasteur was as eager to take back his earlier statements as he was to refute the assertions of those who, like Quatrefages, were not convinced of the contagiousness of pébrine and looked for the occult and mysterious influence of a harmful environment.

Although he was now convinced that the corpuscle was the source of the contagion, he was still unable to define its nature. Rereading the reference works on the question, Pasteur learned that a German investigator in Tübingen, Professor Leydig, had assimilated the corpuscle to a parasite. Leydig had not given any valid proof for his conclusions, but he made a connection between his observations on the silkworm and those concerning a sweet-water fish that seemed to present an identical parasite. Pasteur, who still hesitated, wrote to Leydig to ask for his opinion and described to him the clusters of cells he had seen and the circumstances that had led to contamination. Leydig replied: "On the basis of all my observations, I consider the corpuscles parasitic formations, whatever specific name one may give to them." It is likely that Leydig's letter definitively tipped the scales in favor of the parasite: Pasteur fell in with Leydig's opinion and announced to the world that pébrine was a parasite.

But there was still one problem, namely, how to explain the absence of a linkage between the spots on the silkworm and the corpuscular symptoms. Certain worms, as we have seen, presented spots without being sick. So far, no one had attempted to explain this relative discrepancy. In order to solve the problem, Pasteur used a clever experimental trick: examining broods of healthy silkworms, he segregated those that did not show spots at the time of the fourth metamorphosis and isolated some of them in matchboxes. When the worms began to climb up to their nests, he released the isolated ones and noticed that the solitary caterpillars did not present any spots, while those that had been reared in a group were full of black spots.

From this observation he concluded that the spots could also be caused by accidental injuries sustained as the worms climbed over each other.

In the case of contagion by corpuscles, the animals' skin presented another series of spots, not as numerous as those caused by injuries but otherwise looking very similar. Injuries and corpuscles thus accounted for the so-called pébrine spots, as well as for the discrepancy in their appearance.

"In this connection," Pasteur wrote in his book on the silkworm disease published in 1870, "one cannot help but point out that certain human diseases give rise to spots on the skin. . . . Nor is this the only observation applicable to human pathology that the experiments reported in this book may suggest to receptive minds."[24] Indeed, with our knowledge of Pasteur's subsequent research, "one cannot help but point out" that the scientist was already thinking ahead, beyond the silkworms, to the infectious diseases of humans.

Having demonstrated that the disease was contagious and transmitted by a parasite, and having explained the symptomatic spots, all that remained to be done to complete the investigation was a study of the modes of transmission. In time Pasteur was to show that the disease is transmitted by tainted food, by worm excrement (droppings, as he called it), and by the dust of the nurseries. Corpuscle-tainted mulberry leaves, he pointed out, represent one of the most important sources of contagion. Also, the worms are inoculated with the disease when they pierce each other's skin. The incubation period can be so long that broods of worms issued from contaminated eggs may not come down with pébrine before the upward climb to the nests. On the average it takes thirty days for the animal to become so infested with the parasite that it can no longer spin its cocoon. The observation of corpuscles, the number and size of the cocoons, and the quality of the eggs depend on the stage of the incubation. It is clear, said Duclaux, "that large-scale hatching can only take place if the eggs are pure, and surely they can be pure only if they come from corpuscle-free parents."[25]

The hypothesis of a hereditary disease had also been put forward. Pasteur eventually showed that the infection was indeed congenital and resulted from an infection of the eggs during the gestation period of the diseased female. In short, if pébrine caused Pasteur to stumble so many times, it was because, parasitical at first sight, it was above all endemic. Although the cause of the disease originally did come from the outside, it subsequently proceeded mainly by way of the eggs and the metamorphoses. It forced the scientist to abandon one of his axioms, namely the idea that all disease comes from the outside, and gave a boost to the concept of disease coming from the inside. Pébrine dealt a blow to the theory of airborne contagion, for this illness is transmitted by the egg.

For his first foray into the domain of pathology, Pasteur was thus dealt an unlucky hand, and the path he took was tangled and treacherous. If Pasteur initially minimized the role of the corpuscles and failed to understand their nature, he nonetheless showed remarkable aplomb in rectifying the situation and in the end arrived at a rather accurate description of pébrine, its symptoms, its modes of transmission, its agents, and the possibilities of preventing it. Through his experimentation he was able to arrive at compelling conclusions that others had been unwilling to accept and to make use of existing observations, or in some cases rediscover them. His work brought advances to many fields: hygiene, medicine, microbiology, and the silkworm industry. "I have mastered the corpuscle disease, which before me was considered the only disease affecting sericulture today," Pasteur concluded. "I can impart and prevent it at will."

The Morts-Flats

Yet the problem was not entirely resolved. Just as Pasteur began to feel that he had defeated the scourge by disseminating his method of egg production, new complications arose. He received letters from alarmed silkworm breeders for whom the disease had not abated. In reading this mail and visiting many silkworm nurseries in person, Pasteur became more and more concerned. "He kept us so far from his thoughts," Duclaux tells us, "that we did not understand why he was so worried until the day when he came in and, almost in tears, slumped into a chair exclaiming: 'Nothing has been accomplished, there are two diseases.'"[26]

And indeed, there was a special form of the silkworm disease called the *morts-flats*, or disease of the entrails, or *flacherie*. After the fourth metamorphosis, at the time of voracious feeding called *la grande frèze*, when the healthy worms noisily devoured their mulberry leaves, the diseased ones were indifferent to food. They crawled about on the leaves without biting into them, and sometimes even moved away, looking for a quiet corner in which to die.

Sometimes the dead worms became soft and decomposed, sometimes they remained firm and hard, so that one had to touch them to realize that they were dead. In some cases, the worm laboriously climbed up the branch, slowly spun its cocoon without finishing it, and died without producing a pupa or a moth.

Although flacherie differed from pébrine by the absence of spots, the idea that there were two different diseases was far from generally accepted. Yet Pasteur stated this hypothesis as early as 1865: noting that apparently noncorpuscular worms were affected, he came to formulate, at that rather early stage, the idea that there might be two diseases. Indeed, he even envisaged the possibility that corpuscles in the female moths might weaken the eggs to the point of making them susceptible to flacherie. But the association of the symptoms of flacherie and pébrine was so complex that Pasteur soon revised this first impression and did not pursue the implications of two separate diseases until 1867. At that date, the seriousness of the second illness became evident and threatened to wipe out the first positive results obtained through the new method of controlling the production of eggs. Pasteur was quick to understand not only the existence but especially the importance of the new scourge. Having become an excellent silkworm breeder, he had mastered the art of observing silkworms.

From the very beginning he attributed the poor development of the broods he observed in 1867 to flacherie. These worms had been sluggish in climbing up the branches and their noncorpuscular offspring hatched in the spring had been diseased. This was clear evidence for a disease that could not be attributed to parasites. But if there was no room left for doubt, the facts were becoming more complicated, for the selection of pébrine-free worms showed flacherie to be at least as contagious. It turned out that the batches of eggs affected by the morts-flats disease perished regardless of the circumstances of place or climate.

As he was finishing these observations on a series of individual broods, Pasteur realized that he had to start all over. One scourge was hiding another. His collaborators, among them Duclaux, were eager to set to work on the identification and prevention of another disease. But Pasteur, dejected by what he had discovered, was much less enthusiastic. "We tried our best, of course, to cheer up the master in his discouragement," writes Duclaux. "We told him that if everything had not been done, this did not mean that nothing had been accomplished, only that we had to go back to the job. We were young and we had confidence, not so much in ourselves, but in him. Actually, it was time well spent to watch him wrestle with these difficult problems, constantly searching, sometimes failing to find what he had expected and casting about for a new approach, sometimes triumphant and pressing forward. We did not always know where he wanted to go, for he did not say much. But we tried to guess at it, judging by what had transpired and rectifying our ideas according to what we were able to learn about his."[27]

The completed identification of the first disease made it easier to identify the second. In examining a diseased silkworm it became possible fairly quickly to assign to flacherie symptoms that had earlier been ascribed to pébrine. As for the other diseases of the silkworm, muscardine and grasserie, they were well known and did not interfere. The disaster for the silkworm industry seemed to be essentially caused by pébrine and flacherie. The task at hand, then, was to find out what caused flacherie and how it was propagated.

Silkworm breeders had long been aware of the circumstances that favored it, among them thundery weather with warm and humid ocean breezes, which seemed to impede the perspiration of the worms. Flacherie was also seen in poorly ventilated silkworm sheds, which were prevalent in the region of Perpignan, where the breeding chambers were housed in rooms with only one window. The period between the fourth metamorphosis and the upward climb, when the worm consumed large quantities of wet food, also favored the development of the disease. But these notions were still rather indistinct. They did not give rise to even vague rules, such as to avoid letting sick worms have their feeding frenzy in humid breeding chambers.

Looking for a germ as the possible cause of the disease — for this remained his fundamental hypothesis — Pasteur believed that he was working in a territory closer to something he already knew, the world of bacteria. The sense of smell is well known as a tool frequently used by chemists. Pasteur was no exception to this rule, which is why in these circumstances he made use of the skills he had acquired in dealing with vapors in a laboratory. Whenever he entered a silkworm nursery affected by the morts-flats disease, he was soon struck by a sharp and unpleasant smell that was strongest near the beds of diseased worms. In trying to identify it, he found that this smell reminded him of that of the volatile fatty acids released in the course of the fermentations he had studied. With brilliant intuition, he immediately made the connection between what his sense of smell told him and the facts he had earlier described, concluding that he was once again in the presence of the chemistry of fermentation.

What was the origin of this possible fermentation? Pasteur remembered a few observations formulated by Laffemas, valet to Henri IV, who had reported the death of silkworms fed on wet mulberry leaves. Laffemas had done so much to foster French sericulture that he was rewarded for his services by being appointed supervisor of mulberry plantings for all of France.

The wet mulberry leaf, that was the idea. The first hypothesis to be considered was that this food, being ground up and wet, might itself be the object of fermentation. If this was indeed the case, one had to find out where and how this new fermentative chemistry took place in the diseased worm. The simplest way was to assume that it took place in the enormous digestive tract that occupied a preponderant place in the caterpillar's anatomy. Using a scalpel to make a more thorough analysis of this intuition, Pasteur now opened a silkworm affected by flacherie and noticed under the distended and transparent envelope of the stomach little bubbles of gas that struck the walls of the organ as they rose.

When, following his usual procedure, Pasteur ground up the suspect organ, he believed that he recognized the responsible germ, a motile microorganism he considered akin to a bacterium. "In order to be certain that death in such a flacherie is essentially due to an alteration of the digestive functions arising as a result of fermentation, it is useful to compare the state of the matter contained in the intestinal tract of the diseased worms with that of triturated mulberry leaves left to stand in a more or less closed glass recipient at temperatures occurring in the months of May and June. It is then easy to recognize that in morts-flats disease the intestinal tract behaves like a tube made out of a mineral, glass for example, into which ground-up mulberry leaves have been placed. The same organisms are found in both places; and in both places one also finds either a series of the productions mentioned earlier or several others associated with the release of more or less abundant quantities of gas."[28]

Pasteur believed that he was seeing—frequently in diseased worms, more rarely in pupae and moths—certain of the microorganisms he had encountered in studying fermentation and spontaneous generation. Since the diagnosis seemed fairly easy to make, he even recommended a microscopic method of detecting the presence of a bacterial infection in moths chosen for reproduction, suggesting that following the laying of the eggs, the ground-up intestines of any suspect female be examined under the microscope.

Today we know that the problem is more complex than Pasteur realized at the time. While the disease is indeed partly of bacterial origin, viruses, which could not be detected by Pasteur's microscope, might well make the worms vulnerable to the bacterial invader. To be sure, Pasteur had neither the arguments nor the means to suspect this other infection. But since the bacterium was what he saw, he had to explain where it came from, why it developed, and what mechanisms of contagion were in play.

Moreover, Pasteur initially believed that the germ had no other habitat than the intestine of the diseased worm. Yet it was obvious that flacherie could break out sporadically. What, under these circumstances, could be considered the source of the germ? Pasteur was in a better position than anyone to answer this question. Had he not learned from the alleged spontaneous generations that the germ lives in the air? In order to demonstrate the applicability of this phenomenon to what he was seeing, all he had to do was to leave ground-up mulberry leaves out in the air in warm and humid weather and wait to see if they became contaminated and fermented. The intestine simply served as a test tube that allowed the bacteria to multiply. But the bacterium did need a favorable medium. The mulberry leaf, stripped from the branch, torn up, and stained by drops of sap that had oozed from the branch or from its own stalk, and in addition left to lie in the open air, allowed for such fermentation. Conversely, feeding the silkworms "in the Turkish fashion," that is, on the branch, so that the leaf remained intact, minimized this possibility. The harvested mulberry leaves, which were sometimes left for hours tightly packed together in sacks, were likely to heat up and ferment. "Place a drop of sap on a sheet of glass and cover it with an inverted drinking glass whose inner walls you have moistened in order to keep the sap from evaporating, and within 24 hours, in April and May temperatures, you will see the drop fill up with organisms, mainly vibrios."[29]

The last step was to assert that this was an infectious, and for that reason contagious disease. This is what Pasteur set out to establish by means of experiments similar to those he had used for pébrine. Mulberry leaves dipped into infusions and brush-painted with the ground-up intestine of worms that had died from flacherie, with droppings from diseased worms, or with dust from silkworm nurseries affected by the morts-flats disease were fed to healthy worms to spread the contagion. They did contract the disease. Bacteria multiplied in the intestine of healthy worms. Droppings of the tainted worms contaminated the others. On the soil of the nurseries, which was analyzed, the fragile bacteria usually did not live long. But sometimes they did develop there sufficiently to contaminate healthy worms and cause the leaves piled up in the breeding chambers to ferment.

In doing these experiments, Pasteur also noted down one observation that was to prove important for his future work, even though he did not immediately recognize its implications: if one infects the worms by feeding them fermenting leaves, death ensues in eight to fourteen days. If one uses the excrement of these diseased worms to infect new ones, death will occur sooner. Was this not an indication that the germ does not always have the same virulence and that it can become more aggressive after a first passage through an organism?

But the fact remained that the fermentation of the mulberry leaves and the more or less considerable multiplication of bacteria of varying aggressivity did not explain everything. And this is where the silkworm came in. Its role was to be more or less receptive to the other disease, for the atmospheric conditions (heat, humidity, ventilation) modulated its per-spiration; the spacing, the quantity, and the quality of the feedings had an influence on the fermentation, and all of this constituted a set of factors that did or did not favor the development of the disease. Furthermore, even though the flacherie bacillus was suspended in the atmosphere, at least in the Midi, it was not always present. A healthy silkworm could therefore defend itself against its onslaught, especially if the fermentation was just beginning when it swallowed the bacillus, or if the bacillus was isolated. But whenever the number (or the aggressivity) of the bacteria increased, or when the worm was weakened, or when the contagion was brought in through a different process (a sting, for example), the worm was more likely to become diseased.

It should be pointed out that the description of the multiple causes that facilitate the proliferation of the microbes responsible for flacherie, whether they be bacteria or viruses, was less important than the fact that Pasteur had established that this was indeed an infection and that he had attempted to prevent it. On the basis of his finding, he proposed a series of hygienic measures, including better ventilation for the nurseries, scrubbing the floors, careful management of the silkworms' food, the picking and conservation of the mulberry leaves, and preventing heat and humidity from pervading the atmosphere of the breeding chambers. Before outlining these measures, he went so far as to look for the most judicious prescrip-tions in Chinese works on the subject, for he knew that flacherie had not sprung up yesterday. In these works he found the most simple, practical, and precise methods imaginable for controlling the temperature of the breeding chambers: "The woman who takes care of the silkworms must wear a simple unlined dress. She will regulate the temperature of the prem-ises according to the sensations she herself feels. If she feels cold, she is bound to judge that the silkworms are cold and will therefore stoke up the fire, and if she feels hot, she will conclude that the worms as well are too hot, and so she will lower the fire as much as is necessary."[30] Stagnant and heavy air and air that is hard to breathe or imbued with unhealthy odors were to be condemned. Smoking in the presence of the silkworm had long been considered harmful by Chinese and Japanese authors. These judicious precepts were meant as much for Pasteur's collaborators as for the industrial silk producers.

Nor was this all; for as he was looking for further preventive measures, Pasteur made use of certain observations. When little Marie-Louise raised silkworms in front of the fireplace in the dining room at Pont-Gisquet, he noticed that her broods were always healthy. He attributed this to the draft coming from the duct; as a result he recommended ventilation. Visiting neighboring silkworm nurseries, he identified such problems as excessive humidity, breeding chambers that were too large, and poorly kept beds, whereupon he pointed out that such shortcomings favored infection. But he also thought about other factors that should be avoided, such as exposure to excessive heat from the sun, the strong wind of the Midi, or food that was no longer fresh.

Because a contagious disease results from several factors that come together to facilitate the multiplication of germs, Pasteur discovered hygiene and the possibility of deploying preventive measures. Indeed, if pébrine had indicated to Pasteur that a germ can be responsible for an epidemic, flacherie revealed to him the importance of hygiene in combating microbes. When all is said and done, pébrine and flacherie opened Pasteur's eyes to the rules of contagion and the role of hygiene in infectious pathology.

The Hucksters in the Temple

The silkworm breeders did not easily accept these discoveries. To be sure, Pasteur received benevolent help, loyalty, and sometimes enthusiasm from a few regional leaders of the silk industry. Lachadenède, president of the agricultural association of Alès, Despeyroux, professor at the collège of the same town, Siran, a pharmacist at Grenoble, the Comte de Rodez, director of the silkworm nursery of Canges, all were both zealous and intelligent about applying Pasteur's procedures for producing healthy eggs in a most conscientious manner.

The mayor of Saint-Hyppolyte-du-Fort, Bénigne Jeanjean, was one of his most steadfast supporters. Pasteur often corresponded with him. Jeanjean kept him informed of the quality of the local egg production and spread the word about his precepts in his community. The Comte de Rodez became involved in experimentation and was the first to show that the males have very little influence in the propagation of the disease. Pasteur was generous to those who followed him and asked the minister to reward his faithful early supporters with the Légion d'honneur. He felt that this was the least one could do to foster the principles of combating a scourge that had endangered the economy of southern France for twenty years.

But there were those who did not agree. Pasteur indefatigably pointed out to them that the first major experimental test carried out according to his indications had prepared 2,511 eggs guaranteed free of pébrine. All of these had properly hatched and produced harvests unknown even in times of prosperity in six departments: Basses-Alpes, Hautes-Alpes, Alpes-Maritimes, Var, Hérault, and Vaucluse. To be sure, there had been rather numerous failures in Isère, Ardèche, and Gard, but these were almost always due to the morts-flats disease, which meant that developing measures to combat the combination of the two diseases was the only way to be confident about future harvests.

Why did he always have to persuade, when the results seemed to speak for themselves? Essentially because the naysayers produced a flood of reports and criticisms. They said that the disaster continued unabated, that Pasteur's results were insignificant, and that the same numbers of silkworms were still dying of pébrine and flacherie. They expressed doubt that Pasteur had discovered the disease and called his method of egg production difficult and delicate. Moreover, misprints — which may have been intentional — in local newspapers, particularly the *Moniteur des soies*, sometimes falsified the results of the harvests.

In reality, there were many reasons for these criticisms: habit, or especially the desire to try out their own remedies caused some breeders to apply Pasteur's methods improperly. But if this accounts for the reticence of a few isolated breeders, the political campaign against Pasteur was mainly conducted by the merchants who traded in silkworm eggs. The disaster had fostered the development of a network of importers, and those who were part of it had become rich by trading with the Far East. The trade in Japanese silkworm eggs, in particular, was ruined by Pasteur's procedure, which offered eggs of better quality for a cheaper price.

The greed and jealousy of some was compounded by the stupidity of others, to Pasteur's great exasperation. Thus he wrote to Marquis de Binard in October 1868:

Monsieur le marquis, I have read a first letter to the editor you published in the *Moniteur des soies* concerning the research I have been pursuing for the last four years in the South of France, as well as a second, equally sharp letter which appeared in the October 5 issue of M. Baral's *Journal de l'agriculture*. Permit me to tell you, Monsieur le marquis, without acrimony and with all the deference due to your sense of honor, which I grant you as much as you profess to grant me mine, that you do not know the first thing about my research, its results, the firm principles it has established, and the practical importance it has already acquired. You have not read most of my publications, and as for those that have come into your hands, you have not understood them. . . . Please do read these humble studies with just a fraction of the care I have put into pursuing them, and when you have thought about them and understood them, I would ask you to write again to the *Moniteur des soies* about your impressions and your criticism. If I am pleased with my pupil's progress, I will debate these matters with him. As things stand now, we would be fighting with unequal arms, which would not be worthy of a true gentleman. With the greatest respect I am, Monsieur le marquis, your very humble servant.[31]

Elsewhere, Pasteur responded to a report which claimed that the disasters continued unabated and that the eggs obtained by following his indications had not been as successful as expected: "Considering the matter from the scientific point of view, it seems to me that in an investigation that is as delicate as establishing the causes of an epizootic and the means of preventing it, there is merit in having delineated the questions and the alternatives with as much clarity as I have done. For it is still true that a well-posed question is already half resolved. It would be an undue underestimation of the difficulties of an investigation of this kind not to agree that I must have conducted it with exceptional precision, considering that I predicted the outcome that actually occurred . . . with quasi-mathematical accuracy."[32]

But the dealers in silkworm eggs continued to spread the most mendacious rumors. Pasteur's father-in-law, Charles Laurent, became concerned and wrote to his daughter Marie in a letter dated Lyon, 6 June 1868: "I have to tell you that rumors have been spread here to the effect that the scant success of the worms raised by Pasteur's methods has upset the population of your region to the point that he was forced to leave Alès precipitously, assaulted by a hail of stones that the local inhabitants threw at him."[33] These insinuations left their mark on people's minds — other people's, that is, but not that of Pasteur, who, sure of himself, wrote to the Minister of Agriculture in December 1868: "In the report I had the honor of submitting to Your Excellency under the date of 5 August last, I laid out the proofs that permit me to conclude that sericulture can henceforth avail itself of a proven and practical method to avoid the ravages of the more disastrous of two diseases that cause such severe damage to silkworms. No doubt it will take time to propagate the preventive technique I have introduced. It is always difficult for new practices to gain the acceptance of people who have an interest in the matter; indeed, in the beginning they usually inspire envy in some and defiance in the public at large."[34]

Half-Paralyzed in Body but Not in Mind

On 19 October 1868, Pasteur was to report to the Academy of Sciences on the work on silkworms of the Italian scientist Salimbeni. On the morning of the session, he did not feel well and noticed a tingling sensation over the entire left side of his body. Rather than lying down, he insisted on going to the Quai de Conti. Feeling uneasy about it, Marie accompanied him as far as the steps of the Institut. Pasteur stayed for the entire session and then walked home with Balard and Sainte-Claire Deville. After dinner, he went to bed early, but his malaise soon worsened. He was no longer able to speak or to move the limbs on his left side. Urgently summoned to his bedside, Pasteur's personal physicians, Godélier and Gueneau de Mussy, called in the famous Andral of the Academy of Medicine, who prescribed the application of leeches behind the ears. He acknowledged that he was baffled by this attack of hemiplegia, which was very different from anything he had ever seen. As soon as the blood flowed, Pasteur felt better; his speech returned and some movement was possible. But the improvement was short-lived; once the effect of the leeches had worn off, the situation suddenly deteriorated. The slightest movement became difficult, and the hemiplegia became complete within twenty-four hours, yet Pasteur remained conscious and lucid, a fact that the physicians found unusual. On 21 October, Godélier noted after having examined his patient: "Active mind. Would have liked to talk science."[35] Periods of calm and agitation were to alternate during the next few days.

Pasteur's faithful followers immediately rushed to his bedside: Sainte-Claire Deville, Dumas, Bertin, Gernez, Duclaux, Raulin, and Didon, the latest assistant to be hired. All of Paris's prominent scientists came to find out how he was. The Emperor and the Empress sent a lackey to inquire about the patient's condition.

Over the next few days, there was a slight improvement. His speech returned first. A week after the attack, while Gernez was taking his turn by his bedside, Pasteur undertook to dictate to him a note for the Academy; his disciple did not have to change a single word of the text, which outlined an ingenious procedure for preventing flacherie. Yet at that point Pasteur also asked Marie to let the Emperor know that he was dying "filled with regret that he had not done enough to honor his reign." Sincerity? Pride? Opportunism?

But Pasteur did not die; he began to recover. In order to help him bear the enforced immobility of the first weeks, his supporters and friends took turns reading to him: One chose Pascal's *Pensées*, another Bossuet's treatise *Of the Knowledge of God and Ourselves*; nor did modern works make any concession to frivolity, whether they were a biography of Jenner or a book on Jansenism by the very earnest Silvestre de Sacy.

Meanwhile, Pasteur's greatest worry was the laboratory that Napoleon III had promised to have built for him in the rue d'Ulm; he regularly inquired about the progress of the work. From the windows of the dining room, which overlooked the garden of the Ecole normale, he could see the building site. The workmen had disappeared at the first sign of his illness. He let it be known through General Favé, who came to inquire about him, that he was being buried too soon. Napoleon III and Victor Duruy apologized for this haste and orders were given to resume the construction.

From his invalid's chair he followed the construction of the new building. His lucidity had not flagged for a moment, and he knew that he would remain handicapped. And indeed, he was never to recover completely from the paralysis of the left half of his body, losing the use of his left hand to the point that he had to depend on his collaborators for the most delicate experiments. Fortunately, his right hand remained agile. He also moved with difficulty and therefore requested that an apartment be built above the laboratory, which would make it easier for him to work there after hours. But the building under construction was situated above some quarries and the addition of an extra story would have entailed considerable work to strengthen the foundations. The government was unwilling to face this additional expense.

For twenty years, Pasteur therefore had to cross the courtyard of the Ecole normale to reach his laboratory. Generations of Normaliens thus saw the familiar figure of the scientist, invariably wearing his skull-cap, sometimes of black silk and sometimes of wool in a pattern of small black-and-white squares.

By January 1869, Pasteur was able to walk again and went to recuperate in the south, at Saint-Hyppolyte-du-Fort. He decided to return to this region with some of his faithful co-workers in order to resume work on a study that he considered unfinished: "Monsieur le ministre, it is necessary for me to go south one last time to complete my observations on the diseases present in the silkworms," he wrote on 9 January. "Unfortunately my health has been seriously compromised by a grave and recent accident, so that I require special care and must therefore surround myself with more aids than in the past. I greatly regret this inconvenience, which is bound to create difficulties for the administration."[36]

On 18 January, three months to the day after the stroke, he had himself transported to the gare de Lyon. His wife and little Marie-Louise accompanied him. Installed in a specially arranged compartment where he could lie down, he traveled as far as Tarascon, from where a coach took him to a townhouse in Saint-Hyppolyte-du-Fort, located close to a silkworm nursery. This is where he settled. The state of his health did not permit him to go to Pont-Gisquet. "In this region, where people seek protection only from the heat," Gernez remembered later, "all he could find was a cold, poorly arranged and poorly furnished house."[37]

Maillot and Gernez, who had come to help out, improvised a laboratory. From his armchair or his bed, Pasteur dispensed advice, guidance, and criticism. "The operations, whose phases we followed under the microscope," Gernez wrote, "coincided with his predictions in every point, and he was pleased that he had not given up."[38] In Paris, some people criticized him for making the trip, considering him careless about his health, while others admired his courage. Jean-Baptiste Dumas and Marshal Vaillant sent him letters of support. Pasteur was moving too fast. As soon as he felt a little better, he tried to walk, fell on the stone floor of the cold house and suffered bruises that were "very slow to heal and painful." But soon he confided to Dumas:

> Today the aftereffects of this accident have disappeared completely, and I feel as I did three weeks ago. Progress on moving the arm and the leg seems to have resumed, but all of this is excessively slow. One of these days I will have recourse to electricity, on the advice of Dr. Godélier. . . . As for my head, it is still perfectly sane. Here is how I spend all my days: in the morning my three young friends come to see me and I assign the work for the day. I get up at

noon, having lunched in bed, having the newspapers read to me or dictated a few letters. If the weather is nice, I spend an hour or two down in the little garden of the house where we live. . . . Before dinner, which we — my wife, my little daughter, and I — take by ourselves in order to avoid the fatigue of conversation, my young collaborators come to report on their investigations. Around seven or half-past seven, I feel so tired that I think I will be able to sleep for twelve hours at a stretch; but around midnight I invariably wake up and only go back to sleep toward morning for an hour or two.[39]

As the time for the hatching of the silkworms approached in the spring of 1869, Pasteur sought to accumulate proofs for the validity of his method, for some of the breeders still had doubts. Most notably the members of the Silk Commission of Lyon, who ran an experimental silkworm nursery of their own, still did not consider the process reliable. It was often said that one should not have too much faith in the "micrographs," which is an indication that people were still afraid of the microscope.

"Our Commission," its secretary had written at the end of the previous year, "considers that the testing of the corpuscles provides a useful indication to take into consideration, but that its results cannot be presented as a fact from which absolute conclusions can be drawn." In his characteristic tone that allowed for no rejoinder, Pasteur asserted that the reliability of his products was in fact total. In March, the Lyon Commission asked Pasteur to furnish it with a brood of healthy eggs in view of one last test. Pasteur did better than that. He offered different broods, some of which would succeed, while others would come down with pébrine, flacherie, or a combination of the two diseases. "It seems to me," Pasteur added, "that comparing these different groups would be more apt to enlighten the Commission's judgment of the soundness of the principles I have established than limiting its observations to one or several batches of eggs labeled healthy."[40] He was vindicated in every respect. The predictions he had made for the different batches turned out to be correct. Every box of eggs showed the expected result: one or the other disease or no disease.

At the same period, Maillot traveled to Corsica, where a silkworm breeder of Vescovato, near Bastia, had asked for help. He took along some good eggs to demonstrate the quality of the Pasteur method.

As soon as the weather turned warm, the group returned to Pont-Gisquet, to that calm retreat surrounded by mulberry trees and silkworms where, as Pasteur saw it, everything invited one to work. From here, Pasteur's students were sent out on assignments, transformed into veritable apostles. Duclaux, who had taken charge of the broods at Pont-Gisquet, went to the Cévennes to supervise the egg production of the latest selections. Gernez went to Basses-Alpes to verify the results obtained with the eggs Pasteur had produced the previous year and then went on to the Paillerols estate at Digne. Only Raulin stayed at Pont-Gisquet to complete the study of the mode of counting the cases of flacherie.

In the midst of this campaign, Pasteur received a letter from the Minister of Agriculture, asking him to examine three batches of eggs that a silkworm breeder in Corrèze, a woman who was known for the good management of her nursery, had sent to the ministry to testify to the success of her own method. Thinking that her method was better than Pasteur's and wishing to have it officially adopted, Mlle Victorine Amat had sent these samples of eggs to Paris. The minister asked Pasteur to test them and to submit his detailed report as soon as possible.

Pasteur's reply came a few days later: "Monsieur le ministre, following the instructions of your dispatch of 20 April of this year, I have tested the three fresh lots of eggs you have sent to me. . . . These three kinds of eggs are very bad. If they are raised in breeding chambers, even in very small broods, they will all perish of the corpuscle disease. . . . As far as I am concerned, I am so sure of the accuracy of my judgment that I will not even take the trouble to verify it by raising the samples you have sent to me. I have thrown them into the river."[41]

Hemiplegia had thus by no means dulled Pasteur's quickness to respond and his vehemence in stating his convictions, even though those who came to see him during the long evenings of spring and summer found him peacefully sitting in his armchair in the shade of an orange tree, dictating to Marie a paper for the Academy of Sciences, a reply to letters from silkworm breeders in the Midi, or a pamphlet against some journal of sericulture.

At such times, the work would be interrupted. The conversation might turn to science and the time it takes for discoveries to become familiar to people's minds. Had it not taken Parmentier fifteen years to prove that the potato was safe to eat? And did he not have to wait until Louis XVI had the good idea to wear the pale purple flower in his buttonhole? Pasteur concluded sententiously: "One has to be quick to be useful!"[42]

The letter he sent at about the same time to a certain Achille Vogué, who had asked for a few handwritten words for his collection of autographs, testifies to the same desire to go straight to the heart of the matter, as well as to an uncanny skill in promoting himself.

You are insistent, Monsieur, about obtaining a few handwritten lines from me. I humbly beg you to excuse me for not answering the first two letters you wrote to me about a year ago expressing the same request. I was no doubt wrong to consider them a bit extravagant, considering that I do not have the honor of knowing you. Since you are so keen on autographs, I hope you will realize that this note should be of inestimable value to you. I have not written by hand to anyone for the last five months. On 19 October of last year, I suffered a cerebral congestion, and the resulting paralysis is far from having disappeared. This is what happened to me as a result of the immoderate amount of work I had done over the last five years in my effort to discover a means to prevent the disease that is decimating the silkworms. I will add, to the shame or the glory of human nature, that despite the urgent pleas of my family and my friends, I have wanted to come here to finish this project. This will explain to you why this little note is dated from a small town in the Cévennes. Louis Pasteur, Member of the Academy of Sciences, On assignment at Saint-Hyppolyte-du Fort (Gard).[43]

To Her Majesty the Empress

While the debate over the method of producing healthy eggs continued in the Midi, Pasteur had found a prime ally in Paris: Marshal Vaillant. Minister of the Imperial Household, this old soldier, who was also a member of the Institut and of the Imperial and Central Society of Agriculture, had set up a small silkworm nursery in his office in the Tuileries. This experimental nursery, which he ran according to Pasteur's principles, allowed him to appreciate the usefulness of the method, for he had followed the successive stages in the life of healthy caterpillars: selected eggs, worms without spots, and finally superb yellow-and-white cocoons. He let the moths fly out of his windows toward the Place du Carousel.

Convinced that this was an effective technique, and intent, as much for the sake of the silk industry as for Pasteur, to put an end to sterile quarrels once and for all, Vaillant had the idea of an experiment that would be sure to overcome the last reticence, namely, to apply the method at an imperial silkworm nursery.

It so happened that the crown owned a property, called Villa Vicentina, in Illyria, a few kilometers from Trieste. Princess Elisa, a sister of Napoleon I, had peacefully lived there after the fall of the Empire and had left it to her daughter, the Princess Bacciochi, who had willed it to Napoleon III. Vines and especially mulberry trees were grown on this huge estate, but for many years pébrine and flacherie had ruined the production of silk, as they had done everywhere else.

As part of his functions, Vaillant was responsible for the management of the imperial properties, and it was incumbent upon him not to leave the estate unproductive. Hence the idea of applying Pasteur's method to it and proposing to the breeders to start over with his select eggs. With the consent of the local manager, he therefore asked Pasteur to send to the administrator of the crown's agricultural establishments one hundred ounces of select eggs for distribution at Vicentina. One hundred ounces was a very large quantity, considering that each ounce could yield thirty kilograms of cocoons. At the same time, Vaillant proposed that Pasteur live at the villa so that he could supervise the experiments.

It was October. Elisa's villa, a white, two-story house overlooking spacious lawns and stands of trees in a park of sixty hectares surely was an ideal place for a convalescent. Three weeks later Pasteur, who had briefly gone back to Paris, was on his way there. He went with Marie and his two children; for once, Jean-Baptiste (who was almost eighteen) was included. Since Pasteur was still weak, the trip was made in small stages; the first stop was Alès, where the party had to stop to pick up the eggs. On 25 November, the family finally arrived at Villa Vicentina. Pasteur was immediately delighted with his reception and the beauty of the country. Autumn on the Adriatic. The Pasteurs moved in for a long stay. Raulin was expected to join the family a few weeks later to take charge of the distribution of the eggs and supervise the broods that were being raised.

It was to be a period of calm, although it was not altogether devoted to leisure. For while he was waiting for the eggs to hatch, Pasteur went back to work and composed a synthetic work on his experiments concerning the silkworms. Making use of publications, notes, historical documents, reports to the Senate, press articles, and letters, he dictated to Marie page after page of what soon became a thick book. Pasteur kept to a strict schedule of writing, so that by February the manuscript and galley proofs went back and forth between Paris and Villa Vicentina. The days of winter and spring thus slipped by, all devoted to these literary tasks. They were punctuated by daily outings, either on foot or more often in a small horse-drawn cart that had been bought and repainted in blue with red trim. Aside from these excursions and the visits of a few neighbors, there were few distractions at Villa Vicentina. The children were learning Italian or revising their notions of history: "In one of the town's libraries," Pasteur wrote to Marshal Vaillant,

"I am looking for all the works about Napoleon I in order to move my children's hearts by the great examples of glory and devotion."[44] Jean-Baptiste was working, not without difficulty, on correspondence courses in law.

Surrounded by his loved ones, whom he could enjoy almost for the first time, Pasteur discovered that he had a gentle and tranquil family. But it is true that he was by no means completely recovered. "How I am?" he wrote to Sainte-Claire Deville in March 1870. "I am alive, that is all I can say. Can I hope to return to normal? Yes and no. When my head is clear, I forget the past and make plans for the future. I see myself surrounded by collaborators in a beautiful, spacious laboratory. I walk from one to the other to check on their experiments and I even indulge in the idea that some day I will recover the use of my hand. But sometimes my head is muddled and heavy, and then I despair, and all I want is to go off to die in some hidden corner."[45]

Soon it was hatching time for the silkworms. Pasteur divided the eggs among the different breeders on the estate, setting aside twenty-four ounces with which to raise one large brood under his personal care. Only one incident troubled the raising of these silkworms. The manager of the estate did not want to waste one carton of Japanese eggs he had left and had them taken to the market to be sold. Furious at the thought that these dubious eggs, if mixed with those he had so painstakingly selected, might compromise the harvests and thereby minimize the results of his method, Pasteur summoned the manager. In a grand theatrical gesture, he forbade him ever to reappear in his presence and sat down to write to Marshal Vaillant. Vaillant, who knew Pasteur and his reactions, suggested that he insert in the local papers a notice to the effect that "M. Pasteur guarantees only those eggs that he has raised himself and those he has personally given to the breeders."[46]

Actually, the breeders soon found out about Pasteur's method of egg production. That year's harvest of cocoons was better than anything that had been seen in many years. It netted a profit, after expenses, of 22,000 francs, a considerable sum for the period. Vaillant wrote to Pasteur that the emperor was amazed and delighted and thinking about rewarding him as he deserved. He was thinking about doing what had been done for two of Pasteur's illustrious colleagues, Jean-Baptiste Dumas and Claude Bernard, namely, to appoint him to the Senate. Sainte-Claire Deville, who was considered one of the principal contenders for this post, removed himself from the competition and became one of Pasteur's most ardent supporters.

On 27 July, Pasteur's name was inscribed on the list of promotions, along with that of Emile Augier, who, after Mérimée and Sainte-Beuve, was called upon to represent French literature in the Senate. The decree indicated that Pasteur had been appointed "for services rendered to science through his excellent work."

In the summer of 1870, Pasteur left Villa Vicentina. On his return trip he crossed through northern Italy and Austria, where silkworm breeders were beginning to obtain good results with his method of egg production. Convinced of its effectiveness, the Italian breeders even sought to improve it. It was thus for his work on the silkworm, almost as much — perhaps more — as for fermentation that Pasteur became known in Italy.

It soon became clear that Villa Vicentina represented his last encounter with the Second Empire. His stay at this imperial estate was the symbolic high point of an investigation undertaken at the behest of the government of Napoleon III. It is therefore no coincidence that he dedicated his *Etudes sur la maladie des vers à soie* [Studies on the Disease of the Silkworm] to Empress Eugénie: "To Her Majesty the Empress, homage born of deepest gratitude and high admiration for her lofty spirit and her generous heart." And the dedication, like those of Racine's tragedies, developed into a veritable letter: "Madame, in dedicating these studies to Your Majesty, I am fulfilling a duty. . . . The Empress, touched by the misery brought in the wake of the malady that began to decimate the silkworms fifteen years ago, ruining one of the most prosperous rural industries of France, deigned to take an interest in my first observations and to urge me to pursue them, pointing out to me that science is never grander than when it sets out to widen the scope of its beneficent applications. At that time I made to Her Majesty a promise that I have endeavored to fulfill by five years of persevering research. . . . I am, with the most profound respect, Madame, Your very humble, very obedient, and very faithful servant. Louis Pasteur, Member of the Academy of Sciences."[47]

The victory over the diseases of the silkworms, or rather the discovery of the means to prevent them, had rewarded five years of effort. Despite widespread criticism, jealousies, and calumny, the new method of egg production saved sericulture. Emile Roux later wrote of Pasteur's book on the silkworm that it was a veritable guide for anyone who undertook to study contagious diseases. Pasteur was aware of this and pointed it out to the physicians. He never failed to say to those who came to work in his laboratory, chosen by him to collaborate in his study of infection in animals: "Read the *Etudes sur la maladie des vers à soie*, for I think that it will be a good preparation for the work we are about to undertake."

Even after he had embarked on new and entirely different studies, Pasteur continued to follow the progress of his method in sericulture. For a long time he remained in correspondence with the men who had defended it in its early stages, the pioneers of French silkworm-egg production. His research also gave him other ideas for industrial applications, for Pasteur was struck by the fact that a microscopic parasite could endanger a whole sector of industry. He thought, for instance, about using the murderous power of microbes to get rid of harmful insects. At the time of the great invasion of phylloxera, he proposed the use of germs to combat certain parasites that ravaged the harvests. In this manner, Pasteur discovered the principles of bacteriological warfare.

More important than these applications, and beyond them, Pasteur's work on the silkworms served above all to introduce him to animal biology. On the borderline between the chemistry of fermentation and the culture broth, the dropsy of the moth gave microbiology one of its first credentials in the study of transmissible diseases. The caterpillar of Alès led Pasteur from microbiology to veterinary science and to medicine.

Portraits of Pasteur's parents, drawn in pastels by Louis himself

Marie Pasteur with her daughter Camille

The Pasteur home at Arbois

Louis Pasteur as a student at the
Ecole normale supérieure

Diagram of a crystal, drawn by
Louis Pasteur

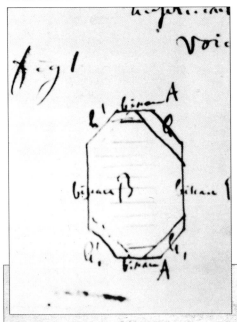

The buildings of the
Ecole normale supérieure ca. 1846

Jean-Baptiste Dumas in 1860

Jean-Baptiste Biot ca. 1850–1860

Pierre-Auguste Bertin in 1874

Louis Pasteur shortly after his arrival
in Strasbourg in 1852

Louis and Marie Pasteur at Pont-Gisquet in 1868,
at the time of the silkworm study

Louis Pasteur in the early 1860s.
Photograph by Pierre Petit.

The Tourtel Brewery at Tantonville
in 1873

The laboratory at Pont-Gisquet, Alès

Sheep being vaccinated against anthrax at Pouilly-le-Fort in 1881

The Chamberland autoclave

Swan-necked flasks from the
Pasteur laboratory

Extermination of rabbits by inoculation with chicken cholera in the
Champagne region, 1887

Vase created by Emile Gallé for Pasteur's jubilee in 1892

Photograph of Louis Pasteur "taken at night" by Liebert ca. 1885

Letter from Louis Pasteur to Joseph Meister, dated 27 November 1885. Pasteur commends the boy for improvements in handwriting, spelling, and even reasoning.

Joseph Meister, photographed at Schlestadt in 1885 by Mabille.

Illustrations on this page copyright Roland Dreyfus.

Jean-Baptiste Jupille in 1885

Group of Russians from Smolensk, at Hôtel-Dieu in 1886 during treatment for rabies

Alexandre Yersin

Emile Duclaux working on silkworms
in 1868

Emile Roux

Adrien Loir, Pasteur's nephew and an assistant

Eli Metchnikoff (*left*) in
his laboratory

Albert Calmette in Saigon, 1882

Louis Pasteur's inaugural lecture at the Académie française, 27 April 1882

Louis Pasteur surrounded by his collaborators, 1884

"The Good Shepherd."
This caricature and play on
Pasteur's name makes reference to
his work on anthrax. It appeared
in *Le Charivari* on 27 April 1882.

This cartoon appeared in *Le Rire*
in 1895. The caption reads:
"Student — What are you doing,
Master? Pasteur — I am vaccinating myself against orders
from Prussia."

L'ÉLÈVE. — Que faites-vous, Maître?
M. PASTEUR. — Je me vaccine contre les ordres prussiens.
(*Floh*, Vienne.)

Main building of the Institut Pasteur
ca. 1900

Louis Pasteur's bedroom

Pasteur in his study

Last photograph of
Louis Pasteur, taken in
the garden of the Institut

Pasteur with his family in the garden of the Arbois house, summer 1892

State funeral for Pasteur at Notre-Dame de Paris, 5 October 1895

Pasteur's burial crypt, at the
Institut Pasteur, 1888

Illustrations copyright Institut Pasteur except those on the eighth page.

INDUSTRIAL PASTEURIZATION

In July 1863, Ildephonse Favé, aide-de-camp to the Emperor, suggested to Napoleon III that he commission Pasteur to study the diseases of wine. For some years now, many wines had been subject to alteration, which caused great hardship to the wine trade. Different diseases gave them a sour, bitter, or insipid taste or an oily look. The situation was such that a few months after the signing of the free-trade treaty between France and Great Britain, an English wholesale merchant could write: "Surprise is expressed in France that the commercial treaty has not brought greater expansion of the English trade in French wines. The reason for this is rather simple. In the beginning we eagerly greeted the arrival of these wines, but we soon made the sad experience that this trade caused great losses and endless trouble because of the diseases to which they are subject."

Favé's approach was straightforward and did not become embroiled in useless formalities, for the urgent nature of the problem demanded the requisitioning of a scientist. "Having acquainted the Emperor with the results of your research on putrefaction, I spoke to him about the question of wine and its diseases. The Emperor is firmly convinced that it would be of the highest importance that you turn your attention in this direction at the time of the grape harvest. His Majesty has agreed to aid you in this effort and to give instructions to this effect to public functionaries who might be called upon to provide you with the needed facilities."[1]

Because of his research on fermentations, Pasteur was indeed well informed about the problems involved in the conservation of wine. Flattered that his work had attracted interest at the pinnacle of power, he accepted the imperial request but refused all financial aid: "Please permit me to devote the modest compensation I receive as a member of the academy, a sum I have set aside for this purpose, to undertake inspections of the fermentation practices in the different wine-producing localities and to study the diseased wines on the spot as they become available, as well as to gather the observations and the wishes of competent practitioners in this field." In truth, the initial diagnosis was rather alarming: Pasteur estimated that French table wines were diseased to such an extent that, despite the denials of the owner-producers, "there may not be a single winery in France, whether rich or poor, where some portions of the wine have not suffered greater or lesser alteration."[2]

Evening Course at Compiègne

Two years later, in the autumn of 1865, when Pasteur had already started numerous investigations and advanced some hypotheses, Napoleon III sent word that he would be most interested in being kept informed of his research and invited him to spend a week at the château of Compiègne.

A royal residence whose origin went back to the earliest days of the French monarchy, and a hunting lodge favored by Louis XV, Compiègne was above all a showcase for the grandeur and luxury of the Empire, or rather of its emperors. For it was here that Napoleon I had married Marie-Louise of Austria and that later Napoleon III fell in love with beautiful Mlle de Montijo, Countess of Teba, who had been invited to be a companion to his mother. When Eugénie marveled at the effect of the dew on the clover when strolling in the park, the Emperor sent her, the next day, an emerald brooch in the form of a clover, studded with diamonds representing the dew. In memory of the early days of their love, the sovereigns were sentimentally attached to Compiègne. It was their official vacation residence.

The custom of "Compiègne Days" went back to 1856. Every year, the court took up residence there for a month and a half in autumn. Series of invitations were organized according to an immutable ritual. Every series involved some one hundred guests accompanied by their servants. At such times the château became a kind of grand hotel that sheltered some one thousand persons belonging to various categories. The invited groups were for the most part composed of princes, ambassadors, ministers, marshals, and high functionaries, but the Emperor also received on each of these occasions three or four figures from the artistic, literary, or scientific world. The painters Eugène Delacroix, Gustave Doré, Ernest Meissonier, Horace Vernet; the writers Edmond About, Alexandre Dumas the Younger, Théophile Gautier, Paul de Musset, Sainte-Beuve, Jules Sandeau, and Alfred de Vigny had attended. The first scientist to be invited had been Claude Bernard in 1864. Pasteur was received in 1865. Milne-Edwards and Chevreul received the call only later.

Napoleon III had commissioned Pasteur's investigations on wine, and some people in the imperial entourage were interested in spontaneous

generation. Thus there was nothing surprising about the invitation, although it did testify to the real importance the government attached to men
of science.

To go to "a Compiègne" was a flattering prospect, but also a daunting
obligation. One had to have a special wardrobe, for the protocol prescribed
the wearing of the formal French dress suit in the evening. Mérimée made
fun of this custom, which frightened some people: "This morning I saw
Sandeau, who arrived here in the flustered state of a man who has just tried
on knee-breeches for the first time." During the day, one could wear a sports
jacket or dress suit, but one had to dress warmly. The letters of guests shiver
with chilly tales; the guests spent their time "either roasting or freezing" as
they went through interminable hallways from the glacial guest rooms to
the overheated imperial apartments.

For Pasteur, this was in fact the first contact with the imperial household and the court. He went to Compiègne without Marie, but accompanied by his domestic Jean, whom he compared to Don Quixote. In the
late afternoon of Wednesday, 29 November 1865, Pasteur and the other
guests arrived at Compiègne on the imperial train. A four-horse carriage
driven by postilions in powdered wigs took them to the château. The local
people leaned out of their windows or lined the streets to see this parade,
which indeed did not lack a certain panache. The guests were received in the
hall of columns. The arrival caused a good bit of excitement, what with the
confusion of luggage and the apprehension of those who came for the first
time. A concierge directed everyone to his or her room.

Having mounted the grand staircase and crossed the guard room,
Pasteur reached his quarters, which had been chosen by the Emperor's
physician, Lucien Corvisart (a nephew of the great Corvisart, who had
been the physician of Napoleon I). He had to share it with a famous
physiologist of the time, Dr. Longet. "Our apartment," he wrote to Marie,
"is very simple, rather like a two-bedroom apartment in a good Swiss hotel,
with a hallway starting by the front door separating the two bedrooms."[3]

Once the guests had settled in, their presentation took place in the
game room, where the ladies lined up to one side and the men to the other;
the imperial couple arrived from the family salon. Eugénie entered on the
arm of the Emperor, a dazzling beauty whose only adornment was the
diamond "Regent," which she wore in her hair surmounted by a spray of
white ostrich plumes. Then, Pasteur reported, "the Empress left the Emperor's arm and stopped in front of every lady her chamberlain presented to
her and spoke to each one of them. The Emperor did the same thing with

the row of men, but since there were many more of them, and since the big shots were in front of the little shots who had yielded to them, the Emperor shook hands with some of the men he knew without saying very much. But then he went back up the line. The bigshots moved to the back and this time the chamberlain named the persons whom the Emperor did not know."

"Monsieur Pasteur," the chamberlain said as the Emperor moved away from M. Dubois (a fashionable sculptor who later did a bust of Pasteur). The Emperor: "Ah! I have long wished to congratulate you on your fine work." "Sire, it is a great honor for me." "Are you continuing your studies? They are so important." "I endeavor, Sire, to keep to the same road. It is only by pursuing the same studies for a long time that one can hope to see a little better than one's predecessors." "You are so right."[4]

Soon it was time for dinner. The Emperor and the Empress walked the long way to the ballroom, where the table was set. The invited guests formed a procession and crossed the vast antechamber, the guard room, and several empty salons. The Emperor took the center seat of one of the sides of the table, the Empress sat across from him. To the Emperor's right was Lady Cowley, wife of the ambassador of Great Britain, to his left the Duchess of Magenta, spouse of Marshal MacMahon, who was in Algeria where he served as governor general. The duchess, Pasteur noted, was a very young lady who did not lack a certain charm.

The quality of the fare received mixed reviews. One of the regular guests, Princess Metternich, reported that the food was good, copious, but not refined. It took so long to serve the seven courses (rice soup, filet of mackerel, mutton tongue, hare in pepper sauce, sweet *entremets*, artichoke hearts, meringues) to a hundred persons that one finally lost one's appetite. Pasteur was too excited to care.

After the meal, coffee was served in the family salon. Watched over by torches in the form of Indians on malachite stands, the room was furnished with old pieces from the First Empire, whose placement was marked in chalk on the floor. . . . Among them were settees for two or three called *conversations* or *indiscrets*. Pasteur went to greet Viollet-le-Duc, a frequent guest at the palace, since he was restoring the nearby château of Pierre-fonds. Then he was presented to the Empress.

Eugénie knew and appreciated Pasteur's work, much to his delight. She drew the scientist into a lively conversation where they talked of microbes, epidemic diseases, cholera in particular, and of the infections in French wine. Seeing the Emperor approaching to speak to Pasteur, the

Empress exclaimed that she wanted to share in their conversation and drew both men toward the chimney: "Monsieur Pasteur, you can't teach only the Emperor."[5]

Pasteur spoke about everything, even about the complexity of right- and left-handed crystals, and about his beloved tartrates, about which he had many good things to say. The Emperor was very interested and commented in a correct and judicious manner, asking, above all, to see Pasteur's discoveries for himself. Setting up the program for the week, he requested the opportunity to examine samples of spoiled wine under the microscope.

After the Emperor had left the room, the Empress sent her chamberlain to look for Pasteur and asked him to sit next to her on her settee. Also seated there were the Prussian ambassador and the novelist Jules Sandeau. The conversation, which Pasteur considered very important, was about the salaries of professors in public service. When the tea was served, it was midnight. The party ended the moment the Empress retired for the night. But Pasteur was so excited by this extraordinary evening that he barely slept three hours that night.

The next day brought hunting to hounds and the carving of the quarry by torchlight, an impressive spectacle that the sovereigns watched from the balcony of the guard room, the guests from their windows, and the public from under the colonnade.

Pasteur did not forget the imperial request and sent Jean to Paris to fetch some bottles of aged white wine and above all the tube with an asbestos stopper that he had used in his experiments on spontaneous generation. While he was waiting for this material, he asked to visit the wine cellar of the château in order to look for some spoiled wine. But this was a delicate matter, for after all, France's first table took good care of its stocks.

Ready for his demonstration, Pasteur could enjoy the pleasures of the château, where a certain casualness prevailed. When he was not working, Napoleon III went around in his overcoat, "just like a good bourgeois." This amazed Pasteur: "The man is really an extraordinary fellow." When he was in exile, the Emperor had known the complete liberty afforded the guests in English country houses. This had taught him to practice a hospitality that did not constrain the guests. The evening entertainments were organized by the sovereigns. Pasteur attended a performance of *Les Plaideurs* played by actors of the Comédie Française. This was the time when conjurer's seances were all the rage, and so the guests witnessed sleights of hand imitating those of the Davenport Brothers, who claimed that they were spiritists, before Rober-Houdin unmasked them as frauds. There were also improvised concerts or living pictures. After all, Mérimée composed his famous dictation at Compiègne.

In the afternoon, the proper thing to do was to participate in planned activities. The guests played Japanese billiards or quoits. At one of the hunts, Pasteur saw Lord Cowley kill almost three hundred pieces of game — rabbits, hares, pheasants, or deer. He was horrified: "This is extremely curious, but extremely cruel, barbaric! . . . It is just dreadful to see this butchery."[6] Even though he did not know much about architecture, the scientist therefore preferred an excursion to Pierrefonds, about which he had nothing to say, except that it was "gothic." This was the fairy tale castle reconstructed according to Viollet-le-Duc's fantasy, about which Mérimée had said that he wondered whether "the crinolines that have trouble negotiating the spiral staircases may not make Pierrefonds uninhabitable."

In fact, the only thing of interest to Pasteur was the demonstration he had promised. He was determined to shine in the presence of the Emperor and the Empress:

> At five o'clock we go to a charming small salon. Many of today's invited guests are already there. The Empress, wearing a charming cloak of red wool, a kind of dressing gown or coat . . . decorated only with a piece of English lace at the neck, enters, salutes the guests, and moves to a small table bearing the lamp. The little group now takes up its places. We are standing in one of the groups a little farther away. Her chamberlain comes over and invites us to move closer to Her Majesty. The Empress assigns us armchairs and chairs around herself and the little table. . . . The conversation begins with a few words about our day but soon turns to science. To my surprise, the Empress talks to me about the diseases of the silkworms. I tell her what I have seen and what I will attempt next year. Truly, I tell her about everything, even the reproduction of the silkworm eggs. This brought the conversation around to generation, heredity, etc., and the Empress's comments were consistently witty, intelligent, and relevant.[7]

Finally, the day appointed for the imperial experimentation arrives. On Monday

> at four in the afternoon, I go to the Emperor's apartments, taking along my microscope, my book, and my wine samples. I am announced. The Emperor comes to the door and asks me to come in. He is alone. In the back of the office M. Conti is working, but he rises in order to leave. The Emperor asks

him to stay. Then he goes to fetch the Empress, and I begin to show their majesties both my drawings and the objects themselves in the microscope. This goes on for more than an hour (this had been agreed upon and announced early in the morning by the Empress's chamberlain). We then go to the Empress's tea-room, where the guests of the day are waiting. The Empress insists on carrying as many of the objects as she can hold herself, and I follow her at a few seconds' interval, so that the Empress produces the effect you can imagine, transformed as she is into a laboratory helper. . . . People want to see and compare human blood and frog's blood. Before the men can even think about pricking their fingers, the Empress has already shed her blood. Everyone, of course, wants to examine Her Majesty's blood. . . . Oh, I forget to tell you that after the session with the Emperor, I asked him to permit me to publish my letter. He read it and told me: "I will be very pleased to have my name associated with these important discoveries."[8]

And then, just as under the First Empire, everything ends with a dance, the *Boulangère*, which the Emperor himself leads since the dancers are not very good. In the grand ballroom lit by a multitude of chandeliers, two orchestras, one at each end of the room, play music for couples in crinoline and knee breeches. The Prince Imperial, not yet ten years old, takes part in the festivities. But Pasteur does not want to do cross capers and takes refuge with the Empress between Viollet-le-Duc and Lord Dudley, the inventor of a telegraphic cable that was to pass under the Channel and connect Calais with Dover. Clearly smitten by the Empress, Pasteur commented: "I see no one here who is not seduced by her. She puts everyone at ease with her charming affability."[9]

Pasteur was deeply flattered by this stay at Compiègne. A sententious father, he wrote to his son Jean-Baptiste: "The honor of spending a week in the Emperor's company which I have just received will make you understand the rewards of hard work and good conduct. You should therefore also work hard, so that one day, God willing, you may have the same satisfaction."[10]

The Nineteenth-Century Drinker

The book that Pasteur presented to Napoleon III contained the first reports on the diseases of wine and the letter he gave him to read was the dedication to the Emperor that would accompany it. In approaching this question, the scientist had adopted the most classic and the surest of all methods, which is to begin with a historical account of the problem.

In the course of the nineteenth century, the consumption of alcoholic beverages had increased in France; this was due in part to improvements in distilling techniques and in part to the desire to avoid the deleterious effects of dubious drinking water. Indeed, advances in the field of hygiene had revealed that water can transmit diseases, particularly in an urban environment. The fear of water therefore grew along with the advances of microbiology. In the end it was considered preferable to consume more "hygienic" beverages, such as wine or beer.

In France, wine making had undergone many changes in the course of the century. The Revolution had brought vineyards into the hands of peasants, thereby depriving the varietals of the care of an elite of practitioners who had experience with the best methods of cultivation and vinification. The First Empire, with its wars and blockades, did great harm to the export trade. From Louis XVIII to Louis-Philippe, the development of new varietals was encouraged, and improved communications brought about by the gradual buildup of a national network of railroads made it possible to give special incentives to the expansion of areas planted in vines.

At the time of Chaptal, wine production was already the second largest agricultural sector, after the production of cereals. New vineyards were planted around the cities, while in the countryside they took over fields at the expense of other crops. Vineyards were expanding so much in the Hérault that this department alone produced three times as much wine as Portugal. Vines were planted in the departments of Var, Vaucluse, and Bouches-du-Rhône. By 1850, two million hectares of French soil were planted in vines. The map of France was rapidly changing as regional specializations disappeared and wine became the beverage of choice. Considered a tonic, it was also part of the pharmacopoeia. Wine drinking was seen as a sign of social advancement, since it was no longer the privilege of the nobility and the Church. Every good bourgeois owed it to himself to have, if not his own vineyard, then at least his wine cellar. In the countryside a meal without a pitcher of wine on the table would have been unthinkable. Wine was taken to the field by the peasant, to the factory by the worker, to the barracks by the soldier.

But the mass production of wine often entailed a loss of quality. In *Illusions perdues*, Balzac puts these words into the mouth of old man Séchard, who cultivates a vineyard near Angoulême: "What do I care about quality? They can have it, their quality, these high-falutin' gentlemen. For me, quality means gold coins!" The neglect of quality brought a string of diseases of the wine in its wake. Fraudulent practices, such as the "plastering" of wine to accelerate the fermentation, or its coloring, either naturally with log wood or artificially with a coal derivative, made wine susceptible to disease, so that its preservation became difficult. Yet wine had to be stored in the shays in order to age, and it also had to be able to travel for export, sometimes beyond the seas.

As for beer, its consumption was initially limited to the frontier regions of northeastern and eastern France. But the rapid rise in the production of hops and the internal migration of populations soon caused a considerable spread in its use, which rose tenfold over sixty years. As the Alsatian proverb had it: "Water is wet, wine is too dear, I take my pipe and I drink my beer." In the hops-producing regions, beer was the drink of the poorer classes, but elsewhere it was gradually becoming a fashionable beverage favored by the dandies. Brewery restaurants or *brasseries* opened in the cities and attracted a clientele of petits bourgeois eager to try out new things. Banter and discussion thrived over a bock, as Maupassant tells us.

The increase in production could not have occurred without the improvement of industrial manufacturing. The beer industry really came into being only in the nineteenth century in the wake of the scientific discoveries of the era, from the identification of yeast to the biological catalyzers. For beer was reputed to be healthy. It did not contain much alcohol but quite a few nutrients. However, this beverage did not always have the required "hygienic qualities." Unscrupulous brewers had no qualms about offering their customers adulterated products. Instead of malt they used an industrial sugar that introduced toxic impurities and led to counteractive fermentations.

Given this situation, Pasteur had a number of reasons for taking an interest in the consumption of alcoholic beverages. Had not his work on fermentation indicated, demonstrated, and characterized the role of yeast? Did not the cultures of microorganisms he had learned to grow make it possible to control not only the quantity but also the quality of a given product? And was he not, in fact, able to detect the conditions making for a bad fermentation, that is, abnormal proliferations of microbes, bad alcohol, and adulterated ingredients?

Yet it was still widely thought that wine making had nothing to do with science. Any production of alcohol was considered to be an art, despite the studies of Chaptal, who had attempted in vain to bring some rigor to the practices of his time. It took Pasteur — his observations on microorganisms and his strong stand on spontaneous generation — to demonstrate the ways in which germs, parasites, and oxidation affect the alteration of wine. One can only salute the perspicacity of Napoleon III and Favé, for it was an unusual step to ask a microbiologist to look into the diseases of wine.

Wine, it was thought at the time, is constantly working, at least until an equilibrium between the sugar and the ferment has been achieved. Too little ferment, and the wine remains sweet; too little sugar, and the excess ferment affects the taste. Everyone knew that wine can be sour, prickly, and acidic, in short, that it can turn to vinegar. Chaptal called this phenomenon the acescency of wine. On this point Pasteur had his first intuition: instead of believing that this was a natural and unavoidable phenomenon, he felt that it was brought about by contamination with a germ from the outside. That is why, here again, he started his investigation by using the microscope. In the samples of wine he examined he was able to find a microscopic fungus, which he called the flower of vinegar or mycoderma. He thus distinguished two stages of fermentation: the first, due to the flower of wine, is necessary for vinification; the second, which produces the flower of vinegar, results in an undrinkable product. In other words, a way had to be found to stop the flower of vinegar from fighting the flower of wine.

Acescency could only occur through contact with air and could be observed in wines kept in barrels; that is why today bottles of wine are stored lying down rather than standing, for even if the bottle is well corked, the wine continues to be susceptible to this infection if the bottle remains standing.

Pasteur started by testing the great French wines. He compared the wines of Pommard and Volney with a few bottles of ordinary table wine from the Jura. The special bouquet of the Burgundies was present only in the absence of microorganisms. His first advice was for the people of Arbois, for he suggested to the vintners of his hometown to examine the wine at every racking to make sure the flower of wine had formed and to avoid the formation of the flower of vinegar.

As they age, wines can also be affected by other parasites. The grease disease, rare in red wines but frequent in white wines from small vineyards, makes the wine look ropy and oily. This problem can appear in barrels as well as in the most tightly corked bottles. Having identified the fungus responsible for it in a white wine of the Nantes region, Pasteur showed it to be round and arranged in clusters or wreaths. The parasite came from a few grapes that had rotted on the vine.

Then there was the bitter disease, which gave the wine a characteristic "old taste." This was one of the most baneful blows that could befall the wine trade, particularly in old wines from the best vineyards. All red wines are susceptible to this disease. First the wine loses its taste. Cellar masters say that it is going soft [*doucine*]. Then the bouquet changes and becomes bitter. Bitterness is an organic disease of pinot wines that did great harm to Burgundy and even Champagne. "Well," said Pasteur, "suppose we hold in our hands a bitter wine, of whatever kind it may be, we will always see the knotted branchings of a parasite, the parasite of bitterness." Experiments were set up to trace this parasite and follow its proliferation over long periods of time. In particular, Pasteur ordered a series of bottles of Pommard 1848, whose aging he conscientiously monitored over several years. Whenever a bottle remained free of the parasite, the scientist rewarded himself by drinking it with his family.

Lastly, wine often turns and becomes cloudy in very hot summer weather. If one shakes it in a glass, one sees silky waves moving about. The wine has no taste, the barrel bulges out, and moisture seeps out of the jointures of the staves. If the barrel is pierced, the wine spurts out forcefully; at the time this was called "having the pushes." This, it was believed, was caused by the rising of the yeast. The frequency and the extent of this trouble was attributed to negligence during the racking of the wine. This disease was in fact not limited to wine but sometimes also affected cider and beer.

A major producer of Montpellier, having sold a wine he considered to be of good quality, soon found himself accused of having cut it with water. The wine was so tasteless that the accusation seemed founded, and the vintner, anxious to restore his reputation, asked Pasteur for an expert opinion. The scientist immediately showed the presence of filaments floating in the liquid and indicated that they were identical to those that accompanied fermentations in his test tubes. It became clear to Pasteur that the wines of Hérault contained a parasite that he would often find in the wines of the Jura. At the time of the casking, he noticed it in samples of ordinary wines made from all kinds of grapes from Montigny, les Assures, and Arbois. He meticulously noted the number, the form, and the size of the parasitical filaments he had seen under the microscope. This evil must be prevented, and in order to do this "nothing is more sensible than the old custom, the wise experience of our forefathers, who advised us to rack the wine in good time to rid it of the sediment" so that it will not "rise."

In this manner, Pasteur's investigation of parasitism caused him to reflect on the activities of the vintner. The scientist felt that he should not confine himself to the abstractions of his laboratory, and that it was through contact with the realities of daily life that he could understand the complex mechanisms he had set out to elucidate. "Many times in the past I have had occasion to recognize the truth inherent in the practice of the old crafts," wrote Pasteur. "It does happen, of course, that these are truths of legend, tinged with the miraculous; but if you don't mind a bit of the supernatural, and if you look at the facts themselves, you will almost invariably recognize that any practice, provided that it is generally followed, is the fruit of reasoned experience." And he continued: "The oldest writings on wine recommend the month of March for the first racking, and a day when the north wind blows, not the south wind, which brings rain, at least in the Jura. Do not think that this is a prejudice or a blind routine. This is a very old custom. To me, it is rational; wine, especially young wine, is oversaturated with carbonic-acid gas. If the barometric pressure has been very low for several days, the wine must discharge this gas. This will bring up tiny bubbles from the bottom of the barrel and these will be able to bring up with them the smallest solid particles of the deposits. The wine will therefore be less limpid than if it is racked on a breezy day, when the atmospheric pressure tends to increase the solubility of gases in liquids. This, I believe, is the reason for the practice of which I am speaking."[11]

The Laboratory of Arbois

It was at Arbois that Pasteur conducted his first experiments on wine in the summer of 1863. When he was asked to study vinification, he wanted to return to the place of his childhood as a matter of course. It was there, after all, that he had learned about vintners, vineyards, and wine.

Once he had decided to comply with the Emperor's request, he called together three Normaliens, Duclaux, Gernez, and Lechartier, and asked them to go with him. There was no time to lose, for the grapes would not keep. Between August 20 and 30, the team sealed the test flasks and prepared the necessary reagents. The day of departure was to be 1 September. In the end the little team spent three consecutive summers, 1863, 1864, and 1865, at Arbois. The result of these investigations was to be the volume *Etudes sur le vin*, whose outline Pasteur presented to the Emperor during his stay at Compiègne. It was published in 1866.

The small town in the Jura, proud of its red and white wines, was eager to contribute to the research of a native son. The municipal council proposed a locale that would serve as his laboratory. But Pasteur declined the offer, preferring to camp out with his assistants in the back room of a café, for he wanted to remain "independent," as he said. One imagines this rudimentary installation with a microscope next to an old incubator he had sent from Paris. This laboratory gradually acquired tubules, tube holders, and gas-jets built to specifications by the local carpenter, blacksmith, or tinsmith. Standing behind the counter where the wines from the different cellars of the region were studied, compared, and analyzed, Duclaux and his colleagues began their research, not without provoking the curiosity and sometimes the jokes of the small local community.

In the evening there were games of billiards amid the dark uphol-stered furniture of a sitting room decorated with allegorical engravings that were sometimes endangered by the billiard cues. The assistants were lodged on the second floor. It was early to bed and early to rise.

Pasteur did not want to miss any of the stages of wine making. In order to remain completely independent, he therefore decided to buy a vineyard of one hundred square meters in a good location at the outskirts of Arbois along the road to Dole. He cultivated, observed, and studied it: "The grapes are put into a vat that has been taken to the vineyard itself. A worker carefully crushes all the grapes, which are then taken to the vintner and poured into barrels. . . . There they ferment and rest for six weeks to two months. Then the clear wine is racked. The racked wine is placed into barrels which are never completely filled. Because of this latter custom, the wine is always covered with mold."[12]

By preaching the racking, bottling, and conservation of wine at low temperature, Pasteur became the advocate of a new art of viniculture. Wine making became a scientific art, to use a paradoxical expression. "It is desir-able to attain that goal, for wine can rightly be considered the most health-ful, the most hygienic of beverages. That is why, of all the drinks known today, wine is the one human beings prefer to all others whenever they have a chance to become accustomed to it."[13]

In the course of its investigations, the laboratory of Arbois also re-turned to its basic experiments on fermentation and the rotational activity of fermented products. Pasteur wrote in his notebook:

Rotational activity of the must of different kinds of grapes:

— Use the colorless filtered must [i.e., the unfermented grape juice] of very ripe grapes.
— Use a specific quantity of not completely ripe grapes taken from specified plants. A few days later, use grapes from the same plants.

—Compare the first findings with those concerning the musts of unequal ripeness.

—Exact analysis of the quantity of sugar, acid, rotational activity, saline content of the must taken off in the first pressing, the second, the third, the fourth . . . There must be variations.[14]

These very rigorous measurements were made in the context of a more and more precise knowledge of the different vintages: the laboratory was becoming a place to turn to for expert opinion. A vintner unhappy with the quality of his wine, a consumer displeased with a purchase, would send samples to Pasteur. The names of the most prestigious vineyards, Pommard, Volnay, Arbois, and those of the most prestigious varieties, from Gamay to Pinot, were marked on his test tubes. Great wines, local wines, good and bad years, all found their way to his laboratory. Pasteur tested these wines, looked at them in a stem glass or examined them under the microscope for possible parasites; but he also tasted them and meticulously noted their gustatory qualities. Yet as a good oenologist, he kept pleasure and investigation strictly separate, and the notebooks of his experiments do not provide the slightest hint that he might have enjoyed his tastings.

Another problem that interested Pasteur was the role of oxygen in vinification and the aging of wine. This was the time when he was studying the aerobic and anaerobic life of germs. Apparently related to chemistry rather than to biology, the question of the aging of wine seemed to be a matter of the composition of the air that remained in the vats and in the necks of the bottles. The air acted like a frontier zone between the Pommard wine and its cork. Now the traditional view was that oxygen is toxic to wine, and that its quality is best when it has very little contact with the atmosphere. For this reason many barrels were fumigated with sulfuric acid. Discovering these practices, which the vintners passed down from generation to generation, Pasteur reflected:

The grapes are thrown into the crushing vats. . . . Air is present. But its quantity is not large in relation to the volume of grapes. Soon the fermentation begins; at this point there is no more oxygen at all, and the liquid is constantly saturated with carbon dioxide. The wine is racked quickly, spurting out in strong jets, and the barrels are filled immediately. During this operation there is only a brief moment of contact with the air. . . . It is done all at once, and in Burgundy the wine is even protected from the air as much as possible. The bottling is one more occasion when the wine necessarily comes in contact with the air, but it too is very brief. . . . It will be obvious to everyone that air has always been considered the enemy of wine and that all the practices of vinification invite us to adopt this point of view."[15]

Observing these manipulations, Pasteur found it difficult to doubt their effectiveness. And so he owed it to himself to find out just what it was that wine gained or lost through its contact with air. He knew that depriving wine of air was to limit its contact with germs. Yet on the other hand, he was also aware of what fermentation owes to air. He therefore had to decide whether air was beneficial to the wine or to the parasite. This was a question Pasteur was soon able to answer. He noticed that air ages the wine, takes away the acidity of its youth, and mellows it. The wine of Arbois yellows as it ages and takes on that onion-skin hue which, as Duclaux put it "was known to our forefathers as it is known to us, because they knew the value of the vineyard while we know only what it costs."

Here again, Pasteur liked to illustrate his subject. He therefore took a young Arbois wine and deprived it of air. A year later, when he uncorked the bottles, Pasteur noted: "These wines, which are more than a year old, at present still have the same color of new wine with which they started, the same green and tart taste of a young wine, and even the fairly noticeable smell and taste of yeast. In short, it seems to me that they have not aged at all. Conversely, the wine treated in the usual manner is already beginning to age."[16] He concluded that this justifies the use of wooden barrels, which allow for slow and steady aeration. He also remarked that at Clos Vougeot, the paint that seals the casks preserves the vivacity and the greenness of the wine but also makes prompt bottling necessary.

In short, in order to keep a wine young, one must keep it away from oxygen; in order to make it age, one must oxygenate it. However, oxidation must not be carried too far, for it damages the wine. The reds become rancid, that is, yellowish and sweet, whereas the whites turn brown, like madeira. The bottle becomes cloudy. A deposit forms along the glass walls. Darkness contributes to this aging process, which is nothing more than the continued effect of oxygen from the air, a kind of slow respiration until the final bouquet is achieved. In this last phase of his studies on wine, when he sought to understand the role of oxygen and the process of aging, Pasteur returned to being a chemist. Equations to express fermentation once again appeared in his notes. Dark wine cellars, wooden barrels, fumigating, racking, equipment, all the age-old practices of the art of wine making were for the first time analyzed as what they really are, namely, as agents of fermentation. A rigorous explorer of the interactions between living matter and the chemical elements he had first studied in his experiments on tartaric acid, Pasteur now assumed his stature as one of the pioneers of biotechnology.

The Beginnings of Pasteurization

When it came to preserving wine, empiricism, sometimes aided by tradition, had long produced techniques that were by no means fruitless. Resin, aromates, sugar, vinegar, and alcohol itself had their heyday in those Greek and Roman recipes described in the *Georgics* and in the treatises of Cato, Columella, and Pliny. What the ancient authors expressed in poetic terms were in fact chemical processes that vintners adopted without asking too many questions. A good bit of imagination was at work here. Thus, in Chile, pieces of meat were thrown into the casks to protect the wine against the vinegar mold. In order to strengthen the wine fungus against the vinegar fungus, wine makers also added young wine to old wine that was turning. Effective insofar as they survived, these measures were nonetheless unreliable, aside from the fact that they became insufficient in the context of an expanding and triumphant commercialization.

If the techniques of chemistry did not provide answers to all the questions, it was logical to supplement them by those of physics. This was Pasteur's reasoning when, faced with the diseases of wine, which he had analyzed in terms of chemistry, he set out to find preventive measures against spoilage. This was one of the first times that he was led to prevent a microbial infection.

Efforts to preserve wine from all the degenerative processes by which it was threatened had involved chilling and heating. Some wine was frozen before being shipped to San Francisco. This was only partially successful, for certain bottles contracted the "bitterness disease" anyway. Heating the wine was also tried. In certain regions of Spain, it had long been the custom to cook the grape must. Elsewhere, in Greece, the grapes were heated in the sun and the wine was boiled. This heating technique was thus not new when Pasteur decided to study it and put it into practice.

Appert, the pioneer of preservation of food, was one of the first to have insisted on the practicality of this method around the early years of the nineteenth century. Heating, as he described it, was a technique that made it possible to preserve a wide variety of foodstuffs. Was Pasteur unaware of this? Yes and no: "When I published the first findings of my experiments on the possible conservation of wine through heating before bottling, it was obvious that I was only giving a new application to Appert's method, but I was absolutely unaware that Appert had thought of this application long before me."[17] Appert had indeed sent some Beaune wine that he had heated in a water bath to Saint-Domingue and noted that upon arrival its bouquet was superior to that of the wine that had never left the port of Le Havre. But if he had thus inaugurated this method, he had not sought to put it into wider use.

Other experiments had also been carried out in this domain. As early as the 1820s, Jean-Antoine Gervais had used heat, but in order to improve the taste of wine rather than to preserve it. As for Vicomte Alfred de Vergnette de Lamotte, a great landowner in the Côte-d'Or who democratically signed himself Vergnette-Lamotte, he was one of the most fervent advocates of the freezing of wine, and also proposed heating for export, but he furnished neither a scientific explanation nor a rigorous description of what for him was an empirical procedure. These predecessors, then, had indeed discovered or applied the heating method, but none of them had said why and how it must be used to preserve the wine.

Here again, it was Pasteur's merit to bring together scattered and complementary data that scientists had hitherto been unable to interpret. He acknowledged that his own contribution had been above all to provide a rigorous experimental demonstration of the method. The process was indeed quite simple, for it consisted of heating the wine for a few moments without air at between sixty and one hundred degrees centigrade. In order to establish the date of this invention, Pasteur took out an inventor's patent for his process, which he called in all simplicity "pasteurization." This step brought him considerable criticism and triggered a veritable polemic in the scientific press. Those who attacked him criticized the patent for its form as much as for its principle. How, they asked, could this process possibly preserve the wine without affecting its taste?

In the departmental council of the Côte d'Or, Thenard raised the question of the rights of Appert and Vergnette-Lamotte. "In short," he said, "the only inventors are the predecessors of Appert, who dimly perceived this fact, Appert, who established it in his experiments, and M. de Vergnette-Lamotte, who used it to good effect. As for M. Pasteur, his book, which is worthy of his deserved fame and shows considerable originality, presents a rational theory of the process, but he is no more its practical inventor than a man who invents a new theory of the plough, however ingenious, would be the inventor of plowing."[18]

Appert was dead, but Vergnette-Lamotte was very much alive. Pasteur defended himself by launching an attack, supported by Balard. In August 1866, untroubled by the slightest self-doubt, he wrote a letter to the editor of the *Moniteur scientifique* (who had sided with Vergnette-Lamotte), in which he became quite violent in his self-justification:

> You say that M. de Vergnette-Lamotte has the priority of the process he made known on 1 May 1865. . . . M. de Vergnette had the idea of simulating the high temperature to which wines are subjected during long sea voyages by storing wines for two months in an incubator or in an attic during July and August. . . . I have no quarrel, Monsieur, with this process. . . . I suggest you try to tell the wholesalers of Bercy or the Hérault to pile up their bottles of wine under their roofs for two months. They will send you packing, and yet you were able to read in the *Comptes rendus de l'Académie* that vintners of the merit of M. Marès of Montpellier are anticipating that they will experience a considerable improvement of their business if they follow my indications. . . . The mycodermas must be very comfortable under the rooftops in the summer, especially in Burgundy. . . . Wine becomes sick there much more rapidly and easily than in cellars. And now, Monsieur, let us talk about M. de Vergnette's paper of 1850, to which I was the first to do justice in my letter to the *Moniteur vinicole* to the extent that it deserved it, although perhaps in rather too polite a manner. . . . Yet it was so short and so instructive! To put it briefly, there is only one rational way — this is M. de Vergnette speaking — to improve wines that must travel far, and that is to concentrate them by freezing. Perfectly clear! Is that what you call a heating process? It is said that M. de Vergnette was the first person who long ago applied heat to wine and noticed certain effects, among them preservation. This is an error, Monsieur. . . . It was Appert who made such attempts and not M. de Vergnette. . . . You should know, Monsieur, that when you find the words "errors of M. Pasteur" under M. de Vergnette's pen, you will find under mine and on the same subjects, the words "errors of M. de Vergnette," and that between his opinions and mine I will not accept you as a judge. . . . You may therefore continue, Monsieur, your habitual denigration of my work and my person. Your attacks only encourage me to persevere. And to end as you do: A word to the wise is enough.[19]

As usual, the matter was perfectly clear to Pasteur: he, and he alone, was right! The public at large, it should be said, was on his side, for Pasteur's adversaries were not of a stature to sustain a fight. Appert had no heir, Thenard abandoned the cause, and Vergnette-Lamotte was ridiculed. As the years passed, Pasteur's paternity of the methods of heating wine was practically never questioned again.

But the subject of heating wine gave rise to other criticism as well. There were those who claimed that this method affected the taste of the wine and caused it to lose its bouquet. This objection to his process did not come from scientists, but from equally respectable people, the home distillers. In this instance, Pasteur had to fight on a terrain where he was less sure of himself, for here he was not attacked on the grounds of chemistry but of epicureanism.

Pasteur did what he always did when he was under attack: he called for the creation of a commission that would verify his tests. He then nominated a blue-ribbon panel of wine experts, among them Hemmet, the official representative [*syndic*] of the wine brokers of Paris, and Teissonnière, a member of the municipal council of the capital. Thereupon a jury was officially convoked and given the title, "Sub-Commission Charged with Certifying the Findings of M. Pasteur Concerning the Conservation of Wine." Comparative tasting sessions were organized. This competent jury rendered a verdict favorable to Pasteur. It acknowledged that heating did not affect the taste of the wine, provided that it was carried out according to the rules prescribed by the scientist.

In the second edition of his *Etudes sur le vin*, which appeared in 1872, Pasteur added an appendix containing the verdict of a second commission, the "Medical Commission on Wine," which had tasted twenty-four kinds of wine heated in the laboratory of the rue d'Ulm. "These additional tastings," wrote Pasteur, "eliminated the last doubt as to the improvement of wine through heating."[20]

The result was that, even while the controversy continued in the more or less scientific journals, Pasteur himself — as well as his disciple Jules Raulin, the Parisian engineer Charles Tellier, Antonio Pacinotti of Bologna, Rossignol of Orleans, Henri Cochon de Lapparent, the director of naval construction in the Ministry of the Navy, and many others — set out to invent heating apparatuses. It was a contest of ingenuity. No heating method would ever reach such a point of perfection! The Pasteur process, as it was called at the time, became an industrial practice and was adopted throughout the regions of France, sometimes, as in the case of Béziers, on a grand scale.

Backed by industry in this manner, Pasteur was bound to win. The *Moniteur viticole* and its team of journalists hostile to the "scientist" capitulated on the occasion of a tasting, and in the end officially recognized the value of the process. The wine producers' and distributors' associations, headed by a group of senators, shared this opinion following a secret balloting and a quasi-scientific blind tasting of different kinds of wine. *La Suprématie de France*, a brochure written by a group of distinguished oenologists, proclaimed that Pasteur was unquestionably correct.

The renown of the process soon crossed the French frontiers. The agricultural commission of Lombardy successfully tried it, and soon the wines of Hungary and Friuli were also heated. In the United States, a vintner in California wrote to the *Monthly Statistics* in New York that "Pasteur is as popular with the vintners of California as the president of the United States. If he were here, they would appoint him to a big job."[21]

Pasteur committed himself to the heating of wine and to the defense of his process with the same passion he had deployed in combating the theory of spontaneous generation. He would personally rush off to Mèze in the remote department of Hérault to demonstrate that an old method of making wine age that was practiced there was quite different from, indeed did not even foreshadow, his own invention. Then he was off to Toulon, where by orders of the Ministry of the Navy, the frigate *Sybille* was about to set sail for West Africa with barrels of heated wine on board.

The experiments with heating wine succeeded because heating does kill microbes. Yet, applied to wine, pasteurization had neither the success nor the scope Pasteur anticipated. By the end of the century, phylloxera was to do more harm to French wine and wine trade than the diseases of cloudiness and bitterness that Pasteur had been able to control. The practice of heating fell into disuse among the vintners, who experienced far greater distress.

By that time, Pasteur himself was also engaged in different pursuits and therefore did not seek to defend his discovery or to promote it. However, the method of anaerobic heating, which must be carefully controlled as to its duration and temperature if the product is to be preserved without alteration of its taste, was to have a considerable influence, which continues to this day.

Does today's vintner in Arbois know that the pasteurized milk he drinks owes its existence and its name to the conservation of wine? For the heating process was soon to be applied to other foodstuffs, as well as to other beverages, first and foremost to milk and beer.

Vinegar and Its Trade

Having finished his study of wine, Pasteur turned his attention to vinegar. This may look like a logical step, since wine can turn into vinegar when it does not age properly. But the case of vinegar was more complex than it seemed. On the one hand, the protective function of vinegar, as of every alcoholic product, was well known: foodstuffs stored in vinegar kept for a long time, and drinking water was sometimes cut with a little vinegar in order to make it more healthful. On the other hand, it was asserted that vinegar was the result of a fermentation without germs.

Science had not waited for Pasteur to analyze vinegar. The existence of vinegar mold and the "mother of vinegar" had been established. The vinegar mold designated a thin skin, smooth when it is young and wrinkled as it gets older, that forms on the surface of the barrels. If this mold is removed and immersed in the liquid, it forms a viscous and gelatinous mass called the mother of vinegar. But this designation was a misnomer, for it is not the mother that brings forth the vinegar; at most it contains the germs necessary for acescency. In fact it is the mold that furnishes the principle of this fermentation, as Pasteur was to show when he proved that the skin on the surface of the liquid is quite simply a proliferation of microscopic fungi: the *mycoderma aceti*.

Pasteur now sought to understand what the fungus takes from the two milieux with which it is in contact, the oxygen-laden air above it and the alcohol below. He found that the mycoderma takes oxygen out of the air and binds it to the alcohol in order to turn it into vinegar. Commenting on this reaction, Pasteur saw in the oxidizing capacity of the fungus some analogy with the behavior of white blood corpuscles.

If this fungus is placed on a barrel of alcohol, it will spread over the surface like a rapidly expanding pellicle. It sucks in the air and acidifies the alcohol. But if the pellicle is submerged, the reaction stops, and no more vinegar is formed, for the microscopic fungus does not secrete anything; it transforms. It must therefore be kept alive, that is, given the opportunity to absorb oxygen, and not be allowed to drown. When the vinegar mold has run out of alcohol to transform, it continues the reaction by starting the combustion of the vinegar it has just produced. The latter is then broken down into water and carbon dioxide.

The action of this mycoderma is so powerful, rapid, and effective that a mass on the order of one gram is sufficient to transform several kilograms of alcohol. The number of microorganisms is insignificant in relation to the quantity of vinegar produced. Perhaps this was the reason why they had been ignored or contested until Pasteur came to observe them in the field of his microscope.

The fungus only acts in air; and since air stimulates the mycoderma, this also justified the prompt bottling of wine as a means to prevent contamination. On the basis of this centuries-old observation, several methods of vinegar making had been developed.

The so-called Orleans process, in use in the Loiret and the Meurthe, made use of very large and well-aerated barrels containing a mixture of wine and vinegar. Another method, known as the "beechwood shavings method" was practiced in Germany. In this process the wine (or beer) that was to be transformed into vinegar was made to drip onto a bed of beechwood shavings piled up in large barrels; holes in the barrels aerated the contents, and the liquid was gathered after it had repeatedly passed through the wood. According to Liebig, who was fiercely opposed to Pasteur's theories, this operation indicated that the vinegar owed everything to the shavings, which in his opinion acted like a dry rot.

Although it was slow, the Orleans process was easily applicable to all kinds of wine. This was not the case for the German process, for the beer was apt to form yeast and the wine, mold, so that they often sealed the beech shavings and stopped the reaction. That is why Orleans had become the vinegar capital of Europe. The debate over vinegar thus seems to have been a rather important element in the economic rivalry between France and the German states.

Pasteur was sure of the biochemical action of the mycoderma. He considered it a new argument against Liebig and tried to convince him. Not having any beech shavings available, he replaced the wood by a rope. If he ran a trickle of alcohol along this rope, nothing happened. But if he first dipped the rope into mycoderma, the alcohol that ran down became charged with acetic acid that formed upon contact with the air. The liquid that dripped into the recipient was vinegar. This use of the rope yielded vinegar for several weeks. It was a brilliant demonstration.

However, despite several publications, particularly in the *Annales scientifiques de l'Ecole normale supérieure* in 1866, Pasteur failed to convince Liebig. The German scientist, who had consulted the director of one of the largest vinegar factories in Munich, still did not believe in the role of the mycoderma. The manufacturer had seen no trace of it in twenty-five years of vinegar making. What was needed were beech shavings, not mycoderma. He did not think that a fungus could grow on the wood. Yet this was precisely what Pasteur suggested.

For Pasteur now set out to look for mycoderma on the surface of the shavings and asked to be sent some samples. Since that seemed complicated, Pasteur proposed a control experiment: if the shavings were scalded, they should become inactive, for boiling water would kill the fungus, if any. But Liebig was equally sure of his position. He scornfully rejected the French suggestions, just as he had rejected all Pasteurian theories concerning microbial fermentations.

Faced with this obstinacy, Pasteur took out a "biotechnological" patent, asserting that *mycoderma aceti* is the agent in the formation of vinegar. On this basis he went on to study the application of heating techniques to vinegar production.

For vinegar also had its diseases, which were as yet poorly known. When visiting a vinegar factory at Orleans in the early stages of this research, Pasteur noticed in poorly filtered barrels the presence of small microscopic organisms, parasites called eelworms. They proliferated so commonly in vinegar that the vinegar makers were not concerned. Indeed, the Orleans manufacturer thought that the eelworm was necessary to the manufacturing of vinegar.

Pasteur soon found out that this was not the case, and that their presence was a dangerous threat to the life of the mycoderma. For the eelworm needed air. By taking oxygen from the air in order to convert the alcohol into vinegar, the mycoderma deprived the parasites: "The eelworms," Pasteur noted, "unable to breathe and guided by one of those instincts of which animals at every rung on the zoological ladder offer the most amazing examples, take refuge on the walls of the barrel. . . . Only here can these small creatures breathe. But it must be understood that they do not yield to the mycoderma without a fight. On several occasions I have witnessed, under particular circumstances, a kind of struggle between the worms and the plant. As the latter, following the laws of its own development, gradually spread over the surface, the eelworms that had gathered below it, often in bunches, seemed to engage in an effort to pull it into the liquid in the form of crumpled shreds."[22] Once it was submerged, the mycoderma became inactive, so that the oxygen could be used by the alcohol, which no longer turned into vinegar.

Once the nefarious role of the eelworm had become clear, Pasteur proposed a means to eliminate it. All this research came to a glorious conclusion in November 1867, when Pasteur gave a lecture at Orleans at the request of the mayor and the head of the chamber of commerce. With the help of a projector, and before an audience that went far beyond the world of vinegar making, he showed pictures of the mycoderma and the eelworm.

At this moment, he truly acted as the promoter of the Orleans process. Once again, science had come to the rescue of industry.

The Year of the Storm

Edmond About had written in 1860: "The unification of Germany is the most fervent and most heartfelt wish of France, for it loves the German nation with an unselfish friendship. France sees without fear an Italy of 26 million inhabitants to its south, and it would not be afraid to see 32 million Germans founding a great nation on its eastern frontier."[23] The imperial policy chose to believe that a united Germany would respect its borders and that the federated provinces would become a neutral state.

In the spring of 1870, Pasteur, who was passing through Strasbourg, and Sainte-Claire Deville, who was on his way back from Bonn, saw troop movements on the other side of the Rhine. It was the deployment of the Prussians along the frontier. The two Frenchmen barely had time to wonder about it.

Events took a dramatic turn in July: in the famous Ems telegram, a veritable provocation, Bismarck explained to the press how William I had tricked the French ambassador, Benedetti. The insult was deliberate. France was swept away by a wave of patriotism. The French were ready to go to war over this diplomatic slap in the face. On 14 July, the Minister of War, Marshal Leboeuf, declared: "We are ready, arch-ready, and even if the war should last a year, we are prepared down to the last puttee!" In the following days, the Legislative Body approved the appropriations for a war that Emile Ollivier, appointed prime minister a few weeks earlier, accepted, as he said, "with a confident heart."

In actual fact, France was isolated, ill-prepared, and militarily under-equipped. Napoleon III insisted on directing the military operations himself, and in order to ensure the continuity of the government, entrusted the regency of the Empire to Empress Eugénie.

By early August, defeats began to accumulate. Alsace had to be evacuated by Marshal MacMahon. Toul and Strasbourg were defenseless. Bazaine retreated to the outskirts of Metz, where he was soon encircled. The remainder of the French troops, followed more than led by Napoleon III, a sick man, allowed itself to be surrounded in the pocket of Sedan. It was there that the Emperor capitulated and declared himself a prisoner on 2 September. Two days later, the Republic was proclaimed at the Paris city hall after Gambetta and Jules Favre, pressured by groups of workers who had occupied the National Assembly, had called for a vote on the liquidation of the Empire. While the empress fled to England, a government of national defense was formed and accepted by acclamation. In early October, Gambetta used a hot-air balloon to travel from the already besieged Paris to Tours, the provisional seat of the government, where he took charge of the continuation of the struggle.

Pasteur had never made a secret of his sympathy for the imperial regime. Brought up with the legend of the Eagle ("The Emperor was more than a great man!") and hatred for the Bourbons, he was fiercely loyal to the imperial dynasty. In his eyes, the Empire was the most precious guarantee for preserving national dignity.

At the outbreak of hostilities, Pasteur was in Paris, living in the rue d'Ulm. There he witnessed the forming of battalions with conscripted young Normaliens. After a vigil of arms in the laboratory, where he bid farewell to Duruy, Sainte-Claire Deville, and Bertin, he thought about enrolling in the national guard, but he was rejected because of the lingering effects of his hemiplegia.

The school in the rue d'Ulm now began to create facilities for taking care of the wounded behind the lines. Bertin, the director, tried to persuade Pasteur to leave Paris. He considered him unsuited for nursing duties and felt that in the end he would be just one more mouth to feed, considering that the provisioning of Paris was becoming difficult. Realizing himself that he would indeed be useless, Pasteur wrote to the editor of Le Gaulois on 8 August: "I have only one way to express my support for our valiant soldiers who are rising en masse to drive out the foreigner. I am sending you another gift of 100 francs and my legal authorization to collect my salary as member of the Institut for the entire duration of the war. Beyond that, I give to the Fatherland my only son, an eighteen-year-old who is today enlisting as a volunteer."[24]

Jean-Baptiste's military service actually began rather inauspiciously. He had barely signed up when he came down with a case of typhoid that pinned him to a hospital bed at Val-de-Grâce. For Pasteur, the anguish of the national crisis was compounded by this illness, for he was particularly frightened of typhoid, which had already claimed two of his children. He did everything he could to make sure that his son received the best possible care.

The news of the capitulation at Sedan and the Emperor's abdication surprised Pasteur while he was still in Paris. On 5 September, he wrote to Marshal Vaillant that he was broken with grief and asked the favor of being remembered to the deposed Empress. He wanted to be counted among those, he said, who would forever remember the blessings of the Empire and who believed that the reign of Napoleon III would go down in history as one of the most glorious periods in French history. For Pasteur, the demise of the Empire had an immediate minor but symbolic consequence: the decree by which he had been appointed senator had not yet become official, so that Pasteur's name would not be listed among the senators of the Empire. This also meant that the scientist would not receive the pension for a position in which he had never served.

On the same day, Pasteur left Paris with his wife and Marie-Louise to return to Arbois; Jean-Baptiste was recovering in his hospital. In Arbois, Pasteur was very restless; he brooded over his anguish and his powerlessness in the little house of his parents, where his sister and his brother-in-law had gone back to tanning. In his father's room, which he now occupied, the portraits of Napoleon I and the King of Rome were still hanging on the wall. Pasteur spent days when his mind was fallow, much to his chagrin. He had brought his beloved microscope, but it had neither a sheet of glass nor slides.

Listening as the news was given to the villagers by the town crier on the bridge over the Cuisance, he learned about the French defeats and the advance of the enemy. Arbois is close to Switzerland, and so he thought about taking refuge there if the "troops of the barbarians" should continue to forge ahead. He fumed against the inertia of Austria, England, and Russia, who stood by passively. But he also raved against the collapse of the French officer corps. For a moment he put his hopes in Bazaine. But that general disgracefully capitulated at Metz on 27 October. Pasteur wanted France to resist "to the last man, to the last rampart," and he also advocated prolonging the war until the dead of winter, "so that all these vandals will perish of cold, misery, and disease." And he solemnly swore that "each of my studies, to my dying day, will bear the epigraph: *Hatred to Prussia, vengeance! vengeance!*"[25]

Following some successes of the army of the Loire under the command of Chancy, the French were crushed at Loigny in the Beauce on 2 December, ironically the anniversary of Austerlitz. At this terrible moment, the Italian government invited Pasteur to leave France in order to direct a laboratory and a sericultural establishment at Milan. The scientist refused the offer in the name of his patriotic conscience: "I would feel that I committed a crime and deserved to be punished as a deserter if I left my fatherland, especially when my fatherland is suffering misfortune, to pursue a material ease that it is unable to offer me." Shortly thereafter, he received another offer, which would have involved going to Pisa. This time, Pasteur hesitated, and wrote to Raulin: "I confess that I am weakening, and that I might well accept, especially if you were willing to share my exile."[26] But Marie's reluctance gave him second thoughts, and in the end he refused, not without thanking the Italian authorities and congratulating them for "giving proper consideration to science in developing the prosperity of their country."

It was at Arbois that Pasteur heard about the loss of Alsace and Lorraine. He deplored this German peace, which was paid for with the dismantling of France. It was, he said, a new feudal serfdom. But he also predicted that France would be able to recover its identity, just as plants that lose their appearance and their name change again when they are cultivated in the greenhouse. A day would come when the genius of the French race would reappear, for "God alone knows the end of the tribulations he sends us." But this end was not yet at hand. "However much I was avid for news when it could still bring some hope," Pasteur wrote to Raulin, "I reject it now that it continually adds to our humiliation and the disasters that have befallen France. I can barely bring myself to read a few scraps of newspaper."[27]

Suffering from insomnia (he slept only two or three hours a night), Pasteur was again worried about Jean-Baptiste, although he did hope that those troops that had kept up their fighting spirit would put up armed resistance in the mountains. Pasteur preferred having his son go to the front and join the active troops to having him stay in the rear in contact with the wounded. He wrote to his commanding officer to inform him that the young man had just gotten over typhoid; in this letter he stressed the danger of contamination, as if exposure to microbes were a greater threat than the Prussian bullets. Shortly thereafter, Jean-Baptiste was able to return to Lons-le-Saulnier and to sign up, like his cousin Joseph Vichot, with the light infantry of Bourbaki's army.

The Prussians were besieging Paris. From the heights of Châtillon, enemy batteries were bombarding the Left Bank. For Bismarck, might made right. Artillery shells fell onto the rue d'Ulm, where Bertin was still in charge. In January 1871, Pasteur learned through a note from Chevreul that the Muséum d'histoire naturelle had been hit. Even though the shells only destroyed some of the greenhouses, killed a parakeet, and decapitated a stuffed crocodile, there was something symbolic about the incident. For Pasteur, it epitomized the aggression of barbarism against science.

In 1868, the University of Bonn had conferred on Pasteur a degree of Doctor of Medicine *honoris causa* for his "contribution to the knowledge of the history and the generation of microorganisms." At a time when Prussia, forgetting its cultural mission inherited from the Enlightenment, was terrorizing France, Pasteur felt that he could no longer tolerate this honor bestowed yesterday by the enemy of today. On 18 January he returned the diploma to the dean of the University of Bonn: "Today the sight of this parchment is hateful to me, and it offends me to see my name, which you have decorated with the qualification *virum clarissimum*, placed under the auspices of a name that will henceforth be loathed by my country, that of *Guillermus Rex*."[28] His German colleagues responded by sending him "the expression of their contempt." Pasteur had this exchange of correspondence published in the *Courrier de Lyon*. He turned it into a patriotic manifesto: "I have yielded to two French viewpoints, one being that science has no fatherland, and the other that kings are contemptible human beings just like everyone else if they violate the laws of humanity."

On 24 January, given the alarming news of Bourbaki's army, Pasteur decided to leave Arbois with his wife and daughter. He borrowed an old coach to reach Pontarlier. It took three and a half days to travel these sixty kilometers. All around him there were people leaving their homes; there was rain, snow, and above all the distress of the soldiers. What he crossed on his way were the remnants of the eastern army, "in an incredible state of demoralization and rout." "The retreat from Russia," he wrote, "cannot have been more horrendous."

This was the debacle Zola was to evoke. Having taken refuge at Besançon, General Bourbaki attempted to kill himself by shooting himself in the head. Without food, without proper clothing, almost without weapons, the survivors were wandering through the fields and along the roads. Among them was the soldier Jean-Baptiste Pasteur, who had "done his duty before Héricourt" on January 13 and 14. Better off than his cousin Vichot, who was suffering from a bad wound in his leg, Jean-Baptiste was safe and sound. As he turned the corner on a country road on the outskirts of Pontarlier, he was reunited with his family.

From there, the Pasteurs traveled to Geneva, where they spent several days in early February. At that time Pasteur wrote to Bertin, who was still in Paris, "Will France have the necessary courage and the perspicacity to discover the causes of so many setbacks? Above all, will she muster the resolve to reject all the utopias which corrupt the political sense in our country and have in these last months wasted all the resources and paralyzed the strength of the provinces. . . . What will become of science in this shipwreck?"[29] Pasteur felt as if he had aged twenty years. Yet Geneva was to be only one stage of his exodus. After a few days' rest, the family went to Lyon to stay with Pasteur's brother-in-law Loir, dean of the Faculty of Sciences.

At Lyon Pasteur learned that Trochu had capitulated in Paris, defeated by famine. Here he heard accounts of looting, in particular that of the apartment of his colleague the physicist Regnault. Echoes of the uprising of many Communes throughout France, even at Lyon, also came to his attention. At the same time, he felt that it was shameful to show pictures of Versailles, where the government had taken refuge.

Paris was cut off. There had been a general exodus, and Pasteur's colleagues were dispersed: Sainte-Claire Deville was in Gex, Jean-Baptiste Dumas in Geneva. There could be no question of returning to the capital, where, even if the insurrection that Pasteur was to call the "Paris saturnalia" was repressed, calm was not about to be restored. Pasteur therefore lingered at Lyon. Spending the end of the winter there, he fretted both about the political situation and about the approaching egg-laying season of the silk-worms at Villa Vicentina, which had escaped the ravages of war. He even asked Bertin to contact the ushers at the ministry, if they were still working there, to find out about Marshal Vaillant's silkworm nursery.

At Lyon, Jean-Baptiste served as private secretary to General Servier. As for Marie-Louise, who had "prepared for it all winter," she was to celebrate her first communion on 23 March. Clearly, neither the war nor the occupation were reasons to forget one's duty to God or the Church.

In his memoirs, Pasteur's nephew Adrien Loir remembers this forced stay at Lyon: "During the war of 1870–71, Pasteur had a long stay at our house, and I can still see him at our dinner table one evening when the drummer sounded the call to arms for the national guard, which was to assemble on the Place Louis XVI, close to our house. The red flag of the Commune flew over the city hall, and it was the day when Commander Arnaud was murdered, killed by the communards. My father rushed off to don his guardsman's uniform, but Pasteur calmly continued to eat his soup as I watched my mother help her husband put on his sword-belt and kiss him goodbye. To my mind, this was war. It seemed strange to me that Pasteur should not move. For several weeks, I had been proudly sporting on the sleeve of my little fur-trimmed overcoat the stripes of a quartermaster sergeant, which had been sewn by a friend of my mother, the daughter of Captain Cognet."[30] The fact was that Pasteur felt crippled and useless.

Throughout this period, he wrote more than he acted. He sought to find scientific rather than military explanations for the disaster and, going beyond the immediate and tactical reasons for the defeat, gave much thought to the insufficiency of the means and the interest that the sciences had received. This was his leitmotif:

> For a long time the public powers in France have failed to recognize the law of correlation between theoretical science and the life of nations. Victimized, no doubt, by its political instability, France has done nothing to maintain, foster, and develop the advancement of science in our country. Strictly following its established momentum, it lived on its past, believing that it was still producing great scientific discoveries because these had brought material prosperity, but failing to realize that it was imprudently allowing their sources to dry up. At the same time, certain neighboring nations, spurred on by France's achievements, diverted them for their own benefit and made them fertile through a wise combination of effort and sacrifice. . . . While Germany founded more and more universities . . . , France, enervated by its revolutions and constantly occupied by its sterile search for the best form of government, paid only halfhearted attention to its establishments of higher education.[31]

Taine and Renan were soon to arrive at the same conclusions.

In fact, Pasteur was once again full of new scientific projects and deeply frustrated that the situation did not allow him to resume working in his laboratory. He was obsessed by one thought, the wish to return to his experiments.

In April 1871, Pasteur decided to leave Lyon for Clermont-Ferrand. There he joined Emile Duclaux in his house in the rue Montlosier, where he took over a few rooms. Marie and the children soon followed the head of the family. Jean-Baptiste, discharged from the army, was to return to his studies. His father urged him to reflect on the testing by fire he had just undergone. He even insisted that his son take German lessons, for although he cursed the Prussian barbarism, he remained convinced of the superiority of the classic German culture.

As for himself, he tried to return to the experiments he had left unfinished in the last few years. He found a space to raise a colony of silkworms, and his work in this domain still brought him great satisfaction, for at this time the Austrian government awarded him a prize for his research on pébrine. He received this distinction with a sense of modesty that was all his own: "This news has given me the greatest pleasure, without however surprising me, for I expected this decision, which I consider to be in keeping with truth and justice." When he was subsequently informed that he might have to share the prize with another recipient, pride overcame enthusiasm: "I do not hesitate to declare that if the ministry of agriculture should accept this proposition, it would be because its judgment is not sufficiently enlightened."[32]

Soon, as if to establish a connection between the beginnings of pasteurization and the need to restore the grandeur of the fatherland, a new subject came to claim Pasteur's attention: beer.

The Beer of Revenge

At Clermont, Pasteur did not want to take over Duclaux's entire house, and so he set himself up in a corner of the chemistry laboratory of the Faculty of Sciences. His studies on beer started out as the result of circumstances. On the one hand, Pasteur had to find something to occupy his mind so as to forget the pain of exile far from the rue d'Ulm; and on the other his former assistant had specialized in the study of brewer's yeast.

Pasteur was eager to contribute his share to his country's recovery. He thought about the loss of Alsace and Lorraine, where the cultivation and the industrial use of hops, as widespread as they were symbolic, would no longer benefit France. The Germanic power must not be allowed to take over everything; it must be attacked in an area where it hurt. France should create the "beer of revenge," as Pasteur called it.[33] "I was inspired to do this research by our misfortunes," he was to explain later. "I began working on it immediately after the war of 1870 and have ceaselessly pursued it since that time, determined to carry it far enough to bring durable advances to an industry in which Germany is superior to us."[34]

There were two categories of beer, the so-called high fermentation or dark beer and the low fermentation or blond beer. In Pasteur's day, all the English beers — porter, ale, pale ale, stout, bitter, and alestout — were produced by high fermentation. The low fermentation beers had first appeared between 1860 and 1870 in Bavaria, from where they had rapidly spread to Austria, Prussia, Bohemia, and eventually France.

The manufacturing of beer was a delicate process. It called for a large number of operations consisting of two principal stages, the malting and the brewing. The malting owed its name to the transformation into malt of part of the starch contained in the barley corns. Since barley in its natural state contains only an insignificant part of the sugar necessary for the operation, the barley corns were soaked in water; this was called the wetting.

Subsequently the drained barley was spread in a thin layer on the ground, where it germinated for two weeks. This germinated and slowly dried barley was called the malt. It was in the brewing, the second essential stage, that the brewer's skill made a difference. The wort of the malt was successively ground, stirred in water, cooked and then cooled in barrels called coolers. The wort was then allowed to ferment, activated of course by brewer's yeast. In order to eliminate deposits, the beer was then clarified before it was put into bottles, where it became effervescent.

The chemistry of this operation is governed by an enzyme, diastase, which Payen and Persoz had described in 1833 without, however, identifying it as an enzyme. Since it was sensitive to heat, the temperature at which the operation was carried out obviously determined the fermentation of the malt and the quality of the alcohol produced. High temperature and low temperature thus designated the two modes of brewing that characterized the two types of fermentation. The "high" beer was easier to produce, but it did not travel. An English brewer almost went bankrupt when he tried to export his production, for his entire cargo had spoiled before it reached India. The "low" and less alterable beer, by contrast, did not have to be consumed immediately, which is why it was called "keeping beer."

The fact is that, for all the experience of the brewers of Pasteur's time, no one was able to explain precisely what happens in the manufacturing of beer. Scientifically, little progress had been made since the invention of the process, which the ancient Egyptians attributed to the god Osiris when they planted barley on the shores of the Nile twenty centuries before our era. As it spread throughout the world, the recipe was improved, but only in strictly practical terms. At most, people knew that if beer was to be healthful and tasty, it must be clear, but there were wide differences in alcohol content, depending on the country and on local customs.

Beer was not really a new subject for Pasteur. When he was working at Lille, he had already studied the role of yeast and determined the optimal conditions for fermentation. During his enforced stay in Auvergne, he sought to refresh his knowledge by visiting a brewery. Only one firm was active in the region, the Kuhn brewery, which had been handed down from father to son at Chamalières.

A skillful brewer, Kuhn used traditional recipes. His quality control consisted of throwing away the yeast when it was no longer good, that is, when the customers of the nearby cafés came to complain about the quality of the product. Beyond that, he had no explanation for anything. All of his theoretical knowledge came from a frequently reissued manual by the chemist Anselme Payen, secretary of the Société d'agriculture, member of the Council of Hygiene and Health, and member of the Institut. In this already old volume, one could read that beer spoils spontaneously in hot weather, that one must be particularly careful not to substitute too many boxwood leaves for the hops, since they cloud the liquid, and so forth. In short, Payen provided a narrow pragmatism rather than a theory.

Pasteur's first step was to turn away from the empiricism handed down by tradition and to state the problem of beer making in relatively simple terms that made use of the fundamental insights he had gained in studying the fermentations. Successful brewing, he pointed out, demanded the development of a culturing process that guaranteed the purity of the yeast and did not allow any contamination by one of the countless germs with which the air is swarming.

From the very beginning, the results were encouraging. In March 1871, Pasteur wrote to Raulin that he was working at the Kuhn brewery, conducting various types of experiments that allowed him to use barley wort to grow yeasts of a remarkable rapidity of action. He felt that he had already achieved a real economic breakthrough by indicating the means of producing a much larger quantity of yeast than was produced at ordinary breweries. He vaunted the advantages of his new process, whose main innovation concerned the type of fermentation, and which, he hoped, would "have the result of harming those German blackguards, whose superior skills as brewers are unquestioned at this time."[35]

What was the novelty of this method? Habitually, the ferment was passed from brewery to brewery. Brewers who ran out because the yeast had actually or supposedly altered would purchase a new sample of pure yeast in Alsace or in Germany or from another brewery in France. But in the summer the ferment became very fragile because of the heat, so that it could easily alter and therefore needed to be kept in iceboxes. Pasteur now invented a process that eliminated the need for continuous refrigeration and made it possible to produce a low fermentation yeast whenever it was needed. This new brewing technique, which Pasteur had patented, not only yielded an excellent beer but also did not require the heating of the wort, which was a considerable advantage.

In August 1871, Pasteur proudly took his new product to Jean-Baptiste Dumas for testing. "My dear master, I have asked the brewer to send you twelve bottles of my beer so that I can have your evaluation of its taste. . . . I hope that even in comparison with the good beers of the Parisian cafés you will find it very good. When the bottles arrive, store them lying down in a cool cellar. Within a few days the beer will become effervescent. It is important to taste it at that point."[36]

When the political events and the establishment of a conservative Republic under Thiers allowed Pasteur to return to Paris in the middle of the summer of 1871, he decided to continue working on this project. Accordingly, he equipped his laboratory in the rue d'Ulm, where he installed a small brewery with a kettle for cooking the malt and a fermenting vat in the basement of the new building that had been started on orders of the Emperor in 1868. The presence of this equipment did nothing to change Pasteur's attitude toward his work. He was as silent and taciturn as ever, but the studies on beer lent a different cachet to the laboratory. The smell of boiling beer filled the air, and the clinking of glasses and bottles was heard on the occasion of the tastings, always presided over by Bertin, whose high spirits and laughter enlivened these strange scientific rituals.

It was not enough to reform the methods of manufacturing, for it was also necessary to answer the most practical questions of the brewers. Pasteur therefore showed by precise experiments that the alteration of beer always coincided with the development of microorganisms that had nothing to do with the yeast. In order to prove this, he sent for beer from the most reputable cafés in Paris and treated them by the techniques of pasteurization he had developed before the war. The heated beer did not undergo any alteration and kept indefinitely in the cellars of the rue d'Ulm. The unheated samples became cloudy. A fungus could easily be detected under the microscope; it was a threadlike parasite which caused the disease of the beer. As he had done in the case of the silkworms, Pasteur set out to make the microscope into a piece of business equipment, urging brewers to purchase one in order to detect these parasites, which, when mixed with healthy yeast, caused the beer to become cloudy, sour, curdled, ropy, or putrid.

Late in the summer of 1871, as the Parisians once again sat down at the little round tables of their cafés to order a bock and to celebrate the return of law and order, Pasteur took a trip to London. He wanted to check out some larger manufacturers, and he also wanted to inform himself about different processes, since English beer was always produced by high fermentation. He also hoped to promote his own technique of yeast selection, his methods of sterilization, and above all the use of the microscope, the only instrument that made it possible to separate the wheat from the tares. Jean-Baptiste accompanied him on this scientific and commercial expedition.

Making the rounds of the London manufacturers, Pasteur finally found a large-size plant, which produced 500 million hectoliters of beer annually, used a hundred horses, and employed 250 persons. Yet its output was not what it could have been, and Pasteur asserted that his method would improve it. Interrupting his inspection tour, he asked to be given a small piece of the English ferment called Porter's yeast, which was taken from the open channel between the reservoirs for the fermentation barrels. Since he took his microscope with him wherever he went, Pasteur, to the amazement of the English brewers, could not resist the pleasure of giving a microbiology lesson. Bent over his apparatus, he easily identified several parasitic germs for, as he peremptorily declared, "the working of the Porter's yeast leaves a great deal to be desired."[37] The Englishmen, astounded and then annoyed, had to accept the diagnosis. Pasteur continued his examination, and when he gathered up the paste used to purify the newly brewed beer, he found the same parasitic germ. The day's production was sure to be undrinkable. He asked some of the clients whether they had any complaints. They finally admitted that on that very morning they had to send for new yeast, since the one he was looking at was unsatisfactory. But Pasteur did not stop with the first barrel but asked permission to subject all those containing that day's beer to microscopic testing.

Returning a week later, he learned that the plant had lost no time in acquiring a microscope. All the beer yeasts had been discarded and replaced by new ferments. The English brewers had understood the lesson. As a result, Pasteur began to project large-scale experiments: "In any case," he wrote to Raulin, "get ready to start the experiments immediately. We cannot wait until new laboratories are ready; we will work in a brewery in Paris or in the suburbs."[38]

Pasteur was reasoning in rather simple terms, and thought that a germ-free yeast would produce a good beer. But he soon came up against the kindly but sly remarks of Bertin, who as his neighbor in the apartments of the Ecole normale, but especially as a good connoisseur of beer, gave his opinion on the taste of pasteurized beer. Bertin, who owed his fine palate of a beer lover to a long stay in Strasbourg, knew all about the Parisian cafés and their specialties. "First give me a good bock," he told Pasteur, "and then you can tell me about it!"[39]

Becoming a brewer was not a thing one could improvise — as Pasteur soon had to admit to himself. Cartesian though he was, he had to recognize that empirical experience sometimes accomplishes as much as careful prevention. Still, the scientist remained convinced that his studies would mark a turning point in the quality of beer. Bertin was less sanguine when he compared samples of the most famous beers with the products made in the cellar of the rue d'Ulm. While it was true that the commercial beers, when improperly stored, became infected and terrible to drink, Pasteur's brews eventually went flat and lost all flavor. Sterile they were, but also mediocre. The fact is that the complex flavor of the beers appreciated by those who frequented the brasseries had to do with many factors, among them the water, the brewing techniques, and the yeast, as well as its sterilization.

Bertin's critiques were troubling to Pasteur. He probably would soon have forsaken this field of experimentation if he had not imprudently solicited the financial help of a scientific group in whose eyes he felt he had to succeed. Duclaux was later to comment on this hesitation on the part of the head of the laboratory: "He had never mastered this subject, because he was never obsessed by it."[40]

And it is true that Pasteur lacked the talent needed for judging beer. Although he had learned to perceive differences through long training, he had by no means achieved the degree of acuity of a brewer or a regular customer of the Parisian cafés. He therefore sought other ways to make the best of his practical experiments. Accompanied by one of his assistants, Grenet, he spent a week inspecting a brewery at Tantonville in French Lorraine. The Tourtel brothers, famous brewers whose firm is still in existence, wanted to learn all about pasteurized beer and the use of the microscope. These studies made in Lorraine were to be used for a book published in 1875, in which Pasteur treated a number of questions that were often only distantly relevant to brewing, for he used beer making as a pretext to develop theories about alcoholic fermentation, the origin of wine yeast, and even of the transformation of species.

Meanwhile, however, he continued to pursue the practical side of his work. Intent on protecting the industrial potential of his diverse inventions, he included several additions in his original patent. In November 1871, January 1872, October 1872, and March 1873, he patented additional details concerning the manufacturing and conservation processes for an unalterable beer, the relevant machinery, and the ingredients used to produce a brewer's yeast that could safely be transported.

In the end, Pasteur's research was put to the most profitable use in Denmark. In that small country, beer brewing was a flourishing industry; the Danes were producers and consumers, but also exporters. Danish beer had to be able to travel, to cross the seas, sometimes as far as India. It is therefore understandable that Pasteur's books, publications, and experimental demonstrations should have attracted a great deal of attention.

In Copenhagen, Carlsberg beer was produced in a famous brewery, which, founded in 1847 by Jacob Christian Jacobsen, had quickly evolved from a simple manufactory to a large plant. Very young and enterprising, Jacobsen sought to make a quality beer. Unable to find an appropriate yeast, he left his farm in Jutland and went to Munich to obtain pots of yeast from the master brewer Gabriel Seldmar, director of the brewery *Zum Spaten*. On his way back, his precious load almost died. Only by wetting the yeast at every stop was he able to keep the ferment alive. It was during this journey that Jacobsen realized the importance of proper conservation techniques. By royal privilege, he was able to install his storage rooms in the ramparts of Copenhagen and to produce the first Danish beer brewed by the Bavarian method. In 1870, his son Karl took over the business and built a new brewery, the celebrated Ny Carlsberg. At this time Jacobsen and his son became aware of Pasteur's work and used it to improve their manufacturing processes. These advances in the brewing process increased their profits to the point that the two brewers, father and son, became the leaders in their field and were able to assign some of their earnings to a philanthropical foundation. Aside from a rich museum of sculpture, known as the Ny Carlsberg Glyptotek, where the neoclassical works of Thorwaldsen occupied the place of honor, they also subsidized the University of Copenhagen and above all built and maintained a major laboratory dedicated to research on beer.

Pasteur would not be forgotten. In 1878, Jacobsen had his marble bust placed at the entrance of Pasteur Street, which led to the Ny Carlsberg. The bust was intended as "a commemoration of the services rendered to chemistry, physiology, and brewing through the study of low fermentation." The bust was executed by Paul Dubois, a fashionable sculptor whom Pasteur had met during his stay at the court at Compiègne, and the scientist received a copy as a gift.

By way of thanking him, Pasteur gave Jacobsen access to his latest invention, a bisque filter he had developed with his assistant Charles Chamberland and which yielded a microbe-free water. In 1884, when Pasteur attended an international medical convention at Copenhagen in the presence of the king and queen of Denmark, he used this opportunity to visit the brewery. By this time, it no longer exported four million, but two hundred million hectoliters of beer per year. The entire Carlsberg plant was run according to the principles taken from Pasteur's studies. The head of the laboratory, Hansen, used different kinds of yeast to give different flavors to his beers, using multiplication techniques he had learned in Paris.

Less than fifteen years after the tragic defeat of 1870, Pasteur had thus indeed succeeded in creating the "beer of revenge." But it was a properly thought-out revenge, a far cry from narrow nationalism and grand patriotic gestures. Looking at an industry he considered inadequate and weak, he wanted to use his expertise to contribute to the recovery of the state. During the French Revolution, he pointed out, Lavoisier, Chaptal, and Berthollet had taught new means of extracting saltpeter, manufacturing steel, and making gunpowder, and Monge had shown how to found canons. The role of the scientist in a national crisis was clear to Pasteur: he was called upon to apply the results of basic research to an emergency situation.

Sainte-Claire Deville expressed a generally accepted idea when he asserted that France had been crushed by technology rather than by tactics. Pasteur went even further: seeking to go beyond a strict distinction between science and warfare, he predicted that economic war would become the new form of the struggle for power and control over people. When he realized that his inventions enabled him to control the microbes that could contaminate wine and beer, he therefore felt that he was handing a weapon to his country.

At the same time, Pasteur could not help but think about other applications for his methods. He had seen the wounded of the battlefields of 1870, he had smelled the atrocious stench of gangrene, and he had recognized these mutilations and these putrefactions as aspects of the same principle he had sought to defeat in order to save fermented beverages or silkworms. "Seeing that beer and wine," he wrote, "undergo profound alterations because these liquids have given shelter to microscopic organisms which, having invaded them fortuitously and without being seen, have rapidly multiplied within them, how can one help being obsessed by the thought that phenomena of the same kind can and must sometimes occur in humans and in animals?"[41]

The Second Empire had crumbled like a house of cards, but this did not slow down the research Napoleon III had asked Pasteur to carry out. On the contrary, it led to technical and even commercial advances rich in implications — and profits — for the future. Indeed, the defeat seems to have spurred the scientist on to greater combativeness and oriented him toward new endeavors, from the needs of industry to the struggle to save lives. Henceforth, Pasteur's laboratory was to have ties not only to the factory and the manufacturing plant but also to the hospital.

IO

A Chemist among the Doctors

Over the centuries, medicine has always been ahead of the precise knowledge of the mechanisms of the body and its cells, for the healer came before the scientist. In treating the patient, the premodern physician considered the symptoms and looked for empirical solutions. After the turning points of the Enlightenment and the Revolution, medicine endeavored to establish principles that would place medical practices on a rational footing. "Medicine is the physiology of the sick person," Magendie proclaimed at the beginning of the nineteenth century, while Lamarck, in 1815, invented the word *biology* to underline the role of the environment in the life processes. The idea that one must understand before one can heal and that a good practitioner must know how the disturbed functions are supposed to work is not as obvious as one might think. For a long time, the lancet was kept apart from the microscope.

It took the teachings of Pasteur and Claude Bernard, who shared a bench at the Academy of Medicine, to usher in a new mentality. Yet the conceptual contributions of microbiology conflicted with the ways of a self-assured and autocratic medical caste, so that more than a quarter-century had to pass before they made inroads into medical and surgical habits and practices. As late as 1883, Michel Peter asserted that a physician has no need to clutter his mind with the science of a chemist, a physicist, or a physiologist; turning to Pasteur in a discussion about typhoid, he added: "What do I care about your microbes? If you have seen one, you have seen them all." In response, Pasteur denounced what he considered a "medical blasphemy,"[1] insisting on the need to redefine the roles of the medical practitioners: In treating the patient, the physician is no longer isolated within a dogma; he must not only heal but prevent illness; prevention and hygiene are individual as much as collective matters. However, it took a long time for his words to be heeded.

Pasteur and Medical Biology

Even before Pasteur had experimented in the medical field, his studies on germs and their effects had laid the foundations for practices from which the surgeons should have been the first to benefit. Yet these men, skillful wielders of the scalpel who were convinced of the prime importance of anatomy, were slow to understand that if their efforts frequently ended in failure it was because they carried with them, on their instruments and their dressings, germs that sowed infection and death. Several generations passed before they understood that autopsies and the handling of anatomical specimens are incompatible with good operating practices and that it is more important to be clean than to be quick.

Practicing asepsis and antisepsis was among the first medical lessons to be learned from Pasteur's studies on the chemistry of fermentation and so-called spontaneous generation. Asepsis designates the set of means deployed to prevent contamination by infectious agents. Sterilized dressings, washing of hands, and wearing sterile gloves and gowns are means used today to block the introduction of microbes into the hospital environment and during surgical and obstetrical interventions. Antisepsis is the technique designed to eliminate microbes and prevent their penetration at the level of the wound. Before there were antibiotics, disinfectants such as iodized alcohol, carbolic acid, or silver sulfate were used as antiseptics. Asepsis prevents microbes, antisepsis destroys them. By discussing, discovering, and going beyond the notions of asepsis and antisepsis, Pasteur came to write the first chapters of his medical revolution.

In the last stages of his studies on beer, Pasteur stated some principles concerning the causes of infectious disease. He had long stressed the logical connection between his work on fermentation and putrefaction and the study of microbial infections, for the research on wine he had conducted at Arbois had already given him an insight into this biological congruence. When he showed that grape juice does not ferment in a sterile milieu, he concluded from this in a flash of insight: "May we not believe by analogy that the day will come when easily applied preventive measures will stop the scourges whose sudden appearance devastates and terrifies entire populations, such as the yellow fever that has recently invaded Senegal and the Mississippi Valley or the plague that has raged on the banks of the Volga?"[2]

But perhaps one has to go back even further in Pasteur's thinking and his notes to find traces of his first medical preoccupations. Did he not already speak of sick or wounded crystals in his writings on asymmetry? Did he not at first interpret racemic acid as a sign of disease in the grape? One should therefore not think that Pasteur changed directions when he took on the epizootic of the silkworm, but rather think of it as an extension of his earlier work; for him, the parasite that he found in the diseased silkworm moth was a new version of the mysteries of yeast and the must of grapes: "Depending on a person's constitution and temperament," he wrote at that time, "epidemic diseases most often affect those who are susceptible to them."[3] Thus, as Pasteur advanced into the chemistry of fermentation, his goal became increasingly clear.

Medical concerns, then, were not new to Pasteur when he decided to study infectious diseases in humans. Actually, the idea was not as original as one might think. In the seventeenth century, the physicist Robert Boyle had already asserted that "he who would explore to the very depth the nature of ferments and of fermentation will be much more capable than anyone else of providing a correct explanation of morbid phenomena."[4] All the treatises of the eighteenth and nineteenth centuries noted similarities between fermentation and disease: "What happens in the fermentation of wine," wrote the chemist Nicolas Lémery in 1713, "can be very helpful for explaining certain diseases, particularly smallpox."[5] However, since at the time the cause of the fermentations was far from understood, the arguments proposed for this analogy were often quite erroneous. "Children are more susceptible to this disease than adults because their blood is more similar to must. . . . People usually have smallpox only once in life, just as must ferments only once." Under this assumption, the exterior signs of inflammation, that is, abscesses, swelling, and rash, caused observers to assimilate the disease to the bubbling of a fermentation. In accordance with Liebig's ideas, diseases were described as "resulting from inert movements of substances undergoing alteration." Others thought that their origin and development was as "spontaneous" as that of the fermentations. Medical theories thus conformed to the current understanding of fermentation.

When Pasteur revolutionized the science of his era by discovering the germs and their role, it was only natural that he should become interested in medicine and hygiene. In studying airborne germs, he had, after all, sought to understand what happened when they entered an internal environment, particularly in the presence of blood and injured tissue. Always intent on fostering the application of his discoveries, he had very soon perceived their practical importance. Under the Second Empire, he called attention to the potential interest of his research for public health and pointed out that great social prestige would come to the Emperor and his government if they adopted this point of view. "The applications of my ideas are immense," he wrote to Favé in 1863. "I am ready to approach that great mystery of the putrid diseases, which constantly occupy my mind."[6]

Yet, if the evidence was so simple and the subject so important, why had it taken Pasteur so long to go to work on the role of the microbe in human pathology? At the time of the debate over spontaneous generation, a physician, Gilbert Déclat, gave a lecture on the role of the microbes and praised Pasteur's work on fermentation and their importance for under-standing disease. At the end of the lecture, Pasteur, who did not know Déclat, went up to thank him, but also to offer his criticism: "The argu-ments with which you have supported my theories are very ingenious, but lacking in rigor. Analogy cannot serve as proof."[7] Pasteur said this because he was convinced that it was useless to keep the controversy going with hypotheses, even if they were well devised. The only thing that counted was experimentation.

Involved in other studies, hard-pressed by the diseases of silkworms and wine, slowed down by the war of 1870 and by his stroke, Pasteur had neither the means nor the time to attack the complexities of pathology. There was also another obstacle, which was perhaps more important than these outward difficulties: Pasteur was not a physician. In October 1873, he wrote to Déclat: "How I wish I had the health and the special knowledge I need to launch myself wholeheartedly into the experimental study of one of our contagious diseases."[8]

Actually, Pasteur's work in experimental physiology had been re-warded as early as 1859 by a prize he received from the hands of Claude Bernard on behalf of the Academy of Sciences. Such a distinction, coming from the founder of experimental physiology, amounted to an official rec-ognition, and Pasteur himself did not hesitate to speak of "physiological chemistry" in connection with his studies. But the practitioners of the nine-teenth century, while they might benefit from the microscope, for example, and from all the other technical advances in the biomedical field, continued to be clinicians first and foremost. In 1876, a fourth of the French medical profession consisted of health officers. Medical culture was encrusted with prejudices perpetuated in the name of tradition and common sense. An ex-perimenter was irritating because of his propensity for questioning every-thing. Journals like *La France médicale*, *La Gazette médicale de Paris*, or *La Semaine médicale*, surgeons like Chassaignac, Péan, or Després, professors like Jolly at Toulouse, Wiart at Caen, or Béchamp, dean of the Catholic faculty of Lille, with their clientele of generalists, had little use for the medical revolution implied in the Pasteurian theories on the role of the microbe.

Since 1860, the diffusion of Darwin's concepts of evolution had stressed the importance of heredity. Illness was often interpreted as a morbid predisposition. The enemy within, the "temperament," the terrain, was to be feared more than the assault of bacteria. Actually Pasteur, who had suffered a stroke that had been treated with nothing more than the application of leeches behind the ears, was well aware that there are other mechanisms of morbidity than the microbe. Thus, when the formidable mandarin Hermann Pidoux, coauthor with Armand Trousseau of the most famous medical treatise of his time, dogmatically stated: "Disease exists within us, because of us, through us," Pasteur did not entirely disagree. "This is true for certain diseases," he wrote cautiously, only to add immediately: "I do not think that it is true for all of them."[9]

At the time, several medical theories existed side by side. All the supporters of contagion, or contagionists, greeted Pasteur's discoveries on the role and the reproduction of germs with approval. In the first half of the century, Pierre Bretonneau at Tours and René Laënnec in Paris were already convinced that typhoid and diphtheria are contagious diseases. Other renowned practitioners also suspected that agents of contagion were involved in a number of other infectious diseases, among them measles, scarlet fever, syphilis, scabies, and smallpox. In one of his last clinical courses at the Hôtel-Dieu, Trousseau, who died in 1867, had explicitly made a connection between fermentation and pathology: "This, then, is the grand theory of ferments applied to an organic function. . . . Perhaps it also applies to the morbid miasmas, which perhaps are ferments that, deposited in the body at a given moment and in certain specific circumstances, will manifest themselves by producing a variety of substances. Thus the smallpox ferment will lead to smallpox fermentation, which will give rise to thousands of pustules; the virus [i.e., poison] of glanders will produce its own fermentation, as will the sheep-pox virus. Other viruses seem to act locally; yet they too subsequently modify the entire body; thus we have hospital gangrene, the malignant pustule, or contagious erysipelas. Under these circumstances, should we not agree that the ferment or organized matter of these viruses will be carried to one place by the lancet, to another by the atmosphere or by pieces of dressing?"[10] These prescient words laid the foundation for the rules of hygiene. Certain obstetricians, such as Jules Helot in 1855 and later Stéphane Tarnier, pleaded for draconian measures of cleanliness in maternity clinics in order to prevent the transmission of agents of infection to pregnant women. These hygienist physicians did not wait for the first experiments of Pasteur and his disciples to preach interdiction, quarantine, and disinfection. Suspect wells were closed long before the discovery of the typhoid bacillus.

Before he had even carried out his first medical experiments, Pasteur thus already had ardent defenders among the hygienists, who were dead set against spontaneous generation. During his 1864 lecture at the Sorbonne, the medical students sat in the first row, eager to hear Pasteur disparage the partisans of heterogenesis. In 1867, a number of physicians came to hear him speak at Orleans about the eelworms in vinegar. On that occasion Casimir Davaine took an important step forward when he asserted that the abstract concept of "miasma" should be replaced by the experimental study of bacteria. Precisely at the same time, Jean-Antoine Villemin sought to demonstrate through inoculation that tuberculosis is a transmissible disease.

Nonetheless, these same contagionist clinicians were often loath to adopt a single-cause explanation based exclusively on the microbe. Did not the ubiquity and the latency of germs prove that in the final analysis the disease is triggered by the terrain, that is, the individual body? Moreover, these clinicians distrusted conclusions arrived at only through laboratory experiments, whose conditions were totally different from the circumstances of the daily life of humans and animals; they listened to Edouard Chassaigne, one of the most powerful chiefs of the Lariboisière Hospital, when he pleaded "against that laboratory surgery, which kills a great many animals and saves very few humans," adding, "every finding that comes out of the laboratory must be cautious, must be modest and reserved as long as it has not received the sanction of the surgeon's long and patient research, as long as it has not obtained the kind of clinical investiture without which there can be no true medical science and practice."[11]

It should also be said that Pasteur's preference for the Academy of Sciences, his predilection for collaborators who were graduates of the Ecole normale or veterinarians, the severe discipline of his laboratory, and his sharp language, which did not sit well with the audiences at medical lectures, at first tended to marginalize him, even make him suspect. It would take many years for the medical world to accept the germ theory and to integrate it into its medical practices.

Certain surgeons finally helped Pasteur overcome his timidity concerning human medicine, for as Alphonse Guérin, Just Lucas-Championnière, Stéphane Tarnier, Ulysse Trélat, and Félix Terrier expatiated on their master's hypotheses, they encouraged Pasteur to interpret and vaunt the merits of asepsis and antisepsis and to make a strong case for combating contagion through hygiene.

The Surgeons of the Empire

In the nineteenth century, according to the surgeon Velpeau, a simple pinprick was "a door opened to death."[12] The risks in lancing a whitlow or an abscess were so great that surgeons hesitated to do it. Minor and in themselves harmless surgical interventions became complicated by secondary infections that often brought gangrene, followed by amputation and death. What, under these conditions, can be said about major operations, such as abdominal interventions, in which the surgeon's hand and his lancet plunged deeply into the patient's internal organs?

After the First Empire, surgery seems to have not just stood still, but actually regressed. In earlier times, surgical patients died less frequently, for the surgeons unknowingly practiced antisepsis: In ancient Greece, Hippocrates had written this about surgery: "What is not cured by drugs is cured by the knife, and what is not cured by the knife is cured by fire, unless the evil is incurable." Wounds were indeed cauterized by fire or by pouring boiling liquid onto the bleeding flesh; in this manner, the Greeks had invented disinfectant substances and developed the custom of treating wounds with detergent, desiccative, astringent, and caustic agents that had a definitive antiseptic effect. In those days, wine, vinegar, alum, sun-dried verjuice seeds, verdigris, fig tree bark, nut-gall, and chalk water were put on wounds, and in addition it was recommended to wash them with hot or boiled water. In 1749, a *Treatise of Medicine and Surgery for the Poor* still recommended not to leave a wound exposed to contact with the air. It also cautioned against inserting a finger or a probe into the wound and suggested that it be dressed with a cloth dipped into hot wine or brandy.

The surgeon Dominique Larrey, a baron of the Empire and the most famous physician of Napoleon's Grande Armée, which he followed from the Egyptian campaign to Waterloo, owed his glory in part to the procedures he invented to ease the suffering of the wounded of Napoleon's campaigns. Known for his skill in using fire, boiled compresses, or improvised splints, and in applying warm oil or brandy to wounds inflicted by canon and gunshots, he was called the Ambroise Paré of the battlefield.

But then came Broussais. François Broussais too was a physician of the Grande Armée. A pupil of Bichat and above all of Pinel, he had learned from the latter to classify diseases into genera and species, just like the plants and animals of natural history. Propagating the ideas of the eighteenth-century English physician John Brown, who held that life is maintained only by irritation, he set out to bring down the old dogmas and to reform therapeutics. In 1816, after the shipwreck of the First Empire, Desgenettes had him appointed professor at the Val-de-Grâce [army teaching hospital]. There he launched a campaign that was to convulse the world of medicine. Endowed with a lively imagination and great audacity, this Napoleon of the Val-de-Grâce, this Danton of medicine, as one of his hagiographers called him, invented physiological medicine. Organs and living tissues, he said, are in a state of equilibrium. Any stimulation engenders modifications. If the equilibrium is broken in one sense, a debility results; if it is broken in another sense, there is irritation, that is, "a stimulation of acts of living chemistry," and consequently, inflammation. Under these assumptions, Broussais saw inflammation everywhere, since most diseases lead to the irritation of the tissues. In his eyes, all diseases were local in origin and characterized by the excess or lack of excitation of the injured tissue above and below their normal state. Diseases, therefore, were not entities but perturbations of the normal physiology. Consequently, Broussais banished medications and strongly recommended the use of leeches. This pragmatic medicine did not waste words and dogmas and went straight to the symptom.

The new medical doctrine of inflammation brought the use of poultices and lint into prominence, and more or less unsterile practices made surgery much more hazardous. Surgeons prescribed emollients, relaxants, and putrefying agents. Complications of wounds were attributed to the suppression of pus, and the conclusion was that abundant suppuration was necessary for healing. A crushed finger healed by traditional medicine, that is, by soaking it in a bowl of brandy, could not convince the followers of Broussais, who saw such unexpected cures as a happy result of luck. At the foot of the dying patient's bed stood pans in which poultices were soaking, and decongestants were prepared from rye flour, malt, marsh mallow, turnips, carrots, onions, and ground tanner's bark. Old and summarily washed hospital sheets were cut up into lint, for lint was considered the indispensable component in the healing of suppurating wounds. The cloth had to be sufficiently fine and above all used, for if it was too new it would form lumps. Finely ground to form dusters, pledgets, balls, or wads, lint was considered preferable to hard, carded cotton, sponge, dried leaves, or straw. How often were these dressings stored near the wards, the dissecting room, and even the latrines?

Also applied to wounds were poorly prepared ointments that were left to age in dusty antechambers. These greasy powders and creams, which according to Broussais's theories were the principal agents in combating inflammation, were also the perfect vehicles for germs. Intended to counteract the miasmas, they were debilitating and repulsive, and they transmitted microbes.

Meanwhile, a few surgeons did experiment with dressings of alcohol, water laced with alcohol, or camphorated alcohol. Léon le Fort, head of surgery at the Cochin Hospital, forbade the use of sponges and demanded rigorous cleanliness and regular washing of hands of his students. But he was an isolated figure, and even though he had reduced the level of mortality in amputations by a third, his colleagues had no idea that one of the main secrets in overcoming infection was a new way of dressing wounds and performing operations.

Try to imagine an operation at that time. It would take place in a small room stuffed with furniture, or else in a large amphitheater. Paradoxically, the needs of science had created a close proximity between the operating room and the dissecting room. Surgeons went from one to the other all the more easily as it was considered highly informative to compare the symptoms of the living with the organic lesions of the cadavers; this was the anatomo-clinical method advocated by Morgagni a century earlier. Students in filthy gowns who participated in the interventions often had in their pockets anatomical specimens they had picked up in the dissecting room. Large groups of spectators, family members, and students often watched operations, huddled in a corner. There was constant coming and going; all kinds of people crossed through the room, where the lighting was poor. The windows, which provided the only light during the day, cast lateral shadows. At night, it was difficult to see at all, for the candles and alcohol lamps were placed far from the operating field, and the surgeon and his assistant cut off what little light there was. If it was an abdominal operation, it would take place in almost total darkness. When the patient was brought in on a stretcher, he or she had to be put to sleep. General anesthesia first came into use in 1864, after William Thomas Green Morton had demonstrated the effectiveness of ether at the Massachusetts General Hospital. Although the method had spread rapidly, the technique remained rudimentary in many instances. A student, often the youngest one, and sometimes even the surgeon's coachman, moistened a cloth with the anesthetic fluid. Almost always there was a danger of explosion or intoxication, for the ether or chloroform was used near the gas lamps. Since the anesthetic was often improperly dosed, the patient frequently awakened and hampered the intervention by impulsive movements. This would badly frighten the onlookers, who were called upon to help. The surgeon shouted to cover the moaning while the family or the students held the patient down.

The surgeon operated in a black suit, looking rather like one of the butchers at the slaughterhouse. His old morning coat was stained with blood and pus. The skill of the surgeon was measured by the state of his smock; the more soaked and stained it was, the prouder he was of it. There was no point in keeping clean; indeed cleanliness would have been considered a sign of preciosity in this masculine job. The surgeon rolled his sleeves up to the elbow. Some, like Péan, protected themselves by tying a towel around their necks. The instruments were often the property of the operator. At the decisive moment, he would pull one from his pocket and wipe it on his handkerchief. If, during the intervention, the surgeon needed to have his hands free, he held the instrument between his teeth. The operating nurse handed him the thread, which she rolled from a bobbin attached to her chest, cutting pieces that were hanging down without giving any thought to microbes. Sometimes the surgeon brought his own sponges, lint, and bandages. The important thing was to work quickly and adroitly to avoid all mishaps. After the operation, the patients were taken back to decrepit wards without running water or central drainage. One can easily understand how difficult it would be to fight these old established methods at a time when the art of the operator was essentially a matter of skill and speed. A surgeon did not hesitate to make the blood spurt, to place his finger on the severed vessel, to cut quickly and deeply into the tissue, unconcerned about those around him, the screams of the spectators, and sometimes the awake patient, about dust, stench, and filth. Wounds almost always suppurated; it was nothing to be concerned about — until the patient died.

During the war of 1870, the catastrophic consequences of these surgical practices reached their dismal high point. Of 13,173 amputations of different kinds performed in the French military hospitals, 10,006 ended in the death of the amputee. One therefore had to think twice before considering an operation that almost always amounted to a death sentence. At the same time, the lint made from sheets soiled by the pus of the last dying man, dirty dressings and sponges continued to be used to dress wounds and wipe off running sores. Wards for the wounded and the amputees — whether in the field hospitals that turned into veritable death houses or in the luxurious lobby of the Grand Hotel that became an improvised surgical unit during the siege of Paris — remained imprinted in the memories of all the medical students of that era: all of them described how, as soon as one entered, one was assaulted by an acrid and putrid smell that impregnated one's clothes. Following the operation, every patient suppurated, to the point that the surgeon Landouzy said that the pus seemed to germinate, as if it had been sown by the surgeon. Little did he know how truly he spoke.

In France, a surgeon by the name of Alphonse Guérin came to suppose by the end of the war of 1870 that infection might be caused by the germs of fermentation described by Pasteur. He wrote:

> I believed more firmly than ever that miasmas emanating from the pus of the wounded were the real cause of the dreadful illness to which, to my dismay, I saw the wounded succumb. . . . It then occurred to me that the miasmas whose existence I had assumed because I had no other way to account for the production of purulent infections, and which were known to me only through their deleterious influence, might well be animate corpuscles of the kind that Pasteur had seen in the air. . . . If miasmas were ferments, I would be able to protect the wounded from their baneful influence by filtering the air as Pasteur had done. . . . I therefore devised the cotton-wool dressing and had the satisfaction of seeing my predictions come true.[13]

What Guérin did after having painted the wound with phenolized water or camphorated alcohol, was to apply increasingly thick layers of cotton-wool. He then covered this dressing with bandages of new and stiff linen which served as compresses. Wrapped in this package, as he put it himself, the cotton-wool dressing remained in place for about twenty days. Applied to the wounded of the Commune at the Saint-Louis Hospital, this procedure yielded remarkable results. Between March and June 1871, nineteen of the thirty-four operative patients whose wounds were dressed with cotton-wool survived. When Guérin became surgeon at the Hôtel-Dieu in 1873, he continued to apply this procedure there and consistently obtained encouraging results.

Semmelweis and Asepsis

Guérin's method was actually the application of the extraordinary intuitions of a surgeon of Hungarian origin practicing in Vienna, Ignaz Semmelweis. Pasteur never alluded to Semmelweis, of whose existence he seems to have been unaware. But there is nothing surprising about that, for the Kafkaesque destiny of this pioneer of asepsis constantly wavered between the tragic and the melodramatic, and his studies were not widely known. In France, it was only in 1924 that a young physician with a literary bent and a fascination for extreme cases became interested in Semmelweis and wrote his thesis about him; the name of this young practitioner who saw himself as a doctor to the poor was Louis Ferdinand Destouches, better known under the name of Céline.

Semmelweis was the first to understand the mechanism by which wounds become infected. Even before Pasteur, he proved that bacteria cause putrefaction and demonstrated that it is possible to prevent postoperative suppuration by requiring surgeons to wash their hands. But, perhaps because "his discovery was too great for the strength of his genius," as Céline was to write, and surely because his character caused him to compound awkwardness with obstinacy, Semmelweis was incapable of convincing the world.

On 27 February 1846, Ignaz Semmelweis, who had just taken up his post as assistant in Professor Johann Klein's maternity ward A of the General Hospital of Vienna, was on duty. This meant that the maternity clinic B, directed by Professor Franz Bartsch, was closed, for women in labor were admitted to the clinic where an assistant was on duty. On that day, Semmelweis admitted a woman who, having gone into labor in the street, had come to the hospital to be delivered. She begged and pleaded to be admitted to Bartsch's clinic and adamantly refused to be delivered in Klein's. It was not the first time that this situation had arisen, and Semmelweis soon found out about the rumors circulating in Vienna: people knew that if the risk of postnatal complications was high in Bartsch's clinic, it was inevitable at Klein's. To enter this clinic was a death sentence; it was better to have the baby in the street. Taking his patients' distress seriously, Semmelweis set out to investigate the causes of the dread infection that struck the new mothers, puerperal fever.

Who was Semmelweis? A young emigré from Buda, born in 1818 as the son of a grocer, he had come to Vienna to study Austrian law. But he never came to like the imperial capital, or the law, which he gave up after a few weeks. Like all contemporary students at loose ends, he was slow to make up his mind. Having given up the law, perhaps he should try medicine. So he signed up for the first available anatomy course in a hospital, attended an autopsy, and came out convinced: he wanted to be a physician.

At that time the Vienna school of medicine was in its heyday. Two men would guide the studies of young Semmelweis. Karl Rokitansky, professor of pathological anatomy, was one of the first to establish a connection between the anatomical examination of organs and the description of pathological symptoms. Explaining and investigating how the alteration of organs modifies their functioning, he brought young clinicians into the dissecting room. In Vienna, the anatomo-pathologist was the specialist in whose field all branches of medicine converged. His comments at the patient's bedside sounded as if he were examining his organs. An affectionate and prestigious teacher, Rokitansky taught Semmelweis to combine clinical observation with the study of histological lesions.

Semmelweis also took the courses of Josef Skoda. Severe and distant, Skoda was a master of clinical science, which he based on auscultation and percussion. An influential figure who enjoyed a great reputation in Vienna, he taught that since it was impossible to cure disease, medicine had to prevent it and therefore must devote itself to the study of epidemic diseases, such as typhoid and cholera. He thus came to place major emphasis on hygiene.

Semmelweis unquestionably owed a great deal to Rokitansky and Skoda, who became the intellectual fathers of his research. But even though he was noticed by his teachers, he was not admitted to their inner circles. Skoda preferred one of his competitors, and Rokitansky steered him toward surgery. Yet Semmelweis became discouraged with this field and abandoned it after two years: "All that is done here seems quite useless to me, and the deaths go on and on. We go on operating without really trying to find out why one patient rather than another succumbs in identical cases." This being the case, Semmelweis turned to obstetrics, and because Klein needed an assistant in his clinic, he went to work there and obtained the necessary certification within two months.

It seemed as if a curse weighed on the two identical obstetrical clinics at the General Hospital of Vienna. In May 1846, 96 percent of the women admitted there died. There was not a glimmer of hope. Nor were the rumors that circulated without foundation: more women died in Klein's clinic than in Bartsch's.

At this point, Semmelweis noticed that in the second clinic only midwives in training performed pelvic examinations and assisted with births, whereas in the first clinic medical students officiated at the births. This observation led him to a practical conclusion, namely, that since death came with the students, they should be banished from the delivery room or, rather, replaced by midwives. When the experiment was tried, Klein's statistics fell in line with those of Bartsch. This was quite upsetting to the two obstetricians, who fired half of the students, especially the foreigners. This preliminary measure brought a lowering of the death rates that lasted for several weeks.

"The cause I am looking for is in our clinic and nowhere else," Semmelweis insisted. He therefore spent his time trying to understand why and how the medical students could transmit disease and death. Observing them in all of their activities, he finally noticed that they blithely went from the dissecting rooms to the cubicles where pregnant women were examined. He was coming close to the truth.

Semmelweis realized this when he learned of the death of a professor of anatomy whom he had greatly admired. Dr. Kolletschka perished of a generalized infection after he had accidentally cut himself during an autopsy. This was the missing link in the chain of his deductions. As he studied the details of the disease and the organic lesions of the autopsied cadaver, Semmelweis understood that puerperal fever and the affection that had killed Kolletschka were one and the same disease entity. Like every other scientist of his time, he did not, of course, know anything about microbes, but he had the intuition that the death of Kolletschka was related to an inoculation with "particles" that had come from the cadaver. These "particles," he surmised, were spread through hospital wards by the hands of students and physicians and transmitted mortal diseases.

At the same time that he discovered the cause of the evil—without, however, being able to produce proof—Semmelweis also invented asepsis. He demanded that physicians and students wash their hands with a chlorine solution until they were clean and the smell of the cadaver disappeared. The application of this method was a genuine success. The mortality rates declined drastically, falling to less than 2 percent in both Klein's and Bartsch's clinics. The matter, then, was perfectly obvious, both as to the nature of the problem and its prevention.

And yet, either because the remedy was too simple or because it had been devised by a young Hungarian immigrant who competed with the pundits of Viennese obstetrics, no one believed in it. One of the main nonbelievers was Klein, on whom Semmelweis was dependent and who seemed to prefer fatalities to a chlorine solution. However, Semmelweis was not completely isolated; Rokitansky and Skoda supported him.

While Semmelweis was convinced of the importance of his theory, he did not find a way to publish it. Timidity, pride, difficulty in expressing himself in German? No doubt it was a combination of all of these things — yet one must also add certain personality problems that led to his complete isolation. He flew into a towering rage whenever he observed the slightest deviation from his rules of cleanliness. He was incapable of being diplomatic with those around him, and Klein was offended by this overly brilliant assistant whom he criticized not only for his theories but also for his insubordination and arrogance. When Semmelweis's contract expired, he therefore refused to renew it. At this point, Bartsch hired the young physician as a supplementary assistant.

Meanwhile Klein succeeded in gathering together at the Faculty of Medicine a number of adversaries of the method developed by his erstwhile collaborator. These men ridiculed Semmelweis's explanations: how could the skillful hands of surgeons possibly transmit such a dread disease? And, in fact, just what was the disease whose only manifestation was the smell of cadaver?

Jealousies and criticism were unleashed at the hospital. The personnel and the students were tired of washing their hands. The most famous German obstetrician, Kiwich, wishing to learn firsthand about the experiments of which he had heard, came to Vienna to observe the odd methods practiced at the maternity clinic of the General Hospital. His verdict was straightforward: he wrote that he had seen nothing that would support Semmelweis's theory.

In 1849, the scope of the debate widened, and the university milieu was rocked by a veritable scandal. To be sure, Semmelweis still had Skoda and Rokitansky to defend him, but he had not yet published the details of the experiments he had carried out, even though he propounded them with messianic fervor. In 1850, three years after he had discovered the cause of puerperal fever, he finally condescended to speaking about the topic before the Vienna Medical Society. But he showed insufficient conviction, and the defense of his friends was not spirited, while his adversaries energetically denied a cadaver-related affection. Semmelweis's paper was not even deemed worthy of inclusion in the society's report.

At this point, Semmelweis abruptly fled Vienna. Was he disappointed in the medical establishment or guided by his old demons of pride? Was he ashamed of his poor German or afraid that he would be labeled an immigrant forever? From one day to the next, without warning or saying goodbye to anyone, he simply left all those who had supported him and who, disappointed and hurt, stopped speaking of him and his theories, except to criticize them.

Having returned to Pest in 1850, he was appointed director of an obstetrical clinic and then to a professorship at the university. In these

capacities he kept applying the same theory, continued to observe its success, and was glad to speak nothing but Hungarian. But this was precisely
what defeated him: Pest was not Vienna, not a capital of European-wide
influence. Before long, no one remembered his work.

In August 1860, when Semmelweis finally decided to publish his
discoveries, he wrapped his thesis into so much bitterness and megalomania
that he lost his credibility: "I am predestined to reveal the truth," he proclaimed like a prophet.[14] In reality, he was already condemned to general
paralysis and dementia. Syphilis in its terminal stage or Alzheimer's? Whatever it was, confinement was inevitable, and when he died in 1865, the
founder of asepsis was a shadow of his former self and his work gradually
slipped into oblivion.

Lister and Antisepsis

Since Semmelweis had not succeeded in communicating his discoveries to the scientific world, it is the Englishman Joseph Lister who must
be officially credited with being the first to apply Pasteur's discoveries to
medicine. He became the founder of modern surgery when he identified
gangrene of the extremities as the first example of a pathological fermentation in humans, thereby establishing a connection between the putrefaction
of necrosed tissue and the action of germs from the outside discovered by
Pasteur.

It was not by coincidence that Lister became acquainted with Pasteurian reflections, for his father, Joseph Jackson Lister, was a wine merchant at Upton (Essex) who, despite an early end to his schooling, was
keenly interested in mathematics and optics. That is how he came to initiate
himself into the techniques of the microscope, and eventually to publish
studies on the form of red corpuscles and the manner in which they are
stacked in cylinders. Interested not only in biological matters, Joseph Jackson Lister also discovered some optical laws concerning the achromatic
alteration of lenses. These would open the doors of the British Royal Society of Optics to him.

Joseph Lister thus grew up in a scientific atmosphere, to which must
be added the puritanism of his mother. He was educated in Quaker schools
but clearly had little use for the Quaker way of life when it came to choosing
a profession. He was not even ten years old when he declared that he would
become a surgeon. His father was delighted to see his son heading for a

scientific profession, and in 1848, when Joseph attained his maturity and enrolled in the university, gave him one of his best microscopes. The son was to put this present to good use, and his first studies charted the direction of his future research. Of these two papers presented to the University College Hospital, London, one concerned gangrene, and the other the use of the microscope in medicine.

A brilliant career was thus off to an early start. At the age of twenty-seven, Lister managed to wrangle a stint of special surgical training with the renowned practitioner James Syme, who shortly thereafter helped him obtain a position as certified surgeon in Edinburgh and, more importantly, accepted him as his son-in-law: Syme's first assistant married Agnes, the chief's daughter.

Simultaneously with his activities as a surgeon, Lister pursued research. Agnes and Joseph had converted their kitchen into a laboratory. On the stone floor they studied the inflammation of frog's feet and the appearance of blood clots. Having examined the different stages of an inflammation under the microscope, Lister wrote to his father, "Yesterday I carried out an experiment that kept me up a whole wonderful night."[15]

His reputation as a research surgeon kept growing at Edinburgh, to the point where he felt somewhat cramped in the historic Scottish capital. He therefore did not hesitate for a moment when he learned that a chair in surgery was becoming vacant at Glasgow. Soon Lister took his lancet, his microscope, his ethics, and his religion to that much more developed and more modern town where he became the third incumbent of a chair of surgery created in 1815. This position allowed him to undertake research on blood.

How did he, a young specialist in hemostasis and inflammation, come to be interested in airborne microbes? Like all surgeons, Lister often faced abscesses that ruined the benefits of the scalpel; he also observed that bone fractures usually healed without complications when they were not open. Infection developed when the skin was torn, so that the bones came into contact with the air. The explanation most often given by the surgeons was that gangrene had something to do with the oxygen in the air. As it entered the wound, it was said, the oxygen corrupted the tissue and caused it to putrefy. This explanation pleased everyone, particularly the operating surgeons, who thus could not be held responsible for the development of abscesses and septicemia. It was credible enough, yet it soon became clear to Lister that it could not by itself account for the phenomenon. How, for instance, could one explain that flesh did not become infected even though the blood brought oxygen to the tissues? Moreover, in certain traumas within the closed thorax, air did penetrate under the skin, and yet no abscess appeared on the broken rib. This led Lister to a different explanation: as a result of his study of coagulation and inflammation, he came to believe that an external agent was involved.

This was the time of Pasteur's first publications on the chemistry of fermentation. In 1865, Thomas Anderson, professor of chemistry at Glasgow University, first read some of the reports Pasteur had presented to the Académie des sciences in the late 1850s. Knowing that Lister was working on putrefaction, he showed them to him, and Lister experienced a flash of insight, in which he immediately understood the potential link between the decomposition of organic matter and postoperative infection. He also did not hesitate to accept the role of bacteria. Won over by Pasteur's explanations and acting as a true scientist, he began by trying to verify and confirm his colleague's findings. With the help of his wife, he set out to replicate the French scientist's experiments in the family laboratory, and his conclusions were indeed confirmations: fermentation and putrefaction cannot occur in sterilized sugar or protein solutions unless germs are introduced from the outside.

For Lister as for Pasteur the ambient air now became one of the principal causes of the propagation of microbes. Since one could not reasonably expect to prevent any contact of the wound with the surrounding milieu, Lister invented a means of destroying bacteria within the wound itself. He felt that it must be possible to kill germs with the help of a substance that would not be toxic to human tissue, and thereby to prevent abscesses despite a septic environment. Here is how he explained his thinking: "When I read Pasteur's article, I said to myself: just as we can destroy lice on the nit-filled head of a child by applying a poison that causes no lesion to the scalp, so I believe that we can apply to a patient's wounds toxic products that will destroy the bacteria without harming the soft parts of this tissue."[16]

Among all the substances traditionally used to treat wounds, certain ones, such as wine and turpentine, unquestionably had an antiseptic effect, yet they also did much to favor infections. The task thus was to develop a harmless product that carried no risk of side effects. Lister explained the circumstances that led him to use carbolic acid: "In 1864, I was very struck by reading a report about the remarkable effect of carbolic acid on sewage water. Adding a very small quantity of carbolic acid to this water removed all putrid smell from the area irrigated by it, and even destroyed the intesti-

nal parasites which usually infected the cattle grazing on these pasturages."
It was a stroke of genius on Lister's part to analyze a recommendation of the
hygienists in biological terms and to understand that the carbolic acid did
not simply remove the smell that testified to putrefaction, but that it also
killed the microbes and could therefore be used to prevent gangrene. He
had just discovered antisepsis.

Could there be a better subject to which to apply this intuition than
an open fracture, in which splintered bone had pierced the bruised skin, and
in which the bleeding and necrosed surface had led to a secondary infection
of the wound? On 12 August 1865, Lister had the first occasion to experiment with a carbolic acid dressing. An eleven-year-old boy who had been
run over by a horse-drawn carriage was brought to the Royal Infirmary of
Glasgow with an open fracture of the tibia. The ragged-edged wound was
not too dirty. Lister washed it with carbolized water, then applied a cotton-
wad soaked in carbolic acid. Over the following weeks, the dressing was
changed several times to permit a new application of carbolic acid. The
treatment was a complete success; no infection occurred.

The same treatment was promptly applied to ten other cases of compound fractures, and all ten cicatrized without difficulty. Since these first
results seemed to be so encouraging, Lister decided to apply his method in
other circumstances as well. He went on to propose carbolic antisepsis for
tubercular abscesses, which often failed to drain properly due to secondary
microbial infections. He also experimented with sutures of silk threads
soaked in carbolic acid, which he successfully used to treat an arterial mal-
formation. More audaciously still, in April 1867 he operated on a tibia that
had improperly knitted after a closed fracture. This was a gamble, for even
though the fracture had not pierced the skin, he now made an incision,
despite the great risk of suppuration that this implied. Here too the carbolic
acid dressing made the operation successful. At this point, Lister was sure of
his method and prescribed it for various types of amputations.

Unlike Semmelweis, who was incapable of communicating his findings, Lister took steps to publicize the success of his treatment. In a series of
five articles published in the *Lancet* in 1867, he announced the invention of
antisepsis and his indications for the treatment of compound fractures and
various cases of suppuration. He again presented his work at a meeting on
9 August 1867 of the British Medical Association, then on 2 May 1868 to
the Medical Society of Glasgow, and finally in his inaugural lecture at the
University of Glasgow on 8 November 1869. In each one of these speeches,
Lister first outlined Pasteur's theories and then described his own findings:

"If we inquire how it is that an external wound communicating with the seat of fracture leads to such grave results, we cannot but conclude that it is by inducing, through access of the atmosphere, decomposition of the blood which is effused in greater or less amount around the fragments and among the interstices of the tissues. . . . Turning now to the question how the atmosphere produces decomposition of organic substances we find that a flood of light has been thrown upon this most important subject by the philosophic researches of M. Pasteur, who has demonstrated by thoroughly convincing evidence that it is not to its oxygen or to any of its gaseous constituents that the air owes this property but to minute particles suspended in it, which are the germs of various low forms of life."[17]

Over the years, Lister brought little modification to his antiseptic technique, which was enacted like a ritual. First he sprayed a cloud of carbolic acid into the surrounding air as well as over the area to be operated on. Then he disinfected the wounds, his hands, and the instruments with carbolized water. Bandages soaked in carbolic acid were placed on the area to be treated, and changing them called for the same ceremonial of filling the atmosphere with spray. Since carbolic acid irritated the skin, he replaced it with increasingly dilute carbolized oil. And since sticking plaster seemed insufficient, the bandages became more complicated; the crust of carbolic acid was protected by a layer of waterproofed silk, on which were placed eight layers of carbolized muslin, the last two of which were isolated by a sheet of gutta-percha. In the end, resin and liquid paraffin were poured over the entire package. In some cases, a sheet of lead was added to ensure the total isolation of the wound.

Despite all these results, Lister encountered rather strong resistance, particularly from English physicians, who watched the success of this adoptive Scot with suspicion. They were suspicious of this eccentric practitioner, who had a laboratory in his home and did research on postoperative infection. In August 1869, when Lister succeeded his father-in-law in the chair of clinical surgery, he began teaching a course in bacteriology that baffled the medical students. They admired him as a teacher, but they did not understand why such a good surgeon should take an interest in microbial cultures and work with a microscope.

Pasteur and Lister were two very different personalities. While Pasteur did not hesitate to provoke his adversaries and to attack them frontally, Lister always kept his composure and ignored the polemics against him, preferring to rely on science and statistics. That is why in January 1870 he published two comparative series of data on amputations, which showed that before antisepsis the mortality was sixteen cases out of thirty-five, or 45 percent, while with antisepsis it was six out of forty, or 15 percent. Despite the eloquence of these figures and the clear reasoning of Lister, the physicians continued to doubt. Over the following nine years, Lister was to become one of the most controversial surgeons in the world.

In fact, outside Glasgow, no practitioner in Great Britain was willing to apply Lister's techniques. It was easier to ignore the progress made than to reform. Where was the surgeon who, having considered rapidity of execution the essence of the art of surgery all his life, would suddenly agree to roll up his sleeves and to waste time soaking his hands in a corrosive liquid, and to spray the air with a substance that obstructed the view, irritated the eyes, and made everyone cough? Carbolic acid reddened one's hands and could lead to eczema. One constantly had to stop to rinse the sponges and to wash one's hands and instruments.

Moreover, certain surgeons who had not adopted antisepsis produced results that were as good as Lister's. Thus Lawson Tait, a surgeon in Birmingham who preferred strict cleanliness to the use of carbolic acid, which he considered too dangerous, washed his hands with soap and boiled water before every operation and also used boiled compresses and carefully cleaned instruments. His results made the Lister method look unnecessarily complicated. The fact is of course that Tait practiced antisepsis without knowing it; incapable of explaining his own method, he used absurd arguments against Lister: "Applying the findings of *in vitro* experiments to *in vivo* phenomena does not make sense." He even went so far as to add peremptorily: "I couldn't care less about germs."[18]

This matter of germs was indeed the main difficulty, and the main grievance against Lister was that he had made Pasteur's theories about microbes his own. Most surgeons still thought that microbes were harmless and secondary parasites that were incapable of playing a role in infection.

In France, this reticence was at least as strong as in England, and many people continued to see pus as a salutary substance; a certain Armand Després for instance claimed that the maggot devoured the vibrio. By 1869, however, a young surgeon, Just Lucas-Championnière, had gone to Glasgow to meet Lister. He returned full of enthusiasm, but his report, published in the *Journal de médecine et de chirurgie pratique*, did not convince anyone. He failed to obtain permission to experiment with Lister's method in France. It was only in 1874 that he was able to bring some progress to medical thinking, for when he was appointed successor to the head of surgery at the Lariboisière Hospital, where all the patients' wounds were suppurating, he was finally in a position to prove the effectiveness of antisepsis.

This first experiment coincided with the beginning of a direct relationship between Lister and Pasteur. On 10 February 1874, the British scientist wrote to his French colleague:

> My dear Sir:
> Allow me to beg your acceptance of a pamphlet, which I sent by the same post, containing an account of some investigations into the subject which you have done so much to elucidate, the germ theory of fermentative changes. I flatter myself that you may read with some interest what I have written on the organisms which you were the first to describe in your *Study on the lactic fermentation*.
> I do not know whether the records of British Surgery ever meet your eye. If so, you will have seen from time to time notices of the antiseptic system of treatment, which I have been labouring for the last nine years to bring to perfection.
> Allow me to take this opportunity to tender you my most cordial thanks for having, by your brilliant researches, demonstrated to me the truth of the germ theory of putrefaction, and thus furnished me with the principle upon which alone the antiseptic system can be carried out. Should you at any time visit Edinburgh it would, I believe, give you sincere gratification to see at our hospital how largely mankind is being benefited by your labours.
> I need hardly add that it would afford me the highest gratification to show you how greatly surgery is indebted to you.
> Forgive the freedom with which a common love of science inspires me, and
> Believe me, with profound respect,
>
> > Yours very sincerely,
> > Joseph Lister.[19]

Such civility and modesty were not always very common among scientists, and Pasteur was rather surprised by the gratitude of a foreign surgeon with whose work he was not really familiar. He replied to Lister on 27 February 1874:

> Dear Sir, your letter was extremely gratifying to me, and I have immediately asked one of my good friends who knows English well to translate for me the

brochure that you were kind enough to send. I had indeed heard of your system of antisepsis . . . but I must admit to my great embarrassment that I was, and still am, rather uninformed about your work, although I have long wished to learn more about it. . . . I am astounded by the precision of your manipulations, and by your perfect understanding of the experimental method, for it is a mystery to me how you can carry out research that demands so much care and constant work while exercising the profession of a surgeon and head of a great hospital establishment. I do not think that another example of such a prodigy could be found in my country. . . . As for the substance of this work, I have no right of course to contest the legitimacy of your conclusions, but if I had the leisure, I would take the liberty of presenting a few critical observations to you. . . . Your method of culturing is remarkable for its rigor and for its understanding of the difficulties inherent in this kind of study, yet I make bold to tell you that I wish it were even more rigorous and that you paid more attention to the purity of your seeds. I would like to continue, but writing quickly tires me. In October 1868, I was struck by a paralysis of the left side, and I still have not entirely recovered from this terrible blow. At any given moment my head can sustain only a limited amount of effort and work. Bending my head, especially while writing, is quite difficult for me. Still, I am sorry to have to leave you.[20]

This moment marked the beginning of France's heroic period of antisepsis, which was to last for ten years, from 1875 to 1885. Lister's method, based as it was on the idea that germs come from the ambient air, attempted to eliminate them from the operative field. It was not long before scientists realized that many more germs were spread by the hands of the operators, by instruments and sponges, and that the air was much less charged with germs than had been thought — as indeed Pasteur had already demonstrated at the time of the polemic over the so-called spontaneous generation. This meant that Lister's technique was most effective in surface operations, but proved much less useful in operations involving deep cuts, such as abdominal interventions. Moreover, the drawbacks denounced by Lister's adversaries were real: antiseptic products did have a caustic effect, especially on the peritoneum that envelops the abdominal viscera. Surgeons therefore gradually came to prefer preventive measures to eliminate all risk of infection, so that it was no longer necessary to disinfect the wounds. Little by little the sterile smock replaced the surgeon's old morning coat and the operating instruments were now sterilized in the autoclave. Asepsis took the place of Lister's antiseptic ritual.

A Day at the Académie de médecine in 1873

Lister's deference in addressing Pasteur and the cautious tone of Pasteur's reply are symptomatic of the gap that separated biology from medicine in the early 1870s. Science and research, the professor and the experimenter, the library and the laboratory, none of these communicated very well with each other. Even within the precincts of the Académie des sciences, Pasteur had to note with regret that certain of his colleagues had not yet taken to the microscope. Fortunately, others sustained his goal of achieving a synthesis of the life sciences and even spurred him on to ever greater efforts. One of them was his old mentor Balard, who in connection with the silkworm disease publicly declared in January 1872: "Can we not hope that by persevering in this course, you will eventually preserve the human species itself from some of those mysterious diseases which may be caused by the germs contained in the air?" To be sure, Pasteur himself was eager to broach new experimental investigations and to launch himself into research concerning the most sacred subject of all, human life; yet as he approached the age of fifty, he felt tired, overwhelmed by unnecessary obligations, and concerned about his own health.

At the end of October 1872, he therefore applied for permission to retire. To this end, he went straight to the top of the hierarchy and wrote to Adolphe Thiers, the president of the Republic, via Jules Simon, at the time Minister of Public Education.

> Serving as Professor at the Faculty of Sciences of Paris and director of a laboratory of the Ecole pratique des hautes études connected with the Ecole normale supérieure, I suffered, at the age of 45, on 19 October 1868, an attack of hemiplegia that has irremediably compromised my health. My mental faculties have not been affected in any way, but while they allow me to continue to be of some use to the progress of Science, the physical weakness resulting from the persistent paralysis and the rapid fatigue the brain suffers as a result of sustained effort make it impossible for me to resume my teaching functions. Consequently, I find myself forced at the age of fifty to apply for my retirement as a professor, although I wish to keep the laboratory I am directing now, a post that carries a compensation of 4,000 francs.[21]

Further on in his letter, expressing nostalgia or resentment, Pasteur recalled the favors the Empire had bestowed on him: "I can furnish proof that at the Council of Ministers held at Saint-Cloud on 19 July 1869, the Emperor suggested that I be granted a national reward. M. Alfred Leroux, at the time Minister of Agriculture and Commerce, approved this project, to which a year later the government would have added my appointment to the dignity of senator (by a decree that was signed but not promulgated on 27 July 1870)."[22] It may have been tactless to insist in this manner, but the Republic did not hold it against Pasteur and granted his request.

Freed of his obligations at the University, he was ready to go into the medical field. But he lacked the proper qualifications. As a chemist, he might not have much weight in dealing with a world of medical practitioners that he did not know, or perhaps knew only too well as a world that jealously guarded its prerogatives and its special knowledge. In their *Traité élémentaire de thérapeutique*, Trousseau and Pidoux had written, "A chemist who has found the conditions for respiration, digestion, or the action of a particular medication will think that he has found the theory of the body's functioning and its phenomena. They all have this illusion, and the chemists will not cure themselves of it."[23] Pasteur knew all about this reticence and backbiting, but he was willing to disregard them and to put up with the dim view the medical world took of experimental research. But he did need a forum.

In early 1873, a vacancy in the section of members at large in the Académie de médecine presented the opportunity for such a forum. Friends suggested that he become a candidate, which he promptly did. On 25 March 1873, Pasteur was elected, but only with a one-vote majority, by forty-one votes out of seventy-nine cast.

It must be understood, of course, that if the physicians thus admitted a chemist to their society, it was because they wanted him to second them in their own efforts — and surely not to have him dictate new laws to them. But then, Pasteur was already a member of the Académie des sciences, and it seems likely that he received many votes on these grounds. It was a matter of adding an already prestigious name to the roster.

The Académie de médecine had been instituted in 1820 by a royal ordinance of Louis XVIII with the aim of reviving the work and the influence of two societies founded in the previous century, the Académie de chirurgie and the Société royale de médecine, which the Convention had abolished in 1793. A first and rudimentary Académie de médecine had been founded in 1814 at the prompting of the famous Dr. Guillotin, but its existence at that time was ephemeral. Divided into three sections, medicine, surgery, and pharmacy, the new academy was given charge of "all government concerns regarding public health, particularly epidemics, diseases specific to certain regions, epizootics, the propagation of vaccination, and mineral waters." It thus became a legal arm of the government.

Yet in the first forty-five years of its existence, this body contributed little to the prevention of infectious diseases. Not that opportunities were lacking. The plague was alive in the Near East and threatened the Mediterranean Basin. Yellow fever, which was endemic in the Caribbean, had ravaged Barcelona and touched some of Italy's port cities. Cholera had spread throughout Europe and certain French regions, causing thousands of deaths in 1832. Typhoid fever and diphtheria attacked populations with a tenacity against which there seemed to be no defense. Fallacious theories on the causes of epidemics and infections contributed to the perpetuation of dangerous illusions. As the dominant spokesman of official medicine, Broussais thundered against the notion of contagion, suggesting that it was better to speak of "the genius of the epidemic," "epidemicity," or "special medical constitutions." In 1850 one could still hear it said that the presence of a choleric individual is by no means necessary for the propagation of cholera and adds nothing to its power of expansion. It was also felt that even if cholera were contagious, physicians would have the duty not to say so. And since this was the official line, the Academy took steps to abolish all quarantines. Fortunately, there were signs of a change of direction around 1860; so now the Academy began to look into the salubrity of the Paris hospitals. Some members, such as Auguste Nélaton, went along with the evolution of science and emphatically made the point: "The faith by which one has lived must never become a chain. One has done one's duty by it if one has wrapped it into the scarlet shroud in which dead gods are buried."

When Pasteur was received in this learned assembly, the recent debates over the findings of Jean-Antoine Villemin were fresh in everyone's mind. In 1865, this young professor at Val-de-Grâce had presented to a meeting of the Academy a paper on the cause and nature of tuberculosis, in which he suggested on the basis of a series of experiments that it was a contagious and inoculable disease. This proposition was discussed for three years. Hermann Pidoux, resplendent in his blue dress coat with gold buttons, vehemently opposed what he considered an insane idea: "For tuberculosis, a constitutional, a diathetic disease, the terrain is everything, not the seed." The octogenarian Piorry did not believe Villemin's explanations either; he felt that opportunistic infections, particularly septicemia, were exclusively due to poor ventilation in the hospital wards: "Do away with putrid smells and mortality will diminish."

Watching as Pasteur slowly and with measured steps ascended the stairs of the Académie de médecine for the first time on that first Tuesday of April 1873, did any of his new colleagues realize that they were receiving the protagonist of a new medical revolution?

The academy had left its earlier abode in the rue de Poitou in 1850, moving into the former chapel of the Charité in the rue des Saints-Pères, not far from Saint-Germain-des-Prés. This place of worship had been secularized by the Revolution in 1797 and converted into an amphitheater for Corvisart's courses; the architect, who had wished to give it the aspect of a classical temple, had taken his inspiration from the description of the temple of Aesculapius at Epidaurus. Little is left today of these interior appointments typical of the neoclassical fashion, which favored porticoes, colonnades, inscriptions in large solemn lettering, and busts of illustrious predecessors.

Pasteur had to cross a narrow vestibule and then walk a short distance to desk number five, which stood close to the president's table. Not far from him sat Claude Bernard. The great physiologist came to the Academy as an associate member and never missed an opportunity to deride the physicians who maintained that physiology "cannot be of any use to medicine and that it is no more than a luxury science without which one can get along perfectly well." To those who extolled the educational virtues of anatomy over those of physiology, he liked to respond, "Descriptive anatomy is to physiology what geography is to history."[24] When Pasteur took his place next to him in the little hall, Bernard therefore could not help but see in him an ally against these self-assured doctors who always seemed to say, "I have just saved my fellow man."

Spurred on by the discoveries following from his work on the chemistry of fermentation and spontaneous generation, Pasteur did not hesitate to lay out his ideas concerning the germ. He never ceased to defend its existence and to prove its morbid effects. Nor was he shy about appropriating, discussing, and divulging the principles of asepsis and antisepsis, even before he had gone to work on the causes of infectious diseases.

Lessons for Surgeons

Pasteur was not slow to understand how the interpretation of his discoveries had led Lister and his disciples to their success. He had barely become a member of the Academy when Alphonse Guérin, at the time dean of the Faculty of Medicine and in this capacity responsible for creating laboratories for anatomy and pathological chemistry, asked him to come to his clinic to observe the benefits of the use of cotton-wool dressings. Guérin associated abscesses with as yet ill-defined infectious fevers. In his opinion, a purulent infection was a poisoning produced by a miasmic toxic agent, and analogous to malaria. One of his colleagues, the surgeon Verneuil, vehemently opposed this idea. He propounded a theory based on the existence of a local toxin produced by the putrid decomposition and subsequent resorption of tissues.

Leaving the corridors of the Academy, which resounded with these interminable and for the most part sterile oratorical jousts, Pasteur went into the field, to the Hôtel-Dieu, where he most attentively examined Guérin's results and his technique of dressing wounds. It was one of his first visits to a hospital. In the long wards holding row upon row of beds, the dying and the able-bodied were often separated only by a curtain. At their feet stood pans holding body wastes and soiled linen. In these ill-ventilated tunnels divided by a central aisle, peace and quiet were unknown, for there was constant coming and going of students, nurses, service staff, and visitors.

Pasteur observed the scene and examined the bandaged surgical patients. A few months later, when Lister wrote to him to inform him of his method, he immediately made the connection: Guérin's dressings, like those of Lister, were unquestionably responsible for the surgical successes obtained, thanks to the filtering out of the germs by a barrier of cotton-wool and their destruction by the antiseptic solution. Otherwise the germs would contaminate the wound. Reproducing the barrier formed by the dressing with tubes packed with cotton-wool in the laboratory, Pasteur found that this closure did not prevent the passage of air, but that the air that passed through the cotton-wool remained sterile. The closure allowed the aeration of the injured tissue, which he considered essential to proper cicatrization. Better yet, he suggested that the cotton-wool be sterilized by heating.

In order to demonstrate the role of germs in the suppuration of wounds, Pasteur now proposed to cut two wounds on the symmetrical limbs of an animal. One would heal under the cotton-wool dressing, the other, which would be left open, would allow the proliferation of microscopic germs; if a third cut were made under sterile conditions, it should heal promptly. Using cotton-wool was, after all, to apply a procedure that Pasteur had long used to sterilize the air in his glass tubes.

But his lessons were sometimes misunderstood. For the dressing had to be applied with all the care and intelligence implied by the theory. On the occasion of a visit to the Hôtel-Dieu with Félix Larrey, son of the famous surgeon of the Napoleonic period, and Léon Gosselin, Pasteur was surprised and furious to find that a young doctor on duty in Guérin's clinic had bandaged the hand of an injured man without first cleaning it properly. When the visitors lifted the dressing, they were struck by a repulsive smell of pus. Once again, Pasteur delivered a lecture on the germ and on the virtues of the most stringent hygiene.

At the same time, Lucas-Championnière became a champion of antisepsis and asepsis. Defying the opposition of the administration, he introduced the use of nail-cutting scissors, brushes, and soap in the hospital setting. The spraying of carbolic acid à la Lister gave rise to passionate protest; the surgeons hated the watering cans that inundated the operating theaters, forcing them to go around in clogs. Routine, plus carefully learned and widely practiced bad habits, added to what ignorance and fear of innovation did to promote the spreading of microbes. Pasteur was so misunderstood, and the current operating practices were so far removed from his concepts — or indeed disregarded altogether — that he repeatedly had to speak about them at the Academy. He explained how and why the air, the hands, and the instruments of the surgeons had to be sterilized. He advised that the instruments be passed through a flame before they were used, adding, in order to make himself perfectly clear: "By this I mean that the surgical instruments should be subjected to a simple flaming that does not actually heat them. And here is the reason. If one examined a probe under the microscope, one would find on its surface furrows and troughs containing dust particles that cannot be completely removed by the most painstaking washing. Fire, however, will entirely destroy this organic dust. In my laboratory, where I am surrounded by germs of every kind, I therefore never use an instrument without having first passed it through a flame." He also indicated how an infection propagates itself: "This water, this sponge, this lint that you use to clean and cover the wound, deposit there germs which, as you can see, proliferate with great ease in the tissues and would without fail and in very short order lead to the death of the operated on patient if life did not resist the multiplication of germs. . . . If you took a mutton roast, cut into it with a lancet and then inoculated the opening with a cotton-wad exposed to ordinary air in the street, you would easily see the putrefaction of the meat."[25]

Pasteur did not limit himself to Lister's recommendations concerning disinfectants. While he did praise their effectiveness and in passing even found new applications for them — he advised the urologist Guyon, for example, to rinse the bladder with boric acid — he went even further, for he was one of those who sought to go beyond antisepsis to asepsis. Considering the disinfecting of the wound insufficient, Pasteur turned from radical germicide to hygiene and prophylaxis. He thus went straight to the heart of the debate that was to revolutionize medicine in the treatment of the most evident effect of the germ, that is, the abscess. On 1 August 1876, he formulated a revolutionary hypothesis on the formation of pus and consigned it to his notebook of experiments: "One hears about sewing needles passing through the body without causing any harm and about lead bullets lodged for years in this or that part of the body. And yet it is said that a foreign body causes the formation of pus, of an abscess. There is a contradiction here. . . . The idea therefore occurred to me that if the foreign body brings on pus — which, as we have just seen, is not necessarily the case — it must bring in a germ, so that it would be this germ that causes the pus. This idea is bizarre, but nonetheless I mean to test it experimentally."[26] The experiment he proposed consisted of placing under the skin of a guinea pig on the one hand some aged urine in which he had ascertained the presence of chains of corpuscles and on the other a previously flamed foreign body. "Will the chains produce pus while the clean foreign body does not? If that were the case, a huge step would be made in surgery and medicine. It would mean that keeping away pus is a matter of keeping away the germ."

Adopting the basic tenets of asepsis, Pasteur ended by indicating what had to be done: "Only perfectly clean instruments must be used. Hands must be washed after being passed briefly through a flame, which is no more of a discomfort than that felt by the smoker who passes along a piece of live coal. Only sterilized lint, bandages, and sponges must be used, and all water must be boiled before use." And, referring to airborne germs, he continued: "Observation shows us every day that the number of these germs is, as it were, insignificant by comparison with those that exist in dust on the surface of objects or in the most limpid ordinary water." Pasteur outlined every detail of the method of sterilizing test tubes or producing heat-treated cotton-wool; the best way was to place vessels whose openings were closed with a cotton wad into a gas oven that heated the air for half an hour, heating the object to a temperature of about 150 to 250 degrees. Cotton-wool was heat treated inside a tube or in blotting paper.[27]

These experiments with boiled and then cooled and air-deprived water were bound to revive the debate over spontaneous generation. "M. Pasteur tells us that he has been looking for spontaneous generation for twenty years without being able to find it," wrote Poggiale, formerly chief pharmacist at Val-de-Grâce Hospital and a member of the Sanitation Board. "Those, like myself, who have no firm opinion on spontaneous generation, reserve the right to verify, check, and discuss the facts as they come to light, and wherever they may come from."[28]

"What!" Pasteur exclaimed, "I have been working on a subject for twenty years, and I am not supposed to have an opinion? And the right to verify, to check, to discuss, and to ask questions should belong to somebody who does nothing to inform himself, who has done no more than skim our work, his feet propped up against the fireplace in his study? You have no opinion on spontaneous generation, my dear colleague; I can easily believe it, though I am sorry to hear it. . . . Well, I do have one, and it is not based on feeling but on reason, because I have acquired the right to have an opinion by twenty years of diligent work, and it would be wise for any impartial mind to go along with it."[29]

More momentous, and more violent, was the criticism that questioned the concept of secondary infection; in a certain sense, this too lent support to the partisans of spontaneous generation. The surgeon Le Fort, who rejected Lister's and Guérin's operating methods, denied the germ theory, claiming that there was absolutely no room for its applications in the surgical clinic. He maintained that purulent infection in wounds was caused by "the influence of local and generalized phenomena internal rather than external to the patient." "There is no proof for original and spontaneous infection," Pasteur responded. "Water contains germs. The simple washing of a wound with a wet sponge can have dire consequences. The accidental presence of germs under cotton-wool dressings explains their possible proliferation."[30]

Pasteur's views thus collided not only with habits but with opposition grounded in principles, as well as with archaic attitudes that today strike us as well-nigh incredible. Armand Després, for instance, denied the usefulness of both asepsis and antisepsis and, anointing himself the "vestal of French pus," came to the defense of what he considered the best topical remedy for wounds. He remained a defender of soiled dressings to the end of his days. The worst part about this was that he was proud of himself and absolutely sincere. Openly expressing his antipathy for the methods of Pasteur and Lister, he exclaimed, "If the microbial doctrine is correct, why is it that wounds in the mouth, the most septic environment of all, always heal?" Fitting his actions to his thinking, he therefore continued to examine his patients wearing his stained suits, as the greatest masters of the period had done; and, in order to demonstrate his theory, purposely contaminated the drains presented to him by scraping the floor with them. As eloquent as he was opinionated, Armand Després became the ringleader of a number of conservative surgeons who, in the name of traditional surgery, made it a point of honor not to use aseptic methods, and indeed to violate them whenever possible in order to demonstrate their ineffectiveness.

In this debate Pasteur took the position of a vegetalist and a vitalist; in other words, he went after the germ before it had a chance to contaminate. Choosing among the applications developed before or after his theories, he preferred, as he had done in the experiments designed to disprove Musset and Pouchet, the swan-necked flask to the heating of mercury, the sterilization of the vector to the destruction of the germ *in situ*.

Yet asepsis was to gain acceptance more slowly than antisepsis. Appearing in Paris in the late 1870s, it was generally applied only by the end of the 1880s. Ernst von Bergmann in Germany and William S. Halsted in the United States adopted these methods even later in the century, when they began to use the drying cupboard or the autoclave in accordance with Pasteur's findings. The application of asepsis thus went hand in hand with certain advances in hygiene. Moreover, customs varied from country to country; in Germany asepsis was adopted only in the form of boiling, whereas England, which had finally given due recognition to Lister, long remained faithful to antisepsis.

Most surgeons gradually began to use a combination of the two techniques, as Terrier and Tyrolean had done earlier. But this meant extra work, what with cleansing the patient's skin, painting the operating area with a disinfectant, having the surgeon scrub his hands! In order to ease the need for these actions which, no longer considered symbolic, had to be repeated several times in the course of an operation and were thus becoming truly burdensome, Halsted in 1885 introduced the use of the rubber glove, which could be sterilized like the instruments. Overall it was not until 1890 that asepsis, winning out over antisepsis, was adopted everywhere. Learning to use it was to prove as difficult as adopting it, for it was not enough to understand Pasteur's teaching; one had to put it into practice. The surgeon had to learn a new set of gestures and reflexes. Not touching anything that was of doubtful sterility was the mainstay of the new dogma.

But for Pasteur, this was more than a dogma; it may have been the symptom of an obsession. Adrien Loir, his nephew and collaborator, makes much of his preoccupation with hygiene:

The washing of hands was done according to a fixed ritual. In one corner of the laboratory was a deep sink above which a spigot brought in water; to its right was a soap holder. Pasteur had taught me to lather my hands for a long time, then to rinse the soap in plenty of water before putting it back into the dish, so that it would always be clean for anyone who might have to use it. . . . This washing of hands went on constantly throughout the day. . . . Pasteur had a phobia about shaking hands, and that is probably why people found him haughty. He never held out his hand. If by chance a stranger had come to call on him in the laboratory, particularly if it was a physician, and if he had been unable to avoid this time-honored gesture of courtesy, he gave me a slight sign that I knew well, pointing toward the sink with his head, which meant that I was to go to open the spigot.[31]

Science and the Fatherland

Since the collapse of the Empire and the fall of Napoleon III, who died in his English exile in Kent on 9 January 1873, Pasteur had no longer enjoyed close relations with political power. The man who had so admired and then so bitterly regretted the passing of imperial power accommodated as best he could to the successive republican regimes in all their variety. But he did not seek to establish personal relations with any of the officials in the ministries. Jules Simon, after all, was no Victor Duruy.

Following the traumatic events of 1870–71, the economic recovery that lasted until 1873 was comforting to public opinion, which interpreted the acceleration of railroad traffic or the shortage of labor as positive signs. But soon Europe underwent a major economic crisis which also affected France, causing economic stagnation, unemployment, a crisis in the real estate market, industrial bankruptcies, and the decline of agricultural activities. Despite its problems, French society reacted with more than a shiver. Haunted by the memory of the debacle of 1871 and dominated by the ensuing desire for regeneration, influenced also by the positivist thinking of Auguste Comte and his disciples, the republicans were open to new perspectives.

Witnessing the evolution of medical science and of industry, liberal society set out to lay the foundations of a scientific renaissance, which had to begin with a reform of higher education. Nationalism and positivism joined forces to seal a new pact based on their confidence in French science. The laboratories vaunted the ideal of the new scientist, who would be released from all material constraints. In the halls of the universities, for once in the vanguard, the fight was led by the professors. New names became well known: Ernest Lavisse, Marcelin Berthelot, Paul Bert.

The last of these had been a deputy from Yonne since 1872. He was a physiologist who was not afraid to refer to the "soul of the fatherland" of which Ernest Renan had spoken. A former collaborator of Claude Bernard, he had succeeded the great physiologist in his chair at the Sorbonne. In 1874, his fellow deputies chose him to present a national reward to Pasteur. It was the third time that a national reward was given to a scientist: in 1839, Daguerre was honored by Arago in the Chamber of Deputies and by Gay-Lussac in the Chamber of Peers for having perfected Niepce's method of photography; and a similar distinction was bestowed on the engineer Louis-Joseph Vicat, the inventor of modern cement, in 1845.

This time, the Assembly wished to reward a scientist whose work had brought advances to both science and industry: "It is essential to the honor and the interest of nations that the life of great men inspire not only admiration but also envy. . . . The discoveries of M. Pasteur, having first shed new light on the obscure questions of fermentation and the manner in which microscopic creatures appear, have revolutionized certain branches of industry, agriculture, and pathology."[32] In consideration of the services rendered, Paul Bert asked the Assembly to award Pasteur a lifetime annuity of 12,000 francs, a sum that represented "approximately the (fixed and potential) emoluments of the chair in the Sorbonne that M. Pasteur was forced to relinquish because of illness"; in the case of his death, half of this pension was to be paid to his widow. Bert added that the economic and medical implications of Pasteur's discoveries were as yet difficult to assess, so that it would be only fair to increase the amount of the pension over time. The bill was approved by 532 votes against 24. Pasteur's cause commanded a majority that the government would have been hard put to muster on other occasions.

His friends and former students, among them Chappuis, by then rector of the University of Grenoble, Sainte-Claire Deville, and Duclaux, sent their congratulations. Some saw the occasion as a consecration at the end of a great career. Désiré Nisard wrote: "The time has come, dear friend, to apply all your strength to living for your family, for all those who love you, and a little for yourself."

But Pasteur was not the man to stop there, to treat himself at the pinnacle of glory to a peaceful retirement at Arbois. Instead, this official honor, an indication that those who wielded political power had become aware of science, gave him the idea that he should become involved in public affairs.

The first campaign for a republican Senate, established by the Constitution of 1875, was about to begin. Unlike the Senate of the Empire, this body was not an assembly of notables appointed by decree but a political assembly to be elected by limited suffrage and endowed with the same powers as the Chamber of Deputies.

In early January 1876, Pasteur started campaigning for the Senate in his native department of Jura, which was entitled to two senators; he was fifty-three years old. To be sure, it was not too late to launch a political career, especially in the Senate, but Pasteur was an idealist who was not accustomed to intervening in policy debates. He therefore espoused the tenets of a nostalgic conservatism based on respect for the fatherland and for science, on family values and religious traditions. Deeply disappointed that he had not been a member of the Senate of 1870, he probably conformed rather too much to his perennial model, old Jean-Baptiste Dumas, who had always known how to combine a scientific career with political power. Considering himself above party quarrels, Pasteur waited until the last moment, ten days before the deadline, to declare his official candidacy.

He understood very soon that no quarter is given among candidates. Using Lons-le-Saulnier as his headquarters, he lived at the Hotel Europe, where Jean-Baptiste served as his secretary. He conducted an active and passionate campaign. As it was almost everywhere in France, the election was essentially a contest among the bonapartists, the legitimists gathered around the Comte de Chambord, and the republicans of Gambetta and Jules Grévy. The latter were particularly popular in the Jura because Grévy was a native son of the department. There were thus five candidates: two, MM. Tamisier and Thurel, who belonged to the democratic left supported by Grévy; one legitimist deputy, Paul Besson; one bonapartist, General Picard; and Pasteur, who was left with the part of the outsider.

Refusing to cast his lot with Besson, whose reputation had been tarnished in a recent law case, Pasteur fought the battle for law and order by himself. By way of a profession of faith, he communicated three documents to the electorate: a list of his titles and scientific works, followed by the decree of 18 July 1874 concerning his distinguished national reputation; one speech he had given at the lycée of Arbois on the occasion of a distribution of prizes; and a brochure containing his correspondence with the dean of the University of Bonn in 1871.

"What I intend to represent in the Senate is Science in its purity, its dignity, and its independence," was the sum and substance of Pasteur's program. It was a fine formula, but the local newspapers, in particular Jules Grévy's *L'Avenir du Jura* in Dole, were quick to use it against its author. It would be a crime, they said, to interrupt this very important research Pasteur is doing. What a terrible thing it would be if the voters deprived France of such a scientist and sent him to sit in the Senate! The best proof of this is that the candidate himself does not belong to any party; his place is at the Institut, not in the Luxembourg Palace, and be careful not to confuse the two. Pasteur tried to respond that these were the very errors that had led to the defeat of 1870, claiming that as long as science and politics were separate, progress was impossible. "If in the last thirty or forty years the governments and the great constituted bodies of our unhappy France had not utterly neglected the institutions that foster the flourishing of the sciences, Germany would have been defeated. Do you not know that while we are being enervated by politics and its senseless divisiveness, to the satanic joy of our enemies, steam, telegraphy, and a raft of other marvels are transforming other modern societies?"[33]

Such talk was upsetting to both the right and the left, and Pasteur was attacked by Grévy's newspaper as well as by the *Sentinelle du Jura*, where Paul Besson fulminated, rejecting the France of 1792 and equating it with that of 1870. On election eve, Pasteur posted throughout the district a manifesto entitled *Science and the Fatherland*, which was essentially an attack on Grévy: "It is one of France's great misfortunes that there are so many politicians in our assemblies. We seem to be guided by politics with its tiresome and mundane discussions. False appearances! For in reality we are propelled by a few scientific discoveries and their applications. In our century, science is the soul of the prosperity of peoples and the living source of all progress." And, speaking of Germany, he added: "In the last half-century, this rival nation has known how to apply the best part of its resources and its attention to the most noble and most spirited works of the mind and to the progress of Science in its most unselfish form, to the point that the word Germany now evokes teaching and universities."[34]

That very evening, Pasteur participated in a last candidates' forum organized and presided over by Jules Grévy. Looming like the statue of the Commendatore and resplendent in his decorations, Pasteur spoke in defense of the national honor and of the fatherland, declaring himself ready to serve the Republic, since such was the regime, though without prejudging the future in the manner of the *doctrinaires* who, imprisoned by fixed principles, were unable to evolve. "I repudiate all long-term engagement, for the future is unknown to me." To certain rumors alleging that he had benefited from the largesse of Napoleon III, he replied head on: "As for the money I am alleged to have received from the Emperor's private funds, this is a vile calumny. What was not said is that the Emperor died owing me 4,000 francs. These 4,000 francs represented the cost of my and my family's travel to Austria."[35] At this moment, Pasteur had thoroughly forgotten the Empire on which he had once fawned.

But it was all in vain, and Pasteur was wasting his efforts. The verdict delivered by the urns on 30 January 1876 was final. With 62 votes, Pasteur clearly came in last, while the two republicans, Tamisier and Thurel, obtaining a combined total of nearly 1,000 votes, took the two vacant seats. It was a stinging defeat. And who knows: the champion of asepsis did not like to shake hands; perhaps this is what ruined his electoral campaign.

Pasteur always remained bitter about not having been a senator, even if he now hastened to write to his spouse: "I am very, very happy that I lost the election. — Lord, how this success would have embarrassed me! I really felt terrible yesterday when I saw all these delegate types and said to myself: these people are judging my life and my work!"[36] However, he could take comfort in the thought that, although he was not allowed to participate directly in the debate on the reform of higher education, he had in a sense made his voice heard.

Three months after his senatorial misadventure, Pasteur was gratified to learn that the Ministry of Public Education was about to undertake the building of new schools of medicine and pharmacy, to assign new laboratories to the Collège de France, and to enlarge and modernize the Sorbonne. This program of renewal was soon to be completed by a series of social reforms that would bring major changes to the lives of the French people, from grade school to the freedom of association, for the ambition of the Third Republic was boundless!

II

ANTHRAX IN SHEEP AND CHICKENS

I give them experiments, and they respond with speeches!" Pasteur said of his detractors. By 1877 it became evident that as a scientist he could no longer stand by as an outside observer of the research conducted on infectious diseases, and that he himself had to do the experiments to demonstrate the pathogenic role of the microbe.

In this debate, asepsis and antisepsis had figured only indirectly: to be sure, the absence of germs prevented disease, but this did not necessarily mean that they caused it. In his earlier work on the silkworm, Pasteur had been able to produce evidence for a possible transmission of the infection; but because the agent of disease could not be clearly defined, no precise answer could be given to the fundamental question of whether the bacterium was indeed the cause of morbidity.

A direct and definitive demonstration was thus needed more than ever. The beginning of Pasteur's study of the disease of anthrax marked his conversion to experimentation in the medical field. His work on the chemistry of fermentation was now a thing of the past, and a new generation of studies and experiments led the scientist to use the animal as a test object, the bacterium as a source of reflection, and disease as the theme of his research.

Pasteur's microbial theory, which established the link between the causes of fermentation and those of contagion, was to be one of his most revolutionary achievements. On 10 December 1878, at the Académie de médecine, Jean-Baptiste Bouillaud, the man who first described rheumatic fever, declared, "Before M. Pasteur's discovery, physicians knew that in contagious, miasmatic, and zygomatic diseases there was something that fundamentally differentiated them from other illnesses. This something they called contagium, miasma, or virus, and they admitted its existence — without ever seeing it, but with complete and absolute certainty — on the basis of their observations. The great revolution achieved in pathology through the discovery of M. Pasteur is that this eminent scientist has allowed us to touch it with our fingers, or rather see it with our eyes through the lens of the microscope, this *quid divinum* that before him we assumed to exist without ever seeing it."[1]

Before it came to such a tribute, Pasteur had had to fight and to stand up to more than one of his fellow academicians, for being elected to the Academy of Medicine did not mean that he had become a physician. When the company chose a chemist to shed light on its work, it was certainly not planning to treat him like a physician or a surgeon; the idea was to hear a specialist who could discuss related research and compare his experiments with those of his colleagues. At the time, few of them had an inkling that Pasteur would become involved in the debate to the point of monopolizing it and making bold to give advice, even orders, about the art of healing.

It is true that the Academy had looked into the chemist's work. For several decades there had been lively quarrels about the possible correlation between fermentation and contagious disease. There were two opposing camps. The majority attributed these phenomena to the alteration of the organic matter itself, claiming that fermentation, putrefaction, and infection are caused by a chemical force that decomposes or disarranges matter, thereby bringing about the destruction of the tissues and ultimately death. The others supposed that the process is triggered by a living external agent, the *contagium vivum*, which attacks and destroys the tissues of the host it has infested.

This debate brought back the very terms that had also dominated the debate on spontaneous generation in the early 1860s. The first camp purely and simply rejected the microbial theories and denied that yeasts or fermentations played any role in infections. Those who adopted this position were for the most part either unbending reactionaries who swore only by the hazy and outdated ideas of Broussais or, on the contrary, practitioners steeped in advanced ideas who, inspired by Liebig and attracted by the most modern physico-chemical explanations, would have liked to explain the genesis of diseases without having to recur to a living germ. Those of the other camp were more favorably disposed to Pasteur's research inasmuch as it consisted of laboratory studies, but this did not mean that they accepted his experimental incursions into the clinical field: they were perfectly willing to listen to a chemist, provided that he stay in his own territory. Studies involving animals or humans called for titles or diplomas that Pasteur did not have. Pasteur had to become a famous orator before he could make himself heard by the authorities of the Academy.

A Brief History of Contagion

Throughout the ages, the alliance between medicine and religion made it possible not to deal seriously with the question of the origin of disease, for it was easy enough to attribute suffering, epidemics, cures, or fatalities to divine or occult forces. Yet the idea that certain diseases could be contagious, that is, transmitted by lower organisms, repeatedly appeared in the history of science and medicine. A century before the birth of Christ, Varro remarked that swamps predisposed people to certain illnesses and wrote that "the air in these places swarms with tiny and invisible animals which, sucked in by the mouth, penetrate into the body, where they engender diseases." The role of emanations, physical contact, and the sharing of objects or clothing in the transmission of diseases had long been known. Leprosy, plague, cholera, and malaria, all these epidemics that had washed over the world had not failed to produce an increasingly precise body of knowledge on the modes of contagion.

In the early sixteenth century, the appearance and subsequent spread of syphilis were a veritable demonstration of the transmissible nature of certain diseases. At that time the Italian Fracastoro, who coined the term *syphilis*, showed amazing prescience when he asserted that the ambient milieu contains germs of disease that can multiply within the body. But because he lacked the material means to establish the physical nature of these agents, Fracastoro was unable to convince his contemporaries, and the idea that infectious diseases were of microbial origin was stopped in its tracks. A century later, the Jesuit Athanasius Kircher observed a swarming mass of "animalcula" in the blood of plague victims. Their presence, like that of the hitherto invisible creatures that the Dutchman Antony van Leeuwenhoek was just then seeing under his microscope, was considered accidental. Yet all these scattered observations had no lasting influence on the thinking and practices of medicine for almost two centuries: because it was invisible, the infectious agent could not be fitted into the system.

By the end of the eighteenth century, there had thus been little change in the debate that pitted contagionists against anticontagionists. Scientists and eminent practitioners pointed out that the effects of quarantine were not convincing and that a simple change in the weather was sometimes enough to put an end to an epidemic, as was observed in yellow fever. Besides, these epidemic diseases did not affect everyone, and their severity varied in those whom it did affect. Medical science therefore simply put forth empirical explanations taken from Hippocrates's treatises on the role of air, water, and places. The idea of telluric influences as proposed by ancient biology was dusted off and dressed in the hygienist fashion of the moment. Poverty, misery, overcrowding, and malnutrition were indicted; had it not been established, for instance, that scurvy was caused by a nutritional imbalance?

The debate thus involved not only medicine; it was dealing with a true social problem. As early as 1822, at the time of the yellow fever epidemic, the contagionists, who believed that contagion was brought about by direct physical contact between two persons, worked under the aegis of the Board of Health to bring pressure to bear on the government, urging it to institute cordons sanitaires, quarantines, and isolation wards. The anti-contagionists, who thought that the disease had more diffuse causes, were rather closer to the thinking of the Académie de médecine. Opposed to interference of any kind, they supported a policy of laissez-faire, deferred to the business interests of the merchant community in the port cities, and pleaded the cause of a triumphant free market ideology. In the early nineteenth century, the success of Broussais's theories was undoubtedly the main obstacle for the contagionist party. This was the time when students were taught that the most varied morbid species and the most specific infectious diseases can arise fully fledged in accordance with the place, the season, the climate, and the hygienic conditions.

It must be acknowledged that at the time the contagion thesis did not rest on a credible scientific basis and that it was not sufficiently supported by advances in medicine. It had long been difficult to establish the individual characteristics of the different diseases. In the seventeenth century, Thomas Sydenham, dubbed the English Hippocrates, had been one of the first to attempt a classification of the diseases. His work, as well as that of his disciples, who modeled their categories on the classification of plants and animals established by the naturalist Linnaeus, made it possible to recognize and differentiate the circumstances that cause diseases to vary from individual to individual. This was above all a bedside medicine, in which symptomatology was more important than the search for causes. In this context, there could be no question of infectious or contagious disease, and even less of vaccination.

As for the French school of Broussais's opponents, Laënnec, Bretonneau, and his pupil Trousseau, it was mainly interested in basing a new classification of the microbial diseases on the characteristics of the lesions they cause. If in this manner the question of the origin of diseases finally came to the surface, it was still not approached from a contagionist point of view, for the principles applied to it were those of the Italian Giambattista Morgagni, one of the great anatomists of eighteenth-century Padua. According to Morgagni, anatomy was the mainstay of medical science, and dissection its most important technique. He wrote that "symptoms are the outcry of the suffering organs." Understanding the illness, in other words, was achieved by opening the cadaver.

In France, Xavier Bichat was the first to go this route and even beyond it, but he paid a high price, for he died in 1802, at the age of thirty-one, from a pin-prick he suffered while dissecting a cadaver. Looking further than to the organ, he discovered the identity of tissues and cells. He always considered autopsies and dissections the principal means of describing, understanding, and classifying diseases, for in his day the secret of infections was believed to lie in the modification of the anatomy caused by the disease. By the late nineteenth century, this method of investigation involved observations of anatomical cross-sections under the microscope and was known as "pathological anatomy." In seeking to understand disease by opening the body, however, no thought was given to the possible influence of a germ from the outside. Strangely, this approach did not consider the possibility of infection, and it denied the role of microbes. Even if microbes were detected in a diseased location, this finding did not harm the theory, for it held that the reaction of cells to a foreign organism was always more important than the action of such an organism. Disease was thus defined in terms of the changes observed in the injured organ or tissue rather than as the consequence of being injured by a germ.

For all the shortcomings of this reductionist approach, one must recognize that pathological anatomy and cytology did have the merit of bringing the laboratory into the hospital. Yet the very concept of the medical experiment was still far from rigorously defined; thus one can read in one textbook of the period that experiments on rabbits cannot be considered conclusive, for "the rabbit is a neurasthenic animal for which life is a burden, and which is only too glad to get rid of it."

At that time, the germ conceived as an agent of morbidity had by no means acquired widespread recognition. When Liebig and his followers triumphed by minimizing or altogether rejecting the responsibility of microbes in fermentation, Cagniard de La Tour and Schwann were the only scientists to plead — in vain — for the recognition of the deleterious action of microorganisms. The same scenario was repeated in connection with contagion; and it took a long time for the first proofs of the pathogenic role of germs to be accepted. It was not until 1836 that Agostino Bassi of Lodi demonstrated that the muscarine of the silkworm was caused by a parasite. Shortly thereafter, Johann Schönlein identified ringworm and demonstrated that microscopic fungi can be harmful to humans.

By the middle of the nineteenth century, medicine was thus forced to acknowledge that microorganisms, usually parasites, are responsible for certain diseases of the skin or the mucous membranes, such as favus, ringworm, thrush, or scabies. But in doing so, it limited its sight to localized and benign infections without mustering the courage to unify the concept of infectious disease and to recognize the involvement of similar agents in the appearance of such scourges as plague, cholera, and puerperal fever.

Yet cytology and bacteriology should have made their way together. It is paradoxical that their paths should have diverged so much. Looking only at what he had set out to see, be it the tissue or the cell, the anatomist tended to ignore the living germ. Nonetheless certain anatomo-pathologists did look into the role of the microbe in the etiology of pathological lesions. Jacob Henle, to whom we owe the description of the renal tissue, was one of the precursors who pleaded in favor of the microbial theory of infectious disease. His *Handbuch der rationellen Pathologie*, published in the 1840s, produced a synthesis of numerous studies which, together, definitively proved that living agents can indeed cause disease. Henle stressed the fact that the intensity of the infection is out of proportion with the germ that has caused it; moreover, he also produced evidence for the existence of a set period of incubation preceding the appearance of the first symptoms. In many respects, Henle anticipated the postulates later formulated by his student Robert Koch. Thus he wrote that in order to demonstrate the causative role of an infectious agent, one must establish its presence, then isolate and cultivate it, and finally reproduce the disease by inoculating it.

Today it may seem difficult to understand why it was so hard for physicians to identify and accept the pathogenic role of germs. Perhaps they found it impossible to conceive of the idea that profound alterations of the organs could be due to living creatures as small as bacteria. It should be added that this fact, though accepted as true in our day, is by no means fully understood. Many viruses work in ways that still elude our attempts to characterize them.

The idea that living creatures foreign to the body should be able to modify its physiology is thus not self-evident. But if one looks into the matter within a nineteenth-century context, one even has to add that it ran counter to the most advanced scientific thinking of the era, as represented by certain of Broussais's opponents. These men reacted against the hazy concept of vital force and tried to replace it with a physico-chemical perspective. This was the time when the role of light in the functioning of the retina was discovered; and the description of electric and molecular phenomena of the tissues bathed by the liquids of the human body seemed to represent the cutting edge of research. Physiology could be explained by chemistry and physics.

In reality, this approach was comparable to that of Liebig and his disciples in the field of fermentation. If one disregarded the germ, there was no need to think about living processes, and medical or biological science could be reduced to chemical equations and the laws of physics. For the Prussian Rudolf Virchow, the most prominent representative of this school, "disease is not an aberration grafted onto a healthy organism, it is a simple disturbance of health." This amounted to a complete rejection of the microbial theory. It meant that Pasteur would have to do battle against the ideas of Virchow, just as he had had to object to Liebig's views on fermentation.

This opportunity to resume a struggle against Prussian science that had never reached a real conclusion may have been one of the factors that motivated Pasteur to commit himself to a thorough investigation of the role of microbes in infectious disease. But above all, he was aware that this was to pursue a task he had set himself long ago, namely to disprove everything that favored spontaneous generation. "Do you know," he wrote to Dr. Bastian in July 1877, "why it is so important to me to fight and defeat you? It is because you are one of the main adherents to a medical doctrine that I consider extremely harmful to the art of healing, the doctrine of the spontaneity of all disease." Ever since his work on the silkworm, Pasteur was convinced that contagion can kill. He therefore felt that he had no choice, that he had to use his theories concerning the virulence and toxicity of germs *in vivo* in animal and human pathology.

In late 1865, Pasteur was for the first time called upon to deal with a problem in human medicine. A cholera epidemic coming from Egypt had affected Marseille, then Paris. By October, more than two hundred victims per day were counted in the capital. The greatest fear of the authorities was to see a repetition of the hecatomb of 1832. That year, Paris had been devastated by a terrible epidemic, which had left a particularly deep impression because the scourge had not spared anyone: the prime minister himself, Casimir Perier, had died — and it was whispered that what had killed him was not so much the cholera as the debilitating treatment inflicted by Broussais.

And indeed, this diagnosis was very probably correct. In any case, this cruel witticism was a sad and revealing sign of the impotence of curative medicine. In the final analysis, the only attempts to circumscribe the epidemic had been made by the public health authorities, which were willing to sacrifice the populations of the poor neighborhoods, whom they considered condemned from the outset. When calm had returned, the general conclusion therefore was that cholera was not contagious, and that the poor sanitary conditions of the capital were mainly responsible for tens of thousands of casualties. A medical problem had been turned into a social issue.

In 1865, the context was different, and the government's attitude in the face of an urgent and concrete crisis went beyond a strictly hygienist approach. Among other things, it asked Jean-Baptiste Dumas to appoint a committee of three scientists to investigate the infection. Dumas chose Claude Bernard, Sainte-Claire Deville, and Pasteur.

Absorbed in his work on microbial ecology and on the role of the airborne germs that he was just then cultivating in his laboratory in the rue d'Ulm, Pasteur, inspired by his experiments with asbestos filters, proposed a direct approach to the disease. The three scientists went to the attic of the Lariboisière Hospital, located above a cholera ward. There they had an opening cut in one of the ventilation shafts communicating with the ward. With the help of a cooling mix and a ventilator, they filtered this air in order to study its composition. Claude Bernard and Pasteur then went to the ward in order to obtain dust directly gathered from the cholera patients. They had the floor swept, the bedding shaken out, and more sterile flasks filled with air. Their aim was clear: it was to identify a specific germ in order to develop appropriate means to destroy it.

However, because the experiments he had used to combat spontaneous generations were still fresh in his mind, Pasteur looked for these germs in the air instead of examining the patients' blood. Yet blood samples were actually taken at the suggestion of Jean-Baptiste Dumas and Claude Bernard, but they were subjected to chemical rather than bacteriological analyses.

Thus it is clear that Pasteur had not yet had the intuition that in the case of cholera, as in other infectious diseases, it is more difficult to isolate the microbe in a transitional habitat than in the tissues it has injured. That is why these experiments were unsuccessful: the four scientists discovered neither the agent of cholera nor its mode of transmission. The epidemic eventually disappeared of its own accord, and the cholera took its mystery with it.

This experience was to remain the only one of its kind in the 1860s, and it benefited Pasteur only to the extent that it offered him a first opportunity to apply his ideas and his experimental techniques to human medicine. Yet the point should be made that even at this time, all of his endeavors, including those that failed, were informed by his conviction that there was a *contagium vivum*. Before he ever undertook his methodical series of experiments, Pasteur appears to have been convinced of the pathogenic role of the microbe and of its transmission by the atmosphere. He had intuitively understood the mechanism of infectious disease.

Taking On Anthrax

Pasteur's work on the origin of infectious diseases now took a detour through veterinary medicine. In early 1877, he began by looking into anthrax at the request of the Minister of Agriculture, who had been alerted by the departmental council of Eure-et-Loir. Pasteur, who had not long before triumphed over the scourges of the silkworm, would no doubt be able to find a remedy against an ill that was devastating the flocks.

Anthrax was one of the most murderous diseases for domestic animals. It existed in many parts of the world, including Russia, where it was called "Siberian plague." Most French departments, particularly in the central regions, suffered its ruinous effects. Horses, oxen, cows, and sheep were attacked in varying degrees. Victims were counted in the hundreds of thousands.

The situation was particularly bad in the department of Eure-et-Loir. Not one farm escaped the disease and the farmers considered themselves lucky if losses amounted to no more than 5 percent of their livestock. Since the calamity struck without warning, they blamed it on supposedly poisoned pastures, and there was talk of anthrax farms and cursed meadows. The sick animals had fallen under an evil spell, and the rare ones that survived had been saved by a miracle. Occasionally, humans could also be affected: shepherds, butchers, knackers, or tanners were in grave danger from a simple scratch, a small cut when they came in contact with a dead or dying animal.

Where did this ill come from? How was it spread? All the names by which it was called in the eighteenth and nineteenth centuries (evil fire, the black illness, gangrene, anthrax) and the picturesque descriptions that went with them betray the ignorance and the helplessness of the rural practitioners. Since the dead bodies seemed to be burnt or charred, external causes were incriminated, such as the action of the sun, perspiration, summer heat waves. Also accused were stagnant water, the irritating bites of flies and horse-flies, spoiled fodder crawling with insects. Marshes were suspected, particularly in the Sologne. The most generally accepted hypothesis postulated a poisoning by toxic herbs that corrupted the humors.

The disease would strike in a few hours. Falling behind in the flock, the sick animal, its head lowered, breathing heavily and quivering, soon fell down on the grass, soiling it with bloody dejecta. Blood flowed from the mouth and the nostrils. Without a sound, the animal collapsed and often died so quickly that all the hapless shepherd could do was to pronounce it dead. Lying on its back with its legs up in the air, or leaning against a partition in the stable, the sheep presented a huge distended belly. If a knife was thrust into the still warm cadaver, thick, viscous, and blackish blood would ooze out, a kind of ink-colored mush. If an autopsy was performed, it revealed an enormous spleen—hence another name sometimes given to the disease by way of a descriptive foreshortening of a symbolic but also explanatory nature: "spleen blood disease."

The first veterinarians who encountered this disease showed that it could be transmitted, but they did not push their reasoning to the point of speaking of a contagious disease pure and simple. In experiments inspired by those of Magendie, they injected healthy animals with putrefied matter and interpreted the results as poisoning rather than contagion. This erroneous judgment is understandable, considering that poison was known from time immemorial, whereas the murderous effects of bacteria had barely begun to be demonstrated. At Chartres, the capital of the disease, rumor therefore accused a mysterious poison, a telluric miasma that was supposed to intensify during the heat of summer, of killing the animal as it blackened its blood and caused its spleen to bulge.

In the early 1850s, Davaine, who, as Duclaux put it so strikingly, "looked at medicine through the windows of science," was one of the first to notice little rod-shaped microorganisms in anthrax-blood. Casimir-Joseph Davaine, born in 1812, was a physician. A native of Saint-Amand-les-Eaux in the department of Nord, he and his colleague Rayer had carried out an extensive investigation of anthrax in cattle. Their studies, which followed those of Delafond, a veterinarian at the veterinary school of Alfort, and of Pollender and Brauell in Germany, had the merit of drawing attention to microscopic organisms that could be observed in the blood of the dying animals. But this was precisely why they were contested and debated. Davaine had understood that the organisms were not normal constituent parts of the blood, corpuscles or webs of fibrin serum, but foreign particles, which he classified as plants. This insight would not have distinguished Davaine from other illustrious explorers of endemic disease if he had not at the same time asked the essential question: was this a contagious agent or simply a tangible but inoffensive consequence of the disease?

It took nearly thirty years to answer this question. Although it had been stated correctly enough, it did have the unfortunate effect of jumbling the tracks it had exposed. To begin with, since the corpuscle detected in the anthrax blood was assumed to be of plant origin, a few obstinate observers proposed to go so far as to consider it a seed or a spore. This was indeed a perspicacious idea, considering that it was later discovered that a bacterium can sporulate by changing its form. But since the specific anthrax bacillus is small, it was confused with the usual germs of putrefaction and therefore went unnoticed for a long time. Davaine was the first to isolate it, but he did not fully realize the implications of his observation: in a notice about his own work he published shortly thereafter, he did not even mention this discovery — and it was to take him twelve years to recognize its importance.

It was only in 1861, when he became aware of Pasteur's work on fermentation, that Davaine made the connection with what he himself had written earlier. He was struck above all by the description of the butyric ferment, whose cylindrical rod shape resembled that of the corpuscles in the anthrax blood. In 1863, Davaine therefore took another look at his observation of 1850 and gave another paper to the Academy of Sciences on the lethal role of the anthrax bacillus. In his note he explicitly acknowledged Pasteur's work.

As Pasteur pointed out, 1863 was the year in which the experiments carried out in the laboratory of the rue d'Ulm were demonstrating that blood and urine gathered under aseptic conditions remained sterile, proving that there were no pre-existing germs in the internal environment. Spontaneous generation, in other words, did not occur in the human body any more than in swan-necked flasks.

During that same year a physician in Dourdain, the neighbor of a farmer who in one week lost twelve sheep to anthrax, sent Davaine a sample of the blood taken from one of the cadavers. Examination under the microscope once again revealed the immobile and transparent bacteria that were assumed to be specific to anthrax. When he inoculated rabbits with this blood, Davaine saw them die in short order.

One might think that the demonstration was made: here was the bacterium, and here was death; the connection seemed straightforward. But then the blood could also contain other elements than bacteria, and so it was not difficult to contest the hypothesis. Before long, Davaine's theory was indeed contradicted by two professors at Val-de-Grâce, Pierre-François Jaillard and Emile Leplat, who repeated the experiment, using cow's blood that they had shipped in the middle of summer from a knacker's yard. Their finding was quite different: although the rabbits did die from the inoculation of the cursed blood, not a single bacterium was found in their cadavers. The conclusion was obvious: the bacterium was an epiphenomenon. Davaine in turn repeated Jaillard and Leplat's experiments and confirmed their observation, but he proposed a new interpretation: since they had used a cow and not a sheep, Jaillard and Leplat were dealing with a new disease different from anthrax. Under these circumstances, why not call it a cow disease. This, however, was as far as Davaine was able to go.

Some fifteen years later, Koch's experiments became known. Born near Hanover in 1843, Robert Koch was thirty-three years old when he became interested in this problem. At the time he was practicing in Wollstein in the province of Posnan after studying in Göttingen and Berlin, where he had come in contact with the most eminent scientists of the day, among them Jacob Henle. The latter had given him an understanding of the microbe theory and the theoretical and experimental obstacles it was facing. Koch was to reap glory and immortality when he discovered the tuberculosis bacillus in 1882 and the cholera vibrio in 1883, but for the moment he spent his time at the bedside of sick farmers, an occupation that also gave him the opportunity to observe anthrax in flocks of farm animals. In a primitive laboratory he had set up in his home, he tackled this problem by himself, seeking to repeat and supersede Davaine's observations.

In doing so, he first sought to find a culture medium for the anthrax bacteria. It occurred to him to use some drops of the aqueous humor from the eye of an ox or a rabbit. This medium seemed particularly suitable for nourishing the germ, for within a few hours he observed that the rods were becoming longer, forming a kind of tangle of threads that filled the visual field of the microscope. Examining the entire length of the corpuscles, Koch noticed ovoid elements arranged like peas in a pod; these he placed into the category of spores. This observation was close to the descriptions Pasteur had published in his studies of the flacherie of the silkworm. It was thus established that bacteria can sporulate and come back to life from these spores several years later. In his search for the anthrax bacillus, Koch was thus no longer guided by Pasteur's work in the chemistry of fermentation but by his research on the silkworm moth.

When Koch placed a few pieces of spleen taken from diseased animals into the aqueous humor of the eyes, the inoculation with the rods, like that with the spores, produced as severe a disease as when it was transmitted by blood from the spleen. Thus it was possible to reproduce in a living medium what had been observed under the microscope. For the second time, a scientist had come close to the goal, and the problem seemed almost totally solved: the black disease was caused by the *bacillus anthracis*.

At this point, the French physiologist Paul Bert provided new arguments to the opposition, and thereby led Pasteur to join the fray. In January 1877, Bert announced to the Biological Society that it was possible to kill the *bacillus anthracis* in a drop of blood by means of compressed oxygen; yet if one inoculated the remainder of the blood, disease and death would ensue even though no bacterium could be seen. "Bacteridia are therefore neither the cause nor the necessary effect of anthrax disease," he concluded.

It now fell to Pasteur to lift the last doubts and perfect an argumentation that had not been entirely convincing. For his experiments he obtained the assistance of one of his former students at the Ecole normale, Jules Joubert, now professor at the Collège Rollin (today Lycée Jacques-Decour) in Paris. The first problem to be solved was to demonstrate beyond any doubt the role of the bacterium, and then to eliminate any deleterious influence that could be attributed to the blood, to the serum, or even to other microorganisms.

Ever since his studies on beer, Pasteur had surrounded himself by an arsenal of culturing techniques. On the shelves in his laboratory he could find test tubes and media that allowed him to make bacteria proliferate whenever he needed them. He also knew that a germ, even if highly diluted, can multiply sufficiently to invade the preparation if given a favorable environment. It was this property that Pasteur set out to exploit first. The experiment was simple: in order to demonstrate that the bacterium alone transmits the disease, all he had to do was to dilute a drop of anthrax blood and at the same time create conditions favoring microbial proliferation. Through a veterinarian in Chartres, Boutet, Pasteur obtained a sample of the blood of an animal that had recently died of anthrax. One drop taken from this bacterial focus was diluted in urine, a medium that favored its development. After a period of culturing sufficient to permit the germs to reproduce and the blood to become diluted, Pasteur took another drop from this solution and placed it into a second flask of fresh urine, where the incubation was repeated. Then another drop was taken and also diluted, and this process was continued until, at the end of some ten passages and culturings, "the original drop of blood, the one that furnished the first seed, has been drowned in an ocean as it were."

"Only the bacterium," added Duclaux, "has escaped dilution, for it multiplies in each one of the cultures."[2] The culturing was indeed so effective and so rapid that within a few hours the liquid took on a whitish and matted look, as if a wad of carded cotton had been dissolved in it. All that remained was to inject a rabbit with one drop of the product of the last passage. The result was beyond contradiction: this inoculation killed as surely as if anthrax blood had been used. Pasteur had thus performed a conclusive experiment to show that the disease was indeed transmitted by the bacterium. The virulence could be inherent neither in the sticky red blood corpuscles or other constituent parts of the blood, for these were too diluted to interfere, nor in some filterable virus, for if the preparation was filtered, it was no longer lethal. For Pasteur, the matter was perfectly clear: anthrax was indeed a disease caused by a specific bacterium, just as scabies is caused by a mite — the only difference being that in order to see the germ one needs a microscope with its enlarging lenses.

Meanwhile, he still had to explain the experiments of Jaillard and Leplat, as well as Paul Bert's enigma. How could the former have injected guinea pigs with anthrax blood without subsequently finding a trace of the bacillus in the animals, which nonetheless were almost immediately killed by the injection? And why did the latter, having used a procedure that eliminated all visible traces of the germ, continue to see its virulence intact?

More inventive than Davaine, Pasteur refused to go along with the preconceived notion that everything that is happening in the rest of the body also happens in the blood. More imaginative as well, he put forth the hypothesis of a second disease. To begin with, he decided to reproduce the conditions of the experiments supporting Davaine's opponents, namely summer heat and the delayed shipping of the cadavers, in a word, conditions that favored putrefaction. From his earlier research, Pasteur remembered in particular that the decomposition of organic matter is due to specific germs. That is why, accompanied again by the veterinarian Boutet, he went to a knacker's yard near Chartres so that he could personally supervise the taking of anthrax blood. He carefully chose three cadavers whose state of preservation exactly reproduced the experiments of Leplat and Jaillard.

On 13 June 1877, in this knacker's yard near Chartres, Pasteur had before him three cadavers, one sheep that had died sixteen hours earlier, a horse whose death had occurred twenty-four hours before, and a cow that must have been dead for at least two days, since it had been brought from a fairly remote place. Three animal corpses, three time frames, three results from the analysis: the blood of the recently deceased sheep contained many anthrax bacilli; the blood of the horse contained very few, and that of the cow, none at all.

Ever since his groping attempts to understand the silkworm problem, Pasteur knew that two diseases can crisscross and interfere with one another. If this was the case here, he had to produce evidence for a second agent of infection that could not be seen in the blood. As was only logical, Pasteur turned to the blood of the cow, where the anthrax bacillus could not be detected, to find that second agent. He injected a sample of it into the belly of a guinea pig and observed. Within a few hours, the animal presented a distended abdomen and swollen muscles. At the autopsy, Pasteur recorded pockets of gas and a running serosity in the abdomen. Having very carefully removed some of this brackish liquid and examined it under the microscope, Pasteur found in it a set of new microbes, germs of putrefaction. Everything became clear. There was indeed a second disease, this one related to the microbes of putrefaction, anaerobic germs that were paralyzed by air and could be easily observed only in the body of the injected guinea pig.

How could this phenomenon have remained unknown for so long, Pasteur asked himself. Simply because the experimenters were in the habit of examining the blood of cadavers without thinking of looking for the germ elsewhere. "Yet the fact is, not only that the microbes examined here appear last in the blood, but also that in this liquid one of them takes on a very special appearance, a great length that often exceeds the total diameter of the field of vision of the microscope and such translucency that it can easily elude observation. . . . If one takes the trouble to look for it elsewhere, one finds in the tissues and in the serosity of a putrefying animal great masses of these proliferating germs."[3] The second disease was related, then, to a new germ to which Pasteur gave the name "septic vibrio," a term that was to become famous in every language.

In this manner Davaine's theory and the counter-experiments of Leplat and Jaillard were reconciled, for although the septic vibrio was not present in the blood of the sheep that had died most recently, it appeared later in the blood of the animal whose cadaver had for some time lain abandoned on a manure heap.

As for Paul Bert's enigma, it too was solved by Pasteur's discovery. It turned out that the germ whose involvement he had demonstrated was anaerobic. So what had Paul Bert done when he subjected anthrax blood to a stream of oxygen under high pressure? He had, to be sure, suppressed the rapid movements of the septic vibrio through contact with the air, "but this does not mean that the vibrio was killed, for the contact with oxygen transforms it into corpuscles/germs, so that overnight a liquid filled with organized and motile filaments will be reduced to a jumble of extremely weak luminous specks. But if these specks are introduced into the body of a guinea pig or into an appropriate liquid, they reproduce as motile filiform vibrios and the animal dies with all the expected symptoms."[4]

As for characterizing the vibrio and the disease it produces, Pasteur took his clue from an observation made by a Parisian veterinarian, Signol. In December 1875, Signol had sent to the Academy of Sciences a note in which he showed that the blood of healthy animals killed by a blow or asphyxiation, if drawn from the deep veins connected to the intestine, becomes virulent after a few hours. If inoculated, this blood causes a death similar to the kind from anthrax. Commissioned to verify this finding, Pasteur repeated the experiment in January 1876 and found the same phenomenon. But he also detected an additional fact: among the microbes of putrefaction, he noticed a germ that "pushed the blood corpuscles aside in its undulant and creeping movement." At the time, the significance of this observation eluded him, but later, when he was engaged in the search for the origin of the septic vibrio, Pasteur remembered this earlier observation. Now, with that synthetic insight that so often led to his most important conclusions, he stated that the second disease described was the one that is transmitted by putrefied cadavers that are left to rot in the summer heat. Consequently, the septic vibrio "is none other than one of the vibrios of putrefaction. . . . Its germ must exist almost everywhere, and surely also in the matter contained in the intestinal tract. When a cadaver is discarded and if it still contains its intestine, the latter promptly becomes the focus of a putrefaction. At this point, the septic vibrio is bound to spread throughout the serosity, the humors, and the blood of the internal organs of the cadaver."[5]

These discoveries concerning the anthrax bacterium and the septic vibrio thus definitively established the microbe theory of disease. By virtue of their ingeniousness and their precision, Pasteur's experimental techniques had triumphed and shed light on areas where others had gotten lost in an overly sophisticated chiaroscuro. It had now been established once and for all that two different kinds of infection could be caused by two different germs, with one of them, the septic vibrio, developing in case of putrefaction. Beyond these two specific instances, Pasteur also showed that he was able to identify the agents of contagion and describe in every detail the diseases for which they are responsible. Yet, despite the absolute clarity of the demonstration, it was to take years of effort to win over certain minds blinded by prejudice and make them acknowledge that here was the point of departure for a general theory of infectious disease.

Monsieur Colin's Chicken

In the presence of an exemplary demonstration, criticism is not easy, unless one bases it on erroneous or preconceived ideas, like those held by certain of Pasteur's colleagues at the Academy of Medicine. Among these, Gabriel Colin was to prove a fierce opponent of the Pasteurian theory. Professor at the Veterinary School of Maisons-Alfort, he could not bring himself to admit that anthrax is caused by the newly discovered bacillus and believed in the existence of another virulent agent without, however, being able to produce any evidence. This quarrel was to give Pasteur many opportunities to become involved in some of these polemics he rather enjoyed, either in person or through a third person. For Colin was a great talker, who in a monotonous and slow voice kept stating doubts that he took for proofs, talking on and on until finally his colleagues were forced to pay him some attention and to consider his arguments.

In August 1877, the veterinarian Henri Bouley, declaring that despite his fervent support of Pasteur he had been "moved by reading M. Colin," called on Pasteur to give close consideration to the objections that had been raised. Pasteur was at his summer home in Arbois at the time, but this did not keep him from replying vigorously to Bouley. "Oh, how I would like to take literally the honor you are doing me by calling me 'your master' and to teach you a sharp and proper lesson, you man of little faith. . . . Let me tell you quite frankly that you are not sufficiently steeped in the matters about which I have reported, both in my name and in that of M. Joubert, to the Academy of Sciences and the Academy of Medicine. Do you really believe that I would have made these reports if they had needed the confirmations of which you speak, or if the objections raised by M. Colin had been pertinent? You are well aware of my situation in these important controversies; you are well aware that, ignorant as I am of all medical and veterinary knowledge, I would immediately be indicted for presumption if I had the temerity of taking the floor without being armed for combat, struggle, and victory. All of you, veterinarians and physicians, would vie — and justifiably so — in stoning me if I brought specious proofs into your debates."[6]

What were the doubts that bothered M. Colin? The veterinarian took cover behind twelve years of work and more than five hundred experiments on anthrax to assert that he, and he alone, was correct, that there was no other expert but himself in the field, and that Pasteur was mistaken to believe that the bacillus was responsible for anthrax. "M. Colin believes in the presence of a soluble virulent matter because he wants it to be there!"

Pasteur wrote for his part, adding bluntly: "Since none of these five hundred experiments had any bearing on the point at issue, he cannot believe that the experiments of others might be relevant. Yet some do have that character, but he considers them null and void, speaking of them only to distort them or to raise subtle points of dialectic against them."[7] In fact, Colin believed neither in Pasteur's experiments with dilution nor in those with filtration, because he thought that both could alter the microbe. He proposed to relate anthrax to a different agent than the bacterium, for in his own experiments, which relied exclusively on what he could see, he had not detected a single one in the blood of an inoculated animal a few hours after it had been injected with anthrax blood.

Here it was easy for Pasteur to remind him of the need to wait for proliferation to take place before observing the bacteria: "Suppose that there is only one bacteridium in a drop of the blood of an inoculated animal; it would almost be a miracle to happen upon this tiny filament when examining the sample under the microscope. Looking for a new planet would be easier."[8] He also stressed another phenomenon that had given him great trouble in his work on the silkworm, namely incubation, which Colin refused to take into consideration, unable to understand how a disease could be both real and not apparent.

However hard he tried, Pasteur was not completely convincing, despite what he considered his irrefutable explanations. Colin continued to have many supporters among the physicians. On both sides, there was talk of verification, of commissions, of petitions. Pasteur asked the Academy to call upon his adversary to demonstrate the accuracy of his claims; and so did Colin, who urged his friends to demand the verification of the hypothesis of a "mere chemist."

Pasteur was exasperated by these efforts to refute him. "You young people who sit in the upper tiers of this amphitheater," he exclaimed, "and who may represent the hope for our country's medical future, do not come here to look for the excitement of polemics; come to learn about methods. Well, I am giving you as an example of the most detestable of these methods M. Colin's statement that a negative microscopic observation is sufficient to assert that there is . . . an anthrax virulence without any presence of bacteridia in the inoculated material."[9] In appealing to the young in this manner, Pasteur knew what he was doing. If the seasoned masters were accustomed to his emphatic diatribes and listened to them as a matter of courtesy more than of conviction, there was indeed, high up in the amphitheater, a small group of young physicians who did not miss a word of what

Pasteur was saying. It was among them that Pasteur found his most fervent support and that he would recruit the indispensable assistants for his medical experiments.

One day, when the minutes of a session at the Academy suggested that he should repeat the experiments outlined by Colin, Pasteur rose up in indignation: "Last year, I published in collaboration with M. Joubert, professor of physics at the Collège Rollin, two notes dealing with anthrax disease. The new facts I announced at that time have not been as far as I know contradicted by anyone, either in France or abroad. Under these circumstances, I must consider these facts scientifically established, and no one, be it a commission or an individual, has the right to demand other proofs than those we have given. This is a point of scientific law that seems incontestable to me."[10]

Yet this peremptory tone was not enough, and the quarrel was carried to somewhat ridiculous lengths. As soon as Pasteur asserted that anthrax cannot be transmitted to chickens, Colin claimed the opposite. Thereupon Pasteur wrote to him, offering to send him a bacterial nutrient broth that he would like his illustrious colleague to inject into healthy chickens, adding that he would be glad to examine the cadavers, if any. Colin eagerly accepted.

A week later, Pasteur met Colin. "How are my chickens doing?" he asked ironically. The animal was not yet dead, but Colin was sure that it would be within a few days. Pasteur left for his summer place, and on his return, at the first meeting of the Academy, took Colin aside to ask whether the chickens had indeed died. Colin, claiming that the vacation had interrupted things, promised to produce the animal's cadaver in the next few days. Days and weeks passed without news of the chickens, their cadavers, or Colin. A few months later, Colin asked to see Pasteur and, rather crestfallen, told him that he had been mistaken. Anthrax was not transmissible to chickens, and the poor creatures had not died, at least not of the disease, for in the end a dog had gotten into the cage and devoured them.

"Well," Pasteur rejoined, "I am going to prove the contrary." The following Tuesday, 19 March 1878, Pasteur left the rue d'Ulm in frock coat and stovepipe hat, carrying a bird cage. In the cage were three chickens, two alive and one dead. Arriving in the middle of a scheduled meeting, he proudly placed the incongruous cage on the chairman's podium. He then proceeded to explain that the dead chicken had perished of anthrax twenty-nine hours after it had been inoculated with the anthrax bacillus.

However, the development of the disease was due to very special circumstances. If it was indeed correct that chickens were ordinarily resistant to the disease, it was because something protected them. This being the case, Pasteur had had the idea of investigating whether this immunity might be related to their body temperature, which was higher than that of the animals that could contract the disease. With this hypothesis in mind, he had devised the scheme of plunging the chickens into cold water to lower their body temperature. In these conditions, Pasteur declared, the inoculated animal developed the disease and died.

As for the two living chickens, one of them, the gray one, had been plunged into the same bath, which proved that cold water causes neither disease nor death. The other, the black one, was given the bacteria but not the bath and had suffered no ill effects. One more experiment remained to be done, but since the Academy did not hold night sessions, Pasteur performed it in his laboratory. This time, he plunged a chicken into a cold bath, inoculated it, then removed it from the water and raised its temperature before the disease had a chance to kill it. The outcome was in keeping with the hypothesis: the chicken recovered.

Despite all this, Colin refused to give up. Having vowed that he would never again debate Pasteur, he returned to the experiments with the chickens and declared that he would be glad to confirm that they had really died of anthrax bacteria. He felt that it was essential to repeat the experiment, this time with an autopsy and a microscopic examination. "I accept," said Pasteur, "provided that the Academy appoint a scientific commission in which M. Colin will be in charge of drawing up the minutes concerning the conclusions of the experiments." And he added, "I came into your company intending to pursue a program in which I must never allow myself to stumble. And this program, I can give it to you in very few words: For twenty years I have sought to find, and I am still seeking to find, spontaneous generation in the proper sense. God willing, I would be glad to spend another twenty years or more looking for spontaneous generation in transmissible diseases. In these difficult investigations, I will always show severity toward frivolous opposition, but at the same time, I shall always have esteem and gratitude for those who point out to me that I am mistaken."[11]

The commission met over the next few days. On the table lay three dead chickens, which had received doses of bacteria after they had been plunged into cold baths. The meeting room was turned into a laboratory. The scientists rolled up their sleeves. A first chicken was dissected on a slab of marble. In the focal point of the infection, they noticed a serosity. These gentlemen in frock coats crowded around the viewer of the microscope to see the proliferation of beautiful and abundant bacteria scattered throughout the animal's body. Colin threw in the sponge, and at the signing of the minutes recognized that the death was indeed due to the anthrax microbe. He also asked that no autopsy be performed on the two other chickens, which Pasteur generously gave to him, in case he wanted to use them for some additional opposing experiments.

"M. Colin was mistaken," Pasteur concluded. "That is all that matters to me. To begin with, he was wrong not to ask me about the reason for his failure when he tried to produce anthrax in chickens by chilling them. He has of course every right to verify experiments, but by the same token it would also behoove him not to forget certain conclusions before he looks into the work of those whom he intends to contradict."[12] *Vae victis!* When Pasteur won, he rubbed it in. And his use of irony was somewhat lacking in elegance, for some time after this episode he pretended to be astonished that in his new work Colin no longer referred to his former concepts.

The problem of the chilled chickens was, however, more complex than Pasteur realized at the time. Whatever the effect of cold water on the functioning of the organs, the ensuing result might have led to new questions and to an improved understanding of certain mechanisms of immunity. Pasteur had shown how the environment influences the disease and, in so doing, indicated that a pathogenic agent can be present in an organism without causing an illness. But it was only much later that the concept of the "healthy carrier" came into being.

Earthworms against the Flocks

As a result of his discoveries concerning the virulence of the bacillus and his victorious polemic against Colin, Pasteur was considered an anthrax specialist. However, this did not mean that the debate had gone beyond the precincts of the Academy, or the experiments beyond the laboratory in the rue d'Ulm. But, as usual, Pasteur wanted to derive direct and material benefits from his findings. The opportunity to do this arose as early as the summer of 1878, when the Ministry of Agriculture once again commissioned him to undertake a study, this time involving interviews with stock breeders, in order to discover in what ways the anthrax bacterium infested their livestock.

Here was a new riddle to be solved: what was the source of the contagion? Widely divergent hypotheses were circulating on this subject, incriminating such things as spoiled fodder, deficient aeration of the farms, and miasmas in the humid lowlands, but their end result was only one rather vague recommendation, namely, to take the herd to a new grazing pasture whenever an animal had become sick.

In tackling this problem, Pasteur proceeded on the basis of two positively established facts. On the one hand, the disease was transmitted by a bacterium; on the other, it was indeed related to grazing areas. The conclusion was obvious: it had to be shown that the disease develops and maintains itself through foci of infection in which the bacterium proliferates. Before long, Pasteur thought that the spore of the anthrax bacillus might well be mixed with earth and thus infect the grazing livestock. This was particularly easy to imagine because cadavers were usually buried right where the animal had perished, that is to say, in the pasture. The disease could thus take long-lasting hold in an infested area. It was a hypothesis; now it had to be proven. Once again, proof was obtained by observation in the field.

On 16 August 1878, Pasteur officially opened his campaign to investigate the anthrax disease. He established his headquarters near Chartres, at the Saint-Germain farm where he installed his collaborators; he himself did not take up residence there because his other investigations (he was putting the finishing touches on his studies of the silkworm and on the fermentation of grape juice) demanded constant commuting between Paris, Arbois, and Chartres. The team was composed of Charles Chamberland, the current official laboratory assistant, Auguste Vinsot, a young veterinarian and recent graduate of Maisons-Alfort, and finally Emile Roux, a newly minted physician of twenty-five, one of Pasteur's most assiduous listeners at the Académie de médecine. Something of a loner, and without apparent ambition, he was destined soon to become the master's closest collaborator.

"The days when Pasteur came from Chartres, lunch at the hôtel de France did not take long," Roux remembered later. "We would quickly get into the carriage to drive to Saint-Germain, where M. Maunoury had kindly placed his farm and his flock at our disposal. During the drive we discussed the week's trials and the plans for subsequent ones. As soon as he had alighted, Pasteur rushed off to the pastures; standing stock still by the fences, he looked at the batches used for the experiment with his customary sustained attention that took in everything. He could spend hours keeping his eyes fastened on a sheep he believed to be sick; we had to remind him of the time and show him that the towers of Chartres Cathedral were beginning to fade away into the darkness to make him leave. He would talk to farmers and farmhands, and he always paid attention to the opinions of the shepherds, who because of their solitary life give all their attention to their flock and often become discerning observers."[13]

By September — writing from Arbois, where he had gone to supervise the installation of some experimental greenhouses — Pasteur was able to send his first conclusions to the Minister of Agriculture. He and his team had first attempted to infest the sheep by feeding them alfalfa drenched in microbial culture broth — this was the method of contaminated feeding he had used earlier to prove that the diseases of the silkworm were transmissible. But in the case of anthrax, the experiment was rather inconclusive, and the animals seemed determined not to die. The anthrax microbe thus appeared to be less active when ingested by laboratory animals in the rue d'Ulm than when the sheep were grazing in the fields. At this point, Pasteur wondered whether the germs contained in these experimental feedings failed to enter the organism because the grass was too soft. In order to promote the inoculation of the bacterium yet also stay close to the normal conditions in the fields, he therefore had the idea of causing abrasion in the mouths, on the palates and the pharynxes of the sheep. They were now fed a more complex diet, for sharp and cutting leaves were added to their mangers, along with thistles or the bristles of barley spikes. The sheep were penned in fields where the grass was tough and sharp. Under these conditions, the experiment succeeded, and the autopsies of deceased animals showed the characteristic lesions of anthrax. One could therefore conclude that the disease was inoculated when a sheep that had injured its gums and throat swallowed contaminated grass.

The department was full of burial pits where the remains of animals dead of anthrax were piled up, so that the spores of the bacillus proliferated in the soil, particularly on the surface of the stubble. As a result, the slightest and ordinarily insignificant wound could easily bring disease and death. Hence Pasteur's strict injunctions to the farmers: "You must keep the livestock away from such plants as thistles, oat spikes, and straw. Also move away from areas where cadavers have been buried." Eure-et-Loir, then, was not a cursed land, and Pasteur not only vanquished superstitious rumors of sorcery but also established the principles of rural ecology by identifying a polluted area that had been infested with bacteria from buried cadavers.

One might believe that the conditions of transmission had been described in sufficient detail for Pasteur to move on. But one question continued to nag him: how do the anthrax germs come to the surface, considering that the dead animals were buried very deeply in the ground?

One day when he was visiting Chamberland and Roux at the Saint-

Germain farm, he glimpsed a possible explanation. In a recently mown field, he noticed a blackish spot, where the earth had a different color than the surrounding soil. Upon questioning the owner, he learned that this was the place where sheep dead of anthrax had been buried. Intrigued by this answer, Pasteur returned to the spot and, his nose close to the ground, noticed characteristic small mounds of earth. These were the ribbons of soil that the earthworms pile up on the surface when they push the earth out of their tunnels. In a brilliant flash of insight, Pasteur had found the missing link in the chain of transmission of anthrax: "It is the worm that conveys the spore. In burrowing their subterranean tunnel, the worms dig up the area around the pits and then sow and transport the spores they have exhumed."

Impatient to put his hypothesis to the test, he had his assistants collect earthworms on the surface of pits where animals dead of anthrax had been buried several years earlier. He took them back to his laboratory and dissected them in the presence of several invited guests, Henri Bouley, Jean-Antoine Villemin, and Casimir Davaine. From the columns of earth that filled the worms' intestines, Pasteur extracted spores that he identified as anthrax spores. The earthworm was only a temporary host for the bacillus, which it carried from the thick blood of the cadaver to the alfalfa or the thistle on which it settled. A complete explanation for the propagation of anthrax had been found.

At the very time when Pasteur revealed the harmful role of the earthworm as a vector of death, Darwin for his part was interested in the positive and beneficial aspect of its action. In his last book, Darwin showed how the work of the earthworms in opening subterranean tunnels, digging up earth, and bringing it to the surface in their dejecta contributes to draining and aerating the soil. This continual mixing, he pointed out, which enriches the land by bringing in minerals and nitrogen, is highly beneficial to farming. Each in his own way, Pasteur and Darwin thus demonstrated two of the functions of the earthworm, transporting spores and cultivating the soil.

Pasteur's initial observations had of course to be completed, explained, and complemented with additional proofs before they could confidently be used as the source of practical advice. Additional experiments were therefore conducted on a grand scale. In 1880, Pasteur had his assistants collect earth from pits where cows that had died in July 1878 were buried. There the land was rich in humus and the grass was thick, for it was nourished by the decomposing organic matter of the putrefying carcass. But the scientists also found the expected spores — and of course plenty of earthworms.

The dead body that returns to earth does, to be sure, furnish an excellent fertilizer, but it is one that can turn out to be lethal. Pasteur therefore felt justified in laying down rules for the farmers: "You must prevent your animals from grazing in pastures where dead bodies have been buried. Fields where crops are grown must not be used as cemeteries. Grazing or raising forage must not take place on land where dead bodies are buried. . . . Whenever possible, dead animals should be buried in sandy or calcareous, and in any case very poor areas, in a word, in places where earthworms cannot thrive."[14]

Anthrax, a disease of livestock, could thus be redefined as a disease caused by the anthrax bacterium. Equally important were the implications for another science. After the investigation of the pastures around Chartres, hygienic practices could not remain unchanged and had to rely increasingly on the laboratory. New epidemiological concepts could no longer sidestep the germ, an undeniable actor in pathological processes. Henceforth, the success or failure of the hygienists was tied to Pasteur's pipette and to his culture broths. And so was that of the doctors.

Scenes from Family Life

Between the experiments conducted to persuade some and the di-atribes delivered to criticize others, Pasteur had little time for family life. Most afternoons he spent at the two academies, that of the sciences and that of medicine. Accustomed to being assailed by opponents, he never missed an opportunity to respond, even though his old teacher Balard preached moderation and restraint, feeling that a scientist of Pasteur's caliber had no need to measure himself against second-raters, who frequently had not even conducted their own experiments before launching themselves into de-bates. "You should criticize results, not hypotheses," he kept telling Pasteur. Duclaux was of the same opinion when he wrote affectionately, "I can see very well what you stand to lose in these fights: your peace of mind, your time, your health. I fail to see what you can gain by them."[15]

From the meetings of the academies, whether on the quai de Conti or in the rue des Saint-Pères, Pasteur walked back to the rue d'Ulm on foot, or sometimes took a hackney cab, especially if the meeting had been stressful and the limp that had remained after his stroke became worse. He returned to his laboratory to continue working, as well as to find peace and quiet after his oratorical jousts. This was where he spent most of his time. Every morning, in an immutable ritual, he went down to the laboratory around half past eight, returning to the apartment three hours later for lunch with Marie. In the afternoon, if there was no meeting, he continued working. He returned to his apartment in the late afternoon and dictated his notes before going to bed around nine, like the peasant of Franche-Comté that he still was.

In the summer, Pasteur invariably went to Arbois, where he remained until mid-autumn. After the death of his father, Virginie Vichot, his eldest sister, had moved to the ground floor of the house and his brother-in-law Gustave and his sons had taken over the tannery in the building's basement. The second floor was reserved for Pasteur and his family. Three guest rooms had been built in the attic. As the uncontested head of the family, Pasteur occupied the master bedroom, where his father and his daughter Jeanne had died. The adjoining room, to which he had added a fireplace and a balcony, was for Marie. The austere look of the house was unchanged. The furnishings were undistinguished, without style. They came from Pasteur's parents or from his lodgings at Lille or Strasbourg. Plain wallpaper and curtains, some utensils bought at Arbois, and a very few knickknacks brought back from travels through Europe completed the picture.

The only small vanities Pasteur permitted himself were personal souvenirs, such as a few of the pastels he had executed as an adolescent, family portraits, or scenes from the laboratory. His room was a temple to the glory of the great days of the Napoleonic Empire: there was a large medallion of General Bonaparte, the work of the Franc-Comtois sculptor Huegenin; a profile of Napoleon in bronze-covered plaster; near the fireplace, a lithograph of the King of Rome. Other and more intimate relics were displayed on a bureau in Marie's room: a box in Brazilian rosewood that came from Joséphine, and another black box studded with steel nails that had belonged to little Cécile.

Virginie died of an apoplectic attack at the age of sixty-two on 30 July 1880. "Without a painful agony, she died a very gentle death that is enviable for the last moment," Pasteur wrote to his son Jean-Baptiste. Having been present at his sister's demise, Pasteur now was keenly aware of being "the last survivor of the family." In order to take full possession of his little fief in Arbois, he had a new tannery built for his brother-in-law and his nephews a little way downstream on the river. From now on, the family homestead was all his.

Life in Arbois was essentially the same as in Paris, except that Pasteur did not have to go to the academies in the afternoon. However, he only took a short respite after lunch, usually finding someone with whom to play his favorite game, croquet, and again in mid-afternoon, when he took a short walk along the roads or in the vineyards.

From the fields and from the neighboring hillsides, people called out familiarly, "Here you are! Louis, how are you doing?" He knew all the local children, and they all knew him. Back from his walk, he liked to sit on a green bench close to the house, next to some acacias and quince trees that he had planted himself when he was a child. On rainy days, when Marie played cup-and-ball, Louis was not adverse to playing a round of billiards. With a firm grip of the cue, he braced himself on his good leg and won almost every time. Everyone knew that the master hated to lose.

There were few visitors. Beginning in 1875, Amélie Loir, Marie Pasteur's older sister, came to Arbois with her son. Bertin, Pasteur's classmate whom he had known forever, also came every year to bring his daughter, who had gone to school with Marie-Louise. His puns and his good humor always brightened the atmosphere. Also there to spread a little gaiety was Jean-Baptiste. Pasteur showed himself indulgent toward his son's jokes: "Goodness, but you're silly!" he would exclaim as laughter made the watch chain and pince-nez shake on his stomach.

Until 1880, Pasteur always had his sister and brother-in-law and their children for Sunday dinner. The meals were simple and frugal, sometimes barely adequate, for the Pasteur household was run with the strictest economy. "At the Vercels', the dinner guests always left a tip under their plates," recalled an old servant in Arbois. "I never found anything under Pasteur's plate." After Sunday dinner at the Pasteurs', where Arbois wine was of course de rigueur, Gustave Vichot would sing. When he was in good form, Pasteur would chime in with his somewhat shrill voice. He always sang the same little song, which began with

> *One night the devil offered me*
> *Glory and great fortune.*
> *He said to me: Luck is with you.*
> *Choose! But take only one.*
> *— Glory is what I will take . . .*

Almost every Sunday, at a quarter to eight, Pasteur, his missal under the arm and accompanied by Marie, went to hear mass at the church of Saint-Just. Along the way, he often stopped to chat with one person, give advice to another.

In October 1874, Jean-Baptiste married Jeanne Boutroux, a young woman from Orleans who was very distantly related to the philosopher Emile Boutroux, Pasteur's colleague at the Ecole normale. Since, according to the rules, the marriage had to be celebrated at the bride's church, the Cathedral of Sainte-Croix, Pasteur addressed himself directly to the bishop of Orleans, who was none other than the famous Dupanloup of the Académie française, easily the most influential prelate in France. "Monseigneur and distinguished confrère," he wrote without batting an eyelash, referring to his own seat in the Académie des sciences, "Madame Pasteur, a very fervent Catholic, entreats me — and has long done so — to write to Your Grandeur in order to ask if You might be kind enough to bestow the nuptial blessing. . . . Until now I have felt that it would be too bold an undertaking to ask for such a favor. I am finally yielding to her urging, albeit with all kinds of scruples of conscience and because I know that Your Grandeur, accustomed to forgiving many things, will deign to accept my humble apologies." The letter ended with "the homage of a profound respect,"[16] as if the marriage of his son were Pasteur's first step in establishing his candidacy for inclusion among the forty immortals.

Meanwhile, a marriage, however religious, was not enough to secure the future. As his father had done for him, Pasteur fretted about his son's career. He suggested that he seek a position in the administration, more precisely in the Council of State, where he, Pasteur, would be in a position to advance his son's career. "I am extremely anxious for you to take up some serious occupations. There is nothing I desire more than to see you acquire merit and become a citizen who is useful to yourself and to your country. Do not disappoint your mother's hopes and mine. Devote your leisure to reading. You know that my purse will easily open to take care of this additional cultivation of your intelligence." At the same time, Pasteur had few illusions about the material effectiveness of his influence. "Favor in high places is less important than one imagines; in any case, it is a bad way to get in, which leads to nothing but regrets and carries a great many risks. You should therefore be thinking more and more seriously about acquiring merit, personal merit."[17] When all is said and done, all this paternal solicitude yields a rather unflattering portrait of Jean-Baptiste, who seems to have been less than eager to put himself to any trouble and who was glad to let his father take care of all difficulties that might arise.

This went so far that some time after the marriage, in June 1875, Pasteur wrote a rather extraordinary letter to his son's mother-in-law: "Chère Madame, I feel compelled to submit to you some friendly observations concerning your manner of understanding our children's life. . . . How can you expect Jean-Baptiste to work properly and seriously in his little place when you are there, living with them, and when you are invited here, invited there, which often means that they too are invited, so that they are constantly running around. And how agreeable can it be for him not to be alone with his young wife when he comes home after work, or to find that she has gone out? I would ask you to add up the time you have already spent with them since they have set up housekeeping. There are, chère Madame, natural vacation periods, to which even people highly favored by fortune limit themselves. . . . And finally I beg you to consider, chère Madame, that my health demands the most careful management, and that it is most upsetting to me to see you settling in Paris for what you say will be a few days, but which I know will turn into a few weeks. This puts me into a very gloomy state, which must be quite painful for you, but which nonetheless would become worse and worse. Be assured, chère Madame, of my sincere feelings of affection and respect and above all of my desire to be on good terms with you."[18] It would be difficult to be more unpleasant under a veneer of politeness.

On another occasion, Pasteur advised his son not to let himself be governed by his wife, but rather to make her his ally in his strategy of social ascent. "Be sure to work at acquiring merit," he kept telling him. "And Jeanne must constantly encourage you in this quest. The wife makes the husband. Every couple that succeeds in life is served by a warmhearted and energetic woman." It must be said that the scientist had his own concept of the services women were supposed to perform in their homes; he was unexcelled in his ability to make others feel guilty if they left him alone. To his daughter Marie-Louise, who had stayed in Paris, he wrote one day from Arbois: "I am very anxious to see you. . . . I don't even notice that my necktie is not properly tied. It is really because I want to have you here and to enjoy your presence that I ask you to come." With Marie, he was less subtle: "Oh, those shirt buttons. I am still wearing the shirt I wore on the trip. 1st shirt: top button gone. 2nd shirt: hole under the button. 3rd shirt: button sewn on so tightly that I can't grasp it. Oh! Wives! How little you know about your husbands' comforts!"[19]

On 4 November 1879, it was Marie-Louise's turn to be married. She married one of her brother's classmates, René Vallery-Radot, who at the time was beginning to make a name for himself in literature and journalism. A friend of Paul Déroulède, he had published in 1874 his recollections of military life entitled *Journal d'un volontaire d'un an au 10e de ligne*, which was pleasing to Pasteur's nationalist and revanchist bent. The scientist considered this jingoistic writer, who set out to bind all the wounds of the fatherland, an ideal son-in-law. "You brave and talented young people," he wrote to him a few weeks before the marriage, "make them triumph, these principles of yours, in your books, in the theater, in the press, wherever your action can have an effect, for it will be to you that all eyes will turn when the policies of transition will have come to an end and when France will once again be in full possession of herself. The more I think about you, the more you inspire confidence, and indeed hope in me. Because of the sincerity of your tone and the naturalness and the charm of your style, I consider you superior to these two abominable skeptics, Sainte-Beuve and Mérimée!"[20]

This last remark shows that Pasteur had finally gotten over his nostalgia for the Second Empire: calling the senators Sainte-Beuve and Mérimée "abominable skeptics" was to recognize and implicitly denounce the cynicism of Napoleon III's regime. What France needed now were rebuilders who would once again bring morality to literature and politics. Vallery-Radot would be one of them, Pasteur was convinced of it. But his esteem had to be deserved: the day when the novice journalist published a somewhat naughty article about an actress in *Le Figaro*, he was roundly reprimanded by his future father-in-law. "Is it really the author of *Volontaire d'un an* and *Souvenirs littéraires* who has signed these effeminate lines? Excuse my frankness. Real friends sometimes have to be unkind."[21]

Seeing his daughter leave on the arm of a young man, Pasteur could not help reacting . . . just like Jean-Baptiste's mother-in-law, five years earlier. "Everything went well the day of the wedding, which had attracted a large and distinguished crowd," he wrote to his faithful childhood friend Jules Vercel. "No need to add that our young couple is very happy, settled in their little apartment in the rue de Miromesnil. They wisely did not go on a trip. We see them every day, to the great pleasure of my dear wife, to whom this separation has brought great sadness."[22]

In order to alleviate this "great sadness," Pasteur did his best to win over his son-in-law once and for all. Better yet, he eventually made him his first biographer. Vallery-Radot became the anonymous author of *Histoire d'un savant par un ignorant* [History of a Scientist by an Ignorant] published in 1884, while his father-in-law was still alive, before he published his voluminous *Vie de Pasteur* (1900), a veritable hagiographic monument that was constantly republished until the Second World War.

At the time, Pasteur was fifty-seven years old, and his wife had to face the fact that there were no more children at home. Here she was, alone with a taciturn husband who came out of his own thoughts only to dictate to her scientific notes that she did not always understand. But it would never do to refuse, or to protest in any way. Pasteur had never liked to talk, and he was not about to become a chatterbox now that he was aging. He had no gift for punning or for the elegant hues of ironic society-banter, or even a sense of repartee in conversation. But he could speak from a tribune, where one blasted an adversary with the heavy guns of rhetoric.

Above all, Pasteur was incapable of not forging straight ahead, in his life as in his words, and nothing could make him change his course if he was convinced that he was right. He bluntly said what he thought, without holding back. He did, to be sure, have a very strict sense of the proprieties, but he sometimes forgot that he could not speak to an interlocutor in a social setting as he would speak to his confrères at the Academy. Marie would try to stop him, begging: "Louis, don't talk like that," but he went on, churlish and obstinate.

This was the role that fell to Marie, who was totally absorbed by day-to-day concerns, while Louis labored for posterity. It was she who protected him from all material worries, for Pasteur did not deal with money matters. Considering them secondary, he left it to Marie to receive his salary, which was brought directly to the apartment by an accountant. Every morning, she put some money in her husband's pocket; a few francs so that he could take a hackney cab if he went out alone.

The docility shown by Pasteur, who served the government as his father had served the Grande Armée was quite simply that of a zealous functionary. Even at the height of his problems with the Normaliens at the time of the "affaire Sainte-Beuve," he always showed himself respectful toward the hierarchy, accepted the strictures imposed, and did not quibble about his salary. In fact, he always sought to improve the system, trying to make it benefit from the concrete applications of his discoveries. France was indebted to him, for instance, for certain suggestions on improving the road system and for the creation of a corps of government physicians to oversee public sanitation. Because he often walked through the filthy back streets of the Latin Quarter between the Pantheon and the Institut, he understood that all efforts at improving social conditions would remain utopian, and that his own fundamental investigations would remain useless, unless the government took care of sanitation, particularly the evacuation of garbage and polluted water. Without a purification procedure, the canalization system was only a half measure which, in the final analysis, represented a serious danger to rivers and beaches. Not without reason,

Pasteur was distressed to see the Parisians consume great quantities of oysters, imagining that all of them were contaminated.

Pasteur thus found himself compelled to participate in commissions of every kind, even though deep down he was truly content only in the silence of his laboratory, where not one superfluous word was allowed. Those around him understood and respected his taste for solitude, formed barriers to protect him from unwelcome callers, and endeavored to surround his meditation with the outward quiet it required. He had to make his way by himself. He did not allow any criticism to touch him, any discouragement to last. Finding the arguments of his experiments by himself, he then discussed them only with himself, in his own mind.

Pasteur never thought about anything but working, unable to imagine that there could be anything else; he had no use for distractions or transformed them into tedious obligations. Distant but not timid or shy, he no longer felt the need to exchange ideas in order to receive suggestions. He needed no one but himself to nurture his life's work.

12

Culture Broths

Under the Second Empire, a study on the various infections treated in military hospitals during the Crimean War arrived at the following peremptory conclusion: "The specific theory of disease is the great refuge of the weak, untrained, and unstable minds who at present hold sway in the medical profession. There is no such thing as a specific disease; there are only specific conditions of disease."[1]

Twenty years later, in 1875, the situation began to change, and the medical establishment was ready to acknowledge at least an indirect relationship between disease and microorganisms. Those who, convinced by the experiments of Davaine, Koch, and Pasteur, believed that infection follows the germ, felt compelled to bend to new exigencies, namely, those of the diagnosis. However, if uncovering the presence of the microbe in the patient confirmed the bacterial theory, it also required further steps. It was no longer enough to admit the existence of the microbe; it was also necessary to characterize it. Its identification established the specificity of a given disease: henceforth every infection had to be linked to its own germ.

Yet the particular cycles of the anthrax bacillus, which can be a little rod or a spore, confirmed Pasteur in his notion that it is difficult to establish a specific nomenclature of germs. He very often stressed this, particularly in his studies on spontaneous generation or on yeast. As early as 1862, he had insisted on the need for caution on the part of investigators who ventured into these areas: "Would it not have been most inappropriate for me to assign names of genera to the microscopic organisms I encountered in my observations? Aside from the fact that this would have been very difficult for me to do, considering that there is, even today, a great deal of confusion in the designation of these tiny creatures, my work would have lost in clarity; and at the very least I would have moved away from its principal aim, which was to establish the presence or absence of life in a general sense rather than the manifestation of a specific form of life in this or that animal or plant species. I therefore systematically used the vaguest possible denominations."[2]

As his theory of disease became more precise, Pasteur realized that the analysis of germs could no longer be left aside. Since the microorganism looks identical in the blood and in the culture medium, bacteriological culturing became indispensable. The broth revealed the microbe.

The Kitchen in the Rue d'Ulm

The identification of a germ under the microscope was still a difficult and extremely delicate operation, mainly because staining techniques were not yet available. Only a highly trained eye, like Pasteur's, could attempt to isolate and recognize the different bacterial strains. In fact, Pasteur did not want to rely principally on his perspicacity. In order to identify microbes, he developed another procedure, which consisted of finding a culture medium appropriate to their selection. By manipulating the composition of this broth, he succeeded in predicting and favoring the selective growth of specific germs.

Pasteur had become aware of the primordial role of the medium in the cultivation of germs while he was working on the chemistry of fermentation, as well as through his studies on yeast. He had understood that living yeast recycles the components of the dying yeast for its own use. In the case of the production of alcohol by means of brewer's yeast, Thenard had been the first to formulate this hypothesis, although he had been stopped by an observation that did not make sense to him, for he found that the amount of yeast had diminished during the fermentation, and yet he was unable to detect the quantity of nitrogen that should have been produced by the destruction of the yeast cells. Pasteur pursued this line of experimentation and demonstrated that the nitrogen of the dead yeast cells is used by the living yeast for its own synthesis, so that it does not appear in the form of a gas but as an organic compound. If one artificially adds a simple nitrogenous substance or a complex mixture such as gluten or serum casein, the lactic yeast will use the nitrogen for its own development. The same effect is achieved by sugar, in which case the carbon is used by the new yeast cells. All that remained to be done was to find out which other compounds might be necessary for this synthesis. Pasteur discovered that mineral salts were needed. The basic requirements of the culture medium had been found.

This discovery implied that in order to multiply, the germ must be provided with the ingredients necessary for its development. But the opposite is also true, and as a result, determining what contributes to the growth of a germ could be used as a method of identifying bacteria. The culture broth soon became an essential tool for making a bacteriological diagnosis. Pasteur therefore gradually came to adapt his laboratory to these new techniques.

The locale in the rue d'Ulm had started out as a chemical laboratory; it had no facilities for studying bacteria. These had to be invented, and serious thought had to be given to the modes of sterilization, for the culture broth had to be sterile before it was seeded. It should be remembered that the autoclave was not yet in existence and was to appear only in 1884. Before Chamberland's invention, the culturing media were prepared in a vessel called a matrass, whose long neck had been drawn into a point by means of a blowpipe. The culturing media were then taken to a small annex of the laboratory, located nearby in the rue Vauquelin. Here stood large pans filled with water, to which calcium chloride was added in order to raise its boiling point above 100 degrees. After the matrasses were placed into this liquid, a gas flame was lit.

From time to time, there were explosions that caused stoppers to blow out and the boiling liquid to splash up to the ceiling. These risks were unavoidable, and it was Pasteur's good fortune that no one was ever injured in such an incident. After sterilization, the vessels were carried back to the rue d'Ulm and stored in a large cupboard, where Pasteur kept all the varieties of broth he had invented as he set out to identify different germs. Over the years, the number of these vials expanded in keeping with his preoccupations. To be found here were broths made of beef or chicken, yeast water, peptonized water, alkaline broths, neutral media, and others. In this cupboard, each category occupied a particular shelf, and the composition of the medium was indicated in blue ink on a label countersunk into the wood.

Each microorganism was given its own recipe, which Pasteur published. In this manner he described the milieu that favored the growth of both lactic and brewer's yeast, or that which enabled him to cultivate the anthrax bacterium. In the latter case, neutral urine gave excellent results, whereas meat broth suited the germ of chicken cholera. Pasteur also developed a medium for the germ of swine erysipelas, for the pneumococcus, the staphylococcus, the butyric or septic vibrio. Many others were to follow.

Whenever the program of an experiment was established, samples of the basic culture broth were prepared and assigned to each type of germ. This operation was carried out according to a rigorous technique which Pasteur had taught his students, who were not to deviate from it under any circumstances. First, the prepared flask was carried from the cupboard to the weighing room, where it was kept in readiness. The laboratory assistant tiptoed out of the room in order not to stir up the air. After two hours, a time lapse considered sufficient to calm both the air and the liquid, Pasteur was informed that all was ready. "At this point," writes his nephew Adrien Loir, "we would enter the weighing room, still without speaking and with as little movement as possible. I would sit down at the table, and Pasteur took a chair behind me, 50 cm. away from me and a little to the side, so that he could see everything I was doing. On the table stood the wire basket with which today's bacteriologists are still familiar; it was filled with quantities of pointed and sterilized tubes. I would take one of them, break its tip, flame it and place it flat onto an agitator in such a way that the wires did not touch the table. This tube was to be used as a seeding tube, for platinum wire came to the laboratory of the rue d'Ulm only around 1886."[3]

Given his phobia about airborne germs, his fear of imprecise work on the part of his collaborators, and his meticulous sense of order, Pasteur kept close check on everything and left nothing to chance. The weighing room was a sacred place exclusively devoted to seeding operations. Nobody was allowed to enter without special permission. Pasteur did admit some privileged collaborators if they were physicists or chemists, but if possible no physicians, at least until 1886.

For Pasteur was particularly distrustful of clinicians and their outdated methods. He was told how specimens were obtained in the hospital (for bacteriological analyses were beginning to be carried out there too): The resident might carry the culturing tubes in the pocket of his morning coat; he then seeded the pus and distributed it among his containers. When he came to the last tube, he took off the cotton plug and replaced it upside down in order to bring in airborne germs as a control. Hearing this, Pasteur shrugged his shoulders and exclaimed: "I'm glad that these people are not coming here!"[4]

On exceptional occasions, however, he did open his laboratory in the rue d'Ulm to outside observers, namely if he sensed that they could bring him something useful. Thus he agreed to make an exception for young Denys Cochin, the future member of the Paris municipal council who had just completed some experiments on brewer's yeast.

The composition of the culturing milieu was not the only thing to require attention. His experiments in the chemistry of fermentation had given Pasteur an understanding of the importance of the role of oxygen. When his first attempts at cultivating the septic vibrio failed, regardless of the variety of broth used, he therefore put forth the idea that this germ is an exclusively anaerobic organism that is killed by oxygen. Trials carried out in a vacuum or in the presence of carbonic gas were indeed successful, for in these conditions the germs did develop.

These experiments gave him a first idea of how an ordinary germ might become pathogenic. For the septic vibrio is present everywhere; it is found in the soil as well as in the digestive tract. If septicemia is rare, it is because oxygen prevents the proliferation of the germ. Supposing a wound in which the septic vibrio could develop, "the best way to prevent death," Pasteur stated, "would be to constantly bathe the wound with regular aerated water or to bring its surface into contact with atmospheric air. . . . But if in these conditions a single blood clot, a single fragment of dead tissue becomes lodged in a corner of the wound, where it is protected from the oxygen of the air, and if it remains there surrounded by carbonic gas, even over a small surface, the septic germs will immediately, in less than twenty-four hours, give rise to an infinite number of vibrios which, regenerating themselves by division, are capable of producing a fatal septicemia within a short time."[5]

In a few cases, Pasteur was left behind by the imagination of his disciples. Thus, while taking great care to ensure the proper composition and the sterility of the culture medium, he noticed a clever experimental trick his assistant Chamberland had developed without mentioning it to him. Charles Chamberland was working in the Vauquelin annex. One day, when his préparateur was absent, Pasteur went there with Adrien Loir and was astonished to observe the young man's actions. Loir, mechanically following Chamberland's instructions for cultivating the anthrax bacillus, simply deposited on the surface of the broth another microorganism, the *bacillus subtilis*. Pasteur frowned and returned to the rue d'Ulm deep in thought. On the way back, he stopped and, breaking his silence, spoke to the boy: "Why has Chamberland asked you to put these bacilli into the vaccine tubes?" "And then," Loir remembers, "he explained to me what he himself had understood when he saw the precaution taken by Chamberland. By spreading its own film over the surface of the liquid, the *subtilis* bacillus would absorb the oxygen that might affect the virulence of the vaccine. It was a precaution that had not occurred to Pasteur, and Chamberland had not mentioned it to him."[6]

Upon their return to the rue d'Ulm, Pasteur asked Loir to bring him the volumes of his publications. Stopping at a passage where the effect of oxygen on the *mycoderma vini* was expounded at great length, he said to his nephew: "It is all here, you know. I am doing this myself!" And then, quite astonished by what he was finding in his book, he ended by exclaiming: "The stuff that is in here!"

On the basis of these experiments with culture broths, Pasteur set out to search for new therapies. This was a matter of applying to the treatment of diseases what he had learned from the nutritional needs of germs. One day, when he had occasion to observe the treatment of a breast cancer by the application of raw meat, he said to Loir: "Get out your notebook!" and dictated the following note:

> If you break a crystal and put it back into its nutrient liquid, it will repair itself. It takes in nourishment in order to regain its original and normal form, and its general nutrition slows down. What is a cancer, and, more generally, what happens when there is a wound? Here, as in the case of the crystal, nutrition is being deviated, that is to say, the assimilable matter is used first to nourish the diseased part in order to enable it to regain its original and normal strength. . . . Should we, perhaps, attempt to nourish our cells? And if so, should we not, for example, try to place on a cancer or a wound the warm and nourishing blood of an animal, and feed that wound in this manner at regular and frequent intervals? And even, in order to prevent the branching of small vessels that form in a wound in order to bring restorative life to it, why not tie the arteries that feed these branchings, thereby limiting the supply of the body's blood to the still healthy cells of the organism?[7]

The Team

Since the concoction of culture broths required great precision, Pasteur needed préparateurs. However, their number always remained small, for the master felt that what he needed was technical assistance, not company. The laboratory attendants only carried out orders. This was the case with Jean Arconi or Eugène Viala. The latter was twelve years old when Pasteur, who knew his family, brought him from Arbois to put him to work in his laboratory. At the time, Viala could neither read nor write; Pasteur took charge of teaching the young boy and every night corrected the work he had done during the day.

The status of the préparateurs was different. They had real responsibilities, which involved not only feeding the microbes but also monitoring the course of the experiments. Pasteur trained them to obtain what he expected from them: results. As far as he was concerned, there was no such thing as an a priori impossible experiment, and if someone objected that a manipulation was too difficult, all he did was grumble: "That is your business. You will have to figure out a way to do it, just so it is done, and done well."

The préparateurs, almost all of whom were graduates of the Ecole normale supérieure, went back to teaching after their stint in the laboratory. They were selected for Pasteur by his former classmate Pierre Augustin Bertin. A professor of physics and associate director of the school, where he had replaced Pasteur, Bertin chose the candidates from among his own students, most of whom were physicists. In the evening he would go up to the Pasteurs' apartment and interrupt Marie, who, as she did every night, was reading the newspaper *Le Temps* to her husband. They would discuss current experiments and the quality of prospective préparateurs.

Jules Raulin belonged to the first generation, the one that had worked on the chemistry of fermentation. Through his work on the cultivation of a microscopic fungus, *Aspergillus niger*, Raulin made a name for himself in the history of science, where he is known as the inventor of a new culture broth. One of Pasteur's favorite students, he was to complete his career as professor of industrial chemistry at Lyon.

Jules Joubert, by contrast, belonged to the second generation, the one that became involved in medical biology. A professor of physics at one of the Parisian lycées, he came to the rue d'Ulm to test the effect of physical agents on microbes. A physicist first and foremost, he eventually ceased working with Pasteur in order to return to the study of magnetic fields and solenoids. He concluded his career as inspector in the educational system of Paris.

His replacement was Charles Chamberland, who was twenty-seven when he came to the laboratory. Pasteur liked him because of his roots in the Jura, for he was the son of a schoolteacher at Chilly-le-Vignoble near Lons-le-Saunier. Cheerful and fond of the good life, although this did not prevent him from becoming an agrégé in physics, he was of an ingenious turn of mind and loved technical innovations. We owe to him the invention of an oven for sterilizing glassware as well as of the autoclave that bears his name.

The work that came out of the laboratory at the end of the 1870s was usually signed Pasteur, Joubert, and Chamberland. But when the last was appointed associate director of the school, he could no longer be present at the laboratory as often and a new préparateur became necessary. Bertin proposed Louis Thuillier who, at Pasteur's instigation, began studying swine erysipelas. Pasteur was to have a great deal of esteem for Thuillier and was deeply affected by his death in Egypt, where he was sent to study cholera. To replace him, Bertin chose first Pierre Duhem, and then Etienne Houssay. But Pasteur paid no attention to either of them, and they left the laboratory after a few weeks. Duhem eventually became not only a great physicist but above all a historian of science and a remarkable philosopher.

Houssay became dean of the Faculty of Science. As for Perdrix, who succeeded them, he did not leave a deep mark on the rue d'Ulm.

Among all the collaborators of the laboratory, however, one must distinguish three who played a privileged role at Pasteur's side: Duclaux, Roux, and Loir.

Emile Duclaux arrived at the rue d'Ulm in 1862, which was the heyday of the quarrel over spontaneous generation. In 1866, after he had graduated from the Ecole normale, he left Paris for Tours, and then for Clermont-Ferrand, where he taught chemistry. But his successive teaching positions did not keep him from participating in Pasteur's work, assisting him in his silkworm campaigns at Alès and collaborating on the studies of wine.

In 1873, when an irascible professor of physics had slapped the face of the dean of the Faculty of Sciences of Lyon, a vacancy was declared there, and Duclaux obtained the position. He held it until 1878, at which time he returned to Paris, first to the Institute of Meteorology and then to the Faculty of Sciences, where he occupied the chair of biochemistry. Because of his interest in the fermentation of milk and in cheese production, Duclaux was a regular visitor to the rue d'Ulm. In fact, he was the real technical director of the laboratory. A very orderly man, he made sure that everything was in its proper place and that nothing unnecessary for the current experiments was cluttering the premises. The rest of the team went to him when it became necessary to make a decision about which no one dared approach Pasteur. "Pasteur listened attentively to what Duclaux had to say," recalled Loir, "and let himself be influenced by him; whenever Duclaux had come by — always very briefly — Pasteur was calm and went to work. The influence of Duclaux was very different from that exercised by Roux. Roux made him unproductive, whereas the other stimulated him to work."[8]

Roux, of course, was Emile Roux, who turned out to be Pasteur's most ambivalent disciple. In 1878, as Chamberland was about to give an injection to an animal with a syringe, Pasteur stopped him in the middle of his gesture and brusquely asked him: "Do you know how to give an injection?" Hearing his collaborator's embarrassed answer, the master hesitated for a moment and then decided: "There is a young medical student whom Duclaux has presented to me; I don't remember his name, he comes to his laboratory because he wants to take his course at the Sorbonne. He arrives around noon on his way back from the Hôtel-Dieu, where he is a laboratory assistant in Dr. Lionville's clinic. He surely must be used to handling a syringe! Ask this young man to stop by here this afternoon around five and to do this for us."[9]

This is how Roux, who regularly followed the debates in the Académie de médecine and had long admired Pasteur, came to be called to the rue d'Ulm. But when he presented himself at the laboratory, it was almost noon, and the chief had already returned to his apartment for lunch. Since Roux was unable to come back later, he gave the injection without waiting. What a beginning! Pasteur was taken aback that someone had dared do something without him, and that a physician — who was not even a Normalien — had entered his laboratory in his absence. But it soon became obvious that this young student knew what he was doing, and since it was necessary to hire a préparateur who was competent to deal with the biomedical field, Pasteur's choice fell on him.

Until 1886, Roux was the only physician admitted to the laboratory of the rue d'Ulm. He sometimes did not hesitate to oppose the master, despite his respect for hierarchy and the thirty years' age difference between them. When Pasteur had a finding in hand, had built a theory around it, and set out to expound it, Roux never failed to search for a weak spot, to the point of becoming obsessive and tiresome. "This Roux is really a pain," Pasteur groaned. "If you listened to him, he would stop you in everything you are trying to accomplish." Roux did indeed have a very strange temperament. He had understood that quiet and concentration were the key words of the laboratory, and while he often criticized, he was also perpetually vigilant to protect this quiet and to act as a filter between Pasteur and the outside world. By the same token, he was also able to keep away potential rivals and, on occasion, to get rid of them. He was, for example, most displeased with the arrival of a young resident, Robert Wurtz, son of the great chemist, who had been accepted for a training period with Pasteur. Judging him to be insufficiently motivated, Roux lost no time complaining to his master: "These doctors can't do the work, because they have too many things to do. Just give this culture to Wurtz and let him extract from it whatever he wants someplace else!" Dismissed almost on the spot, Robert Wurtz left in a hackney cab holding his culture broths on his knees, having learned the bitter lesson that one entered the laboratory of the rue d'Ulm as one entered the convent: one had to give up everything for it. Some time later, Pasteur went to the Faculty of Medicine, where he met the father of his ex-trainee. Charles Wurtz was working amidst his students in an excited and noisy atmosphere that was in striking contrast to the ambiance Pasteur maintained around himself. "How can you work amid all this commotion?" he exclaimed. "It excites ideas," the other man replied. "It would drive mine away," concluded Pasteur, who always wanted to have the last word.[10]

Later, two more physicians were to appear in Pasteur's entourage, Drs. Grancher and Straus. Using the facilities in the rue Vauquelin, they were never permitted to work in the rue d'Ulm. Joseph Grancher, who liked his comforts, had brought some good easy-chairs, among them a rocking chair, into his laboratory. Pasteur did not understand how one could worry about comfortable seats, for he himself never sat down while working. Having seen this, he shrugged his shoulders, confirmed in his idea that one could not decently allow scientists whose lifestyle was so bad to work in the laboratory of the rue d'Ulm.

As for Adrien Loir, he had the dubious privilege of being Pasteur's nephew. His father, professor of chemistry at the Faculty of Sciences at Lyon, had married Amélie Laurent, Marie's younger sister. After his baccalauréat, Adrien Loir had hoped to attend the military school of Saint-Cyr. But at the time Pasteur needed an omnipresent préparateur, on whom he could count in particular to do the things he was unable to do because of his paralyzed arm. He therefore, without consulting his brother-in-law and his sister-in-law, thought of offering his nephew a position in his laboratory. Pasteur, who in dealing with his family often acted like Napoleon with his brothers, laid down his directives and personally outlined the course of studies he wanted his new pupil to follow. He wanted him to learn glass-blowing, to assimilate the fundamentals of crystallography, and to improve his writing skills by taking a course in calligraphy. Adrien was to serve as both secretary and manipulator.

Once his training period was over, the young man moved into the rue d'Ulm, more precisely into a room above the Vauquelin annex, whose outfitting Pasteur had personally supervised. "He wished to make use of my help in such a way that he would henceforth depend only on himself to carry out the conceptions of his mind. He therefore took complete possession of me. From the first day, I became his tool, the indispensable accessory that he would use as he saw fit without encountering resistance or opposition. It was only after he had constantly had me at his side for some time that it occurred to him that I should obtain some official credentials. He first considered pharmacy, claiming that only pharmacists, like his old teacher Dumas and many others, were capable of becoming good manipulators. Then he thought of making me take courses at the veterinary school of Alfort, but this was too far from the laboratory, and he wanted me to be available all day long. And finally, there was medicine, and this is the path he chose for me."

A photograph taken by Adrien Loir may well summarize the strange relationship between uncle and nephew. Pasteur is posing solemnly, in full evening dress, wearing all his decorations. But the scientist's face is blurred, for Loir had focused the picture on the medal of the Légion d'honneur.

From the Streptococcus to the Staphylococcus

In the course of his work on anthrax Pasteur had learned that recourse to animal testing is indispensable for characterizing an experimental pathology. He therefore decided to complete his experiments by inoculating guinea pigs. This was the very first time that animals were brought to the rue d'Ulm; purchased by Chamberland along the quais of the Seine, these specimens were placed in cages in the center of the laboratory.

Chamberland had also gone to see Lauer, a major supplier of surgical instruments, to acquire a syringe, for the laboratory did not have any. As Loir tells it,

> This Pravaz syringe with a capacity of 1 cm³ had a piston made of two leather disks that were squeezed together by means of a screw located at the end of the piston's stem. These two disks had to be reversed and pushed through a nickel-plated piston that was slightly smaller than the syringe. One had to reverse these two disks with one's finger, and when one had obtained the desired shape, one had to push this piston formed by the two reversed disks through the body of the pump with one's fingernail. It was a lengthy operation, which took more than fifteen minutes. The manipulation was carried out in the presence of Pasteur, who would direct it with his eyes and his voice. He was aware that this piston was dirty, and that the syringe could not be sterilized. That is why he later often asked me to taper a glass tube into a sharp point, for he preferred inoculating the animals with this type of sharp point. The piston made from the marrow of the elder bush came into use only around 1887.

This was the point at which Roux came onto the scene and assumed his full importance. A "professional inoculator," he was the only one who knew how to inject selected germs grown in a culture broth under the shaved and carbolated skin of guinea pigs.

However, by the end of the 1870s, the hospital loomed almost as large as the laboratory in Pasteur's life. It was there that he found a virtually inexhaustible source of microbes. In particular, he studied puerperal fever, which was of course not confined to Austria-Hungary. French maternity wards suffered multiple outbreaks of this disease between 1850 and 1875, and there were catastrophic periods when close to a fourth of the young mothers died in Paris. The obstetrician Stéphane Tarnier was in the forefront of this research; he mounted a campaign aimed at avoiding the sometimes epidemic spread of the disease by using Lister's procedures as well as aseptic methods. Closing the maternity wards and sending the mothers home were no more than temporary measures that did nothing to solve the problem. Yet they had to be taken more than once.

In May 1879, the Académie de médecine placed puerperal fever on the list of its priorities. The gynecologist Hervieux read a tentative report on these epidemics: "I believe, in principle, that the lower organisms involved, be they vibrios, bacteria, rods, or moving bodies, cannot possibly account for puerperal septicemia." But then he added: "I have a terrible fear, which I cannot shake and which the Academy will understand, and that is that I will die before this vibrio has been found." Hervieux then spoke about miasmas. Pasteur was there, sitting on a bench and listening attentively. Suddenly he interrupted the speaker, without leaving his place: "What causes the epidemic is none of these things, it is the physician and his helpers who transport the microbe from a sick woman to a healthy woman."[11] Then, getting out of his seat, he went to the blackboard: "Look here," he said, "this is what it looks like," drawing for his colleagues a microbe in the form of chains, known today as the streptococcus. This sketch was not based on a guess. A few years earlier, while visiting the Pitié Hospital with the surgeon Vulpian, Pasteur had had the opportunity to examine a woman who was about to die of puerperal infection. In the blood and the pus he collected from her, Pasteur found the germ he had seen many times in various kinds of abscesses.

Unlike many others, Hervieux was not a narrow-minded mandarin; he knew how to listen and to see. The very next day he invited his fellow academician to his clinic in the Lariboisière Hospital. Accompanied by Roux and Chamberland, Pasteur went there on several occasions. The three scientists visited every ward and amphitheater with their sterile pipettes, observing, taking samples, giving advice. Pasteur developed the first bacterial cultures using blood samples—today we call them hemocultures—which he obtained by pricking the fingers of infected women with a needle. Blood, vaginal discharges, and autopsies always yielded the same streptococci when these matters were cultured in chicken broth. But Pasteur did not limit himself to isolated observations in Hervieux's wards. He also examined the maternity clinics of the Necker and Cochin hospitals. A few young residents shared their observations with him and sent him samples, asking him to examine them microscopically. In the corridors of the medical schools, people began to talk of this procedure, which allowed them to make rapid diagnoses and reliable prognoses. Thanks to the chicken broth, Pasteur was able to continue growing the germs even after the patient from whom they were taken had died. This method, he felt, was bound to succeed, and of those who did not believe in it, its inventor simply said: "Somehow I will make them go along! Whatever it takes, they will have to come to it."

Bacterial analysis was indeed immensely useful for diagnostic purposes. Thus, when in 1879 Professor Feltz of Nancy informed Pasteur that he had isolated an unknown germ in the course of a fever following on a puerperal septicemia, Pasteur pricked up his ears. He was not convinced: how could this possibly be the same disease if a different germ was present? He asked Feltz to send him specimens and easily recognized a germ in which he was an expert, the anthrax bacillus. In order to convince Feltz, Pasteur decided to give him a clinical demonstration: What could be simpler than to inject guinea pigs with three different preparations, one made from the blood of the young woman of Nancy, and the two others containing bacilli of sheep and cow anthrax, respectively, and then to ship them by railroad? On 13 May at three o'clock in the afternoon, the animals were handed to the conductor. The train stopped at Nancy station on the morning of the fourteenth; later that day the guinea pigs died in Feltz's laboratory. The autopsy was conclusive for everyone, including Feltz! It was learned later that the young victim had lived with a horse trader in a small room next to a stable; what she had died of was undeniably anthrax, and not puerperal fever.

Diagnosis had to lead to therapy. The identification of the streptococcus as the agent of puerperal fever gave Pasteur the opportunity to reiterate the rules of asepsis and antisepsis. He urged the use of boric acid: "It is indicated to keep the outer genital tract almost constantly bathed in a solution of this acid, which the mucous membranes seem to tolerate very well and which impedes the development of microorganisms. . . . At the least, the use of ordinary water for washing the genital tract must be forbidden. What could be simpler than to use only water that has been heated to 115 degrees?"[12] However, asepsis was still not well understood, as the following episode indicates. One day in 1880, Pasteur went to the clinic of his fellow academician Alfred Richet to obtain a specimen. The pipette he was using was of course carefully flamed in the alcohol burner. The next day,

Richet, who had watched the manipulation, decided to pay his dues to the new method. So he flamed his scalpel, let it cool and then, doing what he had always done, passed it twice over his dirty apron, as if to wipe it clean.

The field of investigation that Pasteur had opened was so vast that the smallest detail could become the source of advances; there are blessed periods in the history of medicine when giant steps are taken. Since Pasteur's assistant Chamberland regularly complained of suffering from boils on the front and back of his neck and on his thighs, Pasteur examined him and asked him to lend himself to an experiment, for he had a premonition that he was perhaps dealing with a new source of microbes, or rather with a source of new germs.

It was June, the weather was hot, and under his stiff collar Chamberland was perspiring abundantly. Pasteur broke off the tip of a glass pipette and asked his collaborator to bend his head forward. When a helper had gingerly collected a small amount of pus, he placed it into two culture broths, one made with chicken muscle, the other with yeast. By the next day both liquids had become cloudy and filled with germs; but their appearance differed according to their composition: in the yeast water, the microbes were dispersed, while in the chicken broth they clustered together and stuck to the walls of the receptacle. Pasteur had just revealed the existence of the staphylococcus, the most widespread of our pathogenic germs.

Since Chamberland still presented boils in the subsequent days, he was regularly asked to give additional specimens. Pasteur went so far as to prick his finger in order to obtain blood, which, however, fortunately turned out to be sterile. As Pasteur was very excited about the germ he had discovered, he talked about it to everyone. People told him about identical, and sometimes more severe cases among hospital patients. All kinds of pus were brought to his laboratory; taken from armpits, arms, and backs, from fresh or old boils, it proliferated freely in the broths stored in the cupboard. Every time it turned out to be the same roundish and aerobic, indeed air-seeking germ that had been isolated from Chamberland.

Some time later, in February 1880, the surgeon Marie Lannelongue, a man (despite this first name) who had made his reputation with his work on the bone disease osteomyelitis, asked Pasteur to observe a case at the Trousseau Hospital. Lannelongue had shown that this disease could be cured by trepanning followed by antiseptic lavage and dressings. He was going to operate on a little girl of about twelve, who was suffering from a an obviously inflammatory tumor on the right knee. If it was indeed an infection, Pasteur surely was the best specialist of his day. Having chloroformed the child, Lannelongue made a long incision above the knee. A great deal of pus spurted forth, and was collected by Pasteur, who had come with his broths. Bent over the operating table, he brought out his tubes and went to work. He collected the pus that was running out and also asked the surgeon to drill into the bone so that specimens could be taken from deep inside the bone. Upon his return to the rue d'Ulm, his pockets filled with his precious cargo, Pasteur rushed to his laboratory. A first examination under the microscope showed the pus to be but a confused and undifferentiated mass. Thereupon the liquid was placed into the incubator. A few hours later, microbes began to develop. Observing these developing germs, Pasteur realized that he was in the presence of the same type of bacteria that he had observed in Chamberland's boils. "Osteomyelitis is a boil of the bone marrow," he concluded without hesitation. Pasteur had discovered that the staphylococcus can cause not only skin conditions but also bone disease, and that osteomyelitis, whose cause had hitherto been unknown, is a staphylococcus infection.

Boils and osteomyelitis, two affections caused by one and the same germ. A truly provocative formulation, as Duclaux was to stress: "This was to assimilate a serious disease, and one that was located deep in the tissues, with a superficial and generally slight illness! It was to eliminate the difference between internal and external pathology! I imagine that when he stated this opinion before the Académie de médecine, the physicians and surgeons present must have looked at him over their eyeglasses with surprise and dismay. And yet he was right, and this assertion, though very bold at the time, was a first victory of the laboratory over the clinic."[13]

The Life of Microbes

In September 1879, Pasteur was at Arbois. Walking in the vineyards whose leaves were turning yellow, he thought about the great epidemics. In the last twenty years, his work had brought new ideas into medicine: "Take a microscopic creature living in a given area of Africa, where it exists on plants, animals, even humans, and assume that it is capable of communicating a disease to the white race. If a fortuitous circumstance brought it to Europe, it could trigger an epidemic." A century later, these words are bound to bring to mind the sudden onslaught and the ravages of AIDS.

What interested Pasteur most, however, were the great epidemics of his own time, plague, cholera, and yellow fever. In March 1879, he wrote a note on plague. The previous year, the Academy had appointed a commission to carry out a program of research after an outbreak of that disease in Russia on the left bank of the Volga, in the Astrakhan district, and Pasteur felt compelled to intervene in the debate:

> I was determined never to speak in this Academy unless I could present to it positive and proven facts. I therefore ask to be forgiven if, stimulated by the learned observations of my confrère M. Marey, I have spoken to him of things about which I am ignorant. . . . If I were called upon to study the plague in the places where it holds sway, I would begin by supposing — for at the start of every research project one must let oneself be guided by a preconceived idea — that plague is due to the presence or the development of a microphyte or a microzoan. On that assumption, I would concentrate all my efforts on culturing blood and the divers humors of the body, using blood and humors collected at the end of life or immediately after it with the aim and in the hope of isolating and purifying the infectious organism, and above all with the aim of having it available in a culturing milieu where it would have no association whatsoever with any other substance, whether known or unknown, living or dead.[14]

Following these recommendations, Alexandre Yersin, a pupil and collaborator of Roux, discovered the plague bacillus at Hong Kong fifteen years later.

In September 1881, yellow fever was threatening Bordeaux, for some cases were reported aboard a ship en route from Senegal. Pasteur, accompanied by Roux, rushed to the scene. No experiments could be carried out, for the victims had died during the crossing. The rest of the crew was in quarantine. Hearing that a second vessel, the *Richelieu*, which also carried the disease, was about to arrive, Pasteur decided to stay. He tried to dissipate his boredom at the municipal library, "surrounded by more Littrés than I would be able to consume," as he jokingly wrote to his wife. A young physician, Dr. Talmy, seized the opportunity to introduce himself and to express his desire to help with the experiments. Indeed, he was willing to let himself be confined in the isolation ward with the patients of the *Richelieu*. Negotiations with the health administration took time. Just as the authorization of the minister was obtained, the ship reached the outer harbor. All the sick had died. The body of the last of them had been thrown overboard. So Pasteur had to return to Paris, disappointed that he had not had the opportunity of bringing to light some specific germ "in the body of one of the unfortunate victims of medical ignorance," for, he concluded "a known enemy is already half disarmed."

As for cholera, with which he had already unsuccessfully dealt in 1865, Pasteur came back to it a second time, when an epidemic broke out in 1883 in Cairo. By July, five hundred persons were dying every day. Then the disease spread to Alexandria. Pasteur proposed to the consulting committee on public health to send in a French team. Three volunteers were ready to go: Emile Roux, Edmond Nocard, professor at the veterinary school of Maisons-Alfort, and a physician specializing in parasitic diseases, Isidore Straus. Louis Thuillier was also approached; he initially hesitated, but decided to go along in the end.

Writing from Arbois in August, shortly before the team's departure, Pasteur sent his last words of advice. Everything was meticulously worked out, from the hotel where they should lodge to the cook they should hire, to the details of the work and research to be done: microscopic examination of the stools and the blood of the sick, culturing in a vacuum and in open air, inoculation of guinea pigs. If there were any extra time, Pasteur suggested that they study rinderpest in the field. He added: "It is my earnest wish that during the long voyage to Egypt or during their stay, MM. Straus, Nocard, Roux, and Thuillier will all attentively read my two volumes on the diseases of the silkworm in their entirety. There must be major analogies between cholera and pébrine and flacherie."[15]

The mission was off to a poor start, and the first experiments yielded no results. Then tragedy struck: Thuillier died in Alexandria, felled by an attack of cholera. He was twenty-six years old. Robert Koch, who at the same time was directing a German team, went to his bedside and honored him at his last rites: "M. Koch was one of the pall-bearers. We have embalmed our comrade. He rests in a coffin of soldered zinc," Roux explained.

At Arbois, Pasteur was devastated. It was the first time that one of his assistants had died in the line of duty. Tortured by the memory of Thuillier's hesitations, he tried to justify himself when writing to his old master, Jean-Baptiste Dumas. "The only way to console myself about this death is to think of our dear fatherland and of what he has done for it." And, to his laboratory helper, Eugène Viala, "We must work. That is the only way to console ourselves about this great loss for science and for his friends." All day long, Pasteur kept silent. But in the evening he came out of his deep thought and exclaimed: "I just hope they have not forgotten to take a few drops of blood!" The French biologists were engaged in a race for time with their German counterparts, and Pasteur knew that there was not a moment to lose: surely, the cholera bacillus could not hide much longer. As it happened, it was isolated by Robert Koch's team before the year was out.

In a parallel endeavor, Pasteur was studying animal infections, such as chicken cholera and swine erysipelas. "Poultry yards sometimes experience a disastrous disease that is vulgarly called by the name of chicken cholera. The animal stricken by it becomes very weak and staggers about with its wings hanging down. It succumbs to irresistible somnolence. If forced to open its eyes, it seems to emerge from deep sleep, and soon its lids fall shut again. In most cases death occurs before the animal has moved, and after a silent agony."[16]

Jean-Joseph-Henri Toussaint, professor at the veterinary school of Toulouse, had tried to find in the blood of affected chickens an organism responsible for the disease. However, he was unable to cultivate it, not even in neutral urine, the medium in which the anthrax bacillus had thrived. He therefore wrote to Pasteur and sent him the head of a dead rooster. The first task was to obtain a pure culture. Pasteur went right to work. He found that a solution of brewer's yeast was as unsuitable as urine. Since this was a disease to which gallinacaea were particularly prone, Pasteur had the idea of using a broth made from chicken muscle. This almost immediately proved successful, and before long he easily obtained a culture of pure germs, which proliferated with extraordinary speed. The tiniest drop of this broth spread onto a crumb of bread killed the chicken as surely as the butcher's knife.

Infected excrement constituted the source of contagion. It was without doubt the major cause of all poultry yard epidemics. But Pasteur found one additional cause. When he inoculated the germs of chicken cholera into guinea pigs, these animals did not die, but developed abscesses at the point of inoculation. In the pus that formed there, the same germ was found. This meant that any chicken living in the company of sick but nonsymptomatic guinea pigs would be at risk for infection. This explained a great many epidemics that had been considered punishments sent by Heaven. "How many mysteries in the history of contagion will one day find an even simpler solution than the one I have spoken of here?" Pasteur wondered. Further advances in epidemiology were indeed to uncover a multitude of similar examples in which one animal species can serve as a reservoir of infection for another — as in the case of rabbits for tularemia, monkeys for yellow fever, rodents for plague or typhus.

In November 1882, Pasteur was at Bollène in the Vaucluse. As tactful as ever, he wrote to his son-in-law: "My dear René, it is quite clear that at this point I am thinking about pigs more than about any of you." The reason was that he was engaged in the study of another disease, the red disease or swine erysipelas. Here again, it was a matter of finding the cause. Pasteur and his associates employed their usual techniques: culturing in a medium favorable to the microscopic organism (in this case veal broth), serial dilution of the flask, and finally inoculation into the pig, which died with the symptoms of erysipelas. It took less than two months to identify the agent, demonstrate the mode of culturing it, and prove its virulence. But Pasteur did not stop here and continued to study this germ, which had first been described by Loir and Thuillier under the microscope.

In 1884, after his return from Egypt, Isidore Straus went to Germany, from where he brought back a new instrument, the immersion objective, and a procedure by which germs could be stained. This equipment revolutionized the method of identifying germs. Straus was working in the rue Vauquelin, and communications between the rue d'Ulm and its annex were not ideal. That is why it was Roux who brought the new procedure into the master's laboratory. But at that time the doctor was working independently, and Pasteur was not involved in his research on erysipelas. Wanting to astonish his master, Roux asked Loir for a small amount of microbial culture, stained the germs, and asked Pasteur to come and look at it.

Having observed it under different conditions and with a traditional microscope, not immersed and above all not stained, Pasteur had written that the microbe had the shape of an eight. What he now observed under Roux's microscope baffled him: "What is this rod-shaped microorganism?" Roux pointed to Loir and replied, "It is what this young man is culturing as the erysipelas agent." This was a staggering blow to Pasteur. His first idea was that his assistant had the wrong germ. He started pacing up and down in the laboratory as he always did when something important happened, moaning, "Oh Lord, oh Lord! Adrien, what have you done? Why did I trust you?"[17]

Convinced that Loir had made a mistake, he ordered all the stored stocks to be reexamined and all his notebooks to be checked to find out which of the findings might have been modified. But Loir was sure of himself and asserted that he could not possibly have carried out a faulty maneuver. Pasteur continued to doubt. He sent cables to a large number of veterinarians asking to be notified if more pigs became sick so that he could reexamine them. One reply did come from the Côtes du Nord. Loir was immediately dispatched to take specimens to be studied under the microscope. He sent several slides of blood to the rue d'Ulm. And Roux could triumphantly cable him: "Congratulations on your autopsy at four o'clock in the morning! At the time of the animal's death, the microbe was indeed the same as the one we have here. Please do several autopsies and send the material." When his nephew returned to Paris, Pasteur took him into his arms. "He rarely embraced anyone," Loir concluded, "and so this gesture showed that I had regained his confidence."[18]

The next twenty years, marked by the work of Pasteur and Koch, were to be the golden age of bacteriology. But it was not Pasteur who was to achieve the greatest discoveries. Even though he had developed various techniques for bringing all kinds of germs into evidence, he did not seek to carry out an extensive topological study. The microbes for whose identification he was responsible exclusively concerned the few diseases with which he was confronted by circumstances, by request, or by personal interest.

It was the German school, led by Koch and his disciples, which carried out the systematic studies that would lead to the discovery of most of the bacterial agents of human infectious diseases. The fact is that Pasteur was much more interested in the functioning of microbes than in their precise analysis. The culturing of bacteria and the inoculations of animals were carried out in this spirit, but microscopic examination seemed to Pasteur to be of more limited interest, despite the meticulousness he brought to his descriptions.

Yet Pasteur never failed to observe a function. One day he thus reported on a fortuitous observation and commented in a flash of visionary insight on a fact that was to have an extraordinary impact after the discovery of penicillin, namely, that the proliferation of certain bacteria can inhibit that of others. "Many microbes seem to give rise in their culturing medium to substances apt to harm their own development," Pasteur wrote. "The microbe of swine erysipelas is being cultured in a wide variety of broths. . . . It almost looks as if it very soon encountered a new substance that stops the development of this microbe."[19] This intuition was destined for a great future. It was to prepare the minds of subsequent bacteriologists and chemists for new remedies produced by the secretion of germs that inhibited microbial proliferation.

But Pasteur did not exploit this observation. And yet it could have led him to the discovery of antibiotics fifty years before Fleming.

FROM ONE ACADEMY TO ANOTHER

Seating arrangements in an academy have their importance. At the Académie de médecine, Louis Pasteur sat next to Claude Bernard, with whom he liked to chat. Emile Littré, who belonged to that body as well, was not close to him. But membership in the Académie française was a different matter. Here, in 1881, Pasteur succeeded to the seat of Littré, rather than that of Bernard, who had died in 1878.

When it came to the official eulogy of his predecessor, Pasteur did not hold back his criticism. Littré might be forgiven for having single-handedly written a dictionary that competed with and sometimes contradicted that of the Académie française. But his effort to have the positivist philosophy of Auguste Comte accepted under the Cupola was not something that Pasteur was willing to forget. So he went on the attack, but not directly, for he still had to spare the memory of Claude Bernard, who had also been a fervent positivist.

Major dilemmas were sometimes hidden behind the stately rhetoric of the eulogies for members of the Academy. For all the ostensible gentility of these interminable speeches, the arguments could be as sharp as naked swords, for the Immortals were often engaged in debating fundamental questions.

In his *Introduction à l'étude de la médecine expérimentale*, published in 1865, Claude Bernard had pleaded for a scientific medicine. "The experimental method is that scientific method which proclaims freedom of the mind and of thinking. Shaking off the yoke of philosophy and theology, it also rejects personal scientific authority. This is not a matter of pride and boastfulness; on the contrary, the experimenter expresses his humility by renouncing the claim to personal authority, for he doubts even his own knowledge and subjects human authority to that of observation and the laws of nature."[1]

This free-thinking attitude was at odds with existing ideas and theories. Bernard not only believed in yielding to doubt while nonetheless trying to establish laws; he was also faithful to the materialist principles of medicine. That is why, opposed to all conceptions tending to rehabilitate the old vitalism, he came to question the role of ferments as living beings. A staunch advocate of free thinking, Claude Bernard, along with his school, thus in effect became the ally of those who for many years impeded the fight against microbes and, in a sense, its application at the patients' bedside.

If Claude Bernard thus brought Auguste Comte's thinking into medicine, Pasteur, who opposed the very notion of doubt, went even further. Beyond advocating a scientific medicine, he instigated and established the scientific application of his medical experimentation.

Claude Bernard, the Nonconformist

Pasteur first met Claude Bernard in 1859, in connection with the Montyon Prize in experimental physiology offered by the Académie des sciences. Pasteur was the candidate, Bernard a member of the prize committee. Chosen to report on the work of his fellow scientist, Claude Bernard emphasized Pasteur's experiments, particularly those on the isomers of tartaric acid and on fermentation. In 1857 and 1858, he himself had worked on the generation of microscopic creatures, and when the debate between Pasteur and Pouchet was about to begin, his opinions were not rigidly defined. But this did not prevent him from recognizing Pasteur's originality in this area and proposing him for the prize "on the grounds of the physiological tendencies of his research."

At the time, Claude Bernard was forty-five years old. The son of a vintner in the Beaujolais region, he had spent his youth at Saint-Julien, near Villefranche, dividing his time between working in the vineyards and orchards of the Burgundian hillsides and Latin lessons from the village priest for whom he served at mass. His first contact with medicine occurred when he was working for a powerful pharmacist at Vaise, a suburb of Lyon, not as a préparateur but as a handyman, delivery boy, and floor sweeper.

As he was doing his errands, distributing medicines to veterinarians, Claude Bernard dreamed of breaking out of his mediocre condition. Tempted by literature, he wrote vaudeville shows, spent more and more time on them, read them to his friends, reaped applause . . . and got himself fired by his boss in the pharmacy. So, a five-act play under his arm, he betook himself to Paris, where he encountered the usual disappointments. Wisely deciding to forsake Aristotle for Hippocrates, he became a student at the Faculty of Medicine, where the very strict curriculum of lecture courses was not to his liking. He felt more at ease in a laboratory than at the foot of a lectern.

Claude Bernard qualified for hospital practice in 1837 and then for a residency two years later, in the same class with Alfred Richet. At the Hôtel-Dieu, he worked in the clinic of Magendie, one of the best practitioners of his day, who in medicine and physiology only trusted experience: "I have eyes," he would say, "I do not have ears."[2]

Without indulgence for mere talkers and gratuitous hypotheses, and severe with his students, Magendie noticed Bernard during a difficult dissection and took him on as his assistant for his lectures at the Collège de France. "He did not prepare for his courses," Renan was to recall later, "and allowed his audiences to see him struggle with his doubts and perplexities. Very different from those who take every precaution to avoid the embarrassment that would result from coming too close to an unfamiliar reality, he interrogated nature directly, often without knowing what the answer would be. When he ventured to predict an outcome, the experiment would sometimes say exactly the opposite. In such a case, Magendie would share in his audience's hilarity. He was delighted, for if his system, to which he was not committed, was dented by the experiment, his skepticism — to which he was indeed committed — was confirmed by it." In 1847, when Magendie, tired of teaching, left the Collège de France, he quite naturally called on Claude Bernard to become the acting holder of his chair: "If it is you, I know it will not fall to the distaff side."[3]

For Claude Bernard, this promotion opened an era of discoveries. The young physician had for some time carried out vivisection experiments in a little mezzanine shop in the cour du Commerce-Saint-André, across the street from Dr. Guillotin's house. His first trials went back to 1843, when he had studied the tympanic membrane and tested gastric juices. The path taken by his research reflected that of his thinking, and so he broke new ground in neurophysiology as well as in his work on the digestive and liver functions. By thinking of blood as an interior environment, he opened the way for medical biochemistry and endocrinology. For Bernard, a scientist more than a physician, felt that laboratory work, the foundation of experimental medicine, would turn antiquated therapeutics into sciences. If in his first course at the Collège de France he made the statement: "The scientific medicine I am supposed to teach you does not exist," it was because he himself was about to bring it into existence.

A taciturn man, Bernard never smiled, and seldom commented. He dissected rabbits and guinea pigs wearing his stovepipe hat, a gray and black scarf wound around his neck. Everyone could recognize him by his long wisps of hair and his ample white apron. He constantly moved from one laboratory to the other, conducting endless series of animal experiments at the Sorbonne, the Collège de France, and the Muséum, taking notes on his observations and writing papers of sometimes incredible length for the various academies to which he belonged. One of them came to 227 pages of printed text in two columns.

Claude Bernard's language was less authoritarian than anything that had been seen before, and his manner could not have been less pedagogical. He never advertised his achievements and, unlike many other scientists (Pasteur first and foremost) refused to become a traveling salesman for his theories and findings. On the contrary, he would talk in almost the same tone about his discoveries and his disappointments. Even though he had succeeded Flourens at the Académie française and been recognized throughout Europe as the most important physiologist of his time, Bernard was truly at ease only in his laboratory. "There," said Renan, "amidst the most repulsive sights, breathing the atmosphere of death, his hand deep in blood, he found the most intimate secrets of life. . . . His long fingers that plunged into wounds seemed to be those of an augur in Ancient Rome pursuing mysterious secrets in the entrails of his victims."[4]

Pasteur audited two of Claude Bernard's lecture courses. The first time, between March and June 1860, he followed the lectures Bernard gave at the Sorbonne to young students at the Faculty of Sciences. Pasteur missed practically none of the twenty-six sessions, and in a notebook labeled "Lectures of M. Bernard at the Faculty of Sciences" he covered ninety-six sheets with his tiny handwriting. His notes concerning "the phenomenon of the propagation and development of living beings" cite Bernard's words verbatim, but they can be considered a kind of instant review rather than as simple stenography. Indeed, the lecturer is cited in the third person, and Pasteur notes his own reactions, down to small material details. "I only heard the second part of lecture seven." Only the five lectures on the development of the embryo are not reproduced, probably because the subject was too far from Pasteur's concerns.

During the academic year 1862–63, Pasteur repeated the experience, but in a different setting. He attended Claude Bernard's course on "experimental medicine" at the Collège de France. That winter, Bernard's lectures had a more general and more philosophical character, for they constituted the first stage of his *Introduction à l'étude de la médecine expérimentale.* In his notebooks Pasteur once again recorded whole passages of what was said by Bernard, who on occasion mentioned the discoveries of a young chemist by the name of Louis Pasteur. When that happened, Pasteur, good student that he was, went right on and carefully noted the experiments carried out in the rue d'Ulm by M. Pasteur, as if he were hearing about someone else.

When Pasteur, the eternal student, returned to the benches of the amphitheater, did he realize that while he was recording the master's every word, the latter sometimes thought about things that had very little to do with the rigors of physiology? In an intimate notebook, Claude Bernard left to posterity some confidences that would have made the austere Pasteur's hair stand on end: "I must confess that I was distracted throughout that course. But that's not my fault. Seated a few tiers up was a lady who showed a charming and artistically shoed foot adorned on the left side, the side of the heart, by a bracelet with magnificent precious stones that was tied just above the malleolus. I admit that I had never seen such a thing, and that this bracelet took my breath away. My mind could not get away from it. I was wondering why it was there and not elsewhere. I could not help making all kinds of hypotheses, and meanwhile mistook an aorta for a carotid."[5]

But the relations between Pasteur and Bernard went beyond playing the game of teacher and pupil. The two men were united by mutual respect. Pasteur's senior by about ten years, the physiologist had very actively supported the chemist, particularly in 1862, when Pasteur was a candidate for the Académie des sciences. At that time, Bernard campaigned to have the Alhumbert Prize given to his young colleague in recognition of his work on so-called spontaneous generation. From then on, the two scientists met regularly at the sessions of their academies or in connection with various official functions; thus, in 1869, both Bernard and Pasteur were members of a small group of experts appointed by Napoleon III to produce a report on the state of science teaching in France.

Soon a truly personal relationship sprang up between the two scientists. They helped each other in case of illness. In 1866, Bernard contracted a case of choleriform enteritis; his physicians, Rayer and Davaine, insisted that he leave his laboratory in Paris and take a rest at Saint-Julien, where he was subjected to a regimen whose effects he was pleased to study. But he found this forced retirement difficult to bear, even if it afforded him long hours of meditation that would lead to the writing of his *Introduction à l'étude de la médecine expérimentale*. By way of encouraging his convalescent friend, Pasteur decided to show him his affection by a spectacular homage. To that effect he published in the *Moniteur* of 7 November a long article entitled "Claude Bernard, the Importance of His Writings, His Teaching, and His Method."

In preparation for this article, Pasteur had reread all the writings of the physiologist, particularly those on diabetes, and was forcefully struck by the coherence of a method he considered exemplary. "Seeing with my own eyes the unfolding of so much lasting progress, accomplished with a method of such unerring judgment that at present one could not possibly imagine a more perfect one, I constantly felt the sacred fire of science burning ever brighter in my heart. . . . The experimental demonstrations of Claude Bernard have the clarity and the rigor of those used in the sciences of physics and chemistry." Bernard was overwhelmed by these lines and hastened to thank Pasteur, whom he felt bound to compliment in return. "The admiration you profess to have for me certainly goes two ways." In another letter he wrote to Sainte-Claire Deville, he added: "I received the article Pasteur wrote about me in the *Moniteur*. This article paralyzed the vaso-motor properties of my sympathetic nerve and made me blush to the very whites of my eyes. I was so rattled that I don't even remember what I wrote to Pasteur."[6]

And of course, when Pasteur in turn was stricken, suffering a stroke in 1869, Bernard was one of the first to hasten to the patient's bedside to bring him affection and moral support. This was precisely the comfort Pasteur needed most. Accustomed to fight to have his ideas accepted, he now forced himself to smother his anxieties, to control his feelings in order to appear as sure of himself as ever. With Bernard he could for once open up and even admit his doubts. One day, as they were sitting side by side on the benches of the Académie de médecine, Pasteur looked discouraged and low-spirited. The reticence of the medical profession toward his work was becoming more and more upsetting to him. "How much of my discoveries will remain for future generations?" he wearily asked himself. Bernard knew how to revive his hopes with a few words; he quietly told him an anecdote that sounded like a parable: "Something of you will remain. This morning, my surgeon, M. Gosselin, came to probe my bladder. He brought along a young hospital physician, Guyon, who quotes you as an authority. Gosselin washed his hands after he had probed me. Guyon washed his hands before he did it."

A Posthumous Conflict

In January 1878, Claude Bernard became seriously ill. Confined to bed and looked after by his student Arsène d'Arsonval, he spoke confidentially about the courses he was to give soon at the Muséum and in particular voiced some doubts about Pasteur's theory of fermentation.

According to him, the phenomenon had nothing to do with yeast, since alcohol formed thanks to a soluble ferment; he also denied the role of the bacterium as a living organism. Bernard hinted that he had made discoveries liable to overthrow certain dogmas that had been accepted, in particular by Pasteur, in the field of the chemistry of fermentation. These experiments, he indicated to d'Arsonval, were carried out at Saint-Julien, where he had been in the autumn, at the time of the wine harvest. "It's all in my head, but I am too tired to explain it to you." But at the same time, the physiologist also let it be understood that he had preliminary findings that confirmed his hypothesis. He even hinted that he had started writing a note on the subject.

But death was to prevent Bernard from saying more about it. On 10 February 1878, he succumbed to kidney failure. His disciples, d'Arsonval first of all, but also Paul Bert and Albert Dastre, immediately thought that their first duty was to search in the files of the deceased for notes concerning his last projects. It was d'Arsonval who put his hands on these papers, carefully hidden in a desk in the deceased's room. As he read them, it soon became clear to d'Arsonval that he was not dealing with demonstrated results but only with an outline of work to be done.

What was he to do? Not pay attention to these jottings? But they constituted the master's last thoughts. Publish them? But they were only fragmentary ideas, still at the hypothetical stage. In the end, d'Arsonval and his friends opted for publication and handed the bundle to Marcelin Berthelot. Why did they make this choice? No doubt because Berthelot was one of those who did not believe in the role of living processes in fermentation. Asserting his position as one of Pasteur's most serious opponents, he declared that a distinction had to be made "between the role of microscopic creatures that secrete ferments and that of the ferments themselves." On 20 July, Marcelin Berthelot published Claude Bernard's notes in the *Revue scientifique*, accompanied by a statement to the effect that several friends and students had "thought that it would be of interest for science to preserve the traces of the last preoccupations of this great mind, in however incomplete a form they may have been left to us."[7]

This wealth of precautions only served to draw attention to this publication, which became a veritable slap in the face for Pasteur. Had Claude Bernard not been his friend? Had they not been together at the Institut, sitting side by side? How, then, could he imagine that the physiologist had been secretly trying to find arguments with which to ruin the theories of fermentation without talking about it with his colleague in chemistry? In his memoirs Adrien Loir reports his uncle's stunned reaction upon reading the *Revue scientifique*. "Pasteur heard about it one afternoon and brought the document back to the laboratory. As always when something was seriously wrong, he began to pace up and down in the laboratory, loudly lamenting: "Ah! Mon Dieu! Berthelot's publishing these notes is a disgraceful way to treat Claude Bernard. They were not intended for the public, and the man who forces me to refute them is a scoundrel." And he continued pacing, muttering further imprecations. He finally became so exasperated that he exclaimed: "Why, this man, this man, he is capable of anything! The things he is capable of, this man! He probably cheats on his wife!"[8]

Actually, Pasteur had not yet decided whether he should react publicly. If the notes published by Berthelot were only preliminary jottings, it would no doubt be better not to attach too much importance to them. After all, it was the very method of Claude Bernard to doubt everything, even the most universally accepted statements, in order not to let any error of judgment slip by. Discussing widely held theories was for him the best method to open new avenues of research. And indeed, he owed his principal discoveries in experimental physiology to this sometimes devastating skepticism. Yet, on the other hand, if Bernard had considered the hypotheses contained in his notes bona fide scientific arguments, it became necessary to treat them as such, that is to say, to test them in order to confirm or, preferably, contradict them. For Pasteur did not for a moment doubt the validity of his own conclusions and only hesitated because he felt badly about the idea of a posthumous victory over his friend and colleague. "But then," he reassured himself, "Bernard would have been the first to remind me that scientific truth soars far above the proprieties of friendship, and that it is now my duty freely to discuss his views and opinions."[9]

At the Academy, where Pasteur consulted his fellow members, opinions were divided. Some advised him not to let himself be slowed down by what they considered a return to the past; others thought that out of respect for the memory of Bernard he had to take this posthumous publication seriously, but that he should respond exclusively with facts, since the author was no longer there to debate the issue. This was precisely the point that offended Pasteur's sensitivity and his integrity. He was loath to polemicize against a dead man. By contrast, the idea of crossing swords once again with Berthelot did not displease him at all.

By the following Monday, 22 July, Pasteur had made up his mind. Having obtained the documents in their entirety, he used the ordinary session of the Académie des sciences to expose his doubts. In doing so, he suggested that Berthelot's publication was not faithful to Bernard's thinking and also announced that he would without delay start a new series of experiments on fermentation, not by way of defending his own theories, but in order to continue the work that the deceased physiologist had not had time to bring to a conclusion. The strategy is clear: it was a matter of countering Marcelin Berthelot and at the same time bringing Bernard back into his own camp.

That year, Pasteur's vacation at Arbois was therefore devoted to the experiments mentioned in the posthumous publication. He immediately ordered greenhouses for his Jura vines in order to study fermentation uncontaminated by air. More than ever sure of himself, he did have to face his adversaries' insinuations to the effect that his eagerness to perform additional experiments was proof that he doubted his own theories. "I totally reject this interpretation of my conduct," Pasteur protested in a letter to Jean-Baptiste Dumas of 4 August 1878; "and I have explained myself clearly in my first note of 22 July, where I announced that I was going to perform more experiments solely out of respect for the memory of Bernard. . . . In my opinion, M. Berthelot was quite mistaken when he attributed scientific importance to Bernard's posthumous writing and considered it a polemic against the conclusions of my work."[10]

Eventually, a series of scientific experiments put an end to these quarrels among persons. The findings obtained at Arbois in the summer of 1878 were incontrovertible: Yeast is indispensable for fermentation; unless this microorganism is brought in by the air, it is impossible to produce either wine or beer.

On 25 November, Pasteur published his conclusions in the form of an *Examen critique d'un écrit posthume de Claude Bernard sur la fermentation* [Critical Examination of a Posthumous Article by Claude Bernard on Fermentation]. "The matter of the soluble ferment has been laid to rest; it does not exist," he concluded. "Bernard fell victim to an illusion. Bernard's article amounts to no more than a fruitless attempt to replace well-established facts with the deductions of an ephemeral system. . . . This cannot diminish the glory of our illustrious confrère."[11]

Thus Pasteur found an elegant way out of this posthumous duel: he was in the clear with Claude Bernard, but not with Berthelot who, for his part, was very much alive. The polemic continued until early 1879 with rather acerbic exchanges. Berthelot pointed out that here again Pasteur limited himself to describing phenomena he had already analyzed without arriving at any new discoveries and that he refused, in particular, to consider the hypothesis of a soluble ferment. Pasteur, on the other hand, felt that he had done enough by giving up three months of vacation time in order to refute an esteemed colleague; it seemed to him that he had done his duty and performed a service that Berthelot should have performed before he published Bernard's experiments, namely, test them.

But Berthelot did not leave it at that and renewed his attack, whereupon Pasteur, citing Montaigne, called him "changeable and diverse." They exchanged invectives, they evaded the issues, they schemed and plotted, but in the end they agreed on one point: Claude Bernard had been a great man — and that was the end of the discussion. In order to have the last word, Pasteur did not fail to add: "I dare say that he had this character even before he was born."

There is something pathetic about this dialogue of the deaf. In the course of the controversy, Berthelot completely disregarded the practical aspect of the question that seemed to irritate him. He was only interested in basic science and barricaded himself behind one of Claude Bernard's celebrated formulations: "The scientist always looks for causes." Moreover, Berthelot did not hide the fact that he was not averse to reviving the controversy over certain reigning doctrines that denied the existence of spontaneous generation and advocated a return to vitalism. But he did not counter Pasteur's experiments with any plan of action of his own and contented himself with constructing abstract hypotheses which turned out to be correct!

Compared to him, Pasteur looked fastidious, touchy, rigid, and more sensitive than was reasonable. When the debate was first engaged, he had by no means won his case with his confrères at the Académie de médecine, who were becoming more and more reticent toward his microbial theories of infectious disease. The death of Claude Bernard had deprived him of valuable support, and it is understandable that he was devastated by the publication of a posthumous article that attacked him more or less directly, though on a very different terrain.

The fact is that as far as the question of the fermentations was concerned, differing interpretations of the facts had given rise to a scientific misunderstanding. Pasteur's reasoning was based on the observation of yeasts, Berthelot's on that of enzymes. Pasteur's attitude did not seem to be dictated by concern for basic research, for his work in microbiology had accustomed him to concentrate on those aspects that might be useful to industry and public health. Indeed, this was a constant feature of his scientific work; recall that under the Second Empire already, he had gladly endorsed the directives of the Minister of Public Education: "Instruction in the universities, though of course required to keep pace with the latest scientific theories, must nonetheless, in order to produce useful results and extend its beneficial influence as far as possible, take charge of most of the applications of science to the real needs of the area in which it is practiced."[12]

Science and Medicine

The lesson to be learned from this posthumous conflict ultimately had very little to do with the various theories of fermentation. Much rather, it was a fundamental debate on the experimental method. Because their fields of investigation happened to be contiguous, observers often contrast Claude Bernard and Louis Pasteur, as if called upon to define two kinds of researchers, each with his own motives, values, and reflexes. To be sure, they probably did have two distinct sets of attitudes, but one should take care not to see this conflict as a caricature of scientific research. Moreover, the respective personalities of these two scientists surely contributed to transforming their destinies into highly colored popular images: on the one hand, we have the intransigent Louis Pasteur, a good father and good husband, sure of himself, domineering, and irritating even when he was successful; on the other, there is the Romantic Claude Bernard, tormented, eternally beset by doubts, rejected by his militantly antivivisectionist wife and daughters, looking for a little comfort from his friend Marie Raffalovich. But of course it was not that simple.

It is true that Claude Bernard had developed a practice of experimentation that was completely different from that of his colleague in chemistry. As he himself observed, "M. Pasteur is led by his ideas and wants to subject the facts to them, and I seek to make ideas emerge, without doing them violence, by themselves."[13] Under the influence of Magendie, the physiologist experimented as a matter of principle, without feeling the need to know precisely and at every given moment where he was and where he was going. In his famous red notebook, Claude Bernard thus noted: "Everyone follows his own path. Some have prepared themselves all along and go forward following the furrow they have traced for themselves. But I have come to the scientific field by all kinds of byways and have freed myself of the rules by running through the fields."

From the *Introduction à l'étude de la médecine expérimentale* published in 1865 to the *Principes de médecine expérimentale* unfinished at the time of his death, the theoretical accomplishments of Claude Bernard were considerable. He was indefatigable in his endeavor to rationalize the scientific and even philosophic method. "There is no question that experimentation is more difficult in medicine than in any other science. But for that very reason it has never been as necessary, indeed indispensable, in any other science." The investigator starts with a hypothesis, but only the facts are real; and interpretations, even if one calls them laws, are never more than abstractions. In this manner Bernard postulated a science that would take the place of medicine.

Claude Bernard's principal contribution was to unify anatomy and chemistry through physiology. In this endeavor, his attitude matched the complexity of the situations he encountered, as we can see from the diversity of interpretations and the hypotheses he consigned in his red notebook. He often started with a false assumption, only to find a true fact; at times he also came so close to the truth that he could almost touch it, yet drew back at the last moment. His mind teemed with heterogeneous ideas and suggestions, reflecting a thought process in constant motion that sometimes required tentative formulations. Conceptions appear one by one, although one is not always able to apprehend just how the critical argumentation leads to the decisive conclusion. In a certain manner, Bernard's practice was thus at odds with his theory, for while he used a Romantic formulation when he used the word "leap" to define the moment when the mind soars toward the unknown, nothing in the notebooks on his experiments seems to correspond to such an occurrence, although there must have been some.

But the unique aspect of his genius was his ability to seize upon the unforeseeable. Pasteur has stressed the unexpected character of Bernard's discoveries and pointed out that some of them even surprised their discoverer. The chemist experimented in order to verify his hypotheses; the physiologist made use of chance and was able to launch himself wholeheartedly into a path he could barely discern. "Pasteur wants to lead nature," Bernard had written, "while I allow myself to be led by her; and so I follow her. The observer listens to nature, the experimenter interrogates her. We must do both. We must interrogate nature, ask questions of her, but do not think that she has to respond to your question, just listen, whatever she may say. . . . We must try to understand her, that is all. The scientist is only her secretary and does not dictate his ideas to her. I am nature's secretary. Pasteur and the a priorists want to dictate to her the answers that fit their ideas."[14] All of this can be summarized in the following formulation: Claude Bernard submitted to his experiments, Louis Pasteur was there before his.

Pasteur's modus operandi was indeed completely different. He passionately wanted to overcome all obstacles and accumulate proofs to support a theory that was quite frequently formulated before any experimentation had taken place. This no doubt was why his discoveries were not made by chance; on the contrary, they were often based on an intelligent reorientation of earlier work done by amateurs or scientists working in isolation, who did not know how to promote and defend them. This is how Pasteur reconsidered, developed, and finally appropriated the intuitions of Cagniard de La Tour in the chemistry of fermentation or those of Davaine concerning the microbial theory of disease. On other occasions, Pasteur completed existing notions by establishing revolutionary connections between hitherto separate branches of science. Berzelius and Mitscherlich had founded the chemistry of isomers, and Delafosse and Biot had codified crystallography and optics, but Pasteur was the first to make the connection between these two disciplines that led him to the brilliant and dramatic discovery of molecular dissymmetry.

In working on the vaccines, Pasteur followed the same modus operandi when he reconsidered and expanded the theories of Jenner and their applications. Pasteur's experiments were not so much a matter of inventing anything *ex nihilo* as of assembling and connecting scattered notions in ways that had not been tried or even envisaged before him. His discoveries were due to this uncompromising experimental rigor.

It stands to reason that this amounted to a full-fledged method. Pasteur's philosophy of experimentation can be summarized as a procedure judged a priori logical and then carried out by means and for the purpose of testing and verification. The Pasteurian doctrine primarily came into being as a logical sequence. This is true for each individual experiment, but the same observation can be made if one examines the entire set of research programs developed by Pasteur. In every research adventure, one finds the same attitude in the choice of themes to be pursued, the development of hypotheses, and the stating of conclusions. This is so striking that some may say that Pasteurianism is an eternal series of beginnings.

If Claude Bernard's genius expressed itself in the treatment of the unforeseen, that of Pasteur burst forth in the well-reasoned choice of scientific programs and in the application of his discoveries. His reflections on fermentation unfailingly led Pasteur into the area of pathology. He thus defeated the spontanist beliefs twice, after having twice fought the same battle, led the same crusade.

Pasteur chose to work on the intangible. He had prepared himself to understand, interpret, and discern that which cannot be seized. He began by revolutionizing chemistry when he discovered the differing spatial orientation of molecules that were rigorously identical in every other respect;

but he did not stop there and immediately foresaw the significance of his discovery for the study of the biosphere. In the same manner, his approach to the role of microbes and their reproduction shows the truly phenomenal rigor of his thinking, for although his experiments confronted Pasteur with tremendously complex phenomena, he knew how to divide them in very imaginative ways in order to interpret them. We have seen how he went about unraveling the highly confusing mysteries of the silkworm diseases or the anthrax epidemics. In every case, the experiments were guided by reasoning rather than by perceptions, which can be distorted by illusions. "I have always been afraid to attach too much importance to outward characteristics," Pasteur confided. "I have repeatedly been able to see that apparently distinct forms often belong to the same species and that similar forms may conceal profound differences."

This attitude nonetheless has its limits, as is attested precisely by the conflict that pitted Pasteur against Bernard and Berthelot on the subject of the fermentations. It is quite obvious that in this affair Pasteur was not really interested in chemistry: the idea that a soluble ferment might be responsible for the phenomenon actually seemed conceivable to him, but of minor importance, for he could only think of his own discovery in this field, namely the intervention of a microorganism. In the same manner, Pasteur did not pay sufficient attention to one of the most important concepts put forward by Claude Bernard, that of the internal environment. The notes he had taken on this subject when he audited the physiologist's course did not trigger anything in his imagination; and so he did not push his studies in this area any further, although this might have led him to discover the antibodies or the immune system in general. Yet it is conceivable that his experiments and his reflections might have led him straight in that direction.

If indeed this sometimes overly rigid thinking was a weakness, it was more than compensated for by the second aspect of Pasteur's genius, that is, his remarkable ability to see the application of his discoveries. In this sense he was well and truly the heir to Claude Bernard's physiological pragmatism. Pasteur was neither doctrinaire nor philosophical. What he wanted were immediately paying results and useful applications. His endeavors very soon brought advances to industry and agriculture and revolutionized preventive medicine. To be sure, he limited himself to the most immediate and most easily tested of his discoveries, but when all is said and done, he almost never went back to the fundamentals. There was something of Christopher Columbus in Pasteur: the same obstinacy, the longing for distant shores, the endeavor to join different universes, the exploitation of new resources. His exploration relied on very thorough planning, which enabled him to adapt to the changing circumstances of the universe he was discovering.

Pasteur's life work, whose tenets he bequeathed to his disciples, was the starting point of the new scientific medicine. Because they are adaptable, the Pasteurian methods have a universal character. But it clearly is a practical universality: Pasteur suggested systematic investigations and indicated techniques, whereas Claude Bernard left not methods but a methodology that did not even consider notions of productivity or profitability. This no doubt accounts for Pasteur's immediate impact on the course of the history of humanity.

The physiologist and the chemist were also different with respect to their disciples. Those of Bernard simply continued in his vein, which is not to say that they were without talent. But Pasteur's students became his true heirs, who carried on the master's teachings, studied his methods, laid claim to his models, and elaborated on his explanations. It was an integral part of Pasteur's scientific modus operandi to surround himself with remarkable scientific personalities and to create, beyond his own laboratory, a veritable network of intelligences, built like a very fine grid. From the Faculty of Medicine to the Collège de France, the Pasteurian principles set out to conquer science and the teaching of science. Little by little, conflicts between persons came to be forgotten, and for one Marcelin Berthelot, who continued for a long time to combat Pasteur, there were many disciples of Bernard who joined the ranks of his collaborators. It was Paul Bert, a student and confidant of Bernard, who launched the famous Pasteurian slogan: "One disease, one germ, one vaccine."

All in all, it is perhaps not useful to indulge in endless speculation about this, for to oppose the universal success of the Pasteurian practices to the experimental method of Claude Bernard is like comparing apples and oranges. The two scientists were not speaking to the same constituency, and while both went beyond the confines of their discipline, one transformed the world by transforming public health, whereas the other devoted himself to building a philosophical model.

Yet even today, this opposition reflects two directions pursued by medical research. On Pasteur's side is an exogenous concept, in which the microbe invades the body. On Bernard's side, disease is due to a disturbance of the internal environment. But we must stop opposing these two concepts, for they can coexist and complement each other. Indeed, they not only can, they must. We can no longer make an abstract distinction between external aggression and the internal environment, as has been shown by advances in immunology, discoveries about the viruses, and cancer research.

So the only unresolved question goes far beyond the Pasteur-Bernard debate and truly opens out onto an area of metaphysical anguish that precedes any scientific inquiry. Since the greatest discoveries can be fortuitous as well as planned, the question arises: what is the role of chance? Are we to make use of it or are we to pretend it does not exist?

Naming the Microbes

In the year of Claude Bernard's death, Emile Littré coined the word *microbe* to speak of Pasteur's discoveries. Until then, the "animalcula or microscopic beings" had not had a precise identity. Scientists were just beginning to recognize their actions and their functions, and it was known that they were invisible to the naked eye and that the microscope revealed their multifarious forms, but there was not yet a collective term for them, nor a way to classify them. When Littré proposed *microbe* and asked his colleague Sédillot to obtain the approval of the Académie de médecine for this neologism, he showed that he clearly understood the scientific issues of his time.

The surgeon Charles-Emile Sédillot was one of Pasteur's fervent admirers at the Académie de médecine. Professor at the Val-de-Grâce Hospital, former director of the school of health at Strasbourg, and retired since 1870, he had been named an associate member for his research on the prevention of hospital infections. At the age of sixty-four, he had been one of the first to accept and even encourage Pasteur's studies and their applications in surgery, and it was he who in March 1878 read to the Academy a paper summarizing the positive evolution of the treatment of wounds; operations, he said, now could be performed without major risk and frequently brought cures, while gangrene and amputations were on the wane. His report on the *Influence de M. Pasteur sur les progrès de la chirurgie* [Influence of M. Pasteur on the progress of surgery] concluded with an apologia of the word *microbe*: "These microscopic organisms form an entire world composed of species, families, and varieties whose history, which has barely begun to be written, is already fertile in prospects and findings of the highest importance. The names of these organisms are very numerous and will have to be defined and in part discarded. The word *microbe*, which has the advantage of being shorter and carrying a more general meaning, and of having been approved by my illustrious friend M. Littré, the most competent linguist in France, is the one we will adopt."

At the time, Maximilien-Paul-Emile Littré was not only "the most competent linguist in France," he also had legitimate claims to medical expertise, for if he had abandoned the practice of medicine for literature and lexicology, he continued to be one of the best observers of the scientific world. This double expertise made him a considerable and unanimously respected personality. Gambetta considered him the greatest mind of the century, and Renan wrote of him: "By dint of the colossal learning he had drawn from the most diverse sources and of the sagacity of his mind, Littré had in his day a more complete awareness of life than anyone else."[15]

The author of the famous *Dictionnaire* was born in 1801 into a republican, even Jacobin family. His father, a naval officer during the Revolution, had converted to belles-lettres and learned Greek and Hebrew. He christened his son Maximilien to honor Robespierre, and Paul-Emile in memory of the Roman Republic, but history retained only the name Emile, like a nod to Rousseau. At a very young age, Littré showed an extraordinary curiosity. Wanting to know everything, he began with medicine. Was not France at the time the country of Bichat and of Laënnec, who had discovered auscultation? Soon the young man was admitted to a hospital internship, only to realize that there was a terrible inconsistency between the prestige of French medicine and its day-to-day practice. He was downcast by the existing gaps in teaching and above all by the lack of rational thinking. He felt that medicine, constantly threatened by lurking charlatanism, was still encumbered by notions of the supernatural. Refusing to go along with that game, Littré did not open a practice, did not even defend a thesis and, encouraged by his friend from the Ecole normale, Louis Hachette, who had just founded a publishing house, let himself be tempted by the literary life.

But although he gave up its practice, he did not give up medicine altogether. On the contrary, his scholarly work had the ambition to give it the intellectual framework it was lacking. In this aim he founded, with Charles Bouillaud, the *Journal hebdomadaire de médecine* in 1827 and then, in 1837, the scientific journal *L'Expérience*; he also translated Hippocrates for the publishing house of Baillière and reedited, with Charles Robin, Pierre Nysten's famous *Dictionnaire de médecine*. Parallel to these efforts, Littré wrote numerous articles. In one of these he actually suggested, long before Pasteur, the possibility that contaminating agents might be responsible for cholera epidemics. But he always stopped short of practical applications:

"The goal of science," he proclaimed, "is the truth; the useful comes only later. We must not turn a radiant deity into a humble servant girl."[16]

At the age of forty, having refused a chair in the history of medicine, Littré launched his life's great adventure, the writing of an etymological dictionary of the French language, to whose publication Hachette had committed himself. It was a huge undertaking. The authoritative dictionary of the time, that of the Académie française, did not cite any texts; it was but a skeleton, as Voltaire had already scoffed a century earlier. Now Littré was determined to establish the precise use of each word by citing a few lines from the best authors. This demanded thousands of hours of reading and note taking, not to mention the chore of copy editing! This is what accounts for the extraordinary length of this enterprise; the first sections came out in 1863, and the whole work was not completed until 1873.

The publication of such a *summa* immediately brought fame and glory to Littré and naturally made him appear most competent to name the microscopic organisms Pasteur had detected under his microscope. "To designate the animalcula," he replied to Sédillot, who had consulted him on this question, "I would give preference to 'microbe'; first, as you say, because it is shorter, and then because it keeps the feminine noun *'microbie,'* to designate the state of being microbic." Noting that the word *microbie* could also, given its Greek etymology, mean "short-lived," he added, "This is not truly a Greek expression, but reflects the use that our scientific language makes of Greek roots. . . . My feeling is that it is best not to respond to criticism and to let the word defend itself, which it will no doubt do."[17]

Strangely enough, Littré subsequently forgot his sponsorship. In the *Suppléments au Dictionnaire de la langue française* which he published beginning in 1878, neither *microbe* nor *microbie* are mentioned, only the "microscopic living beings that one only sees in the microscope, also called infusoria or microzoans." Yet the word *microbe* did circulate in the scientific world and gradually infiltrated the columns of the general press, and the expression "microbial or Pasteurian doctrine" was heard with increasing frequency.

In the 1880s, the debate assumed a larger scope. It was now becoming necessary not only to find a name for the microorganisms, but also to define a new type of research and its applications. In 1889, Pasteur asked the linguist Michel Bréal to defend the word *microbie*, which was in competition with the term *bacteriology* coined by the Germans. The French origin of *microbe* was born in France, which was already one sufficient reason to prefer *microbie*; moreover, *bacteriology* seemed overly restrictive, which is why Pasteur insisted and repeated again and again that he would "prefer in the future to use the word *microbie*, which is very general and lends itself to the divisions and subdivisions of the new science with its multiple perspectives, rather than '*bactériologie*,' which is too specific and will lend itself less and less to future discoveries." Scientific usage finally settled the question when it agreed with Pasteur's objections but rejected the neologism he defended. Today we speak of "microbiology."

These semantic as well as epistemological concerns attest to the fact that in the last years of the nineteenth century not only revolutionary practices but also a new medical language came into being. Along with medicine in the narrow sense, new biological disciplines were being born. Pasteur, Sédillot, and all those who were convinced of the need to recognize the patient as an object of science were also persuaded that a simple word can be a weapon in the struggle for progress. The word *microbe* was to make its way around the world, and with it went Pasteur's theories.

Positivism and Free Thought

However, despite their shared views on the microbe, Pasteur and Littré were not made to get along. Great scholars are supposed to respect one another, but they are under no obligation to paper over their differences: Littré was a republican and a Freemason, everything that Pasteur detested most. Above all, Littré was a positivist—and that was even less forgivable.

In that same year, 1878, the Académie française had included the neologism *positivisme* in its dictionary, calling it "a philosophy related to the system of Auguste Comte." Comte, who had died in 1857 after having worked as an assistant teacher at the Ecole polytechnique (even though Arago said of him "He has neither major nor minor mathematical qualifications"), had developed a doctrine which he published piecemeal beginning in 1828. Inherited from Hume, d'Alembert, and Condorcet, and influenced by Saint-Simon and Fourier, positivism set out to reform both society and knowledge, for it wanted to establish politics as a social science by way of encyclopedic and historical investigation. In the manner of most of the contemporary utopias, Comte wanted to create a new social system based on a "religion of humanity," in which everything became a science. He thus considered Descartes's analytical geometry the model of a positive science, since it combined geometrical intuition with algebraic reasoning, a concept with an abstraction. But this ambition to advance science by means of philosophy, or rather to urge the adoption of a scientifically demonstrated philosophy, was rather alarming to scholars, who were accustomed to working in their laboratories without being disturbed. "Auguste Comte's fundamental principle," declared Pasteur, "is to dismiss any metaphysical search for first and final causes, to reduce all ideas and all theories to facts, and to attribute the character of certainty only to demonstrations of experience."[18]

One of Comte's principal shortcomings was the lack of precision and clarity in stating his doctrine. He needed a spokesman to express on his behalf this new view of the world, a person who was knowledgeable about social and political problems as well as about astronomy and biology. It was Emile Littré who took on this task of popularization, for when he discovered the *Cours de philosophie positive*, he felt that here was the unifying system of thought that would allow him to develop his erudition in all directions without scattering his efforts. Auguste Comte therefore became his mentor, and Littré, without fear of being embarrassed by sometimes doubtful assertions, without questioning certain obscure precepts or criticizing anecdotal examples, embraced the positivist religion.

For the followers of Auguste Comte, the scientific quest had no limits. Their desire for unity led them to associate reason and passion, science and feeling. In a series of articles he published in *Le National* in 1844, Littré summarized his convictions as follows: "Philosophy is constituted by all the sciences that purvey knowledge of all existing things."[19]

To publish such a manifesto in a daily newspaper was to introduce an immense public to positivism. But Comte, who suffered from a personality disorder and often showed himself to be arrogant and domineering, was not really satisfied by this turn of events; he preferred to set up his philosophy like a religion, to gather his disciples into a sect, for which he established positivist rites of marriage and sponsorship. Littré therefore gradually distanced himself from the "high priest." The final break came in 1852 when Comte published his *Catéchisme positiviste. Sommaire Exposition de la religion universelle en onze entretiens systématiques entre une femme et un prêtre de l'humanité* [Positivist Catechism. Summary Exposition of the Universal Religion in Eleven Systematic Conversations between a Woman and a Priest of Humanity].

However, Littré separated from Comte, not from positivism. On the contrary, if he judged the master to be unbalanced, he was unshakably loyal to his strict and pure principles. Rigorous to the point of refusing to acknowledge any but scientific knowledge, Littré did not allow for any deviations. True, he left everyone free to imagine the absolute and even allowed his wife to hang a crucifix over the marital bed, but, as far as he was concerned, the idea of God had no scientific foundation. As for the original creation, he did not deny that it might have had a first cause, but, not having found anything that proved it, he did not believe in it either. This meant that positivism could only lead to agnosticism.

If Littré's attitude was not lacking in rigor and even courage, the philosophy he preached nonetheless eventually degenerated into scientism and attracted totalitarian minds who blindly sought to impose their own system. In the name of positivism, the disciples of Claude Bernard and Emile Littré endeavored to put into place a scientific system of morality and politics and asked science to furnish a scale of values which it is altogether incapable of producing. Hence a sometimes unthinking confidence in progress and technology, hence also a radical anticlericalism, both of which have caused some later observers to say that the nineteenth century was "stupid."

Pasteur was prompted to react against positivism both by his innermost convictions and by his day-to-day research. There is no doubt that in his mind Littré constituted a target that was all the more provocative for being seemingly out of reach. For the lexicographer's reputation was unblemished, and indeed the Third Republic had made him into a kind of model of the "secular saint." Yet the scientist of the rue d'Ulm considered this existence perfectly normal and did not think it was an excuse for faulty reasoning. Pasteur wanted to show that the positivists had not understood the meaning of the scientific enterprise. For him it was an exploration, and he believed that one must look for the unknown in what is possible rather than in what has been. That is the experimental method:

> The error of Auguste Comte and M. Littré is to confuse this method with the restricted method of observation. Both unacquainted with experimentation, they use the word *expérience* [which can mean either "experience" or "experiment"] in the sense in which it is used in ordinary conversation, where it has a completely different meaning than in scientific language. In the first case, an "experience" is no more than the simple observation of things and an induction that infers, more or less legitimately, from what has been to what might be. The true "experimental" method must go all the way to incontrovertible proof. . . . In order to judge the value of positivism, my first thought was to look for inventiveness. I did not find it. . . . Since positivism does not offer me a single new idea, it leaves me reticent and distrustful.

Yet even while refuting the positivist philosophy, Pasteur sometimes echoed Auguste Comte. Thus he penned an apologia of experimentation in these unambiguous words: "Experimental science is essentially positivist in the sense that in its conceptions it never concerns itself with the essence of things, the origin of the world or its final destiny."[20] In the same manner, when developing his ideas on the complexity of the organic realm, on the opposition between life and death, on the role of scientific knowledge or the history of science, he seems to have been directly inspired by the very theories he claimed to be fighting. But this is a false controversy; all that needs to be said is that Pasteur was a man of his century and therefore an heir to the Enlightenment. Like most of his contemporaries, friends or adversaries, whether they be Auguste Comte, Claude Bernard, or Emile Littré, he asserted the preeminence of hypotheses, warned against religious or metaphysical prejudices, and willingly abandoned theories that were outdated or useless to industry.

The deeper reasons that set Pasteur against the positivist philosophy were more specific, both in form and in essence. Comte rejected the use of the microscope because it showed dissociated and scattered elements; he also established a hierarchy of the sciences and separated chemistry and biology. Obviously, these were heresies in Pasteur's eyes. More serious still were their disagreements with respect to physiology: Comte, following the naturalist Blainville, assigned special importance to anatomical and topographic information, whereas Pasteur insisted on the insufficiency of morphological research and could never muster an interest in the concept of "internal environment" launched by Claude Bernard.

But even this was not the essential. It was the question of the existence of God that constituted the irreducible fault line that separated Pasteur from positivism. He therefore stated:

> The great and visible gap in the system is the fact that the positivist conception of the world does not deal with the most important positive notion of all, that of the infinite. Beyond that starry vault, what is there? More starry skies, granted. But beyond that? The human mind, impelled by an invisible force, will never cease asking: 'What is there beyond?' . . . It is pointless to reply: beyond there are unlimited spaces, times, and magnitudes. No one will understand these words. Whenever anyone proclaims the existence of the infinite—and no one can escape it—he fills that assertion with more of the supernatural than there is in all the miracles of all religions; for the notion of the infinite has the double characteristic of being inevitable as well as incomprehensible. When this notion takes over our understanding, we can only bow down before it. . . . This notion of the infinite in the world, I see

its inevitable expression everywhere. Because of it, the notion of the supernatural is embedded in every heart. The idea of God is a form of the idea of
the infinite. As long as the mystery of the infinite weighs on human thinking,
temples will be erected for the cult of the infinite, whether the God is called
Brahma, Allah, Jehovah, or Jesus. And on the tiled floors of these temples you
will see humans kneeling, prostrate, deeply immersed in thoughts of the
infinite. . . . Where are the true sources of human dignity, of liberty and
modern democracy, if not in the notion of the infinite before which all men
are equal?[21]

Does this mean that Pasteur, as Renan put it, tried to "carve himself a
God out of the infinite?" Should we see this apologia of the infinite as a
remnant of the spirituality of the Comtois? No doubt. However, if Pasteur
was a man of faith, he also felt that humankind had access to a different
infinite, no longer that of pain-filled contemplation, but that of action, the
only source of great joy: "The Greeks understood the mysterious power of
these hidden things when they created the word *enthusiasm*, literally the
God within."

There can be no doubt that it was to this "daemon of work" that
Pasteur vowed a lifelong cult, and his wife Marie sometimes complained
that she did not see him at mass often enough. In fact, even before the
separation of church and state, Pasteur advocated separation of science and
religion: "In each one of us there are two men, the scientist and the man of
faith or of doubt. These two spheres are separate, and woe to those who
want to make them encroach upon one another in the present state of our
knowledge!" With these words, the man of science simply meant to establish the dividing line between the field of experiments, which was his own,
and that of metaphysics, where he had no particular authority. This was in
itself a form of positivism, but it did not prevent Pasteur from committing
himself to the struggle against free thought.

In 1874, presiding over the award ceremony at the Collège of Arbois,
he clearly stated his position:

I know that the word *free thinker* is written somewhere within our walls as a
challenge and an affront. Do you know what most of the free thinkers want?
Some want the freedom not to think at all and to be fettered by ignorance;
others want the freedom to think badly; and others still, the freedom to be
dominated by what is suggested to them by instinct and to despise all authority and all tradition. Freedom of thought in the Cartesian sense, freedom to
work hard, freedom to pursue research, the right to arrive at such truth as is

accessible to evidence and to conform one's conduct to these exigencies — oh! let us vow a cult to this freedom; for this is what has created modern society in its highest and most fruitful aspects.[22]

Toward the end of his life Pasteur, without disavowing any of his convictions, was to give up these flights of rhetoric that combined polemic with lyricism. He finally understood the futility of these duels between "priest-ridden yes-men" and "wild-eyed priest haters": to be a genius meant knowing how to listen to the silence.

But the time to keep silent and meditate had not yet come. For now he had to fight, and the tiers of the Académie de médecine were among the usual battlefields in that incessant struggle between clericalism and free thinking. It was probably within this chamber that Pasteur's spirituality was most seriously contested. No one had forgotten the quarrel over the so-called spontaneous generations, and the Voltaireans continued to reproach the scientist for having combated a thesis that had made it easy to sidestep the question as to the origin of life and the creation of the universe; indeed, they had still not given up their error. "Spontaneous generation is no longer a hypothesis, it is a philosophical necessity," Pierre Larousse wrote as late as 1874. "It is the only way to explain the creator and the creation and, together with the mutability of organic forms, seems to constitute the two poles that anchor the very axis of life."[23]

On the benches of this auditorium Pasteur thus encountered five of the greatest names in positivism: Claude Bernard, Emile Littré, and Marcelin Berthelot of course, but also Paul Broca, the famous neurologist to whom we owe the first anatomical description of the brain, and Charles Robin, a collaborator of Littré who coined the words *hématies* and *leucocytes* to designate the red and white blood corpuscles. The minutes of the sessions show that after the death of Claude Bernard these scientists withdrew their support from Pasteur. On the occasion of his famous paper of 1880, in which Pasteur described the staphylococcus of the boil as also responsible for osteomyelitis, not one of them spoke to underline the importance of this discovery. To be sure, Littré had no oratorical talent whatsoever and never participated in any discussion, but he might have written one of those articles that carried enormous weight. He did nothing of the sort, and Broca and Robin as well kept quiet.

When it was no longer possible to ignore the involvement of philosophy or religion in the scientific debate, Pasteur openly had to confront the free thinkers, some of whom were even more stubborn than he, and the star performers of that movement were not interested in supporting him in his struggle to advance public health or microbiology. This neutral, not to say hostile, attitude only served to slow down improvements in patient care, and it is no coincidence that a free thinker like Armand Després was militantly active both against the implementation of antisepsis and for the laicization of the hospitals. Yet on the other hand, Pasteur was encouraged by confrères like Stanislas Verneuil or Jean-Baptiste Bouillaud, whose reactionary convictions were known to everyone. Bouillaud, for instance, was not only a practicing Catholic but also a Bonapartist: born in 1796, he had joined the army in 1815 in order to follow Napoleon from the Island of Elba all the way to Waterloo.

In Pasteur's entourage there were also young surgeons who made no secret of their religious faith, men like Octave Terrillon, Charles Monod, or Edouard Quenu. The most famous among them, Just Lucas-Championnière, one of the first to bring Pasteurian theories into the hospital, was the grandson of a leader of the counterrevolution in the Vendée — a fact that his adversaries used to accuse him of Catholic obscurantism.

Fortunately the debate did not pit two monolithic blocs against each other, and here again one must be careful not to draw a farcical picture of the situation. The Catholic Paul-Emile Chauffard, for instance, had logically supported Pasteur's position against spontaneous generation, yet he also refused to accept that microbes can play a role in the appearance of infectious diseases. Thus it was not a matter of spiritualists who were right on one side and free thinkers who were wrong on the other. We must understand that in order to impose his ideas and find ways to put them into practice, Pasteur needed to surround himself with real and objective allies. It was for this reason that he came to combat positivism and agnosticism when they stood in his way.

But today it no longer makes sense to pursue this polemic, as was done until very recently by certain historians of science, who thought and wrote that Pasteur was a genius, though a believer. In retrospect, what is most upsetting about these contentions is not the debate between two ways of seeing the world, which is respectable enough, but the idea that the fate of the sick is often determined by the clash between two dogmatisms.

Immortal!

At the Académie française free thinking was much less acceptable than at the Académie de médecine. In 1863, while the first fascicles of his dictionary were coming out, Littré presented himself as candidate for Biot's seat. Bishops' crosiers were raised on all sides, and the cabal was led by Mgr. Dupanloup, who orchestrated the election of Comte de Carné, a dyed-in-the-wool Orleanist. Eight years later, in 1871, the same scenario was enacted, except that the Second Empire had just collapsed. Thanks to the support of Thiers and Hugo, who had returned from exile, Littré succeeded Abel Villemain, former Minister of Public Education, but the election took place in so tense a climate that the ever watchful Dupanloup resigned from the Academy when he learned of the election result.

Littré was to occupy his seat for ten years; he died on 2 June 1881 at the age of eighty. His death rekindled the polemics, and there were those who expressed astonishment that this hero of free thinking had been commemorated in a religious ceremony, while others whispered that Madame Littré had her husband baptized on his deathbed. In *Le Figaro* Emile Zola painted a portrait of this "secular man" by means of a bold parallel between Hugo, "a kindly old man in his dotage" (but still alive!) and Littré, an "impeccable logician." "I am haunted by these two men. One is my youth, the other my maturity. They affront and exclude one another. Littré has taken over all the space of my reason. Hugo is no longer anything to my ear but a far-away music. Positivism has put in place the gravestone under which Romanticism will sleep forever."

To Pasteur, this contrast between Romanticism and Positivism did not mean anything. Yet Littré's death did not leave him indifferent. The reason was that he had already been approached; it had been suggested that if his candidacy came at the right time, it had a good chance of succeeding. The old lady of the Quai de Conti, as the Académie française was sometimes called, was not intimidated by men of science among the literary lights and was glad to receive them, just as she received the star performers of politics, the bar, or the pulpit. To be sure, Pasteur was already a member of many academies and learned societies, but his ambition could not resist pursuing what many considered the supreme honor: to be one of the Forty was not a small thing!

Marie Pasteur watched all this, ostensibly with a resigned eye. On 4 June 1881, while her husband was busy with his experiments, she wrote to her daughter: "The day when our 25 sheep died so prettily, we also learned of the death of M. Littré, and I saw appearing over the horizon a certain speech that would willy-nilly have to be given. People in high places have already spoken about this to the interested party, who does not seem to be too alarmed by it."[24]

Alarmed? Not in the least. Except perhaps as a matter of coquetry, for Pasteur was resolutely ready: "I will not be able to avoid a candidacy for the Académie française," he explained to his daughter the next day. "I will do what I am told and will be the first to console myself if by some combination of circumstances I should be eliminated. What alarms me is the trouble this is causing and the loss of time involved in writing an acceptance speech."[25]

Pasteur was obviously flattered and did not for a moment consider turning down the solicitations on the grounds that he was a stranger to literary discussions. The only thing that truly worried him was writing a good speech, for he had to eulogize Littré. What a subject! "I feel that I still have too much to do to fight for the advancement of science and tomorrow's discoveries," he confessed to a friend, "not to begrudge the time I shall have to spend preparing a speech on positivism and 'the deepest thinker of the century,' as was said by Gambetta, who can't possibly believe a word of this — it would just be too stupid."[26]

Clearly, Pasteur did not for a moment doubt the outcome of the balloting. If he presented himself, he would win, whoever the other candidates were. Yet the names that were whispered were not those of nonentities: Ferdinand de Lesseps, the man of the Suez Canal; the poet Sully Prudhomme, author of the famous *Vase brisé*, who in 1901 would become the first recipient of the Nobel Prize for literature; or Paul de Saint-Victor, a brilliant journalist, friend of the Goncourt brothers and a protégé of Victor Hugo. For its part, Pasteur's candidacy was presented by Ernest Legouvé, a successful playwright, particularly thanks to his *Adrienne Lecouvreur*. Legouvé took it upon himself to smooth the way for Pasteur and to discourage the most obstreperous adversaries: "A man like you cannot fail," he wrote to Pasteur, "I would even say, cannot fight, for he will find all doors open. The extra-academic, or if you will, supra-academic aspect of your candidacy will necessarily ensure your success."[27] It should be made clear that Legouvé was the grand-uncle of René Vallery-Radot, and that he had even been the best man at Radot's marriage to Marie-Louise Pasteur — this certainly facilitated things!

On 27 June 1881 Pasteur submitted to the perpetual secretary of the Académie française, Camille Doucet, his official candidacy for Littré's seat. The election was to take place in December.

As is usual, the Forty were a curious cocktail of personalities at the time. The oldest members, or "Burgraves," were François Mignet, a liberal historian, Désiré Nisard, a former colleague of Pasteur at the Ecole normale supérieure, and of course Victor Hugo, hale and hearty at almost eighty.

They were the last representatives of the generation of 1830. But among the ranks of writers, both novelists and dramatists, their succession was assured, for here were Emile Augier, the happy author of *Le Gendre de M. Poirier*; Victorien Sardou, who had become known for his *La Famille Benoiton* and was destined to experience the triumphs of *Madame Sans-Gêne* and *La Tosca*; Octave Feuillet, whose sentimental novels were a reaction against the crudeness of the naturalists; Eugène Labiche, who, a member since 1880, was working on his *Voyage de M. Perrichon*; Alexandre Dumas fils, elected in 1874 with the support of Hugo, who is supposed to have said: "Since I did not vote for the father, I wanted at least to vote for the son!"

Also seated under the Cupola were philosophers and historians, foremost among them Ernest Renan, Hyppolyte Taine, and Henri Martin. But the most important influence was wielded by the politicians, and the Academy was an Orleanist stronghold. The duc d'Aumale, a son of Louis-Philippe who in 1843 had defeated the retinue of Abd el-Kader, had been elected in 1871; this was quite a symbol in itself. He was surrounded by a remarkable lineup of former ministers, such as the duc de Broglie, the comte de Falloux, the duc de Noailles, Emile Olivier. As for the scientists, after the death of Claude Bernard their only representative in that illustrious body was Jean-Baptiste Dumas. In his eighties, he was the picture of health—a fact that incidentally discouraged Pasteur from waiting for his demise to try for a seat!

Custom prescribes that every candidate go through the ritual of visiting every member, and Pasteur did not intend to make an exception. Dumas fils would have liked to dispense him: "I forbid him to call on me," he protested. "I will go to see him to thank him for joining us!" Not everyone agreed; Victor Hugo for instance did not like to have his hand forced; when the scientist called on him to present his respects, he grumbled: "What would you say if I tried to get elected to the Académie des sciences?"

Nonetheless, the campaign for the Academy was well run, and potential rivals were discouraged. On 8 December 1881, Pasteur was elected by twenty out of thirty-three votes cast. His most serious competitor, the novelist Victor Cherbuliez, an imitator of George Sand, obtained only eight votes.

But the most difficult part, at least in the mind of the member-elect, was yet to come, namely the speech honoring his predecessor. The intellectual antipathy Pasteur had felt for Littré was compounded by a deep and irreconcilable political disagreement, for the lexicographer was a sworn enemy of Napoleon III; during the Empire he had refused to pass in front of his statues, and in 1870 he was one of the first to call for his deposition.

Amid all these Orleanists and (a few) republicans, Pasteur meant to show his Bonapartist loyalties.

This combination of difficulties no doubt accounts for the rather long interval between his election and his induction, which was more than four months. Preparations took up the entire winter of 1881–82. Every night, as soon as he left his laboratory to go upstairs to his home, Pasteur closed his door and polished his sentences. He knew that not everything can be said under the Cupola, and that underhanded hints and deadly praise must be used. But at the same time he wanted to be fair. If positivism seemed to him to be a dangerous error that had to be opposed, this should not prevent him from recognizing either the personal qualities of his adversary or the validity of his work. He therefore went to Mesnil-le-Roi near Maisons-Lafitte to meet Littré's widow and his daughter and to visit the austere dwelling and the small garden where the "secular saint" had grown his vegetables. There Pasteur tried to understand; this universe steeped in purity and simplicity touched him, but it made him feel even more strongly that the philosopher had gone astray when from the depth of this retreat he had denied the absolute.

In the course of writing his speech, Pasteur came to realize that the exigencies of literature are not very different from those of science. The more rigor one deploys in approaching the truth, the more one becomes aware that reality is complex, elusive, and sometimes discouraging. The scientist's personal notebooks allow us to follow his hesitations and his second thoughts step by step, and one can reconstruct at least three or four successive versions of the speech, which Pasteur dictated to Marie and subsequently corrected. In a first version, he evoked his own work and spoke about it with a modesty that sounds false: "Once again the emotions of my early days arise in me, and I am seized by a sense of inadequacy." He subsequently tried several times to improve this formulation, but this sentence was clearly important to him, so that until the end he persistently and obstinately evoked "the sense of what [he] was lacking." These rhetorical acrobatics[28] no doubt had their hidden usefulness, for they allowed Pasteur to distance himself frankly from his predecessor, as we can see from a very virulent but finally discarded sentence: "I am called upon to trace for you the life of a free thinker whose independence consisted of his self-imposed obligation not to think at all about certain subjects." A great opening for a eulogy!

Fortunately, Pasteur was not without support. However much he wanted to be the only master of his laboratory, when it came to questions concerning the Academy, he was quite willing to submit to more expert advice than his own. He therefore frequently went to consult his two sponsors, Jean-Baptiste Dumas and Désiré Nisard, whose differing judgments reflected their very different ways of life: Dumas was a socialite who lived luxuriously in a townhouse in the rue Saint-Dominique, while Nisard lived in isolation in a modest apartment in the rue de Tournon. Dumas urged the member-elect to present the indictment of positivism in the name of science and the quest for the absolute, while Nisard, with the same cautiousness he had shown when he and Pasteur directed the Ecole normale, scrupulously examined his sentences one by one as an old professor obsessed by the improper use of words or stylistic awkwardness.

On Thursday 27 April 1882, there was high excitement under the Cupola. Receiving Pasteur was not exactly a routine matter. Not all the Forty were present: Hugo was (diplomatically?) ill and had sent his excuses. But his absence was compensated for by the much noted presence of Princess Mathilde, who had come to applaud her faithful courtier. And of course the whole Pasteur family was there, seated in the first row: Marie, Jean-Baptiste and his wife, Marie-Louise and René Vallery-Radot.

At two o'clock in the afternoon, the new member of the Academy arrived, escorted by his two sponsors. Pasteur was in his sixtieth year, but his hair was still black, only the beard white. In his green and gold frock-coat decorated with the sash of the Légion d'honneur, he was an imposing figure. "The most striking thing about him," wrote a journalist, "is the characteristic energy. His features are strongly marked, his eyes are lively, his body is robust. His masculine and clipped speech reveals a man for whom there is no such thing as obstacles or fatigue." The reporter of L'Illustration, for his part, stressed the scientist's origins in the Franche-Comté and recalled the names of his compatriots who had been equally "bull-headed": "Born at Dole, Louis Pasteur belongs to the strong race that in less than one century has given France Victor Hugo, Charles Nodier, Fourier, Prudhon, and Courbet." Others stressed the fact that on the very day when Pasteur succeeded Littré at the Académie française, Darwin, who had died April 12, was buried with great pomp in Westminster Abbey. They stressed the fact that it was Littré who had popularized Darwin's theories on the evolution of species in France, to the point that certain cartoons depicted him as a monkey perched in a tree. Was it a hint of providence that Pasteur's consecration and the ultimate homage to Darwin should coincide? Was this the triumph of the moral order?

The academician-elect did not want to see it this way. The occasion did not call for polemics, but for gratitude, although this was no reason to abandon one's positions. Pasteur spoke standing, in a firm voice, without the slightest hesitation. When he was interrupted several times — by applause — he cast impatient looks at those who dared cut him off. He began by speaking of himself, but only — and this was almost customary — by way of protesting his unworthiness. "I would be embarrassed to be standing here if it were not my duty to pass on to Science itself the impersonal, as it were, honor you have bestowed upon me."[29] In the course of his speech, the evocation of his own research in biology allowed him to approach the subjects that mattered to him: "By proving that to this day life has never shown itself to man as a product of the forces that govern matter, I was able to serve the spiritualist doctrine which, forsaken elsewhere, was at least certain to find a glorious refuge in your ranks." Spiritualism versus Positivism — Pasteur was launched! All he had to do now was to give the "eulogy" of Littré, which he did by refuting one by one all the arguments of Auguste Comte and his principal disciple. In particular, he lashed out against the positivist claim that it was possible to find a scientific foundation for politics — in other words, Pasteur rejected sociology. "The work that M. Littré published in 1879 under the title *Conservation, révolution et positivisme* is filled with the misunderstandings into which positivism had led him. Why should this surprise us? Politics and sociology are sciences in which proof is too difficult to establish. Entirely too large a number of factors must be considered to solve the problems with which they concern themselves. Wherever human passions are involved, the area of the unforeseen is immense."[30] This criticism was firm but not negative. By insisting on man's need to search for the absolute, and by acknowledging the role of the unforeseen, Pasteur rejected in advance every kind of determinism and central planning; like an early Karl Popper, he asserted himself as an irreducible foe of all forms of totalitarianism.

The rules of the Institut prescribed that newcomers be welcomed by the acting director. In April 1882, the name of the director of the Académie française was Ernest Renan. No one was better qualified both to welcome Littré's successor and to respond to Pasteur. The author of the *Life of Jesus* was visibly pleased to arbitrate this high-level debate which, beyond the grave, pitted the most sparkling minds against each other.

Both Renan's work and his personality lay exactly at the pivotal point between spiritualism and positivism. A Breton brought up in the strictest Catholicism, young Renan was studying for the priesthood when at the age of twenty-two he lost his faith in the Church because he discovered the need to subject Christianity and its canonic texts to scientific critique. He there-fore left the seminary of Saint-Sulpice in order to devote himself to philol-ogy, archeology, and comparative theology; in short to all the disciplines that would allow him to discover the unchallengeable foundations of reli-gion. Converted to positivism by his friend Marcelin Berthelot, and a col-league of Claude Bernard at the Collège de France, where he taught He-brew, Ernest Renan became the quasi-official spokesman for skepticism, although he continued to call himself an "ardent believer." It fell to this complex man, who was convinced that good qualities are less useful than shortcomings and who, though placid in appearance, was tortured by the quest for truth, to respond to Pasteur.

With his long hair falling to his collar, his heavy and closely shaven face, and his little paunch, Renan spoke sitting down. "I beg you to be indulgent, Monsieur, toward intellectual endeavors in which one does not have, it is true, the experimental tools that you know how to wield so marvelously, but which nonetheless can create certainty and yield important findings. Allow me to remind you of your great discovery of right- and left-handed acid. In the intellectual realm as well, there are different directions, apparent oppositions that do not exclude a fundamental similarity. There are minds that are impossible to bring together, just as, to use the com-parison that is dear to you, it is impossible to fit one glove into the other. And yet these gloves are equally necessary; they complete one another. Our two hands cannot be superimposed, but they can be joined."[31]

And finally, pointing to the tiers of benches filled with his colleagues in their green frock coats, he concluded: "When you see all the good things that can be learned from seemingly frivolous literature, you will come to think that the discreet doubt, the smile, and the sense of *finesse* of which Pascal speaks also have their value. In this house you will not have to conduct any experiments; but this modest observation, which seems so outrageous to you, will be sufficient to give you some hours of gentle pleasure. We will impart to you our hesitations, you will impart to us your certainty. Above all, you will bring to us your glory, your genius, the fame of your discoveries. Monsieur, we welcome you!"[32]

14

The Wager of Pouilly-le-Fort

As far as Pasteur was concerned, disease was more than a field of observation. It was also a field of battle. He had understood the terrible consequences of microbial infections, but he was not a physician and his knowledge of physiology and pharmacology was limited. In his opinion, prevention was the principal therapeutic measure, and his imagination did not go beyond the arsenal of antiseptic drugs. However, he did realize that certain individuals, once they had had a disease, became resistant to a new onslaught. He had thought long and hard about Jenner's vaccination and been astounded that medical practitioners made so little use of it.

The Académie de médecine began by attempting to differentiate the agent of the vaccine from that of smallpox — which shows that at the time the principle of Jenner's method was not yet understood. Yet the Academy officially recommended vaccination, even though the failure of certain inoculants had given rise to doubt and concern.

Pasteur was obsessed by this complex of questions. It must be possible, he thought, to find a mode of universal protection against infections. Roux relates that when he came to work in the laboratory, Pasteur told him: "We must immunize against the infectious diseases whose viruses we cultivate." In the nineteenth century the medical meaning of the word *immunity* was not yet the dominant one, and the expression "morbid immunity" was used to distinguish it from parliamentary or ecclesiastical immunity. Haunted by this idea, Pasteur engaged in pioneering work, setting up long series of experiments designed to penetrate the secret of how vaccination worked. After many attempts that led nowhere, it was his research on chicken cholera that put Pasteur on the right track. Ever since his controversies with Gabriel Colin, Pasteur had been interested in epizootics that decimated poultry yards, and he was still working in this area with Roux and Chamberland.

It is certain that the fundamental discoveries were made in the second half of 1879, but the accounts of the witnesses differ rather strikingly. Duclaux asserts that in the absence of Pasteur, who spent the summer in Arbois, the culturing of microbes was interrupted, to be resumed at the beginning of the new academic year. But when the work was taken up once again, it was noticed that the bacteria that had been set aside no longer imparted the disease. Yet in an attempt to enhance their virulence, they had been cultured in chicken broth, but this did not make any difference: when inoculated into healthy animals, the germs produced no effect. Rather than kill these chickens under the assumption that they would be useless for further experiments, Pasteur had the idea of injecting them with a new dose of cultured material, this time consisting of young and active microbes whose virulence was not in doubt. He now noticed that the animals exposed earlier to the aged bacteria resisted again, whereas the healthy chickens perished. Another version of the same facts, reconstituted on the basis of Pasteur's archives, gives less credit to chance and makes Roux, more than his chief, the hero. In Pasteur's absence the young physician, convinced that it was possible to alter a microbe's virulence, had the idea of passing a current of oxygen over a culture, and then to compare the effect of the microbe thus modified.

However that may be, when he learned what had been observed in his laboratory, Pasteur was immediately aware that it could be put to very good use. What had been reproduced was Jenner's modus operandi, but it had been done in a manner that gave hope not only of understanding the mechanism of immunity, but above all of reproducing it, that is, of developing a process for producing vaccines. These two ambitions were obviously related, but Pasteur, with his characteristic sagacity, decided to treat them separately and to begin with the second, for which he was better equipped owing to his genius for experimentation.

In February 1880, he communicated to the Académie de médecine one of his most famous notes, *Sur les maladies virulentes et en particulier sur la maladie appellée vulgairement choléra des poules* [On the Virulent Diseases, in Particular the Disease Popularly Called Chicken Cholera], in which, without revealing his procedures, he envisaged producing vaccines. "It would seem superfluous for me to point out the principal consequences of the facts I have had the honor of presenting to this academy. There are two, however, which may be worth mentioning in particular. They are, on the one hand, the hope of obtaining artificial cultures of all viruses; and on the other, an idea of how to find virus-vaccines of the virulent diseases, which have brought and still bring so much desolation to humanity."[1]

Such a declaration did not go unnoticed. Pasteur's adversaries, led by Jules Guérin, criticized him for not communicating the details of his technique and saw to it that in its official report of 27 July, the Academy censored his conduct. Pasteur was both furious and depressed, for this affair broke shortly after his sister Virginie had died at Arbois. He composed several drafts of a letter of resignation. "I have no regrets about separating from this company," he confided to Nisard. "Individual passions take up too much room there." But he was soon persuaded to reconsider his decision, on the condition that Guérin cease his maneuvers against him. "Recalling the many marks of benevolent attention my work has received at the Académie, and also considering the regret I should feel if I left this illustrious company, I am willing to retract my resignation if the Academy decides, by a public vote, that the passage incriminated in my absence by M. Jules Guérin and transcribed by me in this letter, is scientifically absolutely irreproachable."[2]

In point of fact, Jules Guérin's claim was not unjustified, but it was premature, for it was only in August 1883 that Pasteur was in a position to reveal details of his experimental procedure, since by that time he had worked out an entire program that explained attenuated virulence and indicated the means to obtain it. A committed vitalist, Pasteur essentially presented himself as a defender of the microbial theory of disease. He used it to explain pathological mechanisms (the germ as cause of disease) and also to define treatments: measures designed to defeat infection, he felt, would serve to prevent it. Then, in about 1880, he began to combine the antiseptic techniques of the chemist with an authentic biological reagent. Attenuated virulence, which does not confer the disease, became a means of stimulating the body's natural resistance and enhancing its anti-infectious defenses. Pasteur's studies thus marked the birth of a new discipline, whose development he foresaw, although he did not have enough time left to participate fully in it, namely immunology.

When he used the word *vaccinate* to designate the inoculation of germs with attenuated virulence, Pasteur clearly showed what he owed to his predecessors, above all Jenner, as he acknowledged in a speech he gave in London in 1881: "I have given to the term *vaccination* an extension which science, I hope, will adopt as an homage to the merits and the immense services rendered by one of the greatest men England has produced, your own Jenner."[3] This did not in any way diminish the merit of Pasteur, who, thanks to the development of microbiology since 1870, succeeded in replacing the empirical use of vaccines with a rigorous technique. If the Pasteurian vaccination was very different, both in its principles and its theory, from Jenner's use of the cow's udder, it was once again because he was able to explain and test biological experiments. What Pasteur claimed to have discovered was not vaccination but the vaccine.

Previous Vaccinations

In 1880, vaccination was certainly not a novelty. As early as 1801, Jean-Louis Moreau de la Sarthe, "Doctor of Medicine, Assistant Librarian of the School of Medicine of Paris, Professor of Hygiene at the Republican Lycée, Member of the Commission for Vaccination of the Louvre, the Society of Medicine, the Medical Emulation Society, the Philomatic Society, and the Society for the Observation of Man,"[4] pompously stated in a popular best-seller: "Vaccination is the equivalent of the gesture that conferred invulnerability on Achilles when, held by one heel, he was plunged into the River Styx." Despite all his professional titles, this worthy man had resorted to hackneyed mythological metaphors to speak about vaccination: as to all his contemporaries, the principle remained a mystery to him, and he trusted that the progress of Enlightenment would disseminate and bring about the triumph of the Jennerian method. Moreau de la Sarthe spoke as a philosopher rather than as a scientist. After him, the positivists of the nineteenth century did not do much better; they would try at most to take credit for the medical successes of vaccination, but were never able to propose a real interpretation.

Rudimentary forms of vaccination seem to have been practiced since time immemorial. As early as the tenth century, the Chinese performed inoculations that protected against smallpox by placing small pieces of the crusts of smallpox pustules into the nostrils of healthy people. This technique, which seems to have been imported from India, apparently called for a very precise application: the pustule was taken only at certain stages of its development, and the pus was kept warm, sometimes in the armpit, before being used for collective inoculations. No doubt, this was already a (highly empirical) method of attenuating the germs.

At the time, smallpox was one of the most dreaded diseases. It was most probably brought to Europe by the crusaders in the course of the twelfth century. The Spanish, who had already been exposed to it at the time of the Saracen conquest, were among the most severely affected and contributed further to the dissemination of the disease when they in turn contaminated the Incas and eventually all the Indians of North America. At the time, the notion of crimes against humanity was not even imaginable, and the conquerors had few qualms about using the pustule as a biological weapon to wipe out indigenous villages.

The more a disease is feared, the more it appeals to the irrational. In the eighteenth century, all of Europe lavished passionate interest and feverish hope on the different hypotheses that hinted at the possibility of defeating infection by smallpox. It was Lady Mary Wortley Montagu who, at the English court, launched the fashion of using inoculations, which rapidly spread throughout Europe. The wife of the British ambassador to the Ottoman Empire, Lady Mary brought back to London procedures used in the harems of Constantinople and also widely disseminated among the Turkish population. The method was simple and consisted of communicating a slight case of the disease at an early age in order to prevent a more serious illness later. Groups of elderly matrons who specialized in this technique inoculated smallpox pus they had carefully taken from pustules and preserved in a nutshell. The inoculation was done by pricking the patient with the tip of a needle that had been dipped into the pus. "Taking smallpox is considered a kind of entertainment here, as taking the waters is in other countries," Lady Mary wrote to an English friend in 1717.

These letters were published at the end of the century and caused a great stir at the time when Edward Jenner, the son and brother of pastors, born in 1749, was also working on inoculation and proposed a different procedure that he had adapted from an English agricultural custom. Jenner had begun his career as a student with the famous John Hunter at St. George's Hospital in London. But his lively and curious mind gave him a passion for natural history, and so he classified the collections Captain Cook had brought back from the South Pacific in 1791. On his return to his native Berkeley, he pursued his observations and became interested in all kinds of zoological questions: the hibernation of hedgehogs and bats, the life of earthworms, the nesting of cuckoos. The distance between nature and poetry often being small, Jenner also wrote bucolic odes, such as the "Ode to a Robin, Who Announces the Coming of Rain."

Crisscrossing the countryside indefatigably and in every weather, Jenner was above all a dedicated doctor. He liked visiting farms and stables, where he would chat with the farmers. This is how he learned about a detail known to every farmer in England: humans who had contracted cowpox were resistant to smallpox. Jenner summarized his observations as follows: "In this dairy country a great number of cows are kept, and the office of milking is performed indiscriminately by men and maid servants. One of the former having been appointed to apply dressings to the heels of a horse affected with the Grease, and not paying due attention to cleanliness, incautiously bears his part in milking the cows, with some particles of the infectious matter adhering to his fingers. When this is the case, it commonly happens that a disease is communicated to the cows, and from the cows to the dairymaids, which spreads through the farm until most of the cattle and domestics feel its unpleasant consequences. This disease has obtained the name of the Cow Pox. . . . [B]ut what renders the Cow-pox virus so extremely singular is that the person who has been thus affected is forever after secure from the infection of the Small Pox; neither exposure to the variolous effluvia, nor the insertion of the matter into the skin, producing this distemper."[5]

Wishing to give a scientific basis to this simple description, Jenner began his first series of experiments in May 1796. But what he attempted to achieve was not exactly what he had seen, for he hoped to move beyond it. Rather than material taken from an animal, a cow or a horse, he used matter taken from the hand of an infected woman, Sarah Nelmes, to inoculate a child of eight, James Phipps: "I found it difficult to persuade myself that this subject would really be safe from receiving an infection of smallpox; however, when he was inoculated, a few months later, with the smallpox leaven [*levain varioleux*], he proved completely resistant to this counter-test."[6] Jenner proceeded in this manner because he thought that the cows contracted the cowpox through the hands of cowherds suffering from smallpox. According to him, smallpox was somehow transformed into cowpox in the body of the cow. In order to eliminate the obligatory use of the animal for obtaining cowpox material, he had the idea of inoculating from one arm to another, hoping that these successive passages would not increase the virulence of the vaccine material.

Despite these promising experiments, Jenner was rejected by his peers when he attempted to draw conclusions and propose a prophylaxis. His first paper, submitted to the Royal Society in 1797, was turned down. He was told that his studies, in addition to being incomplete, contained too many hypotheses and could only harm his reputation. In such circumstances the same criticism is always trotted out: apparent disorder, lack of rigor, and so forth. Jenner therefore had to rework his paper; he developed his arguments, discussed them as he had been asked to do, and produced precisely the same findings in a pamphlet he published at his own expense in 1798 under the title *An Inquiry into the Causes and Effects of Variolae Vaccinae, a Disease Discovered in Some of the Western Counties of England, Particularly Gloucestershire, and Known by the Name of "Cow Pox."*

This publication was received with considerable misgivings. If the physicians were for the most part ready to accept the idea of variolization, whose effectiveness had been demonstrated by Lady Mary, they were extremely reserved about the idea of a scarification in connection with an animal pustule. They found it intolerable to consider inoculating human beings with the impure humors of inferior creatures, such as bovines or horses. A few of the more open-minded physicians pointed out that, considering that the cow provides some of our most healthful foodstuffs, milk and meat, there is no reason why it should not also bring us the means of healing: "Many medications are provided by animals. We feel no qualms about taking them, because they help us. Why, then, should we refuse vaccine?"[7] The fact is that, more than these debates about principles, the criticism centered on the complicated nature of the procedure. For the technique did spread rather quickly, and in the process some real shortcomings became apparent. Most notable were sometimes severe ulcerations at the point of inoculation and ineffectiveness in certain patients.

In the end, however, the interest of Jenner's discovery was widely recognized. Congratulated by the court and admitted to high society, he became a member of several learned societies and twice, in 1802 and 1806, received major subsidies in view of his finally acknowledged merits.

As one would expect, antivariolous vaccination soon crossed the Channel and was practiced outside of England. Napoleon subjected his army to it in 1805, and in 1809 even promulgated a decree endorsing the new method. In 1811 his son, the King of Rome, was officially vaccinated upon his orders. "The imperial example did more for the vaccine than any law," said Corvisart.

By 1820, thanks to the efforts of the Académie de médecine, vaccination had become a routine practice. However, many people still did not benefit from the practice because of ignorance, fear, or administrative negligence. Organizational problems and especially the collection of the vaccine material were indeed the greatest difficulties. Transmission from arm to arm was at first recommended as the simplest and most effective method. But although it was asserted that the pustule transmits only the smallpox, it was noticed that this method also favored other infections, among them syphilis. In the face of this danger, the Academy decided to go back to vaccine material derived from animals. It now became necessary to appoint commissions that went as far as Italy to locate vaccinia-bearing heifers.

At this time Ernest Chambon, the self-styled "apostle and disseminator of animal vaccine," set up the first vaccination center in Paris's rue Massillon. He was called the "cow man" by the street urchins of Paris, who saw him leading his heifer to various hospitals until the day when, tired of moving about, he created the first *Institut de vaccine* in the rue Ballue. In 1868, a bill proposing the generalized use of glycerine-based vaccine matter was introduced in Parliament. But the law was not passed because the process did not seem reliable. It therefore remained difficult to obtain large quantities of vaccine material. Moreover, the Académie de médecine resumed its discussions when certain theoretical supporters of vaccination, such as Léon Le Fort, refused to make it compulsory in practice because they felt that isolating smallpox victims was more effective. In France the law instituting compulsory smallpox vaccination was not passed until 1902 and not enacted until 1907.

Actually, smallpox is not the only infectious disease that past centuries have tried to prevent in this manner. As early as 1772 Sir Everard Home proposed a similar method of combating measles in his *Principles of Medicine*. "I felt that I would render mankind a great service if I succeeded in making [this disease] more gentle and less dangerous by the same means that the Turks have found of making smallpox less cruel."

As for the Jennerian principle, it had been so well disseminated that long before Pasteur certain scientists had attempted to generalize it, notably by testing its effectiveness in anthrax. A veterinarian of Lyon, Chauvais, remarked that the sheep of Algeria were strangely resistant to inoculation and noted that an animal that had lived through a first infection remains resistant to all further ones. "Preventive inoculations affect the humors," he concluded.

The most important precursor in this area remains Auzias Turenne, who published theoretical writings rather than reports on experiments. A pupil of Geoffroy Saint-Hilaire, this naturalist never ceased to advocate vaccination as a prophylactic measure, long before Pasteur, Davaine, and Koch developed their germ theories. Eight years after his death, which occurred in 1870, his friends collected his research papers, lectures, and publications in a thick volume entitled *La Syphilisation*, for Auzias Turenne had been obsessed with the idea of inoculating human beings with syphilis in order to protect them against it. Such a suggestion was bound to be met with derision, but Pasteur had found entirely new ideas in this volume, as we are told by Loir: "After a dinner at M. d'Andecy's, I talked to Pasteur about Dr. Auzias Turenne, since my host was his testamentary executor and had published his work. In fact, I brought him that big volume with the compliments of M. d'Andecy. Having looked it over, the master told me to ask M. d'Andecy to come to see him. They closeted themselves for more than two hours. Pasteur kept this book at home, not in the laboratory. It was in a special nearby drawer in his desk; he often read in it, and continued to do so for several years. Many ideas were suggested to him by this reading. He copied sentences from it. He never talked about this book, *La Syphilisation*, because people would have made fun of him."[8]

Auzias Turenne's propositions were based on a hypothesis: the soft chancre is to syphilis what vaccine is to smallpox. Having inoculated a monkey with syphilis, thereby performing one of the first experimental infections, Auzias Turenne noticed that if the animal did not succumb, it would never again become infected. He therefore proposed to induce the same refractory state in man by inoculating him with the serosities of the soft chancre. But in the process, Auzias Turenne did more than plagiarize Jenner; by taking on one of the most dreaded diseases of his time, he also produced interesting and rich hypotheses concerning the future of infectious disease.

Long before Pasteur had established the role of germs in contagion, Auzias Turenne already envisaged the role of the viruses, recognizing their capacity for attenuation and hence their ability to confer immunity. He also discovered the existence and the importance of infected individuals who appear to be healthy, the so-called healthy carriers. "If we wanted to persist in seeing syphilis as a divine punishment," he asserted, "this punishment would mainly be inflicted on the innocent and their descendants. . . . One and the same virus can present a variety of modalities, that is, diverse forms or different states of being. . . . Viruses regenerate themselves and become stronger; they also degenerate and become weaker; and this double reciprocal property is fundamental."[9]

Denouncing the medical theories that postulated the spontaneous generation of diseases, Auzias Turenne thundered against the limitations of scientific knowledge. At the same time as Davaine, if not before him, he stated the hypothesis of an infectious origin of anthrax and, above all, proposed an explanation for immunity: "The viruses exhaust the organism when they themselves become exhausted; they destroy it or leave it when it no longer furnishes nourishment. . . . The true physician must know how to use all antagonisms and all resistances. To divide in order to rule is a formula that can be applied to the prophylaxis and the therapeutics of the virulent diseases."[10] Several of these assertions foreshadow Pasteur's research.

Creating Immunity

When analyzing the immunity conferred by the attenuated chicken cholera virus, Pasteur immediately understood that this situation was analogous to Jennerian vaccination. But to him the most important fact was that he was able to manipulate the virulence of the germs, and hence to produce vaccine substances. For the discovery, fortuitous or not, that resistance to chicken cholera was induced by aged bacteriological cultures allowed him to envisage a rigorous experimental study of the process. Assisted by Chamberland and Roux, Pasteur now conceived a series of experiments guided by one reflection, namely that Jenner had transmitted the vaccine from arm to arm; by one observation, namely that cultures aged in air become attenuated; and one technique, namely culturing the germ in successive passages through chicken broth.

In the course of the successive seedings, the germ habitually preserved its virulence, for the experimenter wielded the pipette deftly and reseeded each sample without delay. And indeed, the observation of aged cholera germs showed the importance of the time factor: those that had been made to wait too long lost some of their virulence. Pasteur therefore carried out the passages from one culture broth to the next more slowly in order to demonstrate experimentally what he had imagined, namely that by lengthening the interval between seedings, one reduces the microbe's infectious power. "As this interval becomes greater, one sometimes seems to apprehend certain apparently minor signs suggesting a kind of weakening of the inoculated virus. . . . The different degrees of virulence in successive stages, which hitherto had not been noticeable or at best doubtful, are now translated into considerable effects. . . . In other words, a simple change in the mode of culturing the parasite, nothing more than the lengthening of the intervals between seedings, has given us a method of obtaining progressively lower degrees of virulence, leading to a true vaccinal virus that does not kill, imparts a mild illness, and protects against the deadly disease."[11]

Still, these experiments only indicated a procedure to follow; they did not provide an explanation. Pasteur had sufficient experience to know what kept his germs alive: "The growth of the parasite necessarily takes place in contact with air, because our virus is an aerobic creature, which cannot develop without air. It is therefore only natural to wonder at the outset whether the quality of virulence is not related to contact with oxygen in the air." This question could be answered by experimentation. Pasteur now began to culture the germ in glass tubes sealed with the enameler's lamp containing varying amounts of broth and, hence, oxygen. As the germs multiplied, they depleted the oxygen in the tubes. If kept away from air once the oxygen was exhausted, the microbes no longer multiplied and seemed to be static; however, they did keep their virulence. If the tube was opened five, six, seven, as much as ten months later, the bacteria still killed the animal. But if the same cultures were left in contact with the air, they lost some of their virulence. "The question that occupies us is resolved: it is the oxygen of the air that weakens and extinguishes virulence."[12]

Chicken cholera, which was not a major issue for veterinary medicine, became one for science. But Pasteur, who at the same time was investigating the ravages of anthrax in livestock, was eager to move on to greater things. To develop a vaccine against anthrax: what better use could he make of his discovery? He had barely asked that question before he found a partial answer. In a note of July 1880, Pasteur reported that some sheep that had been pastured above an anthrax-infested burial site had survived inoculation with a virulent culture of anthrax, while another group of sheep, which had remained in the stable, had died. Recalling that chickens who fed on foodstuffs tainted by cholera sometimes developed immunity, Pasteur now wondered whether the surviving sheep of this last experiment did not owe their immunity to an invisible and as yet unidentified infection. Creating an immunity against anthrax was thus a definite possibility; the thing was to find the vaccine.

At this point there was a dramatic turn of events: In the first days of August 1880, Jean-Joseph Henri Toussaint announced that he had succeeded in vaccinating against anthrax. Who was this Toussaint? The son of a lowly carpenter in the Vosges, he had studied at the veterinary school of Lyon, but this soon seemed insufficient. Having become doctor of science and then doctor of medicine, he had attracted attention to the point of being appointed at the age of thirty to the chair of physiology at the veterinary school of Toulouse. Toussaint followed Pasteur's work with the greatest attention, for he dreamed of surpassing his elder. He was encouraged in this ambition by Henri Bouley, who in 1878 asked him to conduct a study of anthrax. That year, between 26 August and 22 September, Toussaint worked in the Beauce, not far from Pasteur and his students. The two teams were in competition, but they did not ignore each other. In fact Pasteur, always sure of himself, did not hesitate to tell his young colleague about some of the conclusions he had drawn from the autopsies of cows that had died of the black disease. Toussaint returned from his mission with a report that appeared shortly after that of the team of the rue d'Ulm and did nothing more than confirm the master's hypotheses.

After this first contact with Pasteur's work, Toussaint pursued his research on anthrax and in particular his search for a vaccine for that disease. On 12 July 1880, he submitted to the Académie de médecine and to the Académie des sciences a sealed envelope, announcing that he had developed a method of vaccinating against anthrax. This, of course, caused an uproar. Some — Colin first of all — were incensed that a young whippersnapper should permit himself to proceed in this manner, for which even Pasteur had recently been criticized. To announce a finding without divulging how it had been arrived at! It just wasn't done. The Academy was not just a desk with secret drawers where one could deposit sealed letters. Toussaint was therefore forced to give in and to reveal his procedure.

This was in August, and Pasteur was vacationing in Arbois. There the content of Toussaint's note was communicated to him by Bouley. Toussaint had vaccinated with blood heated to 55 degrees centigrade, a temperature which, according to him, was sufficient to kill the bacteria; he imagined that the vaccine was related to a "blood charged with unknown principles which must have been formed while the bacterium was alive."

Pasteur replied immediately: "I am in astonishment and admiration before M. Toussaint's discovery; in admiration that it was made and in astonishment that it could be made," he wrote to Bouley. "It flies in the face of all the ideas I had formed on viruses, vaccines, etc. Nothing makes sense to me any longer. Ten times a day I thought about taking the train to Paris. I will not believe this astounding fact until I have seen it with my own eyes, although the observations by which it was established seem unimpeachable. . . . I am too upset to write a longer letter. I have dreamed about this, sleeping and waking, all night long."[13] The fact is that Pasteur was not convinced, for he could not believe that it was possible to vaccinate with killed germs, and even less with bacterial products. However, history was to prove him wrong a few years later, when Roux and Behring showed that it is indeed possible to vaccinate with bacterial toxins. Eventually Roux's pupil Gaston Ramon even developed toxins made harmless by formol, the antitoxins.

Reading about Toussaint's findings, Pasteur could not sit still. He had to go and check them out on the spot. There was no point in continuing this useless vacation. From Arbois Pasteur sent Chamberland detailed and peremptory instructions, asking him to buy some sheep and to reproduce Toussaint's experiment. It soon turned out that the young veterinarian had probably been too quick to shout victory: Pasteur and his team noticed that heat did not kill the anthrax bacteria but only put them to sleep. It did not take them long to recover their virulence. Even treated with carbolic acid, as Toussaint recommended at the time, the vaccine became pathogenic again within a few days. Having been attenuated at Toulouse, the anthrax preparation no longer gave protection by the time it reached Maisons-Alfort; the time spent in transit restored the bacteria to full strength. Raising the temperature was no help; the bacterium had lost its ability to vaccinate.

Pasteur was furious at having gone to this trouble to invalidate a poorly designed experiment. "No, above all in matters of such importance, and when one is on such shaky ground, one must not upset the scholarly community to this degree, especially when the strangeness of the findings calls for the greatest reserve. M. Toussaint was equally rash with respect to septicemia and chicken cholera. To assert with such confidence that these two diseases are completely identical is to know nothing about either one of them. And he tells us that his conclusions are backed by more than 250 experiments! If one boasts of having done 250 experiments, it means that one has done them badly more than 249 times."[14]

Toussaint gave in and recognized that his method was not reliable. Pasteur calmed down and let him know through Bouley that "as a matter of discretion he would abstain from all detailed criticism, leaving it to M. Toussaint to test his own findings." Meanwhile, the problem remained unsolved. Now that Toussaint's hypothesis was eliminated, the anthrax vaccine had yet to be discovered. In the process Pasteur, who had believed that he had found a universal method in oxygenation, encountered a number of serious difficulties. For one thing, anthrax was not chicken cholera, and furthermore, the anthrax bacillus produced spores capable of regenerating themselves at any time. Yet within six months he did succeed in developing a vaccine. The strategy was to prevent the formation of spores while keeping the bacillus alive in order to attenuate its infectious power by means of oxygen. Despite all the criticism he had just formulated, Pasteur borrowed Toussaint's idea about the action of heat when he determined a narrow band of temperature within which the bacterium can grow without sporulating. Here it was possible to attenuate it by means of oxygenation, the method Pasteur had used all along. In short, the vaccine was obtained by growing the microbe at 42 to 43°C for eight days and infusing oxygen into the cultures. In this manner they became harmless to guinea pigs, rabbits, and sheep, the animal species most susceptible to anthrax. "Before its virulence is extinguished," Pasteur concluded, "the anthrax microbe goes through different degrees of attenuation. Moreover, as is also the case for the microbes of chicken cholera, each of the states of attenuated virulence can be reproduced by culturing."[15]

The vaccine for anthrax was mastered in the same manner as that for chicken cholera. When Pasteur communicated these findings in March 1881, he explained the fundamentals of the process: "Seeking to decrease virulence by rational means is to experiment in the hope of using active viruses that can easily be grown in the bodies of animals or humans to prepare virus-vaccines of limited development, which will be capable of preventing the deadly effects of the active viruses."[16]

In fact, the idea of attenuating germs was not as new and original as is often believed. Spontaneous attenuation had frequently been observed, long before Pasteur's work. It had long been noticed that in the course of an epidemic the virus diminishes in virulence and simply exhausts itself. In 1860, the English epidemiologist William Farr had put forward the idea that in epidemics the transmission of germs from person to person attenuates their virulence in these successive passages. Shortly before, the physicist John Tyndall, when studying "the critical behavior of atmospheric particles in relation to putrefaction and infection," had noted that germs differ in appearance and development depending on whether they are fresh or old, dry or humid. After all, does not their uneven distribution in time and space account for the seasonal variation of epidemics, and are not the modifications of germs related to the external environment? Yet, however much it was discussed and even studied, the variation of virulence was neither understood nor harnessed. In his experiments, Pasteur showed that he could domesticate the infectious power, for he not only succeeded in weakening the microbe for a vaccine preparation, he also described other procedures designed to restore the germs to their initial vigor.

Then another question arose: would not the attenuated culture that vaccinates the adult guinea pig be liable to be mortal to the new-born animal? Noting, without commenting, that immunity varies with the age of the vaccinated subject, Pasteur used this information to restore its virulence to an attenuated culture. By passing it through a series of new-born guinea pigs and then inoculating it into older animals, one can reinforce the infectious power of anthrax and cause the bacterium to regain its original virulence. It will then kill guinea pigs three or four days old, then a week, a month or several years old, and even sheep. A very similar procedure was used for chicken cholera. This germ too was restored to its original virulence by being injected into small birds (canaries, sparrows — all species that it killed even when attenuated).

Going from the particular to the general, Pasteur considered these modifications of microbes, induced among other things by air, heat, or the age of the inoculated individual, as one of the factors in the spontaneous spreading of the great epidemics and also in their recurrence. Were not plague and typhus examples of this? Starting with this premise, Pasteur was to put forward an entire series of working hypotheses that laid the foundation of modern epidemiology: innocuous germs in the intestine or on the skin can become pathogenic when the organism is weakened or if the skin covering is impaired; the successive passage of a germ from one creature to another or from one species to another is in many cases liable to bring about new forms of contagion.

In his first bacteriological studies, Pasteur based his thinking on the idea that the microbe has no other role than to induce disease — which implies that its power and its properties are invariable. Convinced that the bacterial species were immutable, he refused to admit that the germ can change its form and its function. In the culture broth, all of them were presumed to proliferate in the same manner. Injected into the animal, the bacterium was expected to induce an equally unequivocal disease, provided that it was inoculated in the identical manner. In this conception, the morbid entity was defined by the microbes; in this sense, and on the experimental level, Pasteur agreed with the oldest concept of disease. When studying the *flacherie* of the silkworms he had, to be sure, had an inkling of a fault line in the system, but this had not significantly modified his way of thinking: as he embarked on his microbial adventure, he still hesitated to weaken the stability of the bacterial species and had soon convinced himself that claims of "transformism" were based on imprecise experimentation.

And now, at the age of sixty, Pasteur was once again facing facts that did not fit in with his concepts. Attenuated virulence conflicted with his biological philosophy. He had to renounce his dogmas and enter the debate on the evolution of the species. Should he, then, interpret attenuated virulence in the Lamarckian manner, as a limited transformation of the species under the influence of the environment, or in the Darwinian manner, as a selection of bacterial strains in the culturing medium? Initially Pasteur chose a hypothesis that was closer to his deepest thinking: "I had persuaded myself (and I don't really know why) that it would be more in keeping with the laws of nature to explain all the instances of attenuation I was observing by a hypothesis postulating the mixture in varying proportions of two viruses, one of them very virulent and the other very attenuated, than by the existence of a virus with progressively variable virulence. Having desperately sought, as it were, to prove this hypothesis of two viruses experimentally, I finally came to the conclusion that it was not valid."[17]

Pasteur thus ended by accepting Lamarck's thesis, a step that incidentally was in keeping with the French scientific attitude of that era, which held that the environment influenced the bacterium and that the germ adjusted to the milieu. Pasteur's discovery threw the taxonomy of bacteria into disarray, since it saw the vaccine as a result of the evolution of the species and created a new function. While in Germany Koch and his school were working very hard at describing the world of the bacteria, Pasteur and his students stressed the transformations and the instability of this world.

The Germans were akin to geographers, the French to historians — which in a sense reflected the cultural traditions of the two nations.

In developing such considerations and observations, Pasteur could not fail to unleash controversies. At the Académie de médecine in particular, his old enemy Jules Guérin, encouraged by two other colleagues, Blot and Depaul, kept prattling on about smallpox and vaccines. Incapable of apprehending the importance of Pasteur's hypotheses, he did his best to confuse the debate by constantly compelling his adversary to explain himself, more as a means of holding him back than from a desire to understand. This interminable bickering was exhausting. By June 1880, Pasteur asked Nisard to mediate, since Guérin was his personal physician: "Your friendship for him and for me makes you wish that the discussion should end. It might be in the interest of science that he be forced to understand better what he should understand and that I should drive him to the wall. However that may be, I continue to think that I should, immediately and proudly, cut short the vain endeavors by which these gentlemen have wanted to push the idea that I cannot deal with medical matters since I am not a physician. By now they and the Academy must have understood that if I sometimes speak of these matters, it is because I believe that I have earned the right and the duty to do so by my own studies."[18]

Three months later, on 5 October, the conflict became more acrimonious. Jules Guérin claimed that human vaccinia virus resulted from the inoculation of animal smallpox into human beings, that it was "made human" by this transmission. Without consideration for the almost eighty years of his tenacious adversary, Pasteur brutally replied to him, "Vaccine is vaccine. Anyone who claims to express the relation between human smallpox and vaccine by speaking only of the vaccine and its relation to cowpox and horse pox, without so much as uttering the word human smallpox, engages in a war of words and equivocates in order to sidestep the real point of the debate." Then, refusing to "respond to the indiscreet, intemperate, and unhealthy curiosity of M. Guérin" (who had called for yet another explanation of the process of preparing the attenuated virus of chicken cholera), Pasteur added: "From now on, both of us will fight; we will see who emerges lame and bruised from this match."[19]

This was to go beyond the bounds of propriety. Jules Guérin was furious and hurled himself against Pasteur; he was held back by Baron Larrey, who stepped between the two men, and the meeting was adjourned amid great tumult. The next day, Guérin sent his seconds to Pasteur and demanded satisfaction. Imperturbable and not eager to fight a duel, Pasteur referred him to two men he called the natural seconds for both of them, the perpetual secretary of the Académie de médecine, Béclard, and the acting secretary, Bergeron, the two editors of the official bulletin of the Academy.

Then the *Journal de médecine et de chirurgie*, directed by Just Lucas-Championnière involved itself in the affair. "For our part, we admire the forbearance of M. Pasteur, who is always depicted as violent and ready to go to war. Here is a scholar who from time to time presents short, substantial, and extremely interesting communications. He is not a physician, yet, guided by his genius, he traces new paths in the most arduous areas of medical science. Instead of receiving the tribute of attention and admiration he deserves, he encounters the frantic opposition of a few individuals of quarrelsome disposition, who are always ready to tear him down after they have listened as little as possible. If he employs a scientific term that not everyone understands, or if he uses some medical expression not quite correctly, he must face the specter of endless speeches designed to show him that everything was for the best in medical science before it had to deal with precise studies and avail itself of the resources of chemistry and experimentation. It seems to us that when he said war of words, M. Pasteur was moderate indeed."[20]

In private, Pasteur was not always "moderate." To Legouvé, who expressed his astonishment at seeing him embroiled in this "ridiculous affair," he replied without mincing his words: "There is no doubt that I should have nothing but contempt for the foolish objections of this mean character who is enraged by progress because he is afraid that his pride will be hurt." However, the conflict died down by itself after an exchange of letters showing that Pasteur had adroitly enlisted the support of the Academy's board of directors. In order to calm everyone down, he agreed to express some regret, on condition that his harsh words for Guérin be faithfully recorded in the minutes of the incriminating session. "If in the heat of the discussion I have used a word or made a judgment that might impugn M. Guérin's reputation, I herewith withdraw them and declare that it was never my intention to offend our learned colleague. In our discussions I had only one concern, namely, the vigorous defense of the correctness of my work."[21] Clearly, Pasteur continued to assert, politely but firmly, that Guérin was a fool. Seeming to step back, he gave himself a better running start.

The Victory at Pouilly-le-Fort

Very quickly Pasteur's findings on anthrax vaccination went beyond the walls of the laboratory of the rue d'Ulm. Outside, they delighted some and astonished others, but most of those who paid attention took a critical or skeptical view of these new veterinary practices. How could such a terrible disease be vanquished by such simple means?

Fortunately, Pasteur could call on the support of influential people, not in the exalted realm of science but in the field of day-to-day practice. One of these was a certain Rossignol, veterinarian at Melun, an influential member of the local agricultural society and editor of the *Revue vétérinaire*. In late March 1881, as soon as he heard of Pasteur's communication on the attenuated virulence of the anthrax bacillus, Rossignol called for decisive testing. These discoveries, he said, must not remain without application; we must not allow findings of such promise for stock breeders to gather dust! The veterinarian felt strongly that academic theories must be put into practice in a farmyard, not in the flask of a laboratory. He therefore took the initiative in organizing a public demonstration.

This undertaking was a public relations bonus for Rossignol. If the experiment failed, he would be the one who had found Pasteur wanting, at a time when Pasteur already enjoyed the stature of a savior of humanity. If it succeeded, he too would have achieved something, for the honor of having thought of this outdoors experiment would go to him — and to all veterinarians. Pasteur must have greatly intrigued his contemporaries if they asked him to agree to a control experiment in the heart of the French countryside.

Actually, Rossignol's initiative was well received by the stock breeders, who stated their willingness to finance the operation, many of them no doubt with the secret thought that they would enjoy a resounding fiasco. In any case, the Melun branch of the Society of French Farmers, informed of the project, immediately agreed to sponsor it, and its president, Baron de La Rochette, went to the rue d'Ulm to ask Pasteur to accept a challenge, namely to conduct a public test of vaccination against anthrax.

On 28 April 1881, Pasteur communicated the experimental protocol he intended to follow. Since the Farmers' Society was placing sixty sheep at his disposal, he proposed to vaccinate twenty-five of them. The vaccination would be carried out in two stages: first an initial injection, then a repeat twelve days later according to a sequence to be worked out by the laboratory. Twenty-five sheep would receive no treatment. Then fifty of the animals would be inoculated with virulent anthrax germs. Pasteur predicted that all of the unvaccinated sheep would perish and that only the twenty-five animals of the first group would resist. As for the ten remaining sheep, nothing would be done to them and they would serve as the control group at all stages of the protocol. When La Rochette suggested that ten cows be included in the experiment, Pasteur accepted this condition as well, although no experiments on this type of vaccination had yet been conducted. But this was an opportunity to make the demonstration even more striking.

As soon as the plan was worked out, Pasteur recalled Chamberland and Roux, who were vacationing at the time. "My dear Roux," he wrote on 28 April, "the vaccination experiments will begin at Melun on 5 May. I very much wish you to return by May 3 or 4 at the latest. This is a big and important event. We must have a serious talk about it before we begin. All my very best."[22]

Pasteur was quite right in speaking of a "big and important event." For when he had publicized his attenuation process and vaunted its universal virtues, he had actually acted somewhat precipitously. His findings were far from being as consistent as he implied. Respectable scientists, such as the German Robert Koch, felt that this time Pasteur had gotten ahead of himself.

In the end, it was Roux and Chamberland who saved the day. These two, in their own research in the rue d'Ulm, had reexamined a process described by Toussaint, which, however, due to poorly controlled experimentation, had not produced any decisive results. The process consisted of making the vaccine by using carbolic acid rather than oxygen to attenuate the bacteria of the black disease. When applied rigorously and in the right conditions, this method had proved more cogent than the one Pasteur himself had described. Convinced of its effectiveness, the master now borrowed from his students a process that they themselves had taken from Toussaint.

It should be noted that Pasteur did not forget everything he owed to Toussaint's intuitions. Far from seeking to relegate him to the shadows, as has sometimes been claimed, he always supported him in the future, sponsoring him in particular in 1883 for the Vaillant prize the Académie des sciences awarded him for his work on anthrax. But for the moment the question of who deserved the greater credit had not yet arisen; the task at hand was to win the wager of Pouilly-le-Fort.

For it was in this small village near Melun that Rossignol had his farm, a large complex characteristic of the Brie region: thick stone walls, several outbuildings, barns and stables around a square courtyard; and all around, fields and pastures in the flat countryside. On 5 May 1881, the day chosen for the first injection, a large crowd invaded the railroad stations of Melun and Cesson. Observers had come from Provins, Fontainebleau, Saint-Germain-Laxis, from Courtry and Gregy, from Fourches and many other small places. They all converged on Pouilly-le-Fort, where the animals to be used for the experiments were penned in an enclosure. They had come to see the sheep and cows, to gain a better understanding of anthrax, or simply in order to get a look at Pasteur. They were farmers in wooden clogs and striped pants, councilors-general with the ribbons of their decorations in their buttonholes, physicians in stove-pipe hats and flowing cravats, pharmacists and veterinarians in sleeveless vests. There were also journalists: Gaston Percheron, the chief editor of the *Journal de la médecine vétérinaire pratique*, or Louis Garnier, of the *Revue vétérinaire*. The press had greatly played up the affair, which had become a matter of regional, even national prestige. Nobody wanted to miss such a show. How often do you see a scientist out of his laboratory?

In the crowd one noticed Baron de La Rochette, who had come in a carriage driven by a coachman in a French frock coat. Smiling at everyone, and proud of an event in keeping with his prestige, the old gentleman looked as if he were inaugurating Marie-Antoinette's sheepfold at Trianon. Rossignol was there too of course, umbrella in one hand and notebook in the other. He felt triply honored: as owner of the farm, as journalist, and above all as the instigator of the debate.

The atmosphere was that of a country fair rather than a laboratory. Off-color jokes were noisily bandied about, words of hope and encouragement were spoken. One veterinarian of Pont-sur-Yonne, Biot, loudly proclaimed his incredulity; his former teacher, who was none other than Colin, had especially asked him to be present at the experiments in order to watch Pasteur's manipulations: "You must watch out," Colin had told him, "for the broth of bacteridian cultures has two parts: an inert upper part and a lower and very active part, where the bacteria have accumulated because, owing to their weight, they have sunk to the bottom of the receptacle. They will inoculate the sheep that are not supposed to die with the upper part of the liquid, whereas the controls will be inoculated with the lower part, which will kill them."[23] Ridiculous to the end, Colin therefore asked Biot to give a vigorous shake to the flask before the last inoculation.

When Pasteur appeared, accompanied by Roux, Chamberland, and Thuillier, the spectators stepped aside, but it took some time for the noise to die down. Finally, the animals could be examined. Two sheep were rejected as too weak to undergo the experiment. Since no other animals were available, they were replaced by two goats, and then the herd was divided into three groups that were taken to a large hangar. There was the bleating of the animals, the buzzing of the crowd, Olympian calm on Pasteur's part.

Now Chamberland and Roux opened the black satchel containing the vaccine culture. As they left his laboratory, Pasteur had given precise instructions to his aides: "Be sure you bring the right flask!" Roux prepared the Pravaz syringe and filled it, while Chamberland and Thuillier presented the animal and marked its ear as soon as it was inoculated. Pasteur looked on impassively, ready to pick up any error. One by one, the animals of the first group were injected in the right thigh, receiving the equivalent of five drops of anthrax culture, which Pasteur called the first vaccine. Then it was the cows' turn. One of them was too frail and was replaced by an ox. This time the vaccination was given in the fat part of the left shoulder, and the animal was marked on the right horn.

After the injections were given, the public followed Pasteur to the front parlor of the farm, which had a few color prints on the walls, an old clock, copper pots, a pair of guns. Before an impromptu audience standing in the back or sitting on the few available benches, with Rossignol and La Rochette right in front, Pasteur improvised. He was glad to abandon the rhetoric of the Academy and to speak to farmers who had doffed their caps and to notables who did not know what to do with their stove-pipe hats. Ever since dealing with the agricultural society of Orleans at the time when he was working on vinegar, he knew how to find persuasive words. They listened in silence. Pasteur spoke of himself, of bacteria, vaccination, and anthrax; he explained his method fervently and logically and ended by inviting his public to return for the second inoculation, which would take place on 17 May.

That day, everything continued to proceed as expected. The crowd was as large as the previous time. Pasteur and his three aides repeated their actions of the first day. However, they were using a new batch of vaccine, prepared with more virulent bacteria. "On Tuesday 31 May," Pasteur wrote to his son-in-law, "the third and last inoculation will take place, and this time it will be given to all fifty sheep and ten cows. I am very confident. . . . If there is a truly clear-cut success, it will be one of the most wonderful achievements of science and its application in this century and consecrate one of the greatest and most productive discoveries."[24]

As the day of the last experiment approached, there was a great deal of discussion in the meetings of the agricultural societies, particularly among veterinarians. We must wait, they all said, but those who believed in a possible effect of the vaccine were definitely in the minority.

On 31 May everyone again assembled at Pouilly-le-Fort. Everyone expressed his opinion. Biot was there first, insisting on being allowed to give a vigorous shake to the culture broth before it was inoculated and demanding, also at Colin's behest, that a more virulent dose than agreed upon be injected. Pasteur gave in and agreed to inject a triple dose. He also agreed when asked to alternate the inoculations between vaccinated and healthy animals so as to give each one an equal chance of surviving or dying. Everything was done by three-thirty. That day there were no speeches. Everyone was to meet again two days hence, on 2 June, to check the results.

At no time did Pasteur waver from his calm; he seemed so sure of himself that even his collaborators were impressed. Yet those who knew him well might have wondered, for when Pasteur knew that he was right, he was not calm but combative and sometimes outright aggressive. Not showing any emotion in the farmyard of Pouilly-le-Fort was probably his own way of dealing with his uncertainty.

On 1 June, Chamberland and Roux returned to Pouilly-le-Fort to check on the state of the animals. Among the unvaccinated, they saw many sick individuals, which stayed apart, refused to eat, and breathed heavily with lowered heads. Some of them actually groaned. If one tried to make these animals stand up, they immediately collapsed because their legs buckled. That evening, when Rossignol was about to leave the farm, there were already three dead bodies.

But among the vaccinated, everything was not perfect. One lamb was slightly feverish; one sheep had a temperature of 40 degrees; another presented a swelling at the point of inoculation; and a fourth one limped. In one corner, a ewe refused to eat. When Chamberland and Roux returned to the rue d'Ulm, Pasteur was pacing in his laboratory and rushed toward them. Anxiously listening to his assistants' report, he became perturbed.

While they were conferring, they received a telegram from Rossignol informing them that one of the vaccinated ewes must be considered lost. Impervious until then, Pasteur was suddenly seized by doubt. "For a few moments," Roux recalled, "his faith wavered, as if the experimental method could betray him."[25] Actually, the young physician bore the brunt of the storm: Pasteur accused him of being responsible for the failure of the experiment, of trying to ruin all of his work, of endangering the future of his laboratory. The scientist was beside himself, and Marie Pasteur had to intervene to calm him down.

No one, of course, slept a wink that night. At nine o'clock on the morning of 2 June, Pasteur received another telegram from Rossignol, announcing that eighteen nonvaccinated animals had died, and that the others were dying. As for the vaccinated ones, they were all on their feet. The telegram ended with the words: "A tremendous success."

"This morning at eight o'clock," Marie Pasteur wrote to her daughter that day, "we were still very preoccupied, waiting for that blessed telegram that might have given us the news of some disaster. Your father did not want anyone to distract him from his worries. At nine o'clock the laboratory was informed, and five minutes later I was told about the telegram. I had a little moment of emotion that made me see all the colors of the rainbow."[26]

There was no longer any question, of course, of Roux going by himself to enjoy that success, and Pasteur had regained his habitual self-assurance, tinted with a bit of arrogance. At two o'clock, when he arrived at Pouilly-le-Fort escorted by Chamberland, Roux, and Thuillier, the farmyard was packed. A delirious crowd received them, screaming: all the unvaccinated sheep are dead or dying, the others are all alive! Pasteur stood straight up in his carriage where the crowd besieged him and addressed his opponents, his friends, and everyone else: intoxicated, triumphant, biblical in his enthusiasm, he cried out: "Here it is! Oh ye of little faith!"[27]

Many people had come to be present at the conclusion of the public test, and they all acclaimed and applauded Pasteur, standing around the bodies of twenty-two of the unvaccinated animals, while not far from there the last three of that group were at the point of death. Even Bouley was there and shared the general enthusiasm. At this point Pasteur wanted to crown his success. Like the winner of a race who runs a victory lap, he asked to have one of the animals autopsied in order to show that it had really died of anthrax. Under the microscope the characteristic germs of the disease could clearly be seen. Even the most incredulous veterinarians yielded. Biot also acknowledged that he had been wrong and even talked of vaccinating himself before inoculating himself with the most virulent anthrax he could find.

It was a triumph, but now that he was sure of himself, Pasteur multiplied the controls, which were more rigorous than ever. Judging that the autopsy was not yet sufficient, for one also had to be certain that the vaccinated animals were out of danger, he reserved his conclusion: "The experiment will not be complete until 5 June, when we have decisive proof for the vaccinated animals." This precision in the conduct of the experiment really convinced the crowd. After Pasteur left amid rousing cheers, the enthusiastic onlookers, led by Rossignol, decided to change the farm's name to "Clos Pasteur."

Nonetheless, Pasteur was right to be cautious. During the night of

4 June the vaccinated ewe that had refused all food died. Suddenly the enthusiasm came to an end, and the opponents of vaccination felt hopeful again. But the immediately performed autopsy left no doubt: the cause of death was not anthrax but a miscarriage, for the animal had been pregnant. Vaccination had nothing to do with it. A sigh of relief! But the sheep were not the only heroes. Among the cows and oxen, the results were identical. If they were not vaccinated, they presented severe edema; if vaccinated they suffered no harm. The commission therefore judged that there was no need for further experiments: the vaccine had been proven effective.

The journalists agreed. The specialized press published a plethora of articles, and Pouilly-le-Fort was mentioned even in the London *Times*, which had sent a correspondent to cover the affair. A traveler from Cape-town who was passing through Paris urgently asked to meet Pasteur; he wanted to suggest to him that he should vaccinate the Tibetan goats of South Africa. The scientist was tempted: "Your father would love to make this long trip," Marie wrote to her daughter, "going by way of Senegal, where he could find some good germs of pernicious fever, but I am trying to moderate his ardor."[28] Pasteur did not go; he was needed too much in France. Moreover, he now had to prepare his campaign to be admitted to the Académie française and respond to all those who solicited his presence, for now that his success was evident, he was in demand everywhere. On 13 June he reported to the Académie des sciences; on the fourteenth he was at the Académie de médecine; on the twenty-first he chaired the opening session of the agronomic congress of the French Agricultural Society.

The government as well was interested in the experiment and wanted to honor Pasteur. But the scientist was adamant: he would accept a decora-tion only if he shared it with Chamberland and Roux. The result was the grand cordon of the Légion d'honneur for him and a red ribbon for his assistants. By making this stipulation — albeit without giving a precise ex-planation — Pasteur acknowledged that he owed a great deal to his disci-ples. For the fact is that the experiment of Pouilly-le-Fort had succeeded thanks to the process of attenuation discovered by Chamberland and Roux. However, it was not until 1883 that this key method was described in two successive notes, and not until January 1890 (a few months before Tous-saint died in complete obscurity) that an article in the *Annales de l'Institut Pasteur* brought to light the role Roux had played in this entire experiment. For the time being Pasteur was the master, and Paul Bert had to intervene to make sure that everyone was rewarded. "There were cordial embraces all around amidst the guinea pigs," wrote Marie.

Pouilly-le-Fort became the center of a cult. But the Clos Pasteur soon returned to anonymity. Today it is a place unnoticed by passers-by, situated near the sea of wheat fields of the Beauce, where a vague marker soberly recalls the first victory of the French vaccine.

PASTEUR NOT ONLY RECEIVED congratulations and honors, he also had to deal with a great influx of requests. Everyone wanted to vaccinate against anthrax, and since the making of the vaccine as well as the inoculations had to be supervised, Pasteur had to be everywhere. Chamberland and Roux also traveled throughout the region, operating at Senlis, Vincennes, Pithiviers, Coulommiers, and other places.

A few weeks later, the government asked Pasteur to represent France at the International Medical Congress in London. The physicians would be surprised that his country was sending a chemist, Pasteur supposed, more in order to be begged to go than because he really hesitated. On 31 July 1881 he crossed the Channel with his son and his son-in-law. "I felt nothing on the boat," he wrote to Marie, "but René and J.-B. came close to the most characteristic symptom of seasickness."[29]

In the immense hall of the St. James Palace, Pasteur received an ovation at the instigation of Sir James Paget, surgeon at St. Bartholomew's Hospital, who was the first to describe an important bone disease. Then, on the occasion of a dinner in the city, he was presented to the Prince of Wales, who called himself "a great friend of France," and to the Crown Prince of Prussia, with whom he had to shake hands, Alsace and Lorraine notwithstanding.

Pasteur's address to the Congress concerned vaccination against chicken cholera and anthrax. It was very well received. But the scientist did more than let himself be applauded and did not miss a single session. When Bastian criticized Lister, the chairman of the session asked Pasteur to intervene, but the latter, who did not understand English very well, had not quite understood just what was being discussed. He told the rest of the story in a letter to his wife: "While thunderous applause greeted my name, I whispered to Dr. Lister, 'what did he say?' 'He said that the microscopic organisms in diseases are produced by the tissues themselves.' That was enough for me, and I took the floor to bear down on him. He will have a hard time recovering from this."[30]

While Pasteur was acclaimed in England, it became clear that in France he had not convinced everyone. To be sure, Colin was still on the scene, vigorously protesting to the Académie de médecine and now claiming that he had been the first to demonstrate that inoculation with anthrax can confer immunity. Above all, however, many farmers were beginning to doubt; in particular, those who had not witnessed the experiment of Pouilly-le-Fort were wondering whether the anthrax with which the animals were inoculated had really been virulent. It was rumored that vaccination did not give complete protection.

Confronted with this skepticism arising from deeply held beliefs, Pasteur proposed to stage another Pouilly-le-Fort. It would take place at Chartres and the animals would be inoculated with the blood of a recently deceased animal. Pasteur felt that using blood rather than culture broth would be more convincing. This second public experiment was as brilliantly successful as the first. Bouley enthusiastically congratulated his confrère. "It is too good to be true, they said! And they counted on the vigor of the natural anthrax to dispose of your method. But today you see them converted!"

Converted as well were the people in the South. When Pasteur traveled to Aubenas in the department of Ardèche, he was received by the entire municipal government: flags decorating the city hall, trumpet blasts, triumphal arch of flowers, speech by the mayor. The whole town was celebrating. Presenting a medal bearing his effigy to Pasteur, the speaker recalled that the scientist had saved the region's silkworm nurseries and silk spinning mills before he had defeated anthrax.

The farmers and the veterinarians of Nîmes also wanted to see the scientist. More medals, speeches, and banquets. But he was also asked to reproduce the experiment of Pouilly-le-Fort. Sheep, oxen, and cows were put at his disposal. From Nîmes, Pasteur traveled to Montpellier, where he gave a speech pleading in favor of misunderstood geniuses. "I would like to see the holders of public authority—government ministers, prefects, rectors, mayors—act as spies in search of public merit, charged with finding all those who bring honor to the fatherland. If this could be accomplished, the destiny of France would be vastly enhanced."[31]

Amid these victories over anthrax, and while millions of doses of the new vaccine were shipped throughout France, came the news of the death of Sainte-Claire Deville. Speaking at his old friend's funeral service, Pasteur addressed him one last time in a lyrical tone that was unusual for him: "The earth that bears us, the air we breathe, the elements of which you liked to ask questions and which were always so willing to answer you, all these would gladly speak to us of you. . . . Ah! I beg of you, do not at this moment look at your grief-stricken wife, your distressed sons. In the face of their pain, you would find it too hard to have left this life! You will wait for them in those divine regions of knowledge and bright light, where everything is now known to you, where you even understand the infinite, that maddening and terrible notion which, though forever closed to man on earth, is the eternal source of all greatness, all justice, and all liberty!"[32]

A Time of Controversy

The experiments of Pouilly-le-Fort, followed by those of Chartres, and later those of southern France, had aroused the enthusiasm of many farmers. At the same time they had permitted Pasteur to reassure himself as to the validity of his method, which he now vigorously propagated. News soon spread beyond France's frontiers, and it was not long before inquiries came from abroad, for everyone wanted to benefit from the vaccine and know how it was prepared.

Yet Pasteur was determined to keep sole control over his process and insisted that all vaccine cultures be prepared in his laboratory. This was not a matter of preventing access to information but rather of wishing to avoid errors. The method of attenuation was delicate. In order to be effective, the vaccine had to be manufactured with the utmost care. If it was too virulent, it could trigger the disease and even lead to the death of certain animals. If it was too attenuated, the immunity it conferred was insufficient. Moreover, the resistance of vaccinated subjects did not last forever and diminished over time. If the demonstration was to be effective, the inoculation of virulent bacteria had to take place soon after the last inoculation with the vaccine.

All these considerations prompted Pasteur to write a memorandum to Gambetta, prime minister of France at the time. He proposed the creation of a manufacture of anthrax vaccines, to be headed by himself and two of his collaborators, Chamberland and Roux. This venture would bring credit and prestige to the state. However, the government, swept out of office by a ministerial crisis in January 1882, did not have time to initiate this project. Therefore the large-scale production of vaccine was soon transferred to the annex of the Ecole normale supérieure in the rue Vauquelin. The doses of vaccine were produced under the direction of Chamberland, who inspected them before they could be shipped.

The greatest interest in the discoveries leading to anthrax vaccination existed outside France's frontiers. In September 1881, Baron von Berg, a great Hungarian landowner, asked Pasteur to conduct a public experiment identical to that of Pouilly-le-Fort on one of his estates near Pest. He pledged to place one hundred cows, sheep, and horses at the experimenters' disposal. Pasteur sent Thuillier to prepare for the event.

September in Budapest. The young man admired the beauty of the Danube, the elegance of the women. He was received with a speech of von Berg: "The wealth, the strength, and the glory of a nation are not its soldiers but those of its children who bring luster to science by their work." The Frenchman was given everything he needed to carry out the experiment. At the veterinary school of Pest, several laboratory rooms had been set up for the first trials. But Thuillier was worried: "The vibrations during the journey," he wrote to Pasteur, "have thinned the deposit of my culture broths so much that it hardly precipitates at all." He therefore had to make a special effort to keep the cultures alive.[33]

Moreover, it did not take Thuillier long to realize that the Hungarians were more interested in the secret of manufacturing the vaccine than in the demonstration of its effects. Economic warfare was already in full swing in central Europe, and Hungary hoped to get hold of the vaccination techniques in order to make Germany and Russia dependent on this new industry. Thuillier was plied with questions, which he eluded. A telegram was sent to Pasteur, whose answer was unequivocal: the vaccine was to be prepared in Paris, and divulging the process was out of the question; Thuillier's mission was limited to carrying out several demonstrations of collective vaccination. It was already a privilege for the Hungarians to be the second nation in the world to profit from the benefits of the technique, and there could be no question of allowing them to develop an unauthorized local industry.

In the end, the two injections were carried out, one at Pest, the other at Kapuvar, within the required time frame. The last injection, that of the virulent bacillus, took place in the presence of the press. The demonstration was a total success. The Hungarian newspapers could not praise it enough, to the point that one landowner wanted Thuillier to vaccinate fifteen thousand sheep!

The success was of course credited to Pasteur, who had never left home. Officially thanked by Austria-Hungary, he now turned to the Russians and offered to carry out the same demonstrations for them. But they were suspicious and preferred sending observers to Paris. The reason was that in the other countries of eastern Europe, Pasteur had not entirely convinced the experts. Throughout the year 1882 he was under heavy attack from Robert Koch and the Berlin School.

Koch and his students were critical of all of Pasteur's work, from culturing in broth, the discovery of the septic vibrio, the experiments with cholera-infected chickens, to the role of earthworms in the spread of anthrax, and, of course, the existence and effectiveness of a vaccine. One of Koch's students, Friedrich Löffler, who was to devote his life to the study of tuberculosis and pneumatology, wrote in a memorandum: "The famous experiment of Pouilly-le-Fort with its surprising result is being received, not without reason, with skepticism." Had not Koch demonstrated that under certain conditions of preparing the vaccine, the anthrax spores survive, not in attenuated form but as vigorous as before?

Pasteur was ready to carry out a counter-valuation. But when the Veterinary School of Berlin asked for vaccine samples from the rue d'Ulm, he refused. He wanted a public experiment, a blue-ribbon commission. And he requested that the vaccination be performed by someone from his own laboratory. So the Berlin ministry of agriculture appointed a commission, whereas Pasteur once again sent Thuillier from Paris.

The vaccine went to Germany through diplomatic channels, for the French did not trust the customs officials. "We must do everything to win our victory at Salamis," Pasteur wrote to the French ambassador. The three necessary injections, the first two with the two vaccines and the last with the virulent bacteria, proved to be rather trying. First of all, owing to the interval between injections, Thuillier had to remain on the spot for a fairly long time. "I hear amazing things about what goes on in Berlin," he wrote. "But if I were not a smoker, I would die of boredom." More importantly, however, the breeds of German livestock were particularly sensitive to anthrax. At Packisch, a few of the vaccinated cows died. "As soon as an animal looks as if its head were drooping, all eyes turn from the animal to me and back again. . . . I can assure you that this little game is not easy to take. And it's even worse when I have to be the chief mourner of the vaccinated creatures that have died by my hand and give a speech on susceptibility at their graveside."[34]

The experiment was of extreme importance, for it was designed to impress Koch, whose doubts were growing. Several times a week, Thuillier cabled Pasteur, and when the master was absent, he communicated with Chamberland, with whom of course he used a very different tone. The most important of all of Thuillier's communications concerned a finding of which he informed Pasteur even before it was published in the medical press: Koch had just discovered the agent of tuberculosis, a rod-shaped bacillus resistant to staining with alcohol or acids. Thanks to their staining techniques, the Germans had become the undisputed masters in the science of observing microbes. Pasteur immediately responded, asking Thuillier to verify the discovery and to go to Koch's laboratory to see it for himself. There was no doubt: there really was a new microbe, the terrible Koch's bacillus, which would confer immortality on its discoverer. The latter, however, did not intend to push his investigations further and left it to others to derive therapeutic solutions from his discoveries, for he did not believe in vaccination.

Meanwhile the demonstration continued in Berlin. Constantly surrounded and closely watched by a commission that verified every last detail, Thuillier had successfully carried out the first two inoculations. Pasteur, who wanted the demonstration to be dazzling, sent a particularly virulent dose for the last inoculation. He wanted all the nonvaccinated animals to drop dead at the same time! On 1 July, the results were eloquent, another success. Yet Koch was still not entirely convinced and Pasteur not really satisfied, for the Germans had not publicly acknowledged their defeat.

Pasteur therefore sought to make his case once and for all at the International Congress of Public Health held in Geneva in August 1882. Setting his sights on obtaining from Koch an official response before an assembly of representatives from all countries, Pasteur retreated to Arbois, where he nervously and painstakingly polished a lengthy address. So absorbed was he by making his argument that he even neglected his daily walks on the Besançon road. Then Marie, in her tiny handwriting, copied the erasure-filled pages her husband was about to read before a cosmopolitan audience.

Entering the great hall of the Congress on 8 August 1882, Pasteur noticed that Koch and his students were sitting in the first row. There was thunderous applause as he mounted the podium. Pasteur was one of the very first to speak. Without wasting any time, he immediately went on the attack. He spoke of the attenuation of chicken cholera and anthrax microbes and recalled the triumphal results of his experiments. Then he added: "Yet however blazingly clear the demonstrated truth, it has not always had the privilege of being easily accepted. I have encountered, both in France and abroad, obstinate objectors." Thereupon, he directly addressed Koch and his disciples, the men who had insinuated that he did not know how to cultivate microbes in their pure state, that he knew nothing about the techniques of recognizing microorganisms, so that his work could not be judged by rigorous standards. They had also condemned the method of inoculation consisting of injecting one or several syringes full of liquid under the skin. And Pasteur continued to speak in more and more detail about the experiment of giving chicken cholera to chickens by simply lowering their body temperature after the inoculation. "Dr. Koch," he said, "who finds nothing remarkable in this experiment, wishes to know whether the chilled chickens that contracted the disease were not chickens capable of getting it naturally. . . . And what does the author think may have led me into error? It is the fact that the alteration would begin with the vaccination. . . . The author does not believe that I operated as I said I did, with eighty chickens in certain of my experiments, because that would have cost too much money. But it is true, for in view of establishing the great fact of the attenuation of virulence, my government allowed me not to worry about the expense."[35]

Koch listened in silence, his gold-rimmed spectacles perched on his nose. He did not want to engage in a public debate. Speaking from the podium, he stated that he preferred responding in writing. Pasteur was disappointed, but had to resign himself to waiting. And he had to wait for three months. Finally, in autumn 1882, Koch published a small pamphlet entitled *On Anthrax Vaccination: Response to a Speech Given at Geneva by M. Pasteur.* Koch dug in his heels, alleging that Pasteur had not brought any new scientific evidence to the Geneva Congress; that the attenuation of the viruses was a myth; and that the only way to make progress in studying diseases was to search for microbes.

This was a new provocation. Pasteur was furious and felt compelled to respond. In January 1883 he published a long memorandum in the *Revue scientifique*. "Response to M. Koch, private counselor to the government of Berlin." Addressing his adversary's accusations one by one, he spoke of the obliging German chickens, recalled his first discoveries concerning the anthrax spores, pointing out how early on they had been made, and even brought up his long-ago altercations with Liebig. "Have you not read, Monsieur, among other things, in the Reports of the Académie des sciences, about my challenge to your illustrious compatriot Liebig in 1871 concerning the great inaccuracy of his theory of fermentation?" Liebig and Koch, two adversaries from enemy Germany![36]

Pasteur then accused Koch of being indebted to French science and refusing to admit it. "You assert," he continued, "that the results observed in the fields of Packisch are unfavorable to vaccination, which is not what the report said. You do not acknowledge that you are mistaken about the very principle of the attenuation of the bacterium." And the indictment continues: "Your pamphlet contains a great many passages in which the imper-

tinence of error, as Pascal put it, is really carried too far."[37] Pasteur re-
trenched himself behind his famous experimental rigor. "However violent
your attacks, Monsieur, they will not prevent me from succeeding. I am also
fully confident that this method of attenuating viruses will have consequences that will help humanity in its struggle against the diseases by which
it is besieged."[38]

Actually, Koch and his school were not the only foreign scholars to
express doubt about Pasteur's assertions and publications. In the spring of
1883, he also had to respond to a contradictory Italian experiment. An
attempt to vaccinate sheep according to Pasteur's principle had failed at
Turin; the vaccinated sheep had died along with the controls. As soon as he
heard the news, Pasteur wrote to the director of the veterinary school of
Piedmont to find out about the origin of the anthrax inoculate. It was
blood, and it had not been injected for more than a day after it was taken. "A
grave scientific error," Pasteur replied. "It is obvious that what was injected
was blood containing both anthrax bacteria and septic vibrios, as was done
in the experiment of Jaillard and Leplat. This means that two infections
were transmitted." The director of the school took offense. He was sure of
himself, for the microscopic examinations had indicated the purity of the
preparation. Six Italian professors protested along with him in the florid
style of Italian rhetoric: "We find it marvelous that Your Illustrious Signoria
should have been able to recognize with such certainty, from Paris, the
disease that claimed so many victims among the vaccinated and unvaccinated animals. . . . We do not think that it is possible for a scholar to assert
that an animal has septicemia if he has not seen it."[39] Not content with
sending this letter to Pasteur, they disseminated it in the press.

Pasteur maintained his assertion, and so did the Italians. The quarrel
went on for a full year. Every time the matter was brought up, Pasteur
responded. In April 1883 he lost patience and proposed a public counter-
experiment under the supervision of the Académie des sciences. "If you
accept," he proposed to the Italians, "I will come to Turin and will show
how blood can at the same time contain and transmit septic vibrios and
anthrax bacteria. You will see that the scientific principle is evident." The
Italians did not respond with an explicit invitation but sent a seventeen-
page letter that reviewed all local experiments in this area and ended by
vaunting the existence of an Italian vaccine.

But Pasteur did not give up. "Once again," he wrote on 9 May, "I have
the honor of asking you to have the kindness to inform me whether you
accept the proposition I made to you on 9 April last, to come to Turin to

place before your eyes proof of the facts I have just recalled. PS. In order not to complicate the debate, I will not go into all the erroneous assertions and citations contained in your letter."[40] While Pasteur, in accordance with his words, was busy preparing for his trip, the Turiners were writing a pamphlet entitled *On the Dogmatism of M. Pasteur*. However, this is as far as the matter went, and Pasteur did not go to Italy, where successful vaccination campaigns were soon to be undertaken.

Thus, between 1881 and 1883, while Pasteur's method was used with ever greater practical success, his theories continued to encounter the skepticism of scientists. Colleagues at the Académie de médecine continued to attack the chemist's work. The Gabriel Colins and the Jules Guérins had their supporters, so that tumultuous sessions took place with exhausting frequency. It was not rare to see Pasteur leave the hall trembling with anger at those who simply did not understand the validity of his experiments. In the end he resigned himself to no longer attending the meetings of this academy.

His absence did not, of course, put an end to the controversies concerning the microbe theory, contagion, and therapeutics. In the first weeks of 1883, the treatment of typhoid was placed on the agenda. In 1870, in the midst of the Franco-Prussian War, a German physician, Brand, had achieved an undeniable success by treating typhoid with cold baths of 20°C. A young physician of Lyon, Dr. Glénard, a prisoner of war at the time, had brought the method back to France, where he had been able to persuade his colleagues at the Croix-Rousse Hospital of its merits. This method in fact resembled the one Pasteur had described in connection with chicken cholera. This no doubt explains why experimentation continued for ten years before the Académie de médecine agreed to conduct an official investigation. Indeed, it was an opportunity to reopen the debate against the bacteria, even though the infectious agent of typhoid had been discovered by Eberth in 1880.

"They aim for the microbe and they shoot the patient!" one speaker exclaimed to loud applause. Many felt that the patients had to be protected against the unexpected danger of this barbaric therapy. In this shouting match against the microbes, one voice was louder than all the rest; it was that of Michel Peter, one of the most influential members of the Academy.

Pasteur and Peter were distantly related through their wives, who shared one grandfather, Joseph Loir, veterinarian in Napoleon's army of Egypt. Loir had been married to the daughter of his commanding officer, the veterinarian-in-chief Giraud, on orders of Bonaparte, who was in the habit of arranging marriages in his entourage. So Joseph returned from the banks of the Nile with a pretty wife — and a hat Napoleon had given him as a souvenir. Naturally his eldest son was called Napoléon, and his grandson was none other than young Adrien Loir. It so happened that this young man was serving his apprenticeship in the laboratory of the rue d'Ulm in the afternoon, while working in Peter's clinic in the morning. An uncomfortable position indeed. Adrien often witnessed quarrels between his two masters, for Pasteur and Peter had as much contact as they had contempt for one another.

Peter told everyone who would listen that the microbial origin of disease was not as satisfactory an explanation as it was said to be: "The discoveries of bacteria are among the curiosities of modern times; admittedly interesting, they have practically no benefits for medicine. They are worth neither the time spent on them nor the fuss made over them. After so much and such laborious research, nothing will be changed in medicine, and all there will be are a few more microbes."[41] There is no need to stress Pasteur's reaction to such talk, particularly since it did a great deal to launch a renewed debate on contagion and the vaccines. At the Academy, every time Bouley tried to praise Pasteur's merits, Peter took the floor and was applauded in the name of traditional medicine. He adopted the stance of the prosecutor and transformed the assembly into a grand jury. "M. Pasteur's excuse is that he is a chemist who, inspired by the desire to be useful, wanted to do something for medicine, to which he is a total stranger."[42]

One can understand that Pasteur preferred not to be present at such meetings. On 3 April 1883, the tone escalated when a letter from Peter was read from the podium. The physician now denounced the "microbicide medications" as homicidal. When Pasteur received the minutes of the session, he was on vacation at Arbois, from where he was conducting an epistolary skirmish with the veterinarians of Turin. He immediately left for Paris, determined to fight back. Because it was vacation time, the gates of the rue d'Ulm were closed, and so Pasteur wrote his response in a tiny room of the Grand Hôtel du Louvre. The next morning he gave a resounding speech at the Academy: if Peter, he said, asserts that nothing is further from the spirit of medicine than our methods of analysis, "he does not realize that every time medicine has grown, it came close in its spirit and its methods to the analytical sciences."[43] Going over the arguments advanced by Peter, who had made a particular point of the German and Italian attacks against the French vaccine, Pasteur added, "Meanwhile, you will permit me to point out that in your vain attempt to combat the discovery of the attenuation of viruses and the work of my laboratory, you have found your weapons abroad. . . . My own patriotism, Monsieur, is such that I should be inconsolable if the great discovery of the attenuation of virus vaccines were not a French discovery!"[44]

This peroration was bound to give rise to a salvo of applause — and Pasteur left the podium satisfied with his performance. But this did not mean that he was finished with Peter. He still had to convince all those who were opposed to or hesitant about his ideas. Henceforth, Pasteur's entire scientific strategy was devoted to promoting vaccination and its diffusion.

More than in these debates, in which invectives were traded with such formulations as "microbial madness," "fanaticism for the microbe," or "fetishism," Pasteur was interested in the progress made by French farmers, who for the most part believed in the vaccines and had more and more inoculations performed, as did the Hungarians, the Russians, the Germans, English, Swiss, and Belgians. All over Europe the practice of inoculation took hold at an accelerating pace. Tens of thousands of sheep were vaccinated in the course of the year 1882. Barely two years after the discovery of attenuated virulence, Pasteur was thus in a position to cite eloquent statistical results. Mortality due to anthrax had gone from 9 percent in natural infections to 1 percent after vaccination. As for mortality due to vaccination, it was negligible. The prophylactic effect of anthrax vaccination was evident, even if the immunizations had to be repeated every spring. Inoculation with the Pasteur vaccine gradually became a normal part of the farmer's life.

Since he had understood that he needed to be on the scene to convince the veterinarians and the farmers, Pasteur became a traveling salesman for his discovery. He traveled all over France to defend his process. He often had to moderate the impatience of those who looked for a single vaccine to cure all diseases. "I have barely finished my experiments on anthrax vaccination, and you ask me to find a cure for sheep-pox. Why not for phylloxera?"[45]

As it happened, Pasteur was not interested in phylloxera (which was just then causing terrible ravages in the French vineyards), but in erysipelas or swine fever. In late 1882, he went to Bollène in the department of Vaucluse, accompanied by Loir and Thuillier, to study the disease and identify the bacillus that causes it. The epidemic of the "red disease" was serious. "Not ten thousand but at least twenty thousand pigs have died, and the malady is even more widespread in the Ardèche," Pasteur explained to his wife. He immediately set to work searching for a vaccine based on attenuated cultures. The experiments were to be continued in the Paris laboratory. "Thuillier will return to Paris right away," Pasteur wrote on 3 December, "because we have just bought ten piglets that he will take with him and nurse, if necessary. . . . Young or old, pigs do not tolerate cold well. We will bury them in straw. They are very young and appealing, and one really grows fond of them."[46]

Pasteur was confident. Yet developing a vaccine turned out to be more difficult than expected. The effectiveness of the product depended on the receptivity of the different strains of pigs. A note on this subject was presented in April 1883 to the academies of sciences and medicine, but no trace of this research can be found in Pasteur's later publications. Does this mean that he became discouraged and turned away from this difficulty? More likely he thought that it was sufficient for him to indicate the path to be followed; and indeed a vaccination process was soon developed on the basis of Pasteur's notes and experiments. After 1886, more than 100,000 pigs were successfully vaccinated in France and Hungary.

Meanwhile, for all the importance of this work on chicken cholera, anthrax, and swine erysipelas, and for all his achievements, controversies, and solutions to experimental problems, Pasteur had not yet approached the subject that was to ensure him a permanent place in the legend of vaccination, the study of rabies. In popular opinion, rabies vaccination is still Pasteur's principal claim to fame. For rabies is a disease of both animals and humans, and he was to study its transmission using animal models before embarking upon experiments on humans. It was to be the culmination of his thinking on vaccination and confirm the possibility of stimulating immunity. But at the same time it was to force him to reexamine his first discoveries on the manufacture of vaccines, as he himself acknowledged: "It became clear that the methods that had served for these viruses (of chicken cholera, saliva, swine erysipelas, and even acute septicemia) were inapplicable or insufficient when it came to rabies. It therefore became necessary to think of new methods."

Pasteur was about to enter a new stage, one in which he would use the attenuated virus no longer to prevent disease but to cure it.

Rabies Must Be Defeated!

R abies, Monsieur, is at this time the object of your studies; you are searching for its microscopic organism, and you will find it; humanity will owe you the extermination of a dreadful disease, along with the end of a sad anomaly, by which I mean the slight uneasiness we always feel at the caresses of the animal in which nature best shows us her benevolent smile."[1] In 1882, when Renan welcomed Pasteur under the cupola of the Académie française, the latter had not yet defeated rabies. In a small ironical barb at the famous free thinker, René Vallery-Radot was to write that when he spoke these words, Renan hoped to become a prophet, for once in his life. As it happened, he was only half a prophet, for if Pasteur actually did triumph over rabies, he was never to see its infectious agent, even though he assumed that it was present in the brains of rabid dogs.

To designate this microbe, which he was unable to show, Pasteur employed the word *virus*. He chose this term because it was more abstract, because it depicted contagion as a harmful force exerted on the organism. As a good Romantic, Pasteur liked metaphors, and this one amounted to a veritable premonition, for he was able to divine the virus because he expected to find it, and he never doubted its existence just because it could not be seen. A discovery of the invisible, the rabies virus was real because it conformed to the principles of contagion.

For Pasteur was indeed dealing with a virus. At the time the word had long been used to designate an infectious agent, a poison, or a miasma. But it was not until 1898, following Martinus Beijerink's studies on the mosaic virus of tobacco and the work of Friedrich Löffler and Paul Frosch on hoof and mouth disease, that the viruses would be identified as microbes that are invisible under the optical microscope and capable of passing through filters that can filter out bacteria. As for the pathogenic agent of rabies, it was only in 1903, eight years after Pasteur's death, that experiments made it possible to incorporate it definitively into the category of "filterable viruses" and that scientists were able to discern optically in the cerebral matter of rabid animals the characteristic inclusions that came to be called "Negri's bodies." The existence of viral particles would be actually proven only in 1963, thanks to the electron microscope.

This makes it all the more remarkable that Pasteur was not stymied for long by this microbe that differed from all those he had pursued until then. Having tried in vain, following his usual method, to cultivate it in order to isolate it, he soon realized that he was attempting the impossible. This being the case, he became obsessed by one idea: to find a way to halt the march of this invisible enemy. It is not the least of Pasteur's merits that he was able to hold on to his conceptions in the face of an exception that seemed to invalidate them.

Rabies was the last step in the biologist's career. Pasteur was finally coming to grips with human pathology. But despite the prevailing popular imagery and the hagiographical accounts, this did not make Pasteur into a physician. It was not he who plunged a syringe into the rabies-stricken patient's abdomen, for he was far above such a gesture; he told others how to do it.

Yet it is true that Pasteur knew how to use popular fears and fantasies to impose a new medicine. He now went beyond the prophylactic approach and invented active immunotherapy, treatment by stimulation of the immune system. His victory over the slavering dog amounted to that of human experimentation as a means of "expanding the frontiers of life."

Myths and Realities

Rabies has always filled the popular imagination with obscure and excessive dread. The savagery of this unknown evil, hidden by the tall trees of the forests, lurking under hedges and around the bend of a sunken path, has left indelible marks in popular culture. Who has never heard of these enraged animals, howling of death with bloodshot eyes, foaming at the mouth, and ready to pounce on the hand that comes close? Rabid beasts — dogs, foxes, or wolves, wild or domestic — roaming the countryside truly became objects of terror and malediction. For their disease could be transmitted to man. Their bite was deadly and the disease terrible. It brought days and weeks of anguish, and an agony preceded by atrocious suffering, for the incubation period was very long. Moreover, and this was perhaps even more disquieting, in addition to the rabies of panting and trembling animals shaken by spasms and as furious as they were aggressive, there was also the silent or "mute" rabies, in which the anxious and withdrawn animal licked the friendly hand that came to comfort it and instilled in it the poison of its infectious slaver.

From Antiquity to the Enlightenment, rabies had been a constant topic of discussion. Populating legends, poems, and stories, it was also prominent in the recipes of apothecaries and in the flummery of charlatans. A specter of death, it was disquieting and frightening. People avoided animals that showed any change of temper, pets that turned into strangers and became aggressive, refusing to eat and especially drink. For the fear of water, or hydrophobia, was one of the obvious signs of the affection, so that communities would place buckets of water at the street corners of villages as a means of identifying dangerous animals. Both learned and folk medicine tried everything to invent new remedies, evacuants, sudorifics, poultices, using every variety of ingredient, among them the burnt hair of a bear, droppings of red roosters, shrew's tails, pieces of swallow's nests, and of course all kinds of different herbs. Pierre Chirac, a famous physician of Louis XIV's time, gave a recipe for an antirabies omelette containing the roots of the dog-rose: one piece of this omelette was to be placed piping hot on the wound while the patient ate the rest. Nine days later, unless he had died, the patient also had to drink theriac dissolved in wine — and then wait, always wait. The anxiety over the incubation period was such that most victims had recourse to spells; magic formulas were transmitted in whispers, and the most ancient superstitions brought comfort to the most devout parishes.

This mythical malady had given rise to few theories. The propositions of Fracastoro, who in the late fifteenth century had seen rabies as a contagious disease transmitted by seeds that could enter the organism only through broken skin, had been succeeded by the mystical doctrines of Paracelsus. Rabies had thus long been considered a supernatural phenomenon because its pathology was so mysterious and its symptomatology so dreadful. These hermetic traditions and unverifiable theories were passed on by medical science until at least the second half of the nineteenth century. In 1855 Littré still echoed them in his *Dictionnaire de médecine*.

Yet even though this malady was the subject of so much discussion and caused such terror in the countryside and in urban back streets, it actually had no quantitative importance. As Duclaux pointed out, there were very few cases of rabies. Only a few hundred deaths from rabies a year were registered in France. In Germany, some simple policing measures had already brought a decline in the incidence of rabies, so that even without an antirabies institute the malady caused fewer ravages in the German Empire than in Paris. In any case, it is certain that for man rabies was a scourge of entirely different scope than cholera, plague, or typhoid.

This being the case, one may wonder about the reasons that prompted Pasteur to take an interest in this infection which, even though it struck popular imagination, remained rare. It has been said that this choice was guided by a childhood memory. Pasteur, we are told, recalled the dread he felt one day when he heard about a rabid wolf that roamed through the region of Arbois biting man and beast. He was not even nine years old at the time. While playing with boys of his age on the Courcelles bridge near his home, he saw a group of men going into the blacksmith's house. They were bringing in a poor devil who had just been bitten by the horrible beast. His torso bare, the man had come to have his bites, which were covered with bloody slaver, cauterized with a red-hot iron. Other victims bitten on hands or face died in horrible pain, suffering from hydrophobia. Eight persons were reported injured in the vicinity of Pasteur's house. The terror of the rabid monster continued to haunt the entire region for a long time.

Yet one should not exaggerate the importance of this episode, however great its emotional impact. At most, it may have contributed to Pasteur's decision; the tragic image of such a memory is not sufficient to be decisive. More likely, one should see this new research project as one of the scientist's last attempts to pry open the mysteries of nature. It was the epic gesture of a man who sought to overcome legends, to fight myths with reality. At the same time, one can also assume that Pasteur purposely chose a disease that struck people's imagination in order to create the lasting publicity he needed to promote his immunization procedures.

Clearly, this is how Roux summarized the question: "This malady is one of those that causes the smallest number of victims among humans. If Pasteur chose it as an object of study, it was above all because the rabies virus has always been regarded as the most subtle and most mysterious of all, and also because to everyone's mind rabies is the most frightening and dreaded malady. . . . He thought that to solve the problem of rabies would be a blessing for humanity and a brilliant triumph for his doctrines."[2]

Pasteur himself was to say the same thing when in 1888 he wrote to Grancher that his fight against rabies had above all allowed him to give the widest possible circulation to his discoveries about vaccination. "Recall that at the time I undertook research on rabies only with the idea of forcing the physicians to pay attention to these new doctrines in case these studies would yield some medical information."[3]

Deadly Saliva

In December 1880, when Pasteur began working on rabies, he was well prepared. A letter dated 1878 indicates that he had been interested in this subject for several years. By 1880, the number of rabid dogs had grown so much that in the space of two months a veterinarian of Paris reported five deaths from rabies. There was thus no lack of experimental subjects either in the kennels of Maisons-Alfort or in the offices of Parisian veterinarians who dealt with this disease.

One of these was Jean Bourrel, who in 1862 had created a clinic near the Canal Saint-Martin, at 7, rue Fontaine-au-Roi. Specializing in the study of rabies, he had assembled a large body of data on the various forms of the disease and even written a book that provided his principal therapeutic prescriptions. According to him, the best preventive treatment was to file down or pull out the incisors of dogs and cats! Nor was this all: in July 1880, his nephew and assistant Pierre-Rose Bourrel was bitten by a rabid dog and succumbed after several days of atrocious suffering. Perhaps it was this family tragedy that prompted Bourrel to contact Pasteur. In December 1880, he sent him two specimens from rabid dogs for examination in his laboratory. By that time Pasteur had become a member of the Central Society of Veterinary Medicine and had just distinguished himself by his work on anthrax and chicken cholera. He therefore appeared to be the best choice of a scientist who might take an interest in rabies; and after all the master of the Ecole normale was bound to know more than a veterinarian of the rue Fontaine-au-Roi.

Also in December 1880, Pasteur was informed by Dr. Marie Lannelongue of the Trousseau Hospital that a child of five, who had been bitten in the face a month earlier, had been admitted to his clinic. Agitation, convulsive movements, furious screams, hydrophobia — there could be no doubt about the diagnosis; it was rabies. After twenty-four hours of delirium, the child died, choked by saliva and mucous matter he was unable to swallow. A few hours after the death, Pasteur took a sample from the body, using a brush that allowed him to take up some of this deadly mucus. On his return to the laboratory, he inoculated it into some rabbits.

In doing so, Pasteur did not innovate. On the contrary, he only proved that he was familiar with the earlier literature. For in fact, Pasteur had a close predecessor in the experimental study of rabies, Pierre-Victor Galtier. In August 1879, this young professor at the Veterinary School of Lyon had communicated to the Academy of Medicine a note that had attracted Pasteur's attention: "Describing the first experimental inoculation of the disease, he showed that rabbits are the animals of choice for developing the rabies of suspect dogs after inoculation."[4]

But the experiment conducted by Pasteur turned out to be baffling: The saliva taken from the dead child killed the rabbits too quickly, within thirty-six hours. The incubation period was too short to be consistent with rabies. Moreover, the rabbits died presenting unusual symptoms, namely lungs filled with serosities. While Marie Lannelongue and his colleague Maurice Raynaud, who also reproduced the experiment, believed that they had obtained a model of experimental rabies and hastened to describe what they believed to be the first case of transmission from man to animal, Pasteur was more cautious and more perspicacious. He realized that his rabbits had died from a malady other than rabies, for it did not give rabies to dogs. Unwilling to draw immediate conclusions, Pasteur examined the individual parts of the experiment and discovered in the blood of the rabbits that had died in the laboratory a microscopic organism in the shape of an eight, which multiplied in chicken broth or Liebig's bouillon. This same microbe was also found in the saliva of patients hospitalized for illnesses completely different from rabies, and the description Pasteur gave of it reveals that what he had discovered was the pneumococcus.

For the scientist's adversaries, led by Michel Peter, this discovery only proved that one cannot trust the microbes. Pasteur claims to be working on rabies, it was said, but he is actually studying a different affection. This was a piece of unfairness typical of Peter, whom Pasteur firmly put in his place: "This is indeed a new disease produced by a new microbe; neither the microbe nor the disease has been described before. This tenacity in research, Monsieur, is the honor of our work, and it was because we, my collaborators and myself, pursued these experimental combinations that we were able to demonstrate that the new disease existed in the buccal mucus of children who had died of the same disease as well as in the saliva of perfectly healthy persons. It was then, and only then, that I had the right to assert that the new disease had no relation with rabies."[5]

If the pneumococcus thus temporarily retained Pasteur's attention, rabies was still his major obsession. In 1881, he could write in his laboratory notebook, "Last December, with the help of Chamberland and Roux, later joined by M. Thuillier, we began the study of rabies."[6]

Pasteur did not imagine for a moment that rabies, transmitted by a bite, might in some way be spontaneous. He had no doubt as to the infectious nature of the disease. Everything — the transmission to man or animal and the variations in virulence depending on species and individuals — pointed in that direction. Yet neither the most patient investigations under the microscope, nor the most ingenious staining of the slides made it possible to isolate the microbe responsible.

In conducting this study, Pasteur made use, as he always did, of earlier studies, not only those of Galtier but also those of Duboué. These first investigations of the disease were extremely helpful to him because they came to opposite conclusions.

In January 1881 Henri Duboué, a hospital physician in Paris who had been publishing papers on rabies since 1879, sent his work to Pasteur. Duboué was particularly interested in the evident localization of the disease in the cerebral matter. "The morbid agent," he wrote most notably, "progresses slowly, in a centripetal direction, from the location of the bite to the *medulla oblongata*, and very rapidly, in a centrifugal direction, from this latter organ to the nerves that originate from it."

As for Pierre-Victor Galtier, he was worried about Pasteur's studies and could easily have done without such a prestigious rival. That is why he hastened to communicate to Henri Bouley some new findings, asking him to communicate them to the Académie de médecine as soon as possible. In early 1881, he asserted that he had found the rabies virus only in saliva and also pointed out that nerve tissue was not contagious. "I have inoculated more than ten times, and always without success, the product obtained by expressing the cerebral substance of the cerebellum and that of the bone marrow of rabid dogs."[7]

On the basis of these opposing scientific statements, Pasteur constructed a formal problem to be solved and also deduced from them outlines of a procedure to be followed. The first step was to localize the rabies virus and to answer the following questions: Is the cerebral substance infected or not? Is the saliva the only place where the virus is present? If Pasteur thus took the path indicated by Duboué and Galtier, it was because he had failed to isolate the rabies virus in saliva and was looking for new ways to inoculate it. He therefore chose the cerebral substance of the rabid dog as his first and principal approach to the disease.

Before we follow Pasteur's experiments step by step, it behooves us to point out that Galtier's contribution to their success has never been sufficiently emphasized. Much later, to be sure, in 1908, he was considered for the Nobel Prize for medicine, but Galtier died a few months before the meeting of the prize committee, which then awarded the prize to Ehrlich and Metchnikoff.

The Flask with Two Openings

Rabies was a challenge to the experimenter. The only practicable way to transmit it was to have a healthy animal bitten by a rabid one. However, this means of contagion was not always reliable. But there was no choice, for experiments could not be carried out without a valid model. In rabbits, guinea pigs, and dogs, the incubation period was long. Several weeks could elapse between the introduction of the presumed virus and the first manifestations of rabies. Moreover, transmission was uncertain and irregular. Whether bitten or inoculated, the animal might not develop the disease. "It is torture for the experimenter," Pasteur confessed in connection with these difficulties, "to be condemned to wait for months on end for the result of an experiment."

By May 1881, however, less than six months after the beginning of his research, he was able to communicate his first findings simultaneously at the Académie des sciences and the Académie de médecine. He had developed an infallible experimental method, consisting of inoculating cerebral matter taken from a rabid dog directly into the skull of a healthy animal through a trepanned opening. When deposited on the surface of the dura mater, the outermost of the meninges surrounding the brain, the nerve tissues taken from a rabid animal transmitted the disease in every case. Moreover, in the animal inoculated in this manner, the first signs of rabies (in either the furious or the mute form) appeared within a much shorter time, only one or two weeks. Death ensued in less than a month.

This procedure might seem to be simple and dictated by the logic of experimentation, but no one had thought of it or tried it before Pasteur and his team. Roux, incidentally, reported that Pasteur himself had briefly held up the implementation of this trial because of his reluctance to experiment on an animal like the dog. Perforating the skull of the animal in order to perform the intracerebral injection was a manner of inoculation he found difficult to accept, even if he was not particularly fond of dogs. It was therefore Roux who, one day when Pasteur was absent, did the deed and performed the first inoculation. The next day, when Pasteur learned the news, he pitied the animal: "Poor beast, no doubt its brain is injured; it must be paralyzed." Roux went down to the basement and came back with the animal, which immediately started gamboling happily about the laboratory. Learning that trepanation was bearable put an end to all of Pasteur's misgivings; from that moment on, the experiment was reproduced many times and yielded consistent results. As for the first trepanned dog, it developed rabies after fourteen days of incubation.

Now that the experimental protocol had been defined in this manner, it was time to turn to the most important task, which was to find a vaccine. To accomplish this, Pasteur thought he could obtain an attenuated virus, as he had done for chicken cholera, anthrax, and swine erysipelas.

But how could one be sure that attenuation had actually occurred when dealing with a virus that either killed or produced no reaction at all? The symptomatology of rabies was complex; its manifestations varied with the area of the brain where the virus settled and replicated. The cerebral substance of an animal affected by the most silent form of rabies could produce the most furious form in the animal that was inoculated with it. There was the same uncertainty about the incubation period: if one contaminated several animals with brain matter from the same rabid dog, the reaction time from one to the other could vary from one to three months and more. Although the mode of intracranial inoculation did shorten the incubation period, it did not take care of everything. Pasteur's first concern was therefore to obtain a disease entity whose appearance and symptoms were completely controlled. In order to study the disease in its various aspects, he chose to adhere to only one fixed point of reference, the incubation period.

Once this decision was made, Pasteur went on to the essential, that is, to the search for a means to inactivate a highly potent virus, for he believed that the acquired immunity would be stronger if it were provoked by a highly virulent germ. In order to heighten the potency of the virus, Pasteur used a stratagem that had already served him in the case of chicken cholera, namely, the method of successive passages. But instead of resorting to the traditional culture broth, he made use of animals for modulating the virus. What interested Pasteur was to find out as much as he could about all the pathogenic forms of the virus; he was not asking questions about why things happened, but only how. In this study of the relation between the host and its parasite, which to this day is the object of the most advanced research, Pasteur realized that he had come to the limits of his art, or rather those of the technical procedures available to him. That is why he did not want to become mired in fundamental questions and instead dealt with their experimental aspects.

In point of fact, he noticed that in the process of inoculating the cerebral substance of a rabid rabbit into another rabbit, and from there into another, and so forth in successive passages, the incubation period became noticeably shorter. It decreased from twelve to ten, and finally nine days. At the twenty-first passage, rabies became symptomatic within eight days. Begun on 19 November 1882, this long series of experiments thus provided Pasteur with an incubation period which, shortened to its extreme limit, was constant and fixed in rabbits. A reduced incubation period spelled intense virulence, as Pasteur was able to verify immediately. When injected into a dog, this rabbit rabies produced much stronger manifestations than the common form of the disease.

Pasteur brought all his rigor and all his patience to bear on this research. He proceeded step by step, as is revealed, for example, by his little notebook of observations for the summer of 1883: "Trial with a tube kept at 33 degrees after five days. 28 June: inoculated two fresh rabbits by trepanation. One becomes rabid on 11 July. The second is still quite well on 23 October. In other words, after four months. It is very odd that one was rabid, and not the other."[8] Nonetheless, in the course of his experimentation, Pasteur eventually succeeded in triggering the infection at will.

Thus, without even having isolated the virus, he was able to use it experimentally. Now he had to carry it all the way to the vaccine: "What use can be made of our recent discovery of the existence and the production of different kinds of rabies, all of which are more violent and more rapidly mortal than the known rabies of dogs? The man of science does not scorn anything he can discover in pure science; but ordinary people, terrified by the very thought of rabies, demand more than scientific curiosities. Surely they would be much more interested to learn about rabies viruses if their virulence could be, on the contrary, attenuated. This would give them hope that we can create rabies viruses to be used as vaccines, as we have done for the chicken cholera virus, the saliva microbe, and swine erysipelas."[9]

Using the virus meant trying to weaken it. In a first phase, Pasteur now tried to obtain attenuated virulence by a Jennerian maneuver. This does not mean that he returned to empiricism, only that he relied on a principle that he had successfully used until then, namely the modification of the milieu in which the germ replicates. Whether artificial (culture broth) or natural (the animal), the milieu exerts an influence over the virulence of the germ. By transforming the environment, one therefore also transforms the parasite, for the germ and its environment change their character together. Since this virus could not be cultivated in the incubator,

Pasteur therefore had the idea of looking for a likely subject — in this case an animal species — to weaken it and leave it nonmalignant. At this point he remembered Jenner's first trials. Jenner had introduced into science the idea that the virulence of a microorganism can vary with the animals in whose organisms it develops and who had been fortunate enough to find the cowpox of cattle to bring his research on human smallpox to a successful completion. Confronted with rabies, Pasteur for his part also experimented with different animal species, rabbits, dogs, guinea pigs, and even monkeys. In doing so, he noticed that the passage from dog to monkey weakened the virus and fixed it in a state of reduced virulence; also the incubation periods became longer, which is a sign of a weakened virus.

Attenuated virulence. Pasteur had found what he was looking for to prevent the affection. In a first phase, he used the virus weakened by this method as a vaccine. If he caused animals inoculated with the monkey virus to be bitten, they usually did not contract rabies. This was the description of the first refractory dog.

In August 1884, the Second International Congress of Medicine was held in Copenhagen. Pasteur represented France; Rudolf Virchow, Germany; Sir James Paget, England. At the opening session, in the presence of the King and Queen of Denmark, the King and Queen of Greece, amid all the participants, Pasteur exclaimed: "Gentlemen, science may not have a fatherland, but the man of science must do the utmost for the glory of his country. In every great scholar you will always find a great patriot."[10]

If Pasteur diligently attended the sessions of the Congress, he also took advantage of his stay in Denmark to go to Carlsberg accompanied by his son Jean-Baptiste, who at the time served as secretary of the French embassy. There he was welcomed by his old friend Jacobsen.

But Pasteur had little time to savor the memories of past accomplishments. At Copenhagen the journalists were eager to interview him about rabies. They were very excited about the refractory dogs. The spontaneous nature of the disease definitively seemed to be a thing of the past. Pasteur's reputation was spreading far and wide; his name and his discoveries concerning rabies were now known not only in the cities of France and other countries, but in the tiniest villages, even in Alsace. Yet when he left Copenhagen for Arbois, where he hoped to have a few days of rest, Pasteur was not satisfied. From his retreat in the Franche-Comté he continued to direct the experiments pursued in Paris by Adrien Loir and Eugène Viala, who took turns performing trepanations and writing out lengthy observations while also watching over the rabid dogs and rabbits.

The first experiments with vaccination did not satisfy Pasteur because he was looking for something other than prophylactic vaccination. He knew that the virus with which he was experimenting was nothing more than a weakened virus and that such a virus could not be truly effective, particularly since its method of utilization required too much time. "It [this method] does not easily lend itself to immediate application, and yet this is a necessary requirement, given the accidental and unforeseen nature of rabid bites." This is why Pasteur now envisaged a new method, one that no longer consisted of preventing infection before any bite had occurred, but of treating it once the germ had entered the organism and before the disease had become symptomatic. For this he needed a vaccine that acted more quickly than the disease itself.

This decision seems to contradict all the principles of vaccination. Could rabies really be treated by a preventive measure? This is precisely where Pasteur had a brilliant intuition. What needed to be done was to stimulate a response of the host organism, one that was stronger and more rapid than the disease that slowly spread from the point of the bite to the vital centers of the nervous system. The virus of the monkey, which was weak to begin with, therefore was not suitable. On the contrary, the virus that had to be used would have to be terrible and extremely virulent to begin with, but later attenuated. It also had to possess two distinct and seemingly incompatible characteristics, namely the rapid action of the strong and the harmlessness of the weak virus. Obtaining a "weak strong one" now became Pasteur's new quest, and his technique was to heighten the virulence of the virus in a first phase and to annihilate it in a second.

By proceeding in this manner, Pasteur distanced himself from the Jennerian method. The problem of rabies inspired him to develop an original, elaborate, and well-thought-out method that was far removed from the traditional principles of Pasteurian vaccination. It was a huge step forward, for here Pasteur left the field of bacteriology, itself still in its infancy, to become the first to venture into that of immunology, which he founded as a new science that would provide the means of understanding and manipulating natural immunity.

"I have arrived at a practical and rapid method that has already provided sufficiently numerous and reliable successes in dogs to give me confidence that it will eventually be generally applicable to all animals and indeed to humans."[11] This is what Pasteur said on 26 October 1885 in a progress report on his work before the Académie des sciences.

Did he accomplish this by himself? The experiments on rabies certainly provide the best insight into the relationship between the master and his disciples within the laboratory. Chamberland and Thuillier (until his death in 1883) carried them out; Loir performed the manipulations. But the man who handled the trepan and the syringe was Roux, whose ambition and imagination now came into their own, so that rabies became a subject of discord between Pasteur and his best pupil. Both were deeply committed to this research; they were working side by side, yet almost independently. This work had become a matter of self-respect: Roux felt that he needed rabies to further his career, while Pasteur was obsessed with his fame for posterity. As it happened, Pasteur in a sense owed the definitive elaboration of the attenuation procedure for the rabies virus to Roux. Yet he made use of his collaborator's discoveries without informing him and without acknowledging his contributions. For a time, a climate of distrust and deep resentments poisoned the relationship between the two men.

It is therefore important to pay close attention to the unfolding of the experiments. In 1884, Pasteur sought to find out how long he could preserve the virus of the nerve disease that he wanted to use as an inoculate. To do this, he used the traditional matrasses for seeding his cultures. As usual, Loir performed the operation while Pasteur, standing behind him, carefully supervised his every move and followed him whenever he left the laboratory to go to the incubation room. Loir reports:

> One day, at Pasteur's request, I went to fetch eight culturing matrasses that I had deposited in the seeding room where they had remained for two hours, which is the normal procedure if you want to make a culture. . . . At the end of this time, I went to get [Pasteur]. He sat down next to me, a little behind me, as he always did. After the seeding, which I performed under his eyes, I left the room, walking in front of him, and deposited the tray with the eight matrasses on the table next to his desk. The labels for them were ready. I pasted them on with the great care that I used in every manipulation if Pasteur was watching. Then we made our short way back from the laboratory to the incubation room which, maintained at 37, was located off the hallway, with me still holding my little tray with both hands, and Pasteur following behind me. I stepped aside when we reached the door of the incubation room and, as usual, he opened it. I deposited the little tray on the left side of the shelves where Pasteur kept his own cultures. When I turned around, I saw Pasteur stopped before a 150 cm³ flask with two necks, one on the top and one at the bottom, designed to cause a draft inside the flask. In this flask one could see, suspended by a string, a piece of a rabbit's spinal marrow. The sight of this flask, placed at eye level, seemed to absorb Pasteur to such a degree that I did not want to

disturb him. To prevent the loss of heat from the incubation room, I closed the door we had left ajar. After a long moment of immobility, he asked me: "Who put this flask there?" — "It can only have been M. Roux," I replied; "this is his rack." He took the flask and walked out into the hallway. Lifting it up, he looked at it in bright daylight for a long time, then he returned to his work station without breathing a word. . . . That afternoon, Roux came in and I heard him call: "Sonny!" (that is what he called me). I saw him standing by the open door of the seeding room. "Who has put these three flasks here?" he asked, pointing to the table. "It was M. Pasteur," I replied. "Did he go into the incubation room?" — "Yes, he did." "Did he see the flask on my shelf?" — "Yes, he did." Not another word was said. Roux took his hat, went down the steps and out into the rue d'Ulm, slamming the door behind him as he always did when he was angry. I had just witnessed a great drama. I only understood this later. I never heard him [Roux] say a word to Pasteur about the flasks, and I don't know if they ever talked about them, but I don't think they did. But from this moment on, rabies became a dead letter for Roux; he stopped working on it and no longer came to the laboratory during the day. . . . It was with these flasks that were sitting on the table of the seeding room that Pasteur obtained the attenuation of the rabies virus.[12]

What exactly had happened? What was it that so forcefully struck Pasteur? It was the flask with two openings used to suspend spinal marrow, an idea that had never occurred to him. He therefore sent his laboratory assistant, Jean Arconi, to the supplier of glassware, asking him to bring back a dozen flasks conceived on the principle of Roux's flask, only larger. Pasteur then asked Loir to bring him pieces of caustic potash, to deposit them in the bottom of the flasks, and then to place sterile cotton into the two openings as a means of drying out the air.

Dried air! Here was the secret that made it possible to attenuate the rabies virus. Drying was actually an age-old process employed to preserve the properties of a natural process. The Incas of Peru used it to preserve their foods, and even today we resort to freeze-drying for keeping vaccines. As for the air, which was filtered and sterilized by the cotton wads, it produced the gradual attenuation of the virulence, for as the spinal cord dried out through contact with the air, the virus became weaker and weaker. Within two weeks the spinal cord of a rabid rabbit had lost its virulence. Conversely, as Pasteur was to point out, if the rabid spinal cord was kept out of the air and in a humid state, "its virulence is preserved and does not undergo a change in intensity, provided it is protected from all alteration by outside microbial agents." In fact, the weakening of the virulence does not occur exactly for the reasons Pasteur supposed; it is a complex process involving many factors, among them the potash used to dry the air, the thickness of the suspended spinal cord, the temperature, and the length of exposure. Pasteur thus made use of an empirical procedure that gave him a minimum of humidity, so that one commentator could later say that he owed his success to the erroneous interpretation of a correct fact. That is to put it rather strongly. Perhaps one should simply stress the curious role that luck plays in all great discoveries.

Pasteur thus had his "weakened strong one," which allowed him to embark on a new course of experimentation.

> Here are the means of making a dog refractory to rabies in a relatively short time. In a series of flasks whose air is maintained in a dry state by pieces of potash deposited at the bottom, one suspends every day a piece of fresh rabid spinal cord from a rabbit killed by rabies, rabies that has broken out after seven days of incubation. Every day as well, one inoculates under the skin of the dog one full Pravaz syringe of sterile broth to which one has added one of these drying spinal cords, beginning with a spinal cord bearing a running number far enough removed from the day of this operation to be sure that this cord is completely nonvirulent. Earlier experiments have made this point clear. In the following days, one operates in the same manner every other day with more recent spinal cords, until one arrives at a last and very virulent cord that has been placed into the flask only a day or two previously. At this point the dog has become refractory to rabies. It can be inoculated with the rabies virus subcutaneously or even on the brain surface by trepanation without causing symptomatic rabies.[13]

So now the procedure had been described. Pasteur was so sure of himself that in March 1885 he wrote to Jules Vercel, his old classmate at Arbois: "I have not yet dared to treat humans bitten by rabid dogs. But the time to do it may not be far off, and I would really like to begin with myself, that is, to inoculate myself with rabies and then stop its effects, for I am beginning to feel very competent and sure of my findings."

Vivisection: Necessity and Protest

Pasteur's laboratory soon learned how to deal with rabies. Clearly, the animal did not replace the culture broth; it became its extension. But it also introduced new experimental difficulties related to the keeping of the animals and especially ethical questions. Moreover, manipulating rabies was not without danger, so that everyone had to take stringent precautions to avoid being bitten or pricking himself accidentally.

The first stage of the experimentation used the slaver itself of the rabid animals. It was surely a daring feat, a true tour de force to collect the slaver of a furious dog foaming at the mouth. René Vallery-Radot recalled with horror the scenes he witnessed in his father-in-law's laboratory.

> M. Pasteur invited me to accompany him and we set out carrying six rabbits in a basket. The two dogs were rabid in the last degree. Especially the bulldog, a huge bulldog, was howling and foaming in its cage. An attendant extended an iron bar toward it; it pounced on it and he had great trouble pulling it from the dog's bleeding fangs. Then one of the rabbits was brought to the cage, and one of the floppy ears of the frightened creature was pushed through the bars. But despite the goading, the dog jumped to the back of the cage and refused to bite. "We absolutely have to inoculate the rabbits with this slaver," said M. Pasteur. Two helpers took a cord with a slip-knot and threw it at the dog as one throws a lasso. The dog was caught and pulled to the edge of the cage. They seized it and tied its jaws together. The dog, choking with rage, its eyes bloodshot, and its body racked by furious spasms, was stretched out on a table, while M. Pasteur, bending a finger's length away over this foaming head, aspirated a few drops of slaver through a thin tube. It was in this veterinarian's basement, and at the sight of this awesome tête-à-tête that I saw M. Pasteur at his greatest.[14]

Working with the brain and the spinal cord was equally delicate. Intracranial injections involved two steps. First, one had to obtain the rabid spinal cords, which means that one had to have available a sufficient number of rabid animals to be able to operate in keeping with the rigorous technique from which Pasteur did not allow the slightest deviation. The rabid animals, which were kept in a special kennel, were killed with every possible precaution against bites. Once the dome of the skull was removed, the brain lay exposed. Before removing a piece of the rabid medulla, the affected area was seared with a glass rod in order to destroy any possible contamination. Then, using a tapered tube flamed beforehand, a small amount of this brain matter was aspirated and then diluted in water or sterilized culture broth. Now began the second phase of the experiment. The mixture was aspirated with the Pravaz syringe and had to be inoculated. At this point, experimental dogs or rabbits were put to sleep and kept on the table. While one of the two operators held down the trussed animal to prevent sudden movements, the other bored a hole with a kind of brace and bit. A large apron that was changed at every operation and a cap covering the head conformed as well as could be done to the rules of asepsis. But although the operators washed their hands, they did not wear gloves. Following the trepanation, the injection was quickly made, directly into the brain. The last step was to close the incision without being bitten when the animal woke up.

Dr. Roux's niece, Marie Cressac, has described the atmosphere of the first experiments:

> [Roux], Chamberland, and Thuillier bent down around a table. A large dog was tied down on it, its muscles contracted and its fangs bared. . . . If the animal, despite all the precautions, had caused them to make a false move, if one of them had cut himself with his scalpel, and if a small piece of the rabid spinal cord had penetrated into the cut, there would have been weeks and weeks filled with the anguished question: will he or will he not come down with rabies? . . . At the beginning of each session, a loaded revolver was placed within their reach. If a terrible accident were to happen to one of them, the more courageous of the two others would put a bullet in his head. . . . They were no longer just "researchers" absorbed in the meticulous work of their laboratory; they were pioneers, adventurers of science.[15]

These serial inoculations transformed the appearance of the laboratory. Dogs were now added to the chickens, rabbits, and guinea pigs. Dogs liable to bite and dogs liable to be bitten filled the laboratory. The healthy ones served as subjects for experiments or as controls, while those that slavered were the agents of rabies. Soon there was not a case of rabies in Paris of which Pasteur was not informed; veterinarians would cable him: "poodles and bulldogs in the throes of an attack . . ." and a pound wagon immediately came to pick up the poor beasts. The basements of the rue d'Ulm filled up with various kinds of cages. At night, the whole neighborhood was haunted by the howling of rabid dogs. The most powerful animals, Great Danes and German shepherds, would tear up the floor of their niches and furiously twist their chains. Litter, excrement, and sometimes blood-stained slaver soiled the floor. Between the maintenance of the premises and the regular checking, at fixed times, of the rabies symptoms, there was no lack of work to be done. But what was lacking was space. The dogs were counted by dozens, and the rabbits were even more numerous. The refractory animals in particular, those that did not die, had to be carefully watched for weeks or even months. It soon became evident that the team could not continue to experiment in the tight quarters of the rue d'Ulm or the rue Vauquelin or in the cellars of the Lycée Rollin, where Pasteur kept a great many chicken cages and kennels. He therefore had to turn to the veterinarians who were willing to collaborate in his experiments. "Is your kennel full?" he inquired of Bourrel on 3 April 1884. "Could you keep a few spaces for me, beginning with 4 niches for four dogs? M. Leblanc has already furnished me room for seven dogs at Montmartre. . . . I am paying him 75 centimes per day and per dog. That is what I would pay you as well."[16] But before long, the veterinarians also ran out of space.

It was therefore imperative to find a new site where a vast kennel could be built. First, the choice fell on the forest of Meudon, an isolated place far from the noise of the city. But if the experimenters considered this a very convenient location, the inhabitants of Meudon, stirred up by their mayor, felt differently about it. The mayor asserted loudly and clearly that he respected and admired Pasteur's work, but insinuated underhandedly that a kennel installed in the forest would constitute a constant threat to his fellow citizens. The deputy of Seine-et-Oise, Léon Say, intervened on behalf of the scientists, but to no avail: the press whipped up a veritable public outcry against the purveyors of rabies!

At this point Pasteur learned of an estate called Villeneuve-l'Etang, located in the park of Saint-Cloud (in the village of Marnes-la-Coquette). This too was a rather remote place, where before the war of 1870 a small château had stood. It had been owned by the Duchess of Angoulême, then by the Duc Decazes, and finally by Napoleon III, who found this discreet annex to the palace of Saint-Cloud very convenient. Although not much was left of the main building, some of the outbuildings, particularly the stables, were still serviceable. The sale of this state-owned domain had been authorized since 1878, but since the area was interspersed with ponds, it was not suitable for real estate development and had not attracted buyers. The government therefore took the property off the market and at the request of the Ministry of Public Education assigned parts of it to serve for Pasteur's experiments on the prevention of contagious diseases.

Before long, Villeneuve-l'Etang became a large kennel. Sixty animals were soon placed under the guard of a former gendarme. These were dogs picked up by the pound wagon whether they were rabid or not, and of all breeds—griffons, bulldogs, fox terriers. Sometimes a padlocked carriage would take them to Paris for intracranial inoculations or experimental bites. All around the kennel and along the walls of the property, Pasteur had hutches and cages put up for the rabbits and the guinea pigs. Instead of having a few rooms of the small château of Villeneuve restored for his own use, he preferred to have the outbuildings converted, and so, in the first sunny days of the summer of 1884, Pasteur and his family moved into the rooms once assigned to the officers of the Emperor's Hundred Guards.

Meanwhile, the walkers in the forest of Meudon were not the only ones to worry about Pasteur's work. Among those who sought to place obstacles in the way of his experiments, there were also some who appealed to the moral order; these were the antivivisectionists.

Opposition to scientific experiments on animals in the name of ethics was not a new thing. But the development of experimental physiology had increased the number of opponents and their virulence; they had now become cohorts that formed battle lines to protect the animals. The influence of these leagues was particularly strong in England, where they endeavored to prevent vivisection at an early date. By 1876, it was not rare for English experimenters to find themselves constrained to come to France if they insisted on inoculating a guinea pig. The same current was taking shape in Germany: in 1881, a society in Leipzig had called for punishing anyone who conducted research on animals with imprisonment and loss of civil rights, while other groups had demanded to be allowed to enter faculties of science and medicine so that they could monitor what was going on in their laboratories. The situation became so alarming that Virchow used the International Congress of Medicine of 1884 to denounce the dangers of this trend in a public statement: "Those who attack vivisection have not the faintest idea of what science is, and even less of the importance and the usefulness of vivisection for the progress of medicine. . . . No one who is more interested in animals than in science and the knowledge of truth is qualified to exert an official control over scientific matters."[17]

In France, the first antivivisectionist attacks had been directed against Claude Bernard. He was criticized for placing cannulas into the stomachs of dogs in order to extract gastric juice. In the forefront of those who combated the methods of the physiologist were his wife and his daughter, who made his life miserable even at home. Nonetheless, Bernard was adamant about the usefulness of experiments on living animals. "You ask me which are the main discoveries we owe to vivisection in order to emphasize them in arguments defending these kinds of study. When it comes to that, one can only cite all the accomplishments of experimental physiology; there is not a single fact that was not the direct and necessary result of a vivisection."[18]

After the death of Claude Bernard, Pasteur became the principal target of the antivivisectionists, and as his renown spread beyond the frontiers, he had enemies as far away as England. In fact, it was in order to defend the French scientist against his own compatriots that Darwin declared in 1881: "Physiology can make no progress if we do away with animal experiments, and I am deeply convinced that to hold back advances in physiology is to commit a crime against humanity."[19] In spite of all these pleas, laboratories continued to be regarded as torture chambers.

In the case of Pasteur, this criticism was particularly unfair, because he was the first to object to inflicting any unnecessary pain on the animals. He simply stated that some acts were indispensable, namely those that were legitimized by science. But for the unconditional defenders of the life of animals, this was already entirely too much, and Pasteur regularly received letters containing protests, insults, even threats; there was talk of "taking revenge for his attempts" on the life of the dogs, rabbits, chickens, and guinea pigs in his laboratory. In 1883, when the government was planning to increase the pension of its most famous scientist from 12,000 to 25,000 francs per year, an antivivisectionist league unleashed a veritable campaign of denigration, which blithely equated vivisection with vaccination. The result was criticized as much as the procedure: "It will have taken almost a century," the president of the league wrote to the minister, "to demonstrate the errors of vaccine, but it will take only a fourth of this time to annihilate the fantastic doctrines of M. Pasteur, which contradict the laws of general biology. . . . Science does not recognize in the great discoveries of M. Pasteur anything but a tissue of dogmatic conceptions and unsuccessful attempts more apt to ruin than to enrich the country that would adopt them."[20] This was also the time when a Society to Combat the Exploitation of Public Ignorance with private ignorance was founded; its main target was also Pasteur.

It is true that these antivivisectionist manifestations, these pseudoscientific arguments, and these hostilities spawned by preconceived notions sometimes impeded the proper functioning of the laboratory and spread false alarms in the public mind; but their denigrating efforts often succeeded only in giving unintended publicity to the vaccine and to Pasteur in particular. The slavering dog, rabies, vivisection, and vaccination became topics of conversation in the sidewalk cafés. Sometimes the violence of the antivivisectionists even gave rise to an inverse reaction, to the point that enthusiastic supporters of vaccination volunteered for experiments on themselves. If experimenting on defenseless animals is offensive to sensitive souls, they said, Pasteur can always continue his experiments with reasonable human beings who state that they are willing to be inoculated. This was the case of a diabetic at the Cochin Hospital, who wrote to Pasteur: "Since this life is a burden to me, I have decided to end it. Work, I can't do it anymore. Hospitals, I don't want them any more. I therefore offer myself to you to serve as a subject in your experiments. Do not think that I am puny and fragile. On the contrary, I am big and strong and I look more robust than I actually am. I am forty-five years old, so I can put up with a lot."[21] Or that of an unknown lady, whom Pasteur firmly turned down: "Again, Madame, I absolutely refuse to inoculate you with a rabies virus. Do you really believe that I would be bold or crazy enough to accept such offers? Judge for yourself. And you even have children! Believe me, you should never push your devotion to mankind and to science this far."[22]

Yet the fact is that Pasteur was indeed not far from thinking that human experiments were the only means of achieving the final triumph of his work. He even thought about it in very precise terms, but he felt that he must not experiment on just anybody, even a volunteer. The risk was still too great. That is why Pasteur envisaged a solution which he communicated to the emperor of Brazil, Pedro II, in a letter dated 22 September 1884.

> I have until now not dared to perform any experiments on humans here, despite my confidence in the result and despite the many opportunities that were offered to me after my last lecture at the Académie des sciences. I am too fearful that a failure would compromise future endeavors. . . . And I feel that even after I will have produced many more examples of prophylaxis against rabies in dogs, my hand would shake if I had moved on to the human race. This is where the high and powerful initiative of a head of state might very usefully be brought to bear for the good of humankind. If I were king or emperor, or even president of the Republic, here is how I would exercise my right of pardon for convicts facing the death penalty. On the eve of the execution, I would give the convict's attorney the choice between immediate death for the convict and an experiment consisting of preventive inoculations of rabies designed to make the subject refractory to rabies. If the condemned convict agreed to these trials, his life would be spared. If this were indeed the case — and I am convinced that it would be — the protection of the society that had condemned the criminal would require that he be placed under surveillance for the rest of his life. All of the condemned would accept. A person condemned to death only fears death.[23]

Discovering today what was said in this letter, one cannot help reading it in the light of the crimes committed by the Nazis in the concentration camps. This of course is not the most pertinent reading, even if certain of Pasteur's contemporaries were at times highly critical of the ethics of his conduct. In one of the lectures on "human experimentation" he gave at the Collège de France, Charles Nicolle, one of Pasteur's last disciples, felt compelled to criticize his master: "If Pasteur had been a physician, if he had been made aware, not only by his natural sensitivity, which was very great, but by the professional knowledge of the practitioner, of the danger to which he was exposing his clients, the dreadful thought of an experimentally induced rabies would no doubt have stopped him. . . . He was possessed of that indomitable temerity that a sacred delirium imparts to the genius. The scientist's conscience smothered the conscience of the man." In the last analysis, beyond the fundamental problem of human experimentation, the most shocking aspect of all this is the hypocrisy of the relation between power and ethics. When he turned to the emperor of Brazil, Pasteur knew that he could be heard; by accepting the principle of the death penalty, which neither he nor most of his contemporaries had ever questioned, he could blithely dismiss his misgivings.

The Meister Year

The year 1885 was the year when the French learned about the discoveries of Elie Metchnikoff concerning the role of phagocytes in the resistance to microbes, when the Austrian Carl Auer invented the gas lamp that bears his name. In France, Emile Zola finished writing *Germinal*, while Waldeck-Rousseau achieved the official recognition of associations that would later become labor unions. Peugeot launched the first velocipedes, the government of Jules Ferry fell over the question of Tonkin, and Victor Hugo died on 22 May. His state funeral on 1 June brought almost a million people into the streets of Paris. From his daughter's balcony on the boulevard Saint-Germain, Pasteur watched the immense procession as it moved in silence to the Pantheon. Perhaps he was thinking that this was the funeral of his main competitor in the field of national glory. After all, they were already talking about naming a street in Lille "rue Pasteur"! But for all those who were working in the laboratory in the rue d'Ulm, 1885 was to be above all the crucial year in the battle against rabies.

In the last several years, a rumor has regularly surfaced in scientific circles, particularly among American historians of science. It is claimed that Pasteur repeatedly tried his vaccine on humans before he achieved a real success, and even that he managed to hush up his failures. An accusation of such seriousness does not call for taking sides, but for reconstituting the facts to the extent that the archives allow us to apprehend them. The notebooks of Pasteur's experiments, which have recently become available to scholarship, do indeed contain information that the scientist's official biographers (beginning with his son-in-law René Vallery-Radot) preferred not to use, if indeed they were aware of it.

As is often the case, the rumor contains a kernel of truth, but this truth has been expanded and distorted owing to the ignorance of those who have spoken of it. It should be said that the experiments on rabies followed so closely upon one another in the months of May, June, and July 1885 that one has to follow Pasteur almost step by step in order to understand just what was happening at this time.

To begin with, one must realize what was at stake here; it was an unbelievably bold idea to envisage inoculating human beings. Pasteur was fully aware of this. He knew that among his close collaborators, Emile Roux was opposed to all experimentation on humans because he felt that research on animals had not yet been carried far enough. And indeed, it should be stressed that in most of his experiments with dogs, Pasteur obtained animals refractory to rabies by means of preventive inoculations; in other words, by the application of classic Jennerian vaccination. Only a few subjects had been successfully immunized after they were contaminated by a rabid bite, which is an entirely different matter. Now it turns out that every time Pasteur was appealed to and decided to inject his antirabies vaccine, he dealt precisely with human beings who had already been bitten and were therefore part of the category for which previous experiments furnished few data. Roux's reticence is understandable.

But at the same time one must keep in mind that, once it became symptomatic, which was not always the case, rabies was fatal; if one did nothing after the victim had suffered a bite deep enough to impart the disease, he or she was sure to die a most awful death. This is why, from the time Pasteur knew, not only that he could make healthy animals refractory to rabies, but also that he could stop the progression of the rabies virus in contaminated subjects, every case of rabies in humans presented him with a terrible dilemma. The choice was between experimenting with a treatment that was not absolutely proven and leaving the patient untreated at the risk of an atrocious death.

On 2 May 1885, Pasteur was alerted by a telegram from the director of the Necker Hospital that a sixty-year-old man had been admitted to Dr. Rigal's clinic.[24] Rabies was suspected, for the patient, who had been bitten some time earlier, presented the symptoms of hydrophobia. Pasteur went to Necker that very day. He first spoke to the concierge, who told him that the patient had come in the day before in a state of high agitation. Then, accompanied by Dr. Rigal, he went to the man's bedside and examined his wounds, which had healed over since the attack. The prognosis was thus sufficiently confirmed to warrant a vaccination. "We inoculated under the right ribs one Pravaz syringe of diluted rabbit medulla," Pasteur recorded in his notebook. What he used was the cerebral substance of a rabbit inoculated on the previous 14 April with an attenuated virus.

Since a single injection was not sufficient, Pasteur planned to give a second one that very evening and several more over the following days. But things did not work out as he had ordered. "On 2 May, at ten o'clock in the evening, the hospital's director, who had consulted the Public Welfare Ministry, forbade all inoculations, because Dr. Rigal was not present; so we did not inject another treatment on 2 May, nor on 3 May or 4 May."[25]

On 8 May, since the patient was feeling well, he was discharged from the hospital. Pasteur was still interested in him, but the administration of Necker Hospital did not seem inclined to cooperate: "I wrote to Dr. Rigal asking him to report on this case to the Board. He replied that he would not make the report as I had asked." According to Pasteur's notebook, he returned to the case two weeks later, inquiring whether the patient continued to feel well. Apparently he did not receive an answer—a matter of medical confidentiality!—and never found out more about it.

Then another case diverted his attention. On 8 June, Pasteur went to Lariboisière Hospital to visit a crossing guard of the Eastern Railroad Company who had been admitted four days earlier, after he had been bitten on the hand by a rabid dog as he worked on a railroad track. But it was already too late. When Pasteur arrived, he was in agony; and he died a few hours later "too quickly to be treated," we read in the notebook.

On 22 June, he was notified that a little girl had been hospitalized at Saint-Denis: "A little girl, a charming child of about ten, bitten some six weeks ago (not clear on which day) in the upper lip when she tried to pull her little dog from under the bed." Notified in the morning, Pasteur reached Saint-Denis in the afternoon. The little patient had been complaining of headaches for two days; her case was alarming. The fact that she had been bitten in the lip was an aggravating factor. In agreement with the treating physician, Dr. Leroy, Pasteur decided to intervene: "At my request, Dr. Leroy injected under the skin, below the lower ribs, one Pravaz syringe of bouillon-diluted spinal cord from one of the rabbit-hosts (inoculated on the seventh day of the passage); the preparation had been preserved since 15 June." As early as possible the next morning, Pasteur was back at the little girl's bedside: "With Adrien Loir I go to see the child at ten o'clock. She dies at half past ten in the morning. The symptoms of constant discharge of mucus had rapidly increased."[26]

Pasteur had been defeated, but he wanted to understand. The child's autopsy was performed on 24 June. He sent Loir to bring him the medulla oblongata, which he used to prepare injections with which to test animals, two guinea pigs, two rabbits, and a monkey. By 11 July, the two guinea pigs were dead, one of the rabbits was sick, and the condition of the monkey was normal. One week later, the sick rabbit had died and the other presented the first symptoms of rabies. By 31 August, only the monkey was still well. The conclusion was inescapable: Pasteur had been called in too late, at a point when his treatment could do nothing to stop the virus from advancing into the child's brain.

Even before this set of observations was concluded, Pasteur experimented further with his vaccine. On 6 July 1885, three persons came to see him, two adults and a child. One of the adults, Theodor Vone, a grocer at Meissengott, a small town near Villé in Alsace, had been bitten on the arm two days earlier by his own dog, which he had thereupon shot. His wound, however, was not serious; the skin had not been broken, since the sleeve of his coat had absorbed the bite. But the grocer's dog had earlier attacked a nine-year-old boy, Joseph Meister, the son of a baker at Steige, a neighboring village of Meissengott. Little Meister and his mother were the two other visitors who came to ask Pasteur to help them. Pasteur quickly examined the boy and immediately wrote in his notebook, "Severely bitten on the middle finger of the right hand, on the thighs, and on the leg by the same rabid dog that tore his trousers, threw him down and would have devoured him if it had not been for the arrival of a mason armed with two iron bars who beat down the dog." Pasteur counted a total of fourteen wounds of alarming seriousness. The boy was clearly threatened by rabies. At the same time the circumstances were in some sense favorable, for on the one hand, the bites were very recent, a fact that surely increased the chances of success of the treatment; and on the other, Joseph and his mother had come directly to Pasteur's residence, so that it might be possible not to hospitalize the boy and to take care of him privately, thereby improving the conditions in which the treatment would be administered and monitored.

All these factors made Pasteur's decision particularly delicate. That is why, before pronouncing himself, he wanted to speak to some physicians in whom he had complete confidence. He therefore turned to one of his confreres in the Académie de médecine, Alfred Vulpian, and to a young doctor, head of the pediatric clinic at the Children's Hospital, Jacques-Joseph Grancher. Vulpian was one of the most respected physicians of his day, but this did not prevent him from following Pasteur's research very closely. In fact he was a member of the highly official Commission on Rabies that had been created by the government at Pasteur's request for the purpose of monitoring and evaluating the current research on the disease. Vulpian considered the case sufficiently severe to justify an attempt to vaccinate. As for Grancher, he combined an excellent knowledge of microbiology with his specialization in pediatrics, so that his opinion was particularly weighty; and he too advised Pasteur to administer the antirabies treatment. This was a good thing, for Pasteur needed a physician to give the injection, to which Roux was adamantly opposed, so much so that later he even refused to sign the protocols of the study.

On the very evening of 6 July, Vulpian and Grancher went with Pasteur to the bedside of Joseph Meister, who had been installed with his mother in a room at the Collège Rollin. Careful examination of the bites, their number and their depth, confirmed them in their opinion: if nothing was done, rabies would surely ensue, and the boy would be doomed. Weighing the pros and the cons, and imagining what was in store for the little patient, Pasteur accepted the advice of the two physicians, and Joseph was inoculated.

"As a result, on 6 July, at eight o'clock in the evening, sixty hours after the bites of 4 July, and in the presence of Drs. Vulpian and Grancher, we inoculated into a fold of skin over young Meister's right hypochondrium half a Pravaz syringe of the spinal cord from a rabbit dead of rabies on 21 June; the cord had since then — that is, for fifteen days — been kept in a flask of dry air." Further on in his report, Pasteur clearly stated that the first inoculated spinal cord was the oldest, and therefore the weakest, having dried for two weeks in the flask with two openings. But the treatment required more inoculations in the following days, and so Pasteur had to plan for injections, at regular intervals, of less and less attenuated spinal chord. In view of these injections, the boy had to be kept under constant observation. Fortunately, the Collège Rollin was not too austere a place; thanks to Pasteur's experiments, the laboratory had become a kind of farm, where the country boy found the chickens and the rabbits to which he was accustomed. Joseph even enjoyed playing with the white mice. "Everything is going well," Pasteur wrote to his son-in-law on 11 July. "The boy sleeps well, his appetite is good, and the inoculated matter is absorbed without leaving a trace. If the child is well three weeks from now, I shall consider the success of the experiment assured. However it turns out, I am planning to send the child back to his mother in Meissengott near Schlestadt on 1 August; however, I shall establish a system of observations to be carried out by the good people there."[27]

Yet as the inoculated matter became more virulent, Pasteur became increasingly uneasy. A propos of the last injection, that of a one-day-old rabid spinal cord to be given on 16 July, Marie Pasteur wrote to her children: "My dear children, this will be another bad night for your father. He cannot come to terms with the idea of applying a measure of last resort to this child. And yet he now has to go through with it. The little fellow continues to feel very well."[28]

That day, at eleven o'clock in the morning, Pasteur had Joseph Meister inoculated with the most virulent rabid spinal cord, taken from a dog and strengthened by numerous passages from rabbit to rabbit. Normally, rabies would become symptomatic seven days after such an inoculation. But in this case, it only served to test whether the vaccination had really taken. All in all, the treatment had taken ten days, and young Meister had been inoculated thirteen times.

Even before he had obtained the ultimate confirmation of his success, on July 18 or 19, Pasteur left Paris, leaving Joseph Meister under the care of Dr. Grancher. This rapid departure seems almost incredible, but Pasteur was no doubt tired out, and perhaps also possessed of supreme self-confidence.

However this may be, on 21 July, Joseph Meister, who had been taken on an outing to the countryside around Paris, sent a short letter to Arbois: "Dear Monsieur Pasteur, I am feeling good and I slep well and I also have good aptit. I had fun in the contriside. I dident like to go bak to Paris."[29] Joseph did not return to Alsace until 27 July, but he continued to feel well and Pasteur was fully reassured.

He had defeated rabies, and nothing was more important to him. That is why in late August, when Léon Say suggested that he run for the National Assembly, he refused without regret: "I might accept if I felt that I no longer have enough energy for my laboratory work. I hope to be equal to some further research projects, and, immediately upon my return to Paris I will have to set up an antirabies agency that will for a time be at the center of my activity. I am in possession of a highly perfected prophylactic method against this terrible disease, a method that is as safe for humans as it is for dogs, and of which your sorely tried department will be the first to benefit."[30] And, for the first time, he publicly stated: "Before leaving for the Jura, I dared treat a poor little boy of nine whom his mother brought to me from Alsace, where on 4 July last he had been thrown down and bitten on both thighs, both legs, and on the hand in circumstances that would have made rabies inevitable. He is still in good health."

The very tone of this declaration clearly shows Pasteur's extreme cautiousness, for he did not reveal his exploit right away. Only certain of his close associates were immediately informed, and in fact it was because of an indiscretion on the part of Léon Say that the *Journal des débats* was the first to spread the news: Joseph Meister was cured by Pasteur; he is forever safe from rabies!

Very quickly, even before Pasteur had communicated his results to the Academies of Sciences and of Medicine, his victory resounded beyond the limits of the scientific domain and turned into an act of patriotic revenge. After all, the Meisters were Alsatians, and as such had become German subjects in 1871. Yet it was Pasteur who had saved the child, not a compatriot of Robert Koch. A letter that Pasteur wrote to Joseph on 27 November 1885 does indicate that efforts had been made to minimize the event: "I understand that the German authorities have made a thorough investigation of your accident of 4 July last. I also seem to understand that they would have been happy to find that the dog that bit you was not rabid. That would have cast doubt on the value of the studies which led me to dare administer to you, for the first time since the beginning of the world, a treatment capable of preventing rabies from breaking out. Envious people are found everywhere, and the German scientists, especially some of them, are jealous of French scientists who outstrip them in the search for the truth."

Pasteur stayed in touch with the boy and always spoke to him in this somewhat strange tone, a mixture of paternal affection and professionalism; he also sent him an inscribed photograph of himself, along with postage stamps to put on his answers. After 1885, Pasteur continued to take an interest in Joseph Meister's education; he worried about his health and his future to the point of opening a savings account for him to cover small expenses. The Meister family accepted these attentions with good grace, and indeed solicited them, asking Pasteur's help in finding a job, first for Father Meister and, when the time had come, for the son. Their aim was to escape the German tutelage. Pasteur did his best. Much later, Joseph Meister was employed as a guard at the Institut Pasteur!

Jean-Baptiste and Louise

Immediately upon his return to Paris in the early fall of 1885, Pasteur found himself confronted with another case of rabies. This time the victim was a fifteen-year-old shepherd, Jean-Baptiste Jupille, who lived not far from Arbois at Villers-Farlay in the department of Jura. Six little shepherd boys who were watching their sheep in a field had been attacked by a roaming dog, its muzzle dripping with slaver. In an effort to allow his fellow shepherds, all younger than himself, to run away, Jupille threw himself in front of the dog, armed with his whip. The animal pounced on his left hand, but the adolescent was able to throw it to the ground and to free his hand. He then used the string of his whip to tie the dog's jaws together before he stunned it with one of his wooden shoes and drowned it in a nearby brook. The dog was dead, but the autopsy confirmed that it had been rabid. Unfortunately, Jupille had suffered several deep bites. Something had to be done immediately.

Pierre-Joseph Perrot, the mayor of Villers-Farlay, who knew Pasteur slightly, immediately wrote to him. Pasteur replied by return mail; he recalled the means by which he had just saved Joseph Meister and agreed to do the same for Jean-Baptiste Jupille. So as to leave nothing unstated, he added all the necessary information:

> Jupille was bitten on the fourteenth of this month. This letter will reach you on the 18th. The boy will be here on the morning of the 20th or the evening of the 19th. By then the bites will already be six days old; those of little Meister were only sixty hours old, and I do not know yet, from my experiments, what is the outer time limit, from the time of the bites, when I can begin the treatment. However, I should tell you that I have made dogs refractory to rabies six and eight days after they have been bitten. The boy should stay with me for ten to twelve days, fifteen at the most. Since he probably is not rich, I shall keep him near me, in a room at the laboratory. He will be under supervision, but he can come and go as he pleases without having to stay in bed. He will only be given a small shot every day, but this will be no more painful than a quick pin-prick. The village will have to cover only the cost of a round-trip ticket. I have borne all the other costs I am mentioning here for little Meister. It is true that I am doing this because these are my very first trials.[31]

As soon as he arrived in Paris, Jupille was taken to his room in the rue Vauquelin in an annex of the Pasteur's laboratory. This treatment too was an incontestable success. By the end of October, Jupille was not only out of danger, he could also return to his native Jura carrying in his pocket a bank book of the Savings Bank, in which the Institut de France was about to deposit the tidy sum of one thousand francs! This was because Pasteur had spoken at the Academy on behalf of the young shepherd, urging it to bestow on him the Montyon Prize for virtue to reward him for his heroic conduct.

Owing to all these circumstances, Jupille's cure caused a much greater stir in the press than that of Meister. It is true that the little Alsatian had been treated during the summer holidays, which certainly kept down the excitement. Jupille, by contrast, immediately achieved a great celebrity: everyone was talking about the attack he had suffered and the treatment he had undergone; everyone praised his courage and his devotion; and the sculptor Truffaud even honored him by creating a bronze group that was subsequently placed in the garden of the Institut Pasteur. That, incidentally, was where Jupille came to live; employed as a concierge (like Meister), he proudly sported his handsome uniform and became the darling of visitors and journalists. Unquestionably he became the most photographed personality of the rue Dutot.

However, as one can easily imagine, it was in scientific circles that Pasteur's audacity and his accomplishments created the greatest stir. On 26 October 1885, Pasteur presented an official report about the treatments he had applied to Joseph Meister and Jean-Baptiste Jupille before the Académie des sciences. He had barely finished speaking when Vulpian asked for the floor: "Rabies, that dread disease against which all therapeutic attempts had failed until now, has finally found its remedy! M. Pasteur, who had no precursor but himself in this endeavor, has been led, by a series of studies pursued without interruption over many years, to create a method of treatment that unfailingly prevents the development of rabies in humans recently bitten by a rabid dog. I say unfailingly, for because of what I have seen in M. Pasteur's laboratory, I have no doubt that this treatment will always be successful if it is properly administered within a few days after a rabid bite."[32]

This enthusiasm, though justified by the successful attempts Pasteur had just carried out, was nonetheless premature. The victory over rabies was too quickly assumed and too quickly established as a fact, for these first two official cases had yet to be properly tested. Here Pasteur's legendary prudence and obsessive rigor broke down, for he could not refuse the request to organize a center for the treatment of rabies in the very laboratory of the rue d'Ulm. People flocked there from all sides as soon as a rabid dog was reported anywhere. By early December 1885, eighty treatments had already been achieved or begun.

Joseph Grancher was in charge of the inoculations, while Eugène Viala took care of the preparations of rabid spinal cord under the general supervision of Pasteur. The spinal cords were prepared in a small room with constant temperature, where the flasks with two openings, closed with cotton stoppers, lined the shelves. In each of the flasks the rabid spinal cord of a rabbit was suspended by a string. At the appropriate time, Viala cut this cord, more or less aged and more or less dried out by the potash, into small pieces with scissors and then inserted the fragments into bottles labeled with that day's date. To prepare the matter to be inoculated, he used a pipette to instill a few drops of veal broth into the bottle before grinding up and mixing the contents. The doses of vaccine corresponding to each patient were set aside in a series of glasses protected from dust by a paper cover. Every manipulation, every test tube, every spinal cord was meticulously checked by Pasteur, who was always at Grancher's side when he operated.

It was the master who answered the patients' questions, although he often advised them to turn directly to Grancher, on the grounds that he did not know too much about medicine. "I am a chemist, I do experiments and try to understand what they tell me," he would say. One can imagine the astonishment families felt when they met this half-paralyzed man with his white hair, his pince-nez, and his skull-cap who proclaimed himself a chemist and told everyone who wanted to hear it that he was not a physician and claimed that all he knew was to do experiments. Surly and humble at the same time, Pasteur was nonetheless able to put an end to the anguish of those whom rabies brought to the rue d'Ulm in search of what they considered a miracle.

But Pasteur did not perform miracles, and he was even to experience a terrible failure with little Louise Pelletier. This child of ten had been bitten on the head on 3 October 1885, but her parents did not bring her to the rue d'Ulm until 9 November, thirty-seven days later. By that time her wounds had not healed over and were suppurating. Pasteur hesitated, for so much time had passed that the case seemed desperate. His scientific self-interest dictated that he refuse to treat the child, for he knew that he would most likely fail, which would be grist for his detractors' mill. But, moved by the parents' insistence and the child's suffering, he let himself be persuaded.

The first injection was given on 9 November, and the treatment was completed on the 16th. Three days later, on the nineteenth, the child was seized by convulsive spasms. Soon she was unable to swallow. Grancher, called to her bedside, did not know what he should try, but already there was nothing left to be done: little Louise died on 6 December, the day when Pasteur went to the funeral of Henri Bouley, one of his most faithful supporters at the Académie des sciences. A terrible day for the scientist, who had to bury a friend and see a child die.

Although Pasteur knew perfectly well that this failure did not disprove the value of his treatment, he feared its exploitation by the press. He wrote to his son-in-law asking him to be extremely discreet: "Any publicity can have the most serious consequences. Quite a few of the eighty persons already treated or in treatment now might be frightened if they heard of the death of little Louise Pelletier. One would have to be unaware of the dread caused by this malady to talk about it at length. I alone have the right to speak of all the circumstances relating to my patients and to provide precise explanations concerning them."[33]

But Pasteur's adversaries did try to exploit the drama. However, they were not really able to shake the public's confidence. They would have been silenced if they could have read the letter Louise Pelletier's father had written to René Vallery-Radot after reading his *Life of Pasteur*. "Among the great men about whose life I have been able to learn, none seems greater to me. I do not see one of them capable, as he was in the case of our dear little girl, of sacrificing long years of work, of endangering a universal reputation as a scientist and knowingly risking a painful failure, out of simple humanity."[34]

Meanwhile, at the very time when Louise Pelletier succumbed, the news of the victory of rabies crossed the oceans. Hitherto, Pasteur's work had been known on the other side of the Atlantic only by a handful of specialists. But in December 1885, it was learned that four American children, the sons of workers at the port of Newark near New York, had been bitten by a rabid dog. Pasteur was asked by telegram if he could treat these children. He accepted on the condition that the victims could immediately embark for France. It took underwriting from the *New York Herald Tribune* to pay for the passage, but the four little Americans arrived in Paris in time to be saved. For several weeks, their adventure made the front pages of the American daily papers. This of course was unexpected and matchless publicity for Pasteur and his theories.

In March 1886, it was Russia's turn to appeal to Pasteur. In the region of Smolensk, a rabid wolf had spread terror along its path before it was beaten to death with hatchets. There were nineteen victims who had suffered particularly horrible wounds. One of them was an orthodox priest, who had been attacked on his way to church services; his upper lip and right cheek had been torn off. Another had his whole face torn by the wolf's fangs. These unfortunate people were immediately sent off to Paris, where they arrived in a terrible state. Five of them were so ill that they had to be rushed to the Hôtel-Dieu. In view of the severity of the injuries and the length of the delay, Pasteur decided to give them double inoculations, one in the morning and one at night. Throughout the treatment, they were a strange sight, these moujiks with their bandaged heads and arms and their big fur greatcoats, lining up to be inoculated one by one. Not one of them spoke French. Three died before the end of the treatment, but the sixteen others returned to their country some time later.

When he learned of this success, Prince Alexander of Oldenburg decided to set up an antirabies laboratory at St. Petersburg, which would make it possible to undertake vaccinations using the Pasteur method. So Adrien Loir set off for St. Petersburg with a cage containing two rabid rabbits on 14 July 1886, while General Boulanger, the Minister of War, paraded down the Champs-Elysées on his black charger.

In this manner, the rabies treatment gradually spread throughout Russia. At Odessa in particular, a brilliant biologist, Elie Metchnikoff, deputy director of the antirabies laboratory, used the technique of desiccating rabbit brains and became one of Pasteur's most fervent followers. Before long, rabies victims were vaccinated in London, Vienna, Jena, Warsaw.

And from everywhere they came to the rue d'Ulm. Grancher could no longer handle everything. Pasteur therefore recruited two young agrégés from the School of Medicine, André Chantemesse, a student of Charcot, and Albert Charrin. A surgeon, Octave Terrillon, who was also a friend of Grancher, joined the team as head of the advanced surgical unit and treated the wounds of those who had been bitten. Little by little, the laboratory of the rue d'Ulm came to look like the premises of a medieval miracle worker.

More than twelve hundred French people, from both France and Algeria, were treated preventively against rabies in Pasteur's laboratory. Pasteur declared defiantly in August 1886 that the treatment had been ineffective in only three cases. Could anyone contradict these facts and these assertions?

16

THE CAPSTONE

By 1886, Pasteur's success in treating increasing numbers of rabies cases had become an incontestable reality. Yet the eloquent numbers he could cite were insufficient to disarm his adversaries, who seized on every minor weakness and constantly kept him on the defensive. "As for "As for the accidents that have occurred in spite of the treatment administered," Pasteur declared in May 1886 before the Academy of Medicine, "they are known to everyone thanks to the publicity that was so eagerly provided by the hostile press. For there is a press hostile to my method, and this is not surprising, considering that extremely spiteful persons can be found even within these walls."[1] By these words Pasteur meant Peter, his intimate enemy and the fiercest of his opponents; Peter who continued to lead a passionate and rather skillful campaign against him.

Whatever its source, the criticism was generally based on two types of argument. If the bitten patient who received the treatment did not develop rabies, the critics claimed that since the disease does not always become symptomatic, this was a spontaneous cure; and if the vaccination failed, they attributed the patient's death either to rabies, which thus had not been defeated, or (worse yet) to the inoculation itself. The conclusion was always the same: whether the result was useless or harmful, Pasteur had not discovered anything.

Peter based his strategy on the fact that the disease was quite rare, a point on which he harped again and again. "I have seen two cases of it in thirty-five years of practicing in hospitals and other public institutions, and all my colleagues in rural or urban hospitals count the cases of human rabies they have observed in single digits rather than by tens (let alone hundreds). In order to magnify the benefits of his method and to mask its failures, it is in M. Pasteur's interest to make the annual mortality from rabies in France look greater than it is. But this is not in the interest of truth. Would you like to know, for instance, how many individuals have died of rabies in Dunkirk over the last twenty-five years? The number of deaths was: one. And do you want to know how many have died of it in that town in the last year, since the Pasteurian method has been used? The number of deaths was: one."[2]

Certain official surveys, however, furnished arguments to Pasteur. Mortality following a rabid bite varied between 40 and 10 percent depending on the region. In the summer of 1885, when Joseph Meister was inoculated, five persons died in Paris after having been bitten by a rabid dog. But of course it was difficult to establish such statistics, given the lack of proof. Did the animals that attacked people only have rabies? Might they not also carry other germs? Autopsies of suspect animals were rare; most of the time the dogs escaped or were killed before there was a chance to assess the risk of rabies. Moreover, there were few specialists to confirm such a diagnosis; veterinarians, even if they were consulted, often had little experience in this area.

All these difficulties were compounded by the possibility that rabies might not become symptomatic, even after a bite. For the appearance of the disease in fact depended on the number, the depth, and especially the location of the lesions. It was clear that lesions on the face were much more dangerous than lesions on the calf or the foot, for in the first case the virus was much closer to the vital nervous centers. But other parameters had to be taken into consideration as well: bleeding, especially if it was abundant, could evacuate the rabid saliva; and treatments such as disinfection or cauterization provided by a simple pharmacist were sometimes sufficient to minimize the risk. And lastly, it was understood that the harmfulness of the virus varied with the animal that carried it, so that the lesions caused by a wolf were vastly more dangerous than those brought about by a dog or a cat.

These diverse circumstances made for such a welter of signs that it became difficult to affirm that a patient would die, as Pasteur claimed, or that he could get better on his own, as Peter maintained. There was no way to test these claims. Good faith was the only plea. That is why, as soon as the first statistics became known, Pasteur asked Vulpian to publicize them: "It is therefore permissible to say that whereas among the persons treated fewer than 1 percent died, 16 percent of the persons who were not treated by the Pasteur method have died. . . . If we deduce the 12 persons who have died in spite of the treatment, there are still 264 individuals who have been saved by the treatment. This is very far indeed from the number of 30, which M. Peter gives us to characterize the annual mortality from rabies in France."[3]

Peter, of course, contested these figures. He published a violent diatribe against all of Pasteur's followers in *Le Figaro*: "Last Monday, M. Vulpian communicated to the Academy of Sciences some figures that singularly simplify the question. . . . It would therefore logically and arithmetically follow that in the last twelve months there have been five times more cases of rabies in France than in previous years! Well, I emphatically state that this is monstrously unlikely, and that this is the mathematical artifice that the Pasteurians use to hold up their system."[4] Going beyond statistics, Peter used Pasteur's failures to accuse him of much more serious things. He altogether rejected the idea that one could attempt to administer to humans an extremely dangerous treatment, which in his opinion had not been sufficiently tested, particularly because it called for repeated inoculations. In speaking of the intervals between injections, Peter exclaimed: "The formulas used by Pasteur are more logarithmic than medical; they are purely empirical and founded on *a priori* experiments carried out on human beings." With vibrant eloquence and high emotion, he ended by appealing to the ethical sense of his confreres: "Do you not think that I must be moved by deep conviction if I thus risk losing what is called popularity and alienating the sympathies of this academy that mean so much to me? But I have come to believe that it would be dangerous to keep silent any longer, and I have accomplished what I believe to be my duty. Whatever comes of it will have to come."[5]

It was Vulpian who responded, for Pasteur was unwell, worn out by these perpetual controversies. Why should he come to the meetings of the Academy of Medicine, only to face another run-in with Peter, that man "made of arrogance, ignorance, and triviality?"[6] The pleas delivered by Vulpian were lengthy and passionate. Although they were frequently interrupted by strong applause, everyone was not won over. Peter stepped up his attacks as more and more patients were treated in the rue d'Ulm. In the course of time, failures assumed as much importance as successes; for even though they were rare, they could be used to discredit the treatment itself. Certain observers, no longer content to harbor doubts or express their skepticism, now began to accuse Pasteur and his disciples. In the corridors of the Academy, the expression "involuntary manslaughter" was sometimes freely used. Grancher in particular was a target, as he was to report later: "I felt that the men around me were becoming increasingly disaffected and embarrassed, not to mention the hidden anger they did not express! One day I was at the Faculty of Medicine to give an examination. I heard a furious voice cry out: 'Yes, M. Pasteur is a murderer!' As I entered, a group of my colleagues silently left the room. And the professor who had spoken was not Peter, who at least had the courage of his conviction. . . . Certain papers of the political and the medical press also led an ardent campaign against Pasteur, and I am not even speaking of certain politicians and the antivivisectionist league. And even in the Parisian lycées, students were divided into Pasteurians and anti-Pasteurians who knocked each other about."[7]

Human Guinea Pigs

Truth to tell, there were many, even among Pasteur's closest associates, who objected that the method of antirabies vaccination had not been sufficiently tested to be applicable to humans. Emile Roux, as we have seen, was the first to distance himself from his master; he rarely missed an opportunity to decry his method. His daily criticism finally impressed the most faithful collaborators, in particular Adrien Loir.

One day in 1886, while Grancher was carrying out an inoculation on a patient stricken by rabies, he accidentally pricked himself in the thigh with some of the rabid spinal cord. Pasteur, who was standing behind him, saw the mishap and therefore ordered his assistant to undergo the curative treatment. Grancher agreed without flinching and immediately set up the program for the series of inoculations that was to begin on the very next day. When Pasteur had left the room, Loir pointed out to Grancher that he was taking a risk. "Do you believe, young man, that I would perform this action every morning if I were not sure that the method works?"[8] Grancher's unambiguous reply testified to his unshakable faith.

The next morning, as soon as the injections had been given to that day's patients, the door closed on Pasteur, Grancher, Viala, and Loir. "Are you ready," asked Pasteur, who had ordered Viala to prepare several extracts of rabid spinal cord. "I am," Grancher replied simply. "In that case," Pasteur continued, "you will begin by inoculating me." "No, Monsieur," replied Grancher, "there is no reason for me to vaccinate you, since you have not been infected." Pasteur now turned to Loir and ordered: "Vaccinate me, then, since Grancher does not want to do it." Loir in turn excused himself, claiming that he could only act on orders from Grancher. But then, moved by a bold impulse, he added that he himself was ready to let himself be inoculated to set an example. Thereupon, Loir took the syringe Viala held out to him and inoculated Grancher. Then the latter, without sterilizing the needle, inoculated both Loir and Viala.[9]

Further inoculations were performed over the next days, always behind closed doors, sheltered from indiscreet eyes. Roux, who knew nothing about the matter, was intrigued, tried to find out what was happening, even peered through the keyhole. He eventually understood that someone in that room was being inoculated, and first thought that it was Pasteur. When he realized that the vaccinations concerned Grancher, Loir, and Viala, he became very angry. He threatened to tell Loir's parents. Loir begged him not to do it because it would cause a scandal. Furious at not being heard and deeply upset by what he considered to be a case of "unprofessional conduct," Roux slammed the door of the laboratory behind him.

In private, Pasteur questioned Loir at length and wanted to find out exactly how he felt. "I had become a subject of observation for him," Loir was to explain later. "Almost every day he invited me upstairs to have lunch with him. Then he questioned me. . . . I had to use my arms to pull myself up on the banister, for my legs felt so heavy that I found it hard to move them; I have never felt anything like it since then. Pasteur noted down everything."[10] This means that Marie too was aware that her nephew was involved in this experiment, yet she did not tell the young man's parents. Then, one Sunday, when the whole Loir family had come to the Pasteurs' for lunch, the scientist himself let out the secret when he innocently asked the young man how he felt after his last inoculation. All conversation stopped, and there was icy silence. Then Madame Loir, sitting next to Adrien, had a fit of nerves, instantly persuading herself that she was seeing her son for the last time, and that he was about to come down with rabies.

Fortunately the experiment did not have any untoward consequences and all ended well. Still, injecting a human being with the nerve tissue of a rabid rabbit was not a harmless act. Not only did these repeated injections inoculate rabies, they could also in and of themselves induce symptoms of paralysis and, even though Pasteur denied it, trigger other organic reactions.

All these complications, both at the experimental and the ethical level, had the direct result of detracting from Pasteur's victory. The enthusiasm of the early days, when the cures of Joseph Meister or Jean-Baptiste Jupille were considered miraculous, was a thing of the past. The repeated assaults of Peter, who knew how to be convincing, and the defiance of Roux, who no longer followed his master, eventually came to change the aura surrounding Pasteur's treatments. He found it difficult to impose his ideas and was unable, in the short run, to modify the mentality of the medical world.

Under the title "Rabies Attacks Medical Systems," the famous satirical periodical *La Lanterne* published a carefully crafted savaging of Pasteur: "Our great theoretician has met some eminent surgeons who were able to put his scientific conquests into practice, but what good has he done for the physicians? Almost none at all. His microbicidal medicine is quite simply a homicidal medicine. . . . We beseech our young scientists not to follow the example of Panurge's sheep by entering a path that would be sterile if they followed it indefinitely."[11]

In the face of these attacks and doubts, Pasteur was as sure of himself as ever, perhaps more so; in any case he became more entrenched in his convictions. But he was tired. He was in his mid-sixties and these battles he had to fight day after day wearied him. Each of his victories cost him dearly. He suffered from dizzy spells and arrhythmia, to the point that he had to give in to the pressure brought by his close associates and his doctors. He finally agreed to take some rest.

In November 1886, Raphael Bischoffsheim, a wealthy patron who was particularly interested in supporting major scientific undertakings (he later commissioned Gustave Eiffel to build the Nice observatory) invited Pasteur to use his villa at Bordighera, the most fashionable spa on the Riviera; the queen of Italy, Sainte-Claire Deville, Léon Say, and Gambetta had lived there earlier as Bischoffsheim's guests.

Pasteur was accompanied by his wife, his daughter Marie-Louise, and his son-in-law René Vallery-Radot and their two children. Jean Baptiste, who served at the French embassy in Rome, was not far away and planned to see his parents often. On his arrival at Nice, Pasteur was received with great fanfare by the area's physicians, who had come to express their support and to escort him to the carriage that would take him across the frontier. Pasteur was delighted with Bordighera: not only could he wander among palm trees, orange trees, and in the rose garden filled with brilliant displays even though it was autumn; he also paid neighborly visits to his confrere of the Institut, Charles Garnier, the architect of the Paris Opera, whose villa was a veritable palace. How could he resist the charms of elegant society around a tea table? How could he not be flattered when Prince Napoleon called on him? Accustomed to the republican austerity of the rue d'Ulm, Pasteur now discovered the attraction of Mediterranean luxury, the sensual delights afforded by wealth. Still, he found it hard to convert to this quiet life. He read all the time and worried about the slowness of the Italian mail: every day he impatiently waited for the hour when the mail brought him letters from Duclaux or Grancher. Often he forgot the paradisiacal landscape around him and mentally returned to the four walls of his laboratory, telling Vulpian how to respond to Peter, and continuing to dictate every last detail to Grancher.

Nor was this a whirlwind tour: having arrived in early December 1886, Pasteur and his family stayed until the end of February 1887. This gave him three months away from the Parisian cabals and intrigues. And this vacation, even if it was a working vacation, could have been even longer if nature had not decided to put an end to it: on Wednesday 23 February, which that year happened to be Ash Wednesday, the entire Côte d'Azur was jolted awake by an earthquake. "Daylight was just beginning to pierce my curtains," Marie Pasteur wrote to her son the next day, "and I was about to get up to receive the ashes. Suddenly I said: 'What is this? It's an earthquake!' 'Yes it is,' said your father. Then the whole house began to creak and shake in the most frightful manner. After about a minute it stopped and the children came into our room, more dead than alive. Then there was another formidable jolt, and everyone was perched on your father's bed, except for me. I was petrified to see this pitiful group."[12]

The Pasteur family had had enough. This paradise was too lively for them, and they lost no time leaving it. Their refuge was Arbois. In Franche-Comté at least the ground did not move!

Until the end of April 1887, Pasteur therefore continued to direct from afar the experiments carried out by Loir and Viala, Grancher's organization of the antirabies vaccinations, and Vulpian's responses to attacks. But he could not stay away from Paris forever. After all, the final battle over rabies was being fought in the capital. And this was no longer a matter of criticism, but of actual accusations. France under the Third Republic seems to have loved high-profile legal proceedings, "*affaires*," and Pasteur was to have his share of them.

The "Affairs"

Behind the arguments of those who criticized Pasteur's findings one often finds personal grievances. This was of no interest to the scientist; all he was concerned about were the possible consequences of such attacks, for he feared that his rare failures could be exploited to influence public opinion and cause it to lose faith in his method of vaccination. That is why, whenever there was a death from rabies in his laboratory, Pasteur felt constrained to defend himself by trying to prove that he was not responsible for it. In so doing, he endeavored not only to minimize the facts but also, as far as possible, to turn them to his advantage.

The first case to cause great stress in the laboratory of the rue d'Ulm was the Jules Rouyer affair, named for a child who had been bitten by a roaming dog on 8 October 1886. Twelve days later, he appeared at the laboratory accompanied by his father and underwent the classic antirabies treatment. Then he went back to his normal life until the end of November; but on the 26th of that month he died, having been hospitalized for a lumbago that had developed after he had received a blow in the lumbar region some time earlier. In view of this complex death, the forensic examiner refused to issue a burial permit.

At the time Pasteur was resting at Bordighera. Roux therefore had to take the matter in hand, despite his initial reticence. The problem was that the child's father, Napoléon Rouyer, had gone to the police commissariat of the arrondissement to lodge a complaint against the Pasteur method, accusing the physicians who had treated his son of having caused his death. The child therefore had to be autopsied.

On the eve of this distressing operation, Loir jumped into a hackney cab and went to the morgue, accompanied by a police commissioner. He had been sent to demand that the autopsy be performed under his eyes and in the presence of Grancher. Roux, for his part, had asked to be given the medulla oblongata afterward, so that he could use it to prepare inoculation material to be injected into rabbits; this was the only way to find out whether the child had really died of rabies.

The next day the autopsy was therefore performed at the morgue in the presence of Loir and Grancher, as well as of two other witnesses, both hostile to the Pasteur method. They were a municipal councillor, Dr. Rueff, and a little doctor in a black frock coat, by the name of Georges Clemenceau. The future Tiger was indeed a fierce opponent of Pasteur. In 1865 he had defended a thesis in medicine entitled *De la génération des éléments atomiques*, in which he had sided with heterogenesis. This means that for twenty years he had been waiting for an opportunity to thwart Pasteur and his methods. At long last the suit brought by Napoléon Rouyer allowed him to defend his convictions.

Four grave faces were looking at the small cadaver: Grancher and Loir on one side, Rueff and Clemenceau on the other. Pasteur's supporters and his adversaries were facing off. They were separated by the death of a child, but little thought was given to Jules Rouyer: what was at stake here was the very future of the vaccines.

Enter the forensic examiner who is to conduct the autopsy. Dr. Brouardel takes off his stove pipe hat, calmly puts down his cigar, pushes up his sleeves, and — scalpel in one hand, saw in the other — begins by opening the body over its entire length. Along the way, he takes an opaque liquid from the bladder and places it into a crystallizer so that it can be analyzed later. It is albumin, which means that the child has probably suffered from renal dysfunction, perhaps due to the blow he has received — this at any rate is the diagnosis Brouardel dictates to his secretary.

He continues his exploration. Having turned the body over, he comes to the renal cavity and notes that the kidneys are enlarged and congested. But this is not what most interests the witnesses; they are waiting for the trepanation with which the forensic examiner will finish. He saws open the skull and plunges his hands into the cavity in order to extract the two hemispheres of the brain with the medulla oblongata. This is the crucial moment. Absolute silence reigns in the amphitheater. The entire brain is deposited in a bowl and handed to Loir, as he has requested. Carrying his precious package under the arm, he quickly leaves the room. A hackney cab is waiting to take him to the rue d'Ulm, where Roux is ready to inoculate two rabbits.

He begins by preparing a paste from the crushed brain, which is then diluted spoonful by spoonful in broth; subsequently, the mixture is aspirated into a pipette and finally injected into the trepanned rabbits. This intervention is carried out behind closed doors, for Roux has refused to have any witness whatsoever.

Two weeks later Eugène Viala, who had monitored the animals, brought the news to Loir. The two inoculated rabbits were paralyzed. This clearly was the first sign of rabies; and it meant that the child had developed the disease in spite of the treatment. Loir rushed off to inform Roux, who was still asleep. Roux immediately came downstairs, assured himself that the rabbits were indeed paralyzed, and ordered Loir to go immediately to fetch Grancher. No doubt about it: the child has died of rabies. In order to understand how this has happened, they would have to reconstitute the protocol recording the preparation of the spinal cords that had served to inoculate Jules Rouyer, but Pasteur had taken the notebooks of his experiments with him. This being the case, no other test was possible.

Grancher and Roux were petrified at the thought of the possible consequences of this case. Even though he had not wanted to participate in the rabies experiments, Roux was keenly aware that Pasteur's entire reputation might be tainted. He could think of only one thing to do: Loir had to go immediately to Bordighera to ask Pasteur for his advice. Loir went off. Pasteur was very surprised to see his nephew appear at the Villa Bischoffsheim, but when he learned of the reason for his visit, he reacted with the greatest calm. His first words expressed pity for the little fellow who had died, but he did not seem unduly concerned about the repercussions of this affair. Laconic and fully confident in his subordinates, he sent Loir back to Paris with the famous notebook on the methods of preparation.

A few days later Brouardel summoned Roux. He wanted to have a private meeting with the physician because he trusted him. He knew that Roux had opposed Pasteur's method, not in principle, but because he did not think it was ready. At the same time Brouardel was well aware of what was at stake in this affair and therefore hoped to gain time until a completely reliable antirabies vaccine was available. "If I do not uphold your position, it is because that would set back the evolution of science by fifty years; and that must be avoided!"[13] he said to Roux. And indeed, Brouardel's testimony was crucial. How much maneuvering room did he have? In his quality as forensic examiner, he had to furnish an expert opinion; he could conclude that Jules Rouyer died either of rabies, or of Pasteur's anti-rabies treatment, or of kidney complications due to the blow he had received; and of course he could also incriminate a combination of these causes. If anything were said about rabies in the verdict, Pasteur could be held responsible; otherwise there would be no grounds for action against him.

Brouardel opted for kidney failure, thereby entirely clearing Pasteur of any responsibility. This was to nip in the bud all the hostile campaigns that would inevitably have been unleashed, and to disarm Clemenceau, Peter, and all those who had never forgiven Pasteur for denouncing spontaneous generation as an error.

On 4 January 1887, the case of Jules Rouyer was on the agenda of the meeting of the Académie de médecine. Brouardel had come to render his verdict. Another item on the agenda was a communication by Roux, also in defense of Pasteur. On 28 November, he declared, the medulla oblongata of the Rouyer child had been brought to the laboratory; two rabbits had been inoculated with matter from this medulla after trepanation; more than forty days after the inoculation, these two rabbits were in good health. This declaration definitively eliminated the suspicion of rabies and confirmed the official version of the child's death as given by Brouardel.

Peter had attentively listened to his confreres. He did not believe a word of what he had heard. He rose and began to argue. In his opinion, albuminuria was not significant in a cadaver, particularly if the death had been caused by rabies. And the fact was that all the findings described for young Rouyer were those of rabies.

But this argumentation, solid and logical as it was, finally came up against Roux's report. If the child's brain matter had not transmitted rabies to the rabbits, it was because it was not infected with the disease. The public prosecutor therefore exonerated Pasteur and his method.

Unless one accuses Adrien Loir of lying when he told this story several years later, one must concede that Roux committed a falsehood in order to prevent a worse outcome. More than the attacks against Pasteur's person, what Roux wanted to prevent at all costs was the rejection of the Pasteurian principles by the scientific community. However that may be, there is little point a century later in berating or attempting to exonerate Roux. On the other hand, it is no doubt important to note that this dubious and distressing episode took place in Pasteur's absence. We do not even know how and how much he learned of the facts that followed Loir's impromptu trip to Bordighera.

AT ABOUT THE SAME TIME, another affair made it necessary for the scientists of the rue d'Ulm to turn into veritable private detectives. This case concerned a British citizen, Joseph Smith. On 4 September 1886, Smith, who lived in Surrey, was bitten on the left hand by a rabid black cat. He was first treated by a local surgeon, who sent him to Paris to undergo the Pasteur treatment. Smith returned home on 9 October, apparently cured or, more precisely, without showing any symptoms of rabies. But a day or two later, he began to complain of violent stomach aches, which justified his admission to St. Thomas's Hospital. There he died a few hours later. He was thirty years old. The unusual circumstances of his death led to an inquest. In his deposition, the resident physician of the hospital testified that the patient complained about stomach aches and weakness in the legs, so that he was treated for an "acute paralysis of a very obscure form." The physician stated that Smith had not died from his bites and that his death could be attributed to natural causes. Shortly thereafter, the court brought in a verdict that seemed to confirm this conclusion. The case appeared to be closed.

Meanwhile the French press, always looking for a scandal, tried to exploit these facts. An important daily, *Le Matin*, headlined: "Client of M. Pasteur Bitten by Rabid Dog. Victim Dies After Being Cured of Rabies." Pasteur, who was still in Paris at this time, took this matter seriously; he was concerned and wanted to find out what had happened. The circumstances of this death troubled him. Since there could be no question of doubting the effectiveness of his vaccine or even envisaging possible side effects, he looked elsewhere for an explanation. Thinking about the patient's condition, he remembered that Smith presented signs of alcoholism. This was confirmed by the treating physician, to whom Pasteur declared in his reply,

"Some day we will no doubt learn that alcoholics are particularly suscepti- ble to rabies." But a hypothesis must be proven, and in order to dispel all lingering doubts, Pasteur asked one of his assistants, Dr. Chantemesse, to find out about Smith's behavior during his stay in Paris.

This was the beginning of something akin to a police investigation, involving among other things interviews with everyone who had come in contact with the victim. Looking into his personality brought some sur- prises. It was learned that Smith was an illegitimate child, that he spoke English and French, but that he did not know how to write. He regularly received sums of money, which he spent in low dives. On the very day when he had come to Paris for his treatment, he had returned to his hotel dead drunk, and the same thing happened on the following days. He would come in so late and in such a state, shouting and staggering, that he was almost thrown out of the hotel. On 20 September he was even more inebri- ated than on the other evenings, so that he had fallen into the Seine and was pulled out at the last moment by some boatmen. He had stayed in bed for three days with fever and vomiting. As soon as he was back on his feet, he resumed his drunkard's life, and even corrupted the hotel's bellhop, taking him along on a drinking spree, so that he eventually lost his job.

In short, Pasteur had had the right idea: Smith was indeed an excep- tional and particularly interesting case, which permitted the scientist to advance the hypothesis that alcoholism creates a hypersensitive terrain for the rabies virus. The research team of the rue d'Ulm therefore could feel relieved on two counts: not only was there no criminal responsibility; they had also made an important step forward in establishing the experimental parameters of the vaccination protocol.

This became a reflex with Pasteur. Whenever questions were raised about a death following antirabies treatment, he launched an investigation. A German newspaper, for instance, claimed that the rabies victims of Smo- lensk had died shortly after their return to Russia. Pasteur immediately cabled the mayor of Beloi, asking him for a certificate and also sent for a photograph of the priest showing the scars of his bites as proof that he was alive and well. Pasteur thus did battle with the journalists who raised doubts about his method. In this he acted not only in his own interest; he was afraid above all that some of the rabies victims would no longer come to the dispensary in the rue d'Ulm. After all, it was an article in *L'Intransigeant* that had kept Ferdinand de Lesseps's coachman from letting himself be inoculated.

However, unscrupulous journalists were not the only ones to place obstacles in Pasteur's way. He also had to deal with hostile experimental scientists, often abroad, who claimed to have scientific proof that antirabies vaccination was dangerous. This was the case, for example, with the experiments of Dr. von Frisch.

Sent to Paris by the Vienna Polyclinic in 1886, Frisch had been asked to familiarize himself with the methods of prophylaxis against rabies in Pasteur's laboratory. Pasteur was glad to welcome his Austrian colleague and to have all the pertinent demonstrations shown to him. When Frisch was recalled to Vienna, Pasteur gave him several rabid rabbits with which to continue his studies. However, in Vienna, Frisch was unable to reproduce Pasteur's results and had no qualms about publishing contradictory data that invalidated the method. This he did first in a series of short articles, then in a volume of 150 pages which asserted that Pasteur had followed the wrong trail.

One of the most famous Austrian physicians, Theodor Billroth, took up the matter and attributed even greater importance to his compatriot's conclusions, declaring that von Frisch's experiments added new luster to the Vienna School, and that Pasteur had finally been brought up short. Billroth asserted that the French scientist, however brilliant his earlier experiments, was altogether mistaken about rabies. He spoke of a disaster and reminded his readers of the criticism voiced by Koch and the Berlin School in the matter of anthrax. A classic case of lumping different things together.

This time, he had gone too far! Time to dig up the hatchet in the scientific community! Pasteur replied with an open letter to the president of the Imperial Medical Society of Vienna, which he published simultaneously in French and German in *La Tribune médicale*. He argued as follows: Von Frisch had not respected the incubation periods and had failed to understand the specificity of experiments on rabbits. Yet the virus vaccine had been developed with this species and therefore could not be used as a reference model: "I have never attempted to vaccinate rabbits against rabies by means of the prophylactic method against rabies; nor have I tried to find a method that would be applicable to that animal species."[14] Inch by inch, Pasteur thus defended his findings and the conclusions he had drawn from them.

The affair became European-wide. From Bordighera Pasteur wrote on 9 February 1887 to the director of a Naples newspaper, *Il Pungolo*, which had supported the Austrian physician: "Dr. von Frisch . . . has not succeeded, I am sorry to say. But I can counter his trials with positive results that will overthrow any negative facts he claims to have obtained." To this Frisch responded, "Positive results prove nothing against negative results. But what makes Pasteur think he can call my results negative? It seems to me that his are negative and mine are positive." Over and over Pasteur made the point that his adversaries had no case, that, given his successful treatment of humans in his dispensary of the rue d'Ulm, he did not have to waste his time on vague contestations concerning rabbits, and that he looked forward to the future with serene confidence: "Here again it is Time, which does not plead for one side or the other but is the infallible judge in the last resort, which will have the last word."[15]

Pasteur fought on all fronts. He kept an especially sharp eye on the German-Italian axis, where the opposition was fiercest. He made sure that every last hostile article, whether in the general press or in specialized journals, received his response. These perpetual polemics shed light on Pasteur's relations with the scientific press. Since the first half of the nineteenth century, scientific journalism had gained an important position, independent of any official institutions. There was a great deal of popularization of intellectual matters, which prospered even in the columns of large-circulation daily papers. As early as 1830, Arago had opened certain meetings of the Académie des sciences to the public; journalists seized this opportunity to involve themselves in scientific controversies and began to proselytize.

In the 1880s, after the victory of Pouilly-le-Fort, most of the scientific journalists were ready to defend the honor and the glory of Pasteur, a veritable monument of the national patrimony. A few of them, however, permitted themselves some reservations. One form of contestation consisted of not attacking directly, but of letting the opponents speak as a matter of equity. Thus the *Moniteur scientifique* or *Cosmos*, which acknowledged the validity of the Pasteurian method, gave space to the critiques of Guérin or Peter. Other journals criticized the fact that Pasteur's research monopolized all the attention; they tried to provide information about other methods, such as chemotherapy, on the grounds that Pasteur was able to prevent but not to cure. They also insisted on the weakness of the statistics used by the Public Health Board. And they advocated the return to more traditional and popular forms of medicine.

In their efforts to diminish Pasteur's merits, some writers also went to great lengths to discover all kinds of unacknowledged precursors. While the biologist Pasteur had indeed a particular gift for synthesizing previous studies, the argumentation of the journalists was often rather hackneyed. Thus *La Science universelle* recalled that the principle of antirabies vaccination was inspired by the immemorial practice of mithridatization against snake venom. There were also political undertones; some republican papers, critical of Pasteur's fidelity to Bonapartism, pointed out that Raspail, the "physician of the poor" had already spoken about the pathogenic role of microbes and praised the antiseptic properties of camphor in his *Histoire naturelle de la santé et de la maladie* published in 1843.

But Pasteur was also a popular figure, even if he rarely shed his reserved manner. He did not often write for the large-circulation newspapers, preferring to make himself heard in communications to the academies, but his pupils and friends, Grancher, Vulpian, or Richet, had become advocates who defended him wherever necessary. They published large numbers of articles of popular science and many brochures that found their way into every community in France. Sensational and highly colored pictures, the so-called *Images d'Epinal*, on the front pages of cheap newspapers were more effective than any explanation; they put the final touches to the golden legend of Pasteur, the man who fought rabies with his bare hands, pulling little children from the fangs of the disease. There is no question that the scientist and his associates had an unexpected sense of drama, which is always the most effective tool and the best propaganda. To the devastating myth of rabies and to the redeeming miracle of its treatment, Pasteur added the emotional appeal of real people, casting Meister and Jupille as the heroes of a new theater that showed science in action. At the same time Peter, the Man Consumed by Envy, was cast into the shadowy realm of the wicked, along with the archaic opponents of vivisection. Pasteur personified the March of Progress, resplendent in the rosettes of the Legion d'honneur. Once again there was talk of the salvation of mankind and of relieving popular suffering. For the first time, medicine made the front page of the newspapers.

Creation of the Institut Pasteur

In order to accomplish his mission, triumph over his adversaries, and impose his method, Pasteur needed to have his hands free. As early as 1885, he had managed to find some funds with which to set up a rabies clinic in his laboratory in the rue d'Ulm, but its autonomy was precarious. As patients began to arrive there, the clinic had to take care of all phases of an emergency service, from first aid to inoculations. The clinic was soon overwhelmed; the preparation of rabid spinal cords and the follow-ups of the patients put great stress on the laboratory; and the patients waiting to be inoculated by an injection into the abdomen had to line up between two doors; there was neither an infirmary worthy of that name nor proper sanitary supervision. Something had to be done.

Pasteur had initially felt that the success of his treatment depended on inoculating the victim as soon as possible; he therefore did not think that a single center of antirabies vaccination would be sufficient. But very soon, that is, following the success with Jupille, who had been inoculated six days after he was bitten, he realized that a delay of a few hours, which allowed the victim to be transported to Paris, would not be prejudicial. This being the case, he had only one idea: to set up a treatment center independent of the laboratory of the Ecole normale.

"It is my intention . . . to found in Paris a model establishment that does not require state funding but would be financed by gifts and international subscriptions. I am confident that a single establishment in Paris would be sufficient, not only for France but for all of Europe, Russia, and even North America."[16] Pasteur wrote these lines on 12 January 1886 in a letter to Comte de Laubespin, a philanthropist who had spontaneously sent him the tidy sum of forty thousand francs.

A few weeks later, on 1 March, in a session that has become famous, Pasteur communicated the entire complex of his findings on rabies to his confreres of the Académie des sciences — and then went on to develop his new idea: "On the basis of the most rigorous statistics, you can see that a large number of persons has already been saved from death. Prophylaxis against rabies after a bite is now a reality. This calls for the creation of an antirabies vaccination establishment."[17]

At this point, Admiral Jurien de la Gravière, who presided over the meeting, rose from his seat and congratulated the scientist, not simply, he said, in the name of the Academy, but in that of all of mankind. Vulpian, more down-to-earth, asked for more information about the means that might be available, and about the goals of this unprecedented institution. Pasteur returned to the podium and continued to dream out loud: "In the Paris establishment we would of course have to train young scientists, who would then take the method to faraway countries. And surely we could also do this for the various regions of Europe, but I repeat that there is no need for this. The success of the operations will moreover be better assured by keeping the number of operators smaller." Dreaming did not keep him from being realistic, which is why he adamantly refused to accept money from the French state or from the city of Paris to make sure that he would remain the master in his own house.

At the end of the meeting the members enthusiastically endorsed the opening of a subscription and appointed a commission. Pasteur would get his center of antirabies treatment. To finance this new institution, a subscription was indeed opened in France and abroad under the direction of a committee headed by Jurien de la Gravière. Funds could be sent to the Banque de France, to public treasurers general or private collection agents, and to tax collectors, all of whom passed them on to the Crédit foncier (of whose board Pasteur was a member). Donations depended on the generosity and the means of each individual; a gendarme sent one franc, a poacher fifty centimes, the emperor of Brazil 1,000 francs, and the czar of Russia 97,839 francs. The Milan newspaper *La Perseverenza* collected 6,000 francs following its first appeal.

The *Journal d'Alsace* and other local papers recalled that Pasteur had held a chair at the University of Strasbourg and that "a poor Alsatian peasant, young Meister of Villé, was the first person to whom Pasteur applied his discovery in order to save him from death."[18] To which Pasteur responded, "It was not without genuine emotion that I read the headlines of the eleven newspapers who wish to open their columns to the subscriptions for the new establishment. I was equally happy and moved when I noticed among the many names of subscribers — and how I wish I could thank all of them individually — that of my young friend Joseph Meister."[19] In Alsace-Lorraine, Pasteur collected 48,365 francs; Germany, with a grand total of 105 francs, showed rather less eagerness. The projected Institut Pasteur may have been part of an effort at European reconciliation through science and medicine, but it is obvious that the memory of the war of 1870 still weighed heavily on people's minds.

"The subscription . . . is going well," Pasteur wrote to his old friend Jules Vercel on 24 June 1886. "The Crédit foncier, which acts as a central collection agency for the subscription, will have taken in between 1,400,000 and 1,500,000 francs. But we will need two or three times as much. A considerable capital must be invested if there is to be a sufficient annual income to pay for materials and personnel."[20] In the effort to raise sufficient funds, no stone was left unturned. Thus, at the suggestion of Charles Richet, the "Scientia" Society organized a gala event to benefit Pasteur's antirabies center, to take place on Tuesday 11 May 1886. In the great hall of the Trocadéro Palace (not the severe building we know today, but the curious Moorish pastiche that Davioud had built for the World's Fair of 1878), Pasteur received a lengthy ovation. Constant Coquelin of the Comédie française recited Eugène Manuel's poem to the glory of great men, "whose genius can drive back death," while Gounod himself conducted his famous *Ave Maria*. In thanking Augusta Holmès, Ambroise Thomas, Jules Massenet, Léo Delibes, and Camille Saint-Saëns, whose works were also heard, Pasteur humbly confided, "Do I dare confess that tonight I heard almost all of you for the first time? I do not think that in all my life I have spent ten evenings at the theater."[21]

Little by little this subscription, which was reminiscent of the one that was opened in 1873 for the construction of Sacré Coeur of Montmartre, became the stuff of legend. In his *Journal*, Jules Renard asserts that in order to obtain what he wanted, Pasteur himself sometimes made calls. He reports a scene about which he had heard from Lucien Guitry: "Pasteur called on the widowed Madame Boucicaut, the owner of the Bon Marché department store. She hesitated to receive him, 'It's an old gentleman,' the maid said. 'Is that the Pasteur with the rabies of dogs?' The maid goes to ask. 'Yes,' says Pasteur. He is ushered in. He explains that he is going to found an institute. Gradually warming up, he speaks more and more clearly and eloquently. 'That is why I have assumed the duty of bothering charitable persons like yourself. The slightest contribution . . .' 'Why, of course!' says Madame Boucicaut, just as embarrassed as Pasteur. Some small talk. Then she takes a checkbook, signs a check, folds it, and hands it to Pasteur. 'Thank you, Madame!' he says, 'You are too kind.' He takes a look at the check and bursts into tears. She too is in tears. The check was for a million francs."

This last figure is probably exaggerated, since the total sum collected by Pasteur was 2,586,680 francs. But in any case it is certain that the scientist was his own patron and that he himself made one of the largest individual contributions: one hundred thousand francs. One may wonder how he was able to raise such a sum. Did this money come from the sale of commercialized vaccines since 1882, or simply from the savings of a household that had been housed rent-free at the Ecole normale for thirty years and whose lifestyle was akin to that of cloistered monks or nuns? Adrien Loir indicates that although Pasteur refused to make a profit on the vaccines used within France, he did claim the earnings from sales abroad for himself and his collaborators. Moreover, we should not forget his major national rewards, especially that of 1874, which was doubled in 1883. Judiciously invested, these could yield large returns.

Meanwhile, more was needed than goodwill and money. Space also had to be found: where should the establishment be set up? Pasteur did not like to move around in the streets of Paris. For a very long time now, his whole life had taken place in a narrow perimeter on the Left Bank, in the neighborhood around the rue d'Ulm, the Quai Conti, and the rue des Saints-Pères. Only very special circumstances — visits to Joseph Meister when he was treated at the Collège Rollin or the inauguration of electric lighting at the Louvre department store — would make him cross the Seine. In a word, Pasteur did not intend to go very far, and it so happened that a site was available on the boulevard de Port-Royal, across the street from the Val-de-Grâce Hospital. With the agreement of Grancher, but without consulting his other collaborators, he therefore planned to settle in the immediate vicinity of the rue d'Ulm. But when Duclaux was informed of this plan, he rebelled. The location, he said, was poorly chosen and above all too small, squeezed in as it would be between Val-de-Grâce and Cochin hospitals. Yet the institute of which Pasteur dreamt should be able to grow freely; it not only needed buildings but also an experimental agricultural station; and if they really wanted to make their mark in the future, bigger plans were surely necessary. A few days later, Duclaux announced that he had found the ideal place: twelve thousand square meters of marshy terrain that could be had cheaply in the rue Dutot in the plain of Grenelle. Pasteur was not happy with this solution but nonetheless went along with it in the end, because Duclaux was right.

Grenelle it was, then. But what would be at Grenelle? What would the antirabies institute be called? Pasteur's laboratory in the rue d'Ulm went by the name "laboratory of physiological chemistry." Here a more striking name was needed. Grancher took the lead in this discussion. The first question was whether it should be *institution* or *institute*; the second solution was chosen because the word evoked the Institut de France and its prestige. But Pasteur also had his say; he greatly admired the "Deaf-Mutes Institution" and a formula like "Rabies Institute" would not displease him, for he did not want his name to be used. But this is not what his disciples had in mind, and in the end they overcame their master's misgivings: "Institut Pasteur" it was to be.

It took more than two years to clear up all difficulties. In addition to scientific controversies, administrative and legal complications now had to be dealt with, not to mention meetings with architects and contractors to discuss the building.

The first stone of the Institute weighed only a few grams and was laid by Duclaux: in January 1887 Duclaux, with the agreement of his master, founded a monthly review called *Annales de l'Institut Pasteur*. When the first issue was published, Pasteur was in Italy, which is why its preface consisted of a "Letter on Rabies" sent to his faithful collaborator from the Riviera.

> My dear Duclaux, very often when we were talking in the laboratory, we regretted that we did not have available to us a publication that would be a little more intimate and a little less solemn than the reports of the Académie des sciences. There were times when we either did not report facts and observations that deserved to see the light of day, or failed to respond to criticisms that would have been easy to put to rest. The focus of research in a laboratory sometimes changes so rapidly, and one can so easily be drawn from one direction to another that one is liable to leave aside useful studies that are actually ready for publication. Scattered facts and whole series of experiments are sometimes sacrificed to the emergence of new ideas. . . . You tell me, my dear Duclaux, that you have decided to launch a monthly collection of articles under the title *Annales de l'Institut Pasteur*. The service you will render will be appreciated by the increasing number of young scholars who are attracted by the field of microbiology. The work of our laboratory will of course be the main focus of your *Annales*, but any studies from the outside that you might accept will motivate all of us to do equally well.[22]

It is obvious that Duclaux, even though he placed himself under the aegis of Pasteur, meant to create an independent review that would operate in a truly original manner. The *Annales* broke with the usual compartmentalization of the scientific disciplines and treated issues of public hygiene as well as social, medical, biological, and industrial topics on the same footing. By concentrating on the study of microbes and their virulence, Duclaux and his authors would be able to cover a wide field of investigation, from contagion to immunity, from enzymes to sewers. The result was that, month after month, the *Annales* contributed to the renewal of medicine without ever studying disease as such, and wrought major changes in public hygiene without focusing their studies on squalor and poverty. But above all, and very clearly, they served as a forum for presenting the findings and the discoveries of the Institut Pasteur, a legitimate place to communicate the happy statistics concerning the antirabies vaccination.

Three months later, it was Grancher's turn to found a specialized review, the *Bulletin médical*, which in particular assumed the task of responding point by point to the unrelenting attacks launched against Pasteur and his methods of vaccination in the *Semaine médicale*.

As for the Institute itself, it was gradually taking shape. The building plot in the rue Dutot was purchased in March 1887 for 430,000 francs — which means that the square meter of constructible surface cost 38 francs, the equivalent of 1,000 in today's francs [or $200], surely a great bargain.

A few weeks later, the definitive statute was drawn up and registered with Maître Etienne-Maurice Guérin, notary in Paris: "The purposes of the Institut Pasteur . . . are the following: (1) the treatment of rabies according to the method developed by M. Pasteur; (2) the study of virulent and contagious diseases." There was no precedent for such a foundation: Pasteur had created an autonomous scientific establishment endowed with a legal identity and dedicated to a threefold mission, for it was to serve as a dispensary for rabies treatment, a center of research on infectious diseases, and a school of advanced studies. It was headed by a director assisted by twelve board members, and an assembly composed of thirty members who proposed a new slate of board members every three years. Pasteur, of course, was the first director, and appointed for life; but his successors would serve for six years. These statutes were endorsed notably by the perpetual secretaries of the five academies that composed the Institut de France, as well as by representatives of the Senate, the Chamber of Deputies, and the Faculty of Medicine. Several bankers, those of the Crédit foncier and the Crédit de France, and some important journalists were also included. In short, the Institut Pasteur was held over the baptismal fonts by the most powerful godparents in France.

On 4 June of the same year, a decree of the President of the Republic, Jules Grévy, declared the Institut Pasteur to be an institution of "public usefulness"; henceforth the Ministry of Commerce and Industry would be held responsible for following its development and its work. Pasteur very prudently made sure that his institute was not placed under any administrative tutelage, but the Council of State demanded and obtained a right of official oversight. But in any case, the fact that Commerce and Industry was chosen, rather than Public Education or Agriculture, clearly reveals the very ambitious hopes that all of France placed in the Institut Pasteur.

RABIES, THEN, had provided the initial impulse for the Institut Pasteur. But on the many afternoons in 1888 when Pasteur went to the rue Dutot to see the walls of the Institut go up in this suburban countryside, he knew that he was building more than just a center for combating rabies. The antirabies dispensary would permit him to create a center for the study and investigation of contagious diseases and microbes. The income from the sale of vaccines — those already discovered in the rue d'Ulm and those that would be developed in the rue Dutot — would give him the means of providing for this revolutionary institution.

In the beginning of his career, Pasteur was fascinated by yeast; it became the leaven that raised his research until it achieved the victories over anthrax and rabies. More than ever, he was faithful to his convictions of those days when he now exhorted his colleagues to make their discoveries pay and to find industrial uses for them. Now that Pasteur had chosen to devote himself to the world of health and hygiene, he created an industry in the service of life. Amid the masons and stonecutters, Pasteur could for a moment forget his rabies patients and watch his dream take shape. He was the world's first scientist to become an enterprise.

The Inauguration

14 November 1888. Rue Dutot is bustling with the excitement of a great day. The former village of Vaugirard, a suburb of Paris only since 1860, shakes off its usual torpor of an outlying area given over to sheds and small workshops. The Institut Pasteur has finally risen out of the ground. It consists of two facing main buildings, connected by a gallery of freestone, bricks, and millstone grit in the purest Louis XIII style of which the nouveaux riches are so fond. One might almost take it for a kind of factory because of the tall chimney, which gives the whole complex an industrial look; the steep roofs of the kennels and hutches and the clumps of shrubbery also evoke provincial manufacturing plants, like those that Pasteur had encountered in the Vosges when he was on the trail of tartaric acid. To the right are the hangars of the carriage manufacture "L'Urbaine." To the left, two or three smallish houses with their vegetable gardens and a little farther down the street the Protestant school of the impasse Fremin, and then another very large factory. Across the street, the fences and the meager vegetation of wastelands are still in evidence. But on the horizon, toward the Seine, one can already see the top of the tower that Gustave Eiffel is finishing in time for the World's Fair of 1889.

That day the "Rabies Palace," as the newspapers call it, is swathed in flags: tricolor flags at all the windows, all along the metal fences. Sporting its Sunday best as well is a ramshackle building right across the street, which pompously calls itself "Grand Hôtel de l'Institut Pasteur." Located amid the market gardens, it offers "room and board for all those who wish to be treated for rabies." A little farther on, a café promises wine and lemonade to the same clientele of vaccine recipients.

Little by little, the Institut Pasteur would nibble away at this territory, and these picturesque establishments would give way to more solid constructions. Some time before Pasteur's death a generous donor, Madame Jules Lebaudy, offered to buy the plots across from the Institute and to create a hospital, where Albert Calmette eventually set up a facility for the treatment of tuberculosis. For their part, Laveran, Mesnil, and Marchoux, experts in tropical diseases, were able to create a special wing where they could improve their techniques. When Duclaux succeeded Pasteur, he was in a position, thanks to the donations of Baroness Hirsh, to undertake the construction of a vast building to house the chemical laboratories.

Since the early morning hours, crowds of curiosity seekers had gathered at the entrance of rue Vaugirard, where they were contained by a double row of policemen, later reinforced by mounted police. People were watching from all the balconies, and even from the roofs, where they were hanging on to the chimneys. To go through the doors of the Institut one had to show identification; 1,200 invitations had been sent out. The Grand Dukes Vladimir and Alexis of Russia were expected to attend, as well as the Prince of Oldenburg.

Pasteur arrived at noon, accompanied by Jean-Baptiste. Leaning on his son's arm, he slowly passed by the musicians of the Garde républicaine who were waiting in dress uniform, with their instruments on the ground, and then ascended the flight of steps to the entrance. To the left was the door of the so-called Salle des Actes, where the busts of the main donors had been placed. Here, to name only a few, Don Pedro of Brazil rubbed shoulders with Madame Boucicaut, Baron de Rothschild with the czar of Russia, not to forget the man who without question was the first benefactor of the Institut, Comte de Laubespin. Here, under a rose-colored coffered ceiling, amid the paneling of imitation wood, and in the tenacious smell of drying paint, Pasteur was about to receive his guests.

Half-past twelve. The scientist stands on the threshold. Across his chest he is wearing the grand cordon of a commander of the Légion d'honneur and the insignia of the Russian Grand Cross of Saint Anne. The first delegation he receives is that of the students; there are about twenty of them, carrying the banners of their respective universities. At one o'clock, the first official coaches arrive at the foot of the steps. First come the members of the Institut de France, Jules Simon, Léon Say, Frédéric Passy, Joseph Bertrand, Comte d'Ormesson, Prince Roland Bonaparte, Henri Wallon,

Victor Duruy, Duc de Broglie. Many of them have been Pasteur's close friends for many years. Also to be recognized are MM. Peyron, director of Public Welfare, and Poubelle, the prefect of the department of Seine who has made a name for himself by his contributions to public sanitation. Then arrives the celebrated explorer Savorgnan de Brazza, followed by the directors of the most important newspapers. Over the next few days the inauguration was to make headlines in all the papers, regardless of tendency or style. There were articles or descriptions in *Le Figaro*, *Le Gaulois*, *Le Journal des débats*, *La Liberté*, *La Patrie*, *Le Radical*, *La République française*, *Le Soir*, *Le Temps*, *L'Intransigeant*, *Le Monde*, *La Nation*, *Le Soleil*, *Le Mot d'ordre*, *La Cocarde*, *La Vie du Peuple*, *L'Echo de Paris*, *L'Estafette*, *Le Gil Blas*, *La Lanterne*, *Le Petit Caporal*, *Voltaire*, and of course the *New York Herald Tribune*.

On the dais draped in red velvet trimmed in gold and surmounted by a bust of the Republic and bundled flags, the principal seats are still empty, but the representatives of the diplomatic corps have already made their entry. The ambassadors of Italy, Turkey, and Brazil represent their sovereigns, the king, the sultan, and the emperor, who have not come in person but have made large contributions in support of Pasteur.

Major figures from the world of science are of course being honored. Collaborators from the early days, such as Jules Raulin or Eugène Maillot, have a chance to meet their successors. In constituting the first team for his Institute, Pasteur had considered it important to avoid administrative unwieldiness, and so he privileged youth and efficiency. He had envisaged five laboratories run by five department heads and fourteen laboratory assistants. Emile Duclaux, who was almost fifty, was the oldest; he would be responsible for general microbiology. Medical biology would be divided into two sections, technical microbiology under the direction of Emile Roux (35), and research, under that of Nicolas Gamaleïa (29), the former director of the rabies institute of Odessa. The laboratory of morphological microbiology had been entrusted to Elie Metchnikov (43), another newcomer from Russia. Joseph Grancher (45) would direct the rabies service with André Chantemesse (37). Charles Chamberland (also 37) would be in charge of vaccine production and sales. These laboratory heads would be able to pursue their own research, although all of them were expected to teach and to train the students admitted to the Institute.

At twenty past one, behind schedule, the President of the Republic finally arrives. It is no longer Jules Grévy, who has had to resign in December 1887 in the wake of the scandal over decorations, but Sadi Carnot. The presidential coach drives through the gate and comes to a halt at the foot of the steps. The music of the National Guard strikes up the *Marseillaise* as the delegation pours into the vestibule. The occupants of the official dais had just begun to become impatient. Soon Pasteur is seated to the right of the President. The seats around them are taken by Charles Floquet, President of the Council of Ministers; Louis Peytral, Minister of Finance; Pierre Legrand, Minister of Commerce; and Jules Viette, Minister of Agriculture. Edouard Lockroy, Minister of Education (and second husband of the widow of one of Victor Hugo's sons) arrives last, with a tardiness that the newspapers will gleefully report.

Time for the inevitable speeches. The president of the board of the Institut Pasteur, Joseph Bertrand, is the only member of the Academies who has donned the green tailcoat; it is he who opens the ceremony. He begins by invoking the names of the deceased masters, Jean-Baptiste Biot, Jérôme Balard, Claude Bernard, and Jean-Baptiste Dumas, who had died in 1884, at the time when Pasteur represented France at the celebration of the tricentennial of Edinburgh University. Then Joseph Grancher speaks in his capacity as Pasteur's first assistant for the rabies experiments; as secretary of the Institute's patronage committee, he too evokes a friend who is no longer there, Vulpian, who had died a year earlier. However, he also remembers to speak of the growing ranks of young scientists whom Pasteur seduces "by the implacable rigor of his dialectic and by the absolute form he sometimes gives to his thought."

Speech follows upon speech; some are prophetic, others rather tedious. The atmosphere is solemn, and while the audience is not the elegant Tout-Paris of a first night at the theater, it is very large, and the messages of support, the homages, and the expressions of gratitude go on for so long that Roux and his very young assistant, twenty-five-year-old Alexandre Yersin, prefer to return to their laboratory, still located in the rue Vauquelin, before the end of the ceremony.

Meanwhile, the hero of the day has not yet been heard from. Pasteur rises laboriously and it is obvious that he is quite frail, almost exhausted. He has told his close collaborators that he feels "defeated by time." Still, he is expected to speak. But Pasteur does not speak, his speech is read by his son. The reason is that a year earlier, on 23 October 1887, on a Sunday morning, while writing letters, he had suffered a second stroke, followed by aphasia. That day he was to lunch at the home of his daughter Marie-Louise. The family was sacred to Pasteur; so there could be no question of missing a visit with his grandchildren. He therefore had himself driven to his daughter's house, where he spent the afternoon in an armchair. By evening his speech had returned. But a few days later there was another attack, which did not clear up as easily. Ever since, his speech had remained blurred and hesitant, to the point that on this 14 November 1888, when the high emotion of the occasion was bound to compound the problem, Pasteur was unwilling to move his audience to pity with the mere specter of his former eloquence.

The address read by Jean-Baptiste is a vibrant plea for teaching and research. Pasteur modestly points out that his name has been inscribed above the gates of the Institut against his will: "And I continue to object to a title that pays homage to a man rather than to a doctrine." Nonetheless he does trace his personal itinerary, his struggles on behalf of the University, the creation of laboratories, and government support for research.

He speaks the words of a sage who imparts to his spiritual heirs his ultimate lesson of scientific rigor: "Always cultivate the spirit of criticism. Once it has been allowed to fail, there is nothing to awaken an idea, nothing to stimulate great things. Without it, nothing will hold up. It will always have the last word. What I am here asking of you, and what you in turn will ask of those whom you will train, is the most difficult thing the inventor has to learn. To believe that one has found an important scientific fact and to be consumed by the desire to announce it, and yet to be constrained to combat this impulse for days, weeks, sometimes years, to endeavor to ruin one's own experiments, and to announce one's discovery only after one has laid to rest all the contrary hypotheses, yes, that is indeed an arduous task. But when after all these efforts one finally achieves certainty, one feels one of the deepest joys it is given to the human soul to experience."[23]

Finally, before he closes the ceremony, the President of the Republic confers decorations. For Pasteur had expressed the wish that the inauguration of his Institute should be a day of reward for his collaborators. Duclaux, Grancher, and Chantemesse receive the Légion d'honneur, as does the architect of the Institute, Félicien Brébant, who has refused to accept any remuneration for his work.

The last fanfares are sounded. A fine drizzle has begun to fall. Pasteur no longer hides his weariness. Leaning on Marie's arm, he walks up to the quarters that have been set aside for him, quarters "fit for a czar," as the porter will tell the journalists. The day has been a success, but Pasteur wonders. His thoughts go back and forth between the microbes that kill and the vaccines that protect, contradictory and fragile symbols of the fate of humankind, always poised between war and peace. He repeats to himself some of the final words of his address, which had almost surprised him when he heard them from the mouth of his own son: "God alone knows."

17

THE LAST LIFE SCIENCES

Yildiz Palace, Constantinople, 5 June.

Illustrious Master, His Imperial Majesty the Sultan, my magnificent master, having learned through public reports that have resounded throughout the enlightened parts of every country that a new beneficent star has again arisen in that glorious France, whose civilization and progress have shed light on all of mankind in modern times, has done me the honor of charging me with setting up, subject to his imperial sanction, an official scientific commission which, once it is selected, will travel to Paris in order to call on you and to ask you on his behalf to receive it kindly and to incorporate it into the ranks of your foreign students by admitting its members to your cosmopolitan labora-tory, thereby enabling them to study the pernicious influence of microbes.[1]

Clearly, the creation of the Institut Pasteur was seen as a major event throughout the world, and applications from those who dreamed of coming to work with Pasteur were pouring in. For in spite of old age and fatigue, in spite also of all the honors he had received, the scientist had no intention of resting on his laurels or retiring to his mauso-leum while he was still alive. He was ready to convey his last messages and to perform the ultimate experiments. Some doors still remained to be opened, and the microbes had not yet yielded all of their secrets. Pasteur's last studies, some of which date from before the opening of the Institute, also had their symbolic value.

Bacteriological Warfare

In November 1887, when the entire press devoted its headlines to the scandal over decorations that would soon force Jules Grévy to resign, Marie Pasteur, as she did every night, was reading the newspapers to her husband, since he had to save his eyesight. A conscientious and meticulous reader, she did not leave out any detail, and therefore did not neglect an advertisement inserted in *Le Temps* by the New South Wales government, which was looking for a means to destroy the rabbits that were multiplying alarmingly in New South Wales and New Zealand.

Ever since the American Civil War had raised the price of wool, many Australian settlers had acquired considerable wealth and become large landed proprietors. Imitating English customs, they developed a passion for hunting, and since every landlord wanted to have his own game reserve, dozens of societies for the acclimatization of game had sprung up. The Australian soil and climate were particularly suitable for hares and rabbits, so that these animals multiplied very rapidly; soon the hunters could no longer control them. They now had to appeal to experts in hopes of destroying these rodents which ravaged the properties. The government foresaw a major calamity and dreamed of a massive extermination campaign; hence the advertisement published in *Le Temps* and most of the world's major dailies.

As Marie calmly read this short paragraph, she had no idea of what it would set in motion. For Pasteur considered the Australian advertisement much more interesting than the troubles of Jules Grévy and his son-in-law Wilson. On the very next day, he therefore summoned Loir in order to dictate to him a letter to the director of *Le Temps*, in which he asked that paper to forward the solution he proposed.

While conducting his experiments on rabies, Pasteur had continued to work on chicken cholera. In particular, he had shown that the microbe responsible for this disease can also infect and kill rabbits. Knowing about the devastating effects of a cholera epidemic in a poultry yard, he felt that it would be possible to create an experimental epidemic in the Australian countryside. In his letter to *Le Temps* he described the death struggle of the cholera-stricken chickens and added, "I imagine that the same thing would happen to the rabbits and that by returning to their burrows to die, they would communicate the disease to others, which would propagate it in their turn. But how would one go about making the first rabbits ingest the destructive disease into their bodies? Nothing is easier. In the vicinity of one of the burrows, I would place a fence surrounding a certain area where the rabbits would come to feed. We have learned from our experiments that it is easy to grow the chicken cholera microbe in a state of perfect purity and in any desired quantity, in any kind of meat broth. One would simply have to pour these microbe-laden liquids on the food of the rabbits, which would soon go off to die here and there, spreading the disease all around. I should add that the parasite of the disease of which I am speaking here is harmless to farm animals, except of course to chickens; but then chickens do not have to live in the open countryside."[2]

The procedure Pasteur envisages here cannot be called anything but "bacteriological warfare," for he intended to use his mastery over microbes to impede the natural evolution of other species. Pasteur versus Darwin! This was not the first time that our scientist had envisaged using the resources of microbiology for correcting nature's mistakes, for he had thought about it as early as 1882 as a means of combating the phylloxera that was ravaging the French vineyards.

For the time being, however, the problem was rabbits. Pasteur's letter to *Le Temps* was published verbatim and read by the widow Pommery, owner of a celebrated Champagne cellar in Reims. She too wanted to exterminate rabbits, for they were digging their burrows under her storage cellars and causing stones falling from the ceilings to break her champagne bottles. In December 1887, Madame Pommery therefore addressed herself to Pasteur, who was enthusiastic about the idea of trying out this method on a small scale. Loir was dispatched to spread the chicken cholera on top of the cellars. It was a complete success. Dozens, perhaps hundreds of dead rabbits were counted. The ground was littered with dead bodies. Rather strangely, the epidemic stopped at the borders of the wine-making estate. Everyone was delighted, Madame Pommery at being rid of her rabbits, Pasteur at having achieved success in a new experimental field.

Having learned of the letter published in *Le Temps* as well as of the success achieved in the Champagne region, the government in Sydney now contacted Pasteur, asking him to send strains of the chicken cholera microbes with instructions on how to use them. Since Pasteur was unwilling to give out his reagents unattended by his collaborators, he sent Adrien Loir on this mission. In February 1888, accompanied by two collaborators, Drs. Germont and Hinds, the young man embarked on a steamer of the Orient Line with his cholera-bearing flasks. It took him several months to reach Australia — and he was to remain for five years.

When Loir finally arrived at the antipodes, he was not sure that the microbe would still be effective after such a long crossing and in such a faraway land. Since he did not have an incubator available, he heated the microbe in his own belt before injecting it into two rabbits. It was a complete success, for the two animals soon died in the bathroom of his hotel room; chicken cholera was as deadly in Australia as it was in France. But a triumph was not at hand, for a great deal of time had passed since the advertisement had appeared in *Le Temps*, and the political situation in Australia had undergone a rapid change. A lobby of rabbit exterminators had sprung up, and it had become more powerful than that of the ruined farmers. A law against the use of the French microbe had been passed by Parliament, and this law also prohibited introducing any possible source of contagion into the Australian territory. This isolationist law was not dictated by any ecological concern but rather by the urgent need not to kill the goose that laid the golden eggs.

Having made his way to Sydney, Loir, in his capacity as special ambassador of the Pasteurian methods, nonetheless asked to be received by the New South Wales Minister of Agriculture. He went to see him accompanied by the French consul, who assured him that he was perfectly fluent in English and agreed to serve as his interpreter. At the end of the interview, the improvised interpreter told him that Australia had no use for Pasteur or for his methods. Loir was crestfallen; he might as well go home. But then, rather than sailing immediately, he gave himself a short vacation. By chance he met an Australian senator and told him how badly he had been received by the government. Very much surprised, the parliamentarian looked into the matter, only to learn that the French consul had completely misunderstood the words of the minister who, far from criticizing Pasteur, had gone on and on in his praise. Wishing to make up for this misunderstanding, the senator went to see Loir and suggested that since he could not use the cholera microbe to destroy the rabbits, he should go to work on anthrax, known in Australia as the Cumberland Disease.

This is how Adrien Loir came to found what can be considered the first overseas Pasteur Institute, near Sydney. Informing his uncle on a regular basis, he was to conduct, under his guidance, research on the transmission of anthrax and in particular on peripneumonia in livestock. In 1891 he also became involved with rabies under the unexpected circumstances of an appearance of Sarah Bernhardt.

The famous actress never traveled without her dogs. But Australia was very strict when it came to quarantine. In order to get around the rules, Loir offered his charming compatriot housing for her two animals in his institute, which was protected and where they would enjoy preferential treatment. By way of thanking him, Sarah Bernhardt granted the young scientist an unsurpassable privilege, that of playing the silent part of her lover in Victorien Sardou's *Fédora*. At the end of the play, the heroine faints and falls into her lover's arms — which is why this part was coveted by all of Sarah Bernhardt's innumerable suitors, to begin with His Royal Highness the Prince of Wales.

Many Australians did not forgive Adrien Loir for holding Sarah Bern-hardt in his arms for a moment, and the most jealous took revenge by bring-ing up the subject of rabies. In his defense, Loir pointed out that the length of the voyage, added to the isolation of the dogs in his institute, constituted a sufficient safeguard, for together they extended far beyond the incubation period of the disease. Yet Pasteur, whom his nephew consulted on this mat-ter, was less categorical: "A dog that leaves Europe after having been bitten by a rabid animal will die during the quarantine imposed upon its arrival in Australia in keeping with the incubation period. However, this rule is not absolute; science knows of incubation periods of one year, even two years and several months for rabies; but these are most unusual exceptions."[3]

In any case, this anecdote shows that the New South Wales govern-ment was extremely concerned about the risk of rabies. It is true that at this point Oceania was still free of this affection. Rabies and other hitherto unknown viruses would eventually be brought to Australia by the means of rapid communication, especially air travel.

The Awakening of Immunology

From bacteriology to vaccination, the Pasteurian doctrine was com-ing into its own. Little by little it conquered the globe. Now a new science, still in its infancy, came upon the scene, for the elderly man who came to check on his latest experiments was inaugurating the field of immunology.

Immunology — the word did not even exist yet. By advancing from the notion of natural resistance to that of acquired resistance, Pasteur had laid the foundation of a theory of immunization. The laboratories of the Institut Pasteur were set up to breed bacteria for making vaccines; the science of the infinitely small came to be called microbiology, a term that was finally judged to be more appropriate because it was more general. Here was the starting point of an enterprise of industrial dimensions, which aimed to modify the natural rhythms of life by means of systematic vaccination.

Pasteur thus had appropriated the Jennerian procedure and given it a new name, but this does not mean that he can be accused of stealing another man's legacy. For he had considerably expanded the empirical practice of inoculation or mithridatization. The techniques of immunization had im-perceptibly led him to the science of immunity.

In his last studies, Pasteur recalled that he had started out as a chemist. First in his laboratory of the rue d'Ulm and then in his Institute, his ultimate experiments indicate that he was trying to understand how the same microbe can either kill a person or stimulate his or her resistance. This is where bacteriology merged into immunology. Pasteur brought these neighboring disciplines together. Understanding the role of the molecules, the toxins, and the antitoxins involved both chemistry and biology.

At the time of his first biomedical research, Pasteur thought that he had found the cause of acquired resistance. His reasoning was based on an analogy with the behavior of microbes in culture broths, which exhaust the milieu in which they proliferate. On this assumption he had developed a physiological theory in which the body constituted a veritable interior environment that would be exhausted by a first inoculation with living but attenuated bacteria. This was a nod to Claude Bernard. In his reflections on chicken cholera, Pasteur explained nonrecurrence in these terms: "Once the affected muscle has been healed and repaired, it somehow becomes impotent to cultivate the microbe, as if the latter, by growing there earlier, had suppressed in the muscle some principle that life is unable to restore to it, and whose absence prevents the development of the small organism. In my opinion this explanation, to which we are led at this time by the most palpable facts, will most probably become generally accepted and turn out to be applicable to all virulent diseases."[4] This was not a new hypothesis. Auzias Turenne had clearly stated it before Pasteur.

But other experiments eventually contradicted this explanation. Pasteur's doubts about Toussaint's findings, in particular, amounted to a kind of self-criticism. In short, he had to look elsewhere. At this point he had the intuition that immunity might be conferred by soluble products secreted by the bacteria and therefore directly related to their presence. The study of chicken cholera showed him that the bacterial filtrate brought on the symptoms of the disease, somnolence, lowered eyelids, drooping wings. The chickens injected in this manner seemed to be as sick as those inoculated with active bacteria. On the basis of this finding Pasteur attempted to use these soluble products as vaccines, for this would eliminate the need for live bacteria. These experimental vaccinations with soluble products (or toxins) took place in 1880. They were a resounding failure.

As a result, the pursuit of toxins was abandoned, but only temporarily. For when Pasteur began his systematic investigation of rabies, he thought about it again. Although this did not appear in the protocol of antirabies vaccination that he himself had devised, Pasteur still believed that there might be a vaccinal matter secreted by the rabies virus. In August 1888, he thought that he had achieved a successful vaccination against rabies by means of a virus that had been heated and thereby made non-virulent. This would have given him a chemically dead vaccine. But it was too early for this conclusion. In fact it was Emile Roux who most succinctly posed the problem and eventually succeeded in solving it in his laboratory at the Institut Pasteur.

As early as December 1887, Roux and Chamberland had published in the *Annales de l'Institut Pasteur* a note in which they asserted that it is possible to confer immunity against septicemia by vaccinating the animal with toxins rather than with the living bacterium as in the past. The paternity of this investigation does seem to belong to Roux, as Pasteur wrote to Chamberland soon after the publication: "In speaking of the experiments connected with this remarkable study, I cannot possibly say that they do not belong entirely to Roux, for they do. . . . My age and my experience give me the right, I believe, to tell you that you should not have allowed your name to appear in the title of the study in question. Roux did wrong out of excessive generosity. . . . You may find me too severe, but as head of this laboratory I have a duty to establish the historical truth of the laboratory's work, especially when it results in a scientific observation that is bound to be considered exceptionally valuable by everyone."[5] These quarrels about paternity aside, the important fact is that Roux had developed Pasteur's intuition concerning toxins and that he had used it to create a new type of vaccine.

From 1889 on, as the aged master no longer experimented but limited himself to evaluating and judging the work that was done under his eyes, Roux began to champion new therapies to combat infection. This research was not only French, for in Germany Emil von Behring was following a parallel path. Roux and Behring eventually invented serotherapy, the passive treatment of infections by the injection of serum taken from vaccinated animals.

In 1888 and 1889, Roux and his assistant Alexandre Yersin were working on diphtheria, a dangerous disease which, identified as early as 1818 by Pierre Bretonneau, killed its victims (mostly children) with paralysis and croup, a laryngitis caused by false asphyxiating membranes. In studying the diphtheria bacillus discovered by the German scientists Klebs and Löffler, they noticed that it was not present in large numbers, a fact from which they concluded that its active principle must be an extremely toxic poison. This assumption led them to isolate a toxin produced by the bacillus with which they were able to reproduce the disease. Now a way had to be found to combat this toxin. In 1891 it was learned that in Germany Behring and his assistant, the Japanese Kitasato, had gone even further with the agent of tetanus which they had discovered, for they had not only succeeded in isolating the tetanus toxin, but had also and above all been able to show that the serum of animals vaccinated with this toxin can neutralize its effect. Better yet: this serum, when injected into another animal, conferred protection against the disease.

Applying this method to diphtheria, Roux developed sera that neutralized the activity of the diphtheria toxin. For the moment, he simply reproduced Behring's findings in a different model, but Roux intended to go much further. He now had the idea of taking his discovery beyond the walls of the laboratory; the next step was to use animal serum for treating humans. The method was both simple and delicate, since the injection of a toxin was indeed hazardous to the animal. In order to produce a serum that neutralized the toxin but did not allow it to act, the toxin was inoculated in successive doses. And since large quantities of serum had to be obtained, the animal used for the vaccination was the horse. The last step consisted of bleeding the animal with a large trocar.

On Christmas night 1893, the first child with diphtheria was inoculated with this protective serum. A first trial, a first success: the child recovered. By 1894 the hôpital des Enfants-Malades had therefore become the temple of serotherapy; the application of the method on a grand scale produced spectacular effects. The mortality figures for diphtheria dropped along with the false membranes. In May of that year Roux was invited to Lille by the local branch of the Association of the Friends of Science to report on his work. Pasteur, though ailing and weary, made the trip to hear his disciple. As he listened to Roux—an impressive figure with his shining eyes and his lean, ascetic face—explaining the cure by transfer of immunity as if it were a straightforward problem in geometry, Pasteur felt transported back forty years. What a long way he had come since he had studied the yeasts! Now others spoke in his place. Yet that evening, who received the applause, the master or the disciple?

In September 1894, Roux presented an important paper to the International Congress of Hygiene and Demography in Budapest. The interest it sparked was such that antidiphtheria serotherapy began to be used around the world. In France, prefects urged the government to set up a program of immunological treatments. *Le Figaro* opened a nationwide subscription to help cure children of the croup. In short order the Institut Pasteur received so many gifts that it was able to build stables, purchase about a hundred horses, immunize them, and establish the use of serotherapy on a grand scale. Within three months, it was in a position to send out fifty thousand doses. "Dear Sir," Pasteur wrote to the editor in chief of the *Figaro*, François Magnard, "you have drawn wide attention to the immense service rendered by my dear collaborator M. Roux; you have asked mothers to contribute to the dissemination of a method that will snatch thousands of children from death. Not only has your appeal been heard, you have also had the infinitely touching idea of placing the children at the head of the subscribers. Their entry into life is marked by a good deed. I thank you more than I can express, on behalf of all my collaborators who are working for Science, for their Country, and for Humanity."[6]

Thus it came about that in the stables of Marnes-la-Coquette, the old horse that had carried Marshal Canrobert into battle at Saint-Privat in 1870 spent its last days giving the serum that would save little diphtheritics. And on fine summer days, when young Louis Pasteur Vallery-Radot, born in 1886, visited his grandfather, he was sometimes allowed to ride this retired war-horse converted to the service of immunology.

The simultaneous discoveries of Roux in France and Behring in Germany had thus brought about an extraordinary therapeutic advance. Almost twenty-five years had passed since Sedan, and the progress of science was such that it was beginning to outweigh patriotic resentments. In 1894, at Roux's request, Behring was officially received at the Institut Pasteur, where he was awarded a medal of honor. But on that day, Pasteur refused to appear: for his part, he could not bring himself to honor a German in his establishment as long as Alsace and Lorraine had not been returned to France.

Beyond these more or less healed-over wounds, it was vitally important that the united efforts of the Institut Pasteur and the German school be brought together in a humoral theory of anti-infectious immunity. However, if it was clear that the "humors" can prevent and cure, no one was able to explain how the immunity conferred in this manner persisted in an organism in which all chemical elements were constantly being renewed.

Thus a great many question marks were left, and the founding of immunology was by no means completed. As a man of the laboratory who was more comfortable with experiments than with theories, Pasteur had foreseen very soon that the humoral theory of immunity could not explain everything. Since all that is alive in an organism is the cell, the research had to be oriented in that direction. This is why, when Pasteur heard about the discoveries of the Russian Metchnikoff concerning the digestion of bacteria by white corpuscles, the so-called phagocytosis, he was immediately won over. He had Metchnikoff's studies published in the very first issue of the

Annales de l'Institut Pasteur and offered him a double position as laboratory director and department head at the Institute.

Who was Elie Metchnikoff? Born near Kharkov in 1845, he was trained as a zoologist and acquired most of his knowledge on his travels to Italy, North Africa, or France. Interested in physiological phenomena akin to digestion, he had observed the manner in which amoebas absorb particles. This is what eventually inspired him to see an analogy between this absorption he had observed and the function of the white corpuscles. He wrote in his recollections:

> One day when the whole family had gone to the circus, I was observing the life of the mobile cells of a transparent starfish larva, when I was suddenly illuminated by a new thought. I had the idea that analogous cells probably served to defend the organism against harmful intruders. Sensing that there was something very interesting in this, I became so excited that I began to walk very fast, even going to the seashore to collect my thoughts. I said to myself that a thorn placed into the body of a starfish larva, which has neither blood vessels nor a nervous system, should be very quickly surrounded by mobile cells, as can be observed in humans when they have a splinter in a finger. No sooner said than done. In the little garden at our house I took several thorns from a rose bush and immediately placed them under the skin of several superb starfish larvae that were as transparent as water. I was so excited that I did not sleep at all that night, waiting for the results of my experiment. The next morning I was overjoyed to find that it had fully succeeded.[7]

While holding a position at the bacteriological station of Odessa, where Nicolas Gamaleïa, who regularly corresponded with Pasteur about antirabies treatments, was also working, Metchnikoff was casting about for a foreign laboratory that would permit him to pursue his research. He hesitated between Berlin and Paris. In Germany, Koch received him in an unfriendly manner, for he did not believe in phagocytosis. So Metchnikoff agreed to join the Institut Pasteur, which had offered him a position. Pasteur received him with enthusiasm, marveling at his microscopic preparations showing the struggle between the white corpuscles and the microbes.

With the "Romantic" Metchnikoff, the tormented and visionary Russia, the Russia of Turgeniev and Tolstoy, came to the Institut Pasteur, along with a strange and new panoply of experimental animals that included such creatures as the axolotl of tropical oceans and even crocodiles. When he was working on cholera, typhoid, and tuberculosis, Metchnikoff constantly did battle to have his work recognized in medicine, and when he went from the struggle against microbes to theories about the destruction of aged tissues, he came to see the immune system as the guardian of our identity, the grave-digger of cells attacked by aging or infection. Metchnikoff's cellular theory was to revolutionize immunology and to compel the proponents of humoral immunity to engage in a fruitful dialogue with those of cellular immunity.

This competition remained in the balance as long as Pasteur was alive. But in the years after the master's death, the cellular theory not only encountered the outspoken hostility of the German school but was also overwhelmed by the dynamism of the chemists and the bacteriologists of the Institut Pasteur. Nonetheless, it is to Pasteur's credit that by opening the doors of his Institute to the Russian zoologist, he was able to make the proponents of the two kinds of immunity cohabit under the same roof of the rue Dutot.

In those days of 1894, the researchers in the aged scientist's orbit had found out all, or almost all that was to be known about the life and the functioning of that strange immune system, which protects the identity and the survival of the individual, manufactures antibodies, and destroys infected cells.

The Pasteur Generation

To be working at the Institut Pasteur was to belong to the Pasteurian family. But in the late nineteenth century, this was also a form of marginality, for it meant refusing medical clients, declining university chairs, turning down hospital appointments. Yet by way of compensation, the Pasteurian had the outstanding privilege of working at the cutting edge of advanced studies, at the crossroads of the newest or most subtle disciplines, that is, microbiology, molecular chemistry, immunology.

Among those who spent time in the rue Dutot between 1888 and 1895, and who were thus in a position to come in contact with the aged master, were Albert Calmette, Alexandre Yersin, Charles Nicolle, and Jules Bordet. Even though some of them had only a brief stay at the Institute, they owed their greatest discoveries to Pasteur's entourage and to his personality. "They kept him informed of the research that was being done," Calmette was to say of Pasteur, "and he made suggestions or commented on the experiments. Madame Pasteur, a true scholar's wife, was interested in everything and knew what everyone was thinking and feeling even before they had expressed it. She saw to it with touching solicitude that no worries and preoccupations darkened the meditations and the dreams pursued with such enthusiasm and confidence by those who worked there. . . . Those days of misery and grandeur, they were the heroic era, the time of the Pasteurian epic. . . . In the seven years that Pasteur spent in his Institute among his collaborators and his students, he had the supreme satisfaction of witnessing some of the most valuable achievements of the experimental science that he himself had brought to its highest peak."[8]

The Pasteurian spirit must take wings, for the Institute was meant to teach future apostles; this was something that Pasteur had clearly indicated at a very early stage: "In the Paris establishment we will have to train young scientists who will carry our method into faraway countries." Exporting science beyond the seas — this mind-set was in keeping with the preoccupations of France's colonial policies at the end of the nineteenth century. In the 1880s, in the wake of the severe criticism surrounding the military expeditions to sub-Saharan Africa or Annam, as well as the setbacks in the Tonkin region, which had brought down the government of Jules Ferry, France had slowed down its colonial efforts. But after 1890, the example of England, which was expanding its empire and exploiting the natural resources of its new territories, prompted the French nationalists to urge their government to resume its policy of colonial expansion. Pasteur became one of the architects of this policy through the work of his disciples. It was during his lifetime, at his instigation, and with the active support of Roux, that the applications of microbiology and advanced vaccination techniques reached what was not yet called the Third World. Thanks to Pasteur, French science became both a humanitarian enterprise and an instrument of conquest.

Albert Calmette came to Roux's laboratory in July 1890. A navy physician, he had gone to sea in 1883, at the age of twenty, to serve in the French Far-Eastern squadron; this was the time of the war against the Black Flags, gangs of Chinese pirates who made murderous incursions into the protectorate of Tonkin. Calmette thus spent a few months in the Bay of Along in the entourage of Admiral Courbet, and took part aboard *La Triumphante* in the naval battles of Fou-tcheou and Tam-sui and also at the capture of Kelung (August–September 1884). A few months later he sailed to Gabon, where Savorgnan de Brazza was preparing to conquer the Congo. Passionately interested in microbiology, Calmette was carrying a microscope that he took along on all of his campaigns. In the floating hospital at Libreville he intently studied the sleeping sickness that was decimating the Africans.

In spite of violent bouts with malaria, he never stopped working. Returning to France in December 1887, he left again for Saint-Pierre-et-Miquelon
three months later. On this cold and wild island, Calmette, who tracked
microbes wherever he thought they existed, identified a germ that produced a bright red substance responsible for a strange infection, codfish
salmonellosis, and was able to show that bacteria were growing in the salt
used to preserve the mounds of codfish that filled the holds of fishing boats
during the Newfoundland fishing season.

Albert Calmette was thus already a full-fledged researcher in 1890,
when he presented this work to Emile Roux. At the Institut Pasteur, everyone took an immediate liking to this young man with his impeccable service
record, who had come to broaden his knowledge. Roux encouraged him to
take the microbiology course that was taught at the Institute and then chose
him for a special training course in his own laboratory. Calmette stayed
there for a time, but then Pasteur, who had also noticed the young scientist's qualifications, wanted to make use of his Indo-Chinese experience. He
therefore sent him to Saigon to set up a laboratory for the preparation of
smallpox and rabies vaccines.

Calmette left without delay. By late January 1891, the new Institut
Pasteur was operational and Calmette could embark upon his own research.
It was in Saigon that he undertook the study of snake venom that has made
him famous. Thanks to a snake charmer who brought him a batch of cobras
captured in a village where they had sought shelter from a flood, Calmette
was able to conduct experiments on their venom glands. This, he hoped,
would lead him to an antivenom serum, which he would obtain on the basis
of the same principles as the sera that were just then being developed from
microbial toxins by Roux and Yersin in their studies of diphtheria. The
analogy between venom and toxins was a correct intuition, and Calmette
succeeded in making chickens resistant to snake venom.

In Paris this discovery was greeted with enthusiasm, and Calmette,
with the support of Pasteur and Roux, obtained the permission of the
military authorities to set up a large-scale production of antivenom sera at
the Institute in the rue Dutot. He now systematically carried out Roux's
experiments on toxins with snake venom and soon prepared the first antivenom sera at the Institut Pasteur of Paris by injecting horses with a series of
doses of venom. Handling cobras and rattlesnakes required a certain courage. One morning, a concierge at the Institut Pasteur (it was none other
than Jupille, who apparently refused to be as brave when it came to snakes
as he had been ten years earlier in facing a rabid dog) came to tell Calmette

that no one dared enter his laboratory because some snakes had gotten out of their cages. So the scientist had to capture his runaway vipers by himself, but some of them escaped to the sewers under the Institute. This is the origin of the old and persistent rumor to the effect that serpents and other reptiles haunt the Parisian sewage system.

Pasteur attentively followed Calmette's work. Shortly before his death he entrusted him with the task of creating and directing an Institut Pasteur at Lille. It was there that Calmette, together with Camille Guérin, was to begin his research on the tuberculosis bacillus, which ended with the identification of the famous BCG (Bacillus Calmette Guérin). In 1914 the scientist's existence was deeply shaken when his brother Gaston, director of the *Figaro* newspaper, was assassinated by Madame Caillaux; nonetheless, Albert Calmette continued to be a tireless worker until his own death in 1933. Under his leadership, the Institut Pasteur of Lille made its influence felt throughout Flanders and played an active role in the industrial and agricultural development of the region.

Calmette may have met a young physician of exactly the same age as himself in the microbiology course that he took at the Institut Pasteur. Alexandre Yersin, born in 1863 in the Swiss canton of Vaud, had come to Paris to prepare his thesis in medicine on the subject of experimental tuberculosis. At the same time he was working in Charles Richet's clinic at the Hôtel-Dieu, where he autopsied the patients who had died of rabies. One day he asked a colleague to take him along to the laboratory in the rue d'Ulm. "The courtyard was filled with people under treatment. . . . One can barely get in, so full is it with people of all nationalities. To the left, by the entrance is a door leading to M. Pasteur's study. By that door stands a little man with a paper in his hand, who calls out the names of the persons who are to enter. . . . The little man turns on us like a bullet and shouts, 'Who are you? What do you want?' That little man is M. Pasteur. . . . As I leave, I thank M. Pasteur for his kind reception. He deigns touch my hand. I was told later that Pasteur is terribly grouchy and that it is quite difficult to see him. He is always afraid that a stranger will steal his secrets."[9]

Yersin met Roux at the Hôtel-Dieu on the occasion of an autopsy. The two men immediately understood each other, and as soon as Yersin had passed his examinations at the Faculty of Medicine in July 1886, he spent all his afternoons in the rue d'Ulm, assisting Roux with his preparations. Pasteur, for his part, asked him to change the patients' dressings. In 1887, at the time when he went to work for Dr. Grancher at the Childrens' Hospital, Yersin became the personal préparateur of Roux, who hired him under a (renewable) one-year contract at a salary of fifty francs a month. Immediately after the creation of the Institut Pasteur, the twenty-five-year-old Yersin thus participated in the first studies on the diphtheria toxin.

But Yersin's roving and passionate temperament found it hard to put up with the routine of the Institute's laboratory benches. The young physician wanted to see the world. That is why he dropped everything and sailed as a simple ship's doctor on a courier ship of the French Far-Eastern Transportation Company. Disembarking in Vietnam, he became an explorer setting out for the unknown wilds of Annam. At Saigon he called on Albert Calmette, who encouraged him and helped him outfit his expeditions. For Yersin was interested in more than public health; he also hoped to contribute to the colonizing effort by providing geographic and ethnological observations. There were no medications in his bags; he took with him only a gun, a microscope, a Chamberland filter, and some canned food.

In 1892 he returned to Paris in search of support and funds for further missions. Roux, who bore him no grudge for having left him, received him with open arms, as did Pasteur, who recommended him to the Minister of Foreign Affairs. On his return to Indochina, Yersin was asked by the governor-general to conduct an on-site study of the requirements for building a road leading deep into the Moï country. This was to be the object of three missions, during which he reached the high plateaux of Liang-Biang, produced a topographical map of the possible access routes toward Cambodia, and studied the way of life of the Moï people.

On one of these expeditions Yersin discovered a place that would be suitable as a health resort; eventually the town of Dalat was built there. Racked with malaria when he returned, he stopped at Saigon and took some rest at Calmette's home, but soon he was off again. In May 1894, just as he was preparing to explore the mountain chain that stretches from northern Tonkin to the Yun-nan plateau, a severe plague epidemic broke out in southern China. By that time Calmette was no longer in Saigon, but the government of Indochina had not forgotten that Yersin had also worked at the Institut Pasteur, and so it sent him to investigate the matter.

The plague was in Hong Kong. Yersin rushed to the scene. By 15 June 1894 he had a small laboratory in place. But the island authorities preferred dealing with a Japanese team led by Kitasato, the former collaborator of Behring, and Yersin was not allowed to autopsy any of the victims. Undaunted, he carried on his research in secret, bribing some English boatmen who had to dispose of the dead bodies. The pus of the buboes obtained in this manner allowed Yersin to identify the plague bacillus on 20 June 1894, just a few days after he had arrived in Hong Kong. Nor did he stop there, for he proved that the epidemic started with the rats whose dead bodies littered the streets of the contaminated neighborhoods. Although he was unable to develop a vaccine, the plague victims could be treated with the antiplague serum he immediately began to produce.

Yersin set up his laboratory at Nha Trang, on the shores of the China Sea. This is where he worked for the rest of his life, in addition to raising crops and animals. He not only built enormous stables for the serum-producing water buffaloes but also cultivated tropical plants. Thus he introduced the rubber tree into Indochina and set up the first Indo-Chinese nurseries of heveas from the Sunda Islands and developed plantations of cinchona trees to make quinine available. Until the end of his life, the director of the Pasteur Institutes of Saigon and Nha Trang worked as a full-time physician and also pursued his research. An old sage who over the years became passionately interested in photography, astronomy, and meteorology, he died on 1 May 1943 in his colonial house on the shores of the spectacular Bay of Nha Trang.

Another outstanding figure among the first Pasteurians was Charles Nicolle. Nicolle barely knew Pasteur personally. He only encountered him once, in 1894. It was in a staircase in the rue Dutot; Pasteur was laboriously coming down, stair by stair, while the twenty-eight-year-old Nicolle bounded up four steps at a time. The young man stepped aside to make room for the old scientist, who looked at him and asked, "Are you working?" "Yes," Nicolle replied. "We must work," Pasteur articulated slowly and then continued on his way.[10] These were the only words ever exchanged between the master and the disciple, but this did not prevent Nicolle from becoming the recipient of the 1928 Nobel Prize in Medicine for his contributions to microbiology.

Nicolle's father was a physician at Rouen and professor of natural history at that city's School of Arts and Science. A supporter of Pouchet, he lived to see two of his sons take Pasteur's side. Maurice, the older, was the first to go to work in the laboratory of the rue Dutot, but his stay was brief, since Roux soon sent him to Constantinople with the mission of founding an Institut Pasteur there. This had become a possibility when Chantemesse, dispatched to study the causes of the recurring cholera epidemics at the request of Sultan Abdul-Hamid II, had suggested that a microbiology laboratory should be created in Turkey.

As for Charles Nicolle, born in 1866, he had passed the difficult competitive entrance examination for an internship in the Parisian hospitals on his first attempt in 1889, having studied for it together with Léon Daudet. He therefore had the opportunity to frequent the literary and artistic milieu of the last years of the century, meeting such figures as Barbey d'Aurevilly, Edmond de Goncourt, or Nadar at the Sunday soirées at the home of Alphonse Daudet.

Nicolle was slightly hard of hearing, and therefore unable to practice the stethoscope kind of medicine; however, he overcame this handicap by turning to biology as it was taught at the Institut Pasteur. Starting out as a préparateur, he came in contact with Metchnikoff, "of whom I would be unable to say," as he was to put it in one of his lectures at the Collège de France, "whether he was a poet or a scientist, for there was something magical about his ideas." He also met Roux, "a man of a cold and logical mind who had no use for flights of imagination, but was possessed of such clear intelligence that this very clarity was as compelling as the most stirring passion."

Nicolle studied with these two masters. After defending his thesis in 1893, he returned to Rouen as professor of pathology and clinical medicine at the School of Medicine. He founded the first sanatorium in the area. But the decisive turn in his career came in 1902, when he was asked to direct the Pasteur Institute of Tunis, which had been founded in 1893 by Adrien Loir after his return from Australia. Charles Nicolle was to turn this modest laboratory, consisting of two tiny and poorly equipped rooms, into one of the most influential scientific centers in the world.

In Tunis, he began by ferreting out and studying a parasitical disease that was rampant in tropical countries, leishmaniasis, establishing that it was transmitted to man by an insect, the phlebotomus, or blood-sucking sand fly. Above all, and this was his major accomplishment, he identified the vector of typhus, which at that time was causing such ravages that on his way to the Tunis hospital he had to step over the bodies of typhus victims who, lying curled up in the streets, were waiting to be admitted. Nicolle noticed that the disease, though extremely contagious, stopped at the doors of the hospital: not one nurse, not one physician came down with it. From this he deduced that the contagion had to do with the filthy rags of the sick. "I asked myself," he wrote, "what happened between the door of the hospital and the ward. What happened was this: the typhus victim was taken out of his clothes and his underwear, shaved and washed. The agent of contagion was therefore something attached to his skin and to his linen, something that water and soap got rid of. It could only be the louse; it was the louse!"

Charles Nicolle was forty-three years old when he made this discovery, but this did not prompt him to leave Tunisia, not even when he was elected to the Collège de France and awarded the Nobel Prize. He continued to work in Tunis, particularly on Malta fever, until his death in 1936. His last wishes reflect how much he felt he owed to those who had trained him: he asked to be buried within the walls of his Institute, dressed in his white smock and the apron worn by the Pasteurians. He had spent his life in accordance with the answer he had given Pasteur one early morning in 1894: he had worked.

The name of the Belgian Jules Bordet, the last of the great Pasteurians of that generation may be least known by the general public. Yet his work provides nourishment for research to this day. Born at Soigies in 1870 as the son of two schoolteachers, Bordet became interested in science at a very young age, turning the attic of the family home into a chemistry laboratory. At sixteen, he became a student at the medical school of the Free University of Brussels. Eight years later he graduated with a fellowship for studying abroad. He decided to use it to go to Paris to perfect his knowledge in microbiology. Coming to the Institut Pasteur in 1894, he worked there with Metchnikoff until 1901. Bordet was a passionate and discreet researcher. Loir said that his blue eyes seemed to reflect the heavens of his dreams. In the twilight years of his life, he actually took up the study of astronomy.

Under the aegis of Metchnikoff, Bordet discovered immunology, but as his mastery and his imagination grew, he turned away from the theses of the Russian scientist and became a strong proponent of the theory of humoral immunity. Having noticed that the serum of vaccinated guinea pigs caused a clumping of cholera vibrios, he isolated the serum components that conferred immunity, the antibodies. This discovery made it possible to understand how the neutralizing sera described by Roux and Behring were actually working: Bordet showed that there are molecules, the antibodies, which recognize the bacterial components called antigens, and that vaccination stimulates the antibodies, thereby making them capable of inhibiting the action of microbes and their toxins. Bordet also identified another component of serum that is sensitive to heat, the complement. It is this component, which gives the antibodies their functional properties and becomes bound to them in the antigen complex, that causes the bacteria to dissolve. Moreover, because they recognize the antigens of bacteria, the antibodies can be used to identify and diagnose infections. Bordet began with typhoid and then went on to plague. Working in his small laboratory at the Institut Pasteur, Bordet thus came to invent the principles of serodiagnosis for microbial infections.

A few years after Pasteur's death, in 1899, Bordet established that the function of the immune system is not limited to reacting to germs. A reaction can be triggered by very different causes than aggression by a pathogenic agent. Red corpuscles or substances from outside the body may be sufficient. This being the case, immunology separated from microbiology and became a science in its own right. Its main purpose became to understand what enables antigens to recognize self and non-self, to distinguish what properly belongs to an individual or a species, and what does not.

It is true that Bordet had worked with Metchnikoff more than with Pasteur; nonetheless, the aged scientist had left his mark on him. In 1901, when he returned to Belgium, Bordet asked and obtained from Pasteur's widow, who held the rights to the name, permission to give his laboratory the name "Institut Pasteur du Brabant." There, in Brussels, he continued his work, which ranged from the discovery of the whooping-cough bacillus (which he and his brother-in-law, Gengou, identified in his daughter's throat), to a study on blood coagulation. When Jules Bordet obtained the Nobel Prize in Medicine in 1919, Pasteur once again shared in the honor.

Calmette, Yersin, Nicolle, and Bordet were not the only ones. Between 1888 and 1895 there were of course other remarkable Pasteurians whose names are part of the history of medicine, men like Etienne Marchou, the founder of the Pasteur Institute of Senegal, or Marcel Mérieux, a student of Roux, who created his own enterprise in Lyon in 1897.

Because they had worked in the rue Dutot or glimpsed Pasteur's silhouette under the trees of his Institute, they all considered themselves his disciples; they became a kind of family. Through the cult of Pasteur and his doctrines, and by propagating the microbial theories and those of the nascent immunology, this family kept the glory of the Institut Pasteur alive and made its patrimony yield ample fruit.

At the same time, the Pasteurians also saw to it that science, medicine, and prevention became everyone's concern, in particular that of the state. Very soon the Pasteurian approach was no longer the exclusive property of the scientists of the rue Dutot; it even spread beyond the confines of the French colonial adventure, even though the suffering populations of Africa and Asia were unquestionably the first to benefit from the overseas Pasteur Institutes.

As Pasteur had said many times, science must serve all of humanity. Health and education, the practical application of the Pasteurian discoveries, and technical cooperation with the Third World, these were the mainstays of the ideal disseminated by Pasteur's disciples.

The Last Breath of Life

In early 1892, Pasteur's health suddenly deteriorated; he became confined to his room.

In May of that year, a committee formed in Denmark announced its intention to celebrate the scientist's seventieth birthday on 27 December of the coming year. It had opened a national subscription to raise the funds needed for the minting of a commemorative medal. The movement soon reached Norway, where a second committee was founded, in this case for the purpose of founding a fellowship that would permit a young biologist to go to France to complete his studies. In Sweden, plans were made for a prize bearing the name of the French scientist; the king headed the list of subscribers.

In France, where the press reported these initiatives, the learned societies did not want to be left behind. The section of medicine and surgery of the Académie des sciences also appointed a committee, to be headed by Joseph Grancher. For the celebration of "M. Pasteur's jubilee" the rector of the Parisian university system opened the doors of the great amphitheater of the Sorbonne, at whose inauguration Pasteur had been present in 1889. Four thousand invitations were sent out. The President of the Republic promised to come.

The ceremony took place on 27 December 1892. In the great amphitheater of the Sorbonne, under the huge painting of Puvis de Chavannes, the first rows and the hemicycle were reserved for the members of the Institut. Along the sides were the seats for the University professors and administrators, and for the representatives of French and foreign learned societies. Further back were the deputations of the Ecole normale supérieure, the Ecole polytechnique, the Ecole centrale, and the Ecole vétérinaire, as well as students of the faculties of medicine and pharmacy.

At mid-morning Pasteur entered through the right-hand door leading to the main stage on the arm of the President of the Republic, Sadi Carnot. Both were wearing the grand cordon of the Légion d'honneur. The aged scientist was greeted by an immense ovation. Then he was led to a small table at the end of the stage, and it was from there that, seated throughout, he listened to the speeches honoring him.

And now the ceremony was under way; it was to be long and tedious. There were speeches and addresses from every department of France, every country in Europe and in the world. Speeches on behalf of kings, presidents, prefects, members of academies. The speakers came from Sweden, from Turkey, from Germany, Italy, Austria-Hungary, Belgium, and else-where. When the English delegation came to the stage, Pasteur slowly rose to embrace Lister.

The rue d'Ulm presented a distinguished gift. The Normaliens had commissioned Emile Gallé to make a unique and extraordinary vessel. By way of "crystallizing Pasteur's thought," the famous glassblower of Nancy had found a way to enclose microbes in transparent crystal. "The infinitely small proliferating in this vessel."

Once again that day, Jean-Baptiste read his father's speech: "Young people, young people, have faith in those sure and powerful methods whose first secrets we are just beginning to understand. All of you, what-ever your future career, do not allow yourselves to be touched by denigrat-ing and sterile skepticism, and do not become discouraged by the sadness of certain hours that a nation has to experience. Live in the serene peace of laboratories and libraries. Ask yourselves first: 'What have I done to acquire knowledge?' And as you advance, ask: 'What have I done for my country?' And this you will do until the moment when you may experience the supreme happiness of thinking that you have in some way contributed to progress and the good of humanity. But to whatever degree life will have favored your efforts, when you approach the great goal, you must be able to say to yourself: "I have done my best."[11]

The last to speak were the people of Dole. Throughout their address, Pasteur shielded his face behind his hand to hide the deep emotion stirred by the memory of his father and his mother and the evocation of his origins in the Franche-Comté.

The ceremony was over. Sadi Carnot embraced Pasteur to thunderous applause. The Republic was embracing Science. This was the time of the Panama Scandal and the first anarchist assassinations. Two years later, the president was assassinated at Lyon.

Pasteur had become a living symbol, embodying both science and France. French-speaking countries in particular considered him a special ambassador. At the proposal of the deputies of the Province of Quebec, the Canadian government gave his name to a county on the Maine border. A few months later, it was Algeria's turn to honor the scientists of the metro-pole; the government general of Algeria decided to name a town in the province of Constantine "Pasteur." "Thanks to you," Pasteur wrote to Jules Cambon, my name will remain attached to this corner of the earth. When in the future a child of this village will ask about the origin of this name, I would hope that the teacher will simply tell him that it was the name of a

Frenchman who had loved France very much. . . . The thought that my name might some day awaken in a child's soul the first stirrings of patriotism makes my heart beat faster."[12]

In his last years, perhaps since he had kept watch at Joseph Meister's sickbed, Pasteur seems to have been obsessed by thoughts of childhood. No doubt, memories of the deathbed scenes of his daughters Jeanne, Cécile, and Camille also came back to haunt him. In memory of these deceased children, the daughter of Marie-Louise and René Vallery-Radot, born in September 1880, had been christened Camille. Pasteur watched her grow with passionate interest. In September 1891, he dictated to his wife a long letter to the little girl; it was a very detailed account of the eruption of Krakatoa. Pasteur had boned up on volcanology to satisfy Camille's curiosity.

In February 1893, Marie wrote to Jules Marcou, Pasteur's friend from the early years at Besançon, who had become one of the most respected geologists in the world: "Dear M. Marcou, your friend Pasteur continues to be fairly well, but he must resign himself to put aside all work that is in any way strenuous. He takes much interest in the work of others. He still enjoys going to the Academies. He sees his grandchildren grow up with great solicitude, and then there are so many things he has to do for his health that he keeps quite busy."[13] Above all, Pasteur was impeded in his mobility. Travel was difficult, but this did not prevent him from moving about between the rue Dutot, Villeneuve L'Etang, and Arbois.

In February 1884, he was saddened by the death of his dear childhood friend Jules Vercel. On 25 October he made one of his last official appearances at the Institut Pasteur when Casimir Perier, the incumbent president of the Republic, came to bestow the insignia of a commander of the Légion d'honneur on Roux. A few days later, on Thursday, 1 November, as he was about to leave for his daily visit to his grandchildren, he became violently ill and fainted. He was carried to his bedroom. He did not regain consciousness until that evening, when he requested that someone stay with him. He remained in bed for three months. Had he suffered another stroke? In any case, he was treated for albuminuria and placed on a milk diet.

His children and grandchildren moved into the apartments at the Institut Pasteur. Shifts for sitting with him were set up. In groups of two, Pasteurians and family members took turns around the clock at the master's bedside. Roux and Chantemesse took over the Sunday night shifts. As soon as he had a free moment, Metchnikoff came to visit Pasteur, whatever the hour. And so did Grancher. By the end of December there as a slight improvement. On 1 January 1895, Pasteur received all of his students and collaborators to wish them a Happy New Year. That day, Alexandre Dumas the Younger also came to salute his confrere of the Académie française. "I wanted to begin the year right," he told Pasteur, unaware that he himself would not live to see its end.

In the following days and weeks, Pasteur gradually improved. In April he insisted that the Normaliens, who were celebrating the hundredth anniversary of their school, should be invited to his home. On this occasion Roux decorated the tables with glass bulbs containing culture broth, test tubes, and swan-necked flasks. That day, Pasteur had himself carried to the laboratory and asked to see the plague bacilli that Yersin had isolated. This was to be his last observation at the microscope.

The spring of 1895 came unusually early. Soon the trees were covered with leaves, and Pasteur enjoyed this renewal of nature whenever he could. He spent long afternoon hours under the Institute's chestnut trees in quiet conversation with his collaborators or with the faithful Charles Chappuis, now rector of the University district of Dijon.

In May he learned that the Berlin Academy of Sciences wished to honor him with the medal of the Prussian Order of Pour le Mérite. The idea that he was to be distinguished by Emperor William II gave Pasteur a burst of energy; his anger was good for him. Alsace and Lorraine were still German! How could he possibly accept a decoration from the victors of 1870? He refused this honor, just as he had once returned his diploma *honoris causa* to the dean of Bonn University. But at the same time he also declined the offer of an ultranationalist, Comte Féry d'Esclands, who wanted to set up a committee to publicize Pasteur's refusal. Tired and worn out as he was, the aged scientist had not lost his keen sense of diplomacy and knew how to drape himself in his patriotism without giving purchase to the extremists.

On 13 June Pasteur left the Institute; having laboriously made his way down the steps to the rue Dutot, he comfortably settled in a carriage that was to take him to Villeneuve-l'Etang, where he was to spend the finest summer days. But he was becoming increasingly weak and had little chance to enjoy the park. His strength ebbed rapidly; soon he could no longer leave his bedroom, where members of his family read to him accounts of Napoleon's last battles or scenes from the life of Saint Vincent de Paul. Another stroke brought on aphasia. He died on 28 September 1895, at twenty minutes past four in the afternoon.

The government immediately ordered a state funeral, while the newspapers published the scientist's last wishes, words of a moving sobriety:

"This is my testament. I leave to my wife as much as I am allowed to leave her under the law. May my children never stray from their duty and continue to give their mother the tenderness she deserves. L. Pasteur. Paris, 29 March 1877. Arbois, 25 August 1880."[14]

An immense funeral procession moved from the Institut Pasteur to Notre-Dame. The new president of the Republic, Félix Faure, was accompanied by all his cabinet ministers; the Grand-Duke Constantine of Russia and Prince Nicolas of Greece were present as well. The people of Paris also followed the hearse, which was surrounded by a detachment of the Garde Républicaine. There was only one sour note: the irascible Edmond de Goncourt wrote in his *Journal*, "The honors rendered to great men — Good Shepherds though they may be [this is a pun on the name Pasteur = Shepherd] — are becoming a bit excessive it seems to me; perhaps they are becoming heirs to what used to belong to God."

In keeping with his family's wishes, Pasteur was not interred in the Pantheon but in a crypt built in the cellars of his Institute, very close to the apartments where he had spent the last years of his life. There he is keeping watch at the doors of the laboratories where his work continues.

In 1940, when the Germans, who had occupied Paris, wanted to visit the tombs of Pasteur and Marie — who died fifteen years after him — they were denied access by a little white-haired man. It was Joseph Meister, the concierge of the Institute and the guardian of the sanctuary. He had survived the war of 1870 and he had survived rabies, but this humiliation was more than he could tolerate, and so he obstinately refused to unlock the gate of the crypt and to allow the descendants of the Prussians to disturb Pasteur's rest. Sinking into deep depression, he locked himself into his little apartment and finally committed suicide.

Pasteurian Science Triumphant

Pasteur was dead, but Pasteurian science was more alive than ever. The revolution he had triggered continued to change the world, while his life became the stuff of legend.

A media phenomenon of the first order, Pasteur was the first scientist to bring questions of experimental medicine, and of medicine in general, to the front pages of newspapers. In this sense he was the principal architect of the transformation of public opinion in medical matters. At the end of the nineteenth century, in a world where laboratories were still few and far between, and where the teaching of science was hemmed in by traditional rules and without prestige, he generated faith in the progress of medicine and created a new type of hero: the Pasteurian scientist. Without this media revolution, the advances of science would surely have been much more timid.

To appreciate the worldwide character of this innovation, one should recall the enthusiasm of the American press in 1886, when Pasteur vaccinated the four rabies victims from New York. The cured children were put on display in a shop window in the Bowery, where 300,000 curiosity seekers paid to get a glimpse of them. On this occasion, every American town, down to the very small ones, celebrated the name of Pasteur and the effectiveness of French science. On the occasion of the hundredth anniversary of the scientist's birth, President Warren Harding wrote to Raymond Poincaré, prime minister of France at the time: "America was one of the first countries to put Pasteur's discoveries into practice. Pasteur belongs to America as much as he belongs to France." As late as 1928, the pharmaceutical company Parke Davis tried to find the first vaccine recipients, hoping to use them in a publicity campaign, while in Chicago one of the four bitten children, by now fifty years old, officiated at the unveiling of a bust of Pasteur. Not even Koch's identification of the tuberculosis bacillus aroused as much emotion.

But of course it was in France that Pasteur continued to be celebrated as a national hero. By attaching extreme importance to the French origin of his discoveries, he had adopted a reflection of Disraeli as his own: "The health of the people is the foundation on which rests all the honor and all the power of the statesman. Attending to the people's health is his foremost duty." Even though he had chosen independence by refusing any administrative tutelage of the state over his Institute, the Republic did its best to make Pasteur into a secular saint, exploiting his image to glorify patriotic devotion to science. A veritable cult came into being, and its liturgy was fixed for a long time by the biography of René Vallery-Radot and the film of Sacha Guitry. Pasteur's face was elevated to the rank of an allegory, for in 1929 he became the first famous man (except for Napoleon III) to be depicted on a postage stamp; and in 1966 he was honored by the Banque de France, which had his portrait engraved on a widely used banknote, the five-franc bill. The symbolism is eloquent, for here Pasteur is no longer only a scientist or a discoverer; he serves to bind people together. Two French ships were christened with his name: the first, launched in 1928, served as a troop ship, while the second, launched in 1968, was a cruise ship.

In 1922, the celebrations commemorating the centenary of his birth marked the high point of the cult of Pasteur. The president of the Republic, Alexandre Millerand, solemnly paid homage to the man "who has left his mark on all who think and all who suffer." And, he continued, "The cult of great men is a principle of our national education. A people that remembers its dead draws secret reserves of strength and hope from these commemorations."

On 27 December 1922, at seven in the morning, the church bells in every town and village in Franche-Comté began to ring as if to celebrate a second Christmas. At Dole itself, the town officials assembled in a procession that moved to the lycée in the rue des Tanneurs, where a sheaf of flowers and some palm branches were deposited at the door of Pasteur's birthplace. At Strasbourg, the liberation of Alsace was celebrated along with the memory of the scientist who "never separated the fatherland from science." In Paris, the festivities took place in May 1923 with such events as the homage of the Ecole normale and a banquet for French and foreign delegations at Versailles.

The Institut Pasteur opened its doors. Visitors were allowed to visit the crypt and the scientist's living quarters. There, an exhibition had been assembled, showing flasks containing racemic crystals, the origin of all subsequent discoveries, silkworm cocoons sickened by pébrine, some of the flasks which, hermetically sealed by Pasteur at the time of the experiments against spontaneous generation, were still sterile — along with a profusion of vials, retorts, test tubes, and round flasks whose labels allowed the viewer to follow the history of all of Pasteur's discoveries. Donations for the support of scientific research were solicited; collection boxes were set up to gather funds for laboratories, "the new temples of science," and one could buy pictures showing "Le bon Pasteur" (the Good Shepherd) against a deep blue background.

Throughout the world, squares and streets were named for him; his portrait was hung in classrooms, in laboratories, in hospitals. Pasteur would have been a hundred, and what was being celebrated was the century of science, the century of Pasteur.

SO MANY SPEECHES about this taciturn man.

From the Second Empire to the Third Republic, and even after his death, Pasteur had known how to use the powers that be for his own purposes, and governments in turn had used him as a symbol to establish their own political identity and to justify their civilizing power. He had acquired glory by developing experimental methods sufficiently new to checkmate his adversaries by dismissing the older experts. This he did with prodigious skill and, at times, considerable unfairness. He meticulously orchestrated the practical triumph of his convictions by setting up appropriate research programs and finding the means to carry them out.

Pasteur thus has the stature of the principal architect of a medical revolution that matches the political and industrial revolutions that had shaken Europe since the end of the eighteenth century. Others, like Claude Bernard, had prepared the ground, but without question it was Pasteur who redesigned the medical landscape in which we have been living and moving for more than a century.

Beyond discovering and mastering germs, he created the field of medical research and immediately gave it its full dimension. "The object of scientific research," he said, "is the improvement of human health." On the basis of this conviction, he sought to unify teaching and research in the pursuit of healing. And indeed, Pasteurian science revealed the medical effectiveness of disciplines that had developed outside the hospital. Medicine as healing and prevention thus entered a new era and became a scientific endeavor. This encounter between the scientist and disease was not fortuitous; it had occurred in the context of specific problems. From the culture broth to the vaccines, from the toxins to serotherapy, the achievements of Pasteur and his disciples led to the patients' bedside. As mortality and morbidity began to decline, post-Pasteurian society discovered that it was possible to walk out of a hospital alive and well. This also meant that medical know-how could be disseminated beyond the circle of initiates.

Pasteur himself went beyond disease. To him, living matter was more than an object of science; he wanted to transform everyday life. More than a technique, pasteurization is a symbol. Sanitation, antisepsis, disinfection, sterilization—from the mass-produced foods to sewage drains, our very reflexes have changed.

In the end, the figure of Pasteur has so many facets that at times it seems impossible to integrate them all into the picture: the scientist at the bedside of suffering humanity, the creator of new biomedical disciplines, the industrial empire-builder of course; but also the nonconformist university professor, the obsessive hygienist who did not want to shake hands, the solitary researcher who nonetheless knew how to find excellent people to work with him.

Yet in a certain sense his path seems to have followed a straight line, for Pasteur never lost sight of a single problem. As a Romantic chemist and the founder of biology, he created the conditions for a new relationship between man and life, and therefore between man and death.

In April 1862, Louis Pasteur was not yet forty years old. So it was a young scientist who wrote a very long letter to the minister in charge of overseeing research, Gustave Rouland, to give an account of his work. He began by declaring that he was "almost embarrassed" to pick up the study at the point where it had been left by Lavoisier; he then recalled his first discoveries concerning the asymmetrical crystals and fermentation, and added:

> I have come to the conclusion that the destruction of organic matter is due mainly to the proliferation of microscopic beings endowed with the special ability to break down complex organic matter, to initiate slow combustion and to bind oxygen, abilities by which these beings become the most active agents of that necessary return to the atmosphere of everything that has had life of which I spoke earlier. I have demonstrated that the atmosphere within which we live constantly carries the germs of these microscopic beings, which are always ready to grow within dead matter, where they can fulfill the destructive role that sustains their own life. And if God had not seen to it that the organic laws governing the mutations of tissues and liquids in living bodies prevent their propagation, at least in the normal conditions of life and health, we would live under the constant threat of being overwhelmed by them. But as soon as the breath of life has been extinguished, there is no part of the plant or animal organism that they cannot use as their food. To sum up, after death has occurred, life reappears in a new form and endowed with new properties.[15]

These lines sum up Pasteur. Written at a decisive juncture in the direction of his entire research, as he was about to turn from fermentation to disease, they reveal his fascination with the continuation of life as it proceeds from destruction to destruction.

Thirty-three years later, on 28 September 1895, when Marie gently separated Pasteur's hands in order to slip a crucifix between his fingers, it was of little importance that this imputrescible piece of boxwood would be buried with the scientist's remains. Humanity continues to dream that his closed fists hold the last specks of dust of his tartaric acid crystals.

CHRONOLOGY

1822   27 December: Louis Pasteur born at 2 A.M. in Dole (Jura), as the third child of Jean-Joseph Pasteur and Jeanne-Etiennette Roqui.

1823   15 January: Pasteur baptized.

1825   10 March: Birth of Pasteur's sister Joséphine.

1826   The Pasteur family moves to Marnoz (Jura), where Jean-Joseph rents a tannery.

       15 January: Birth in Clermont-Ferrand of Marie Laurent, Pasteur's future wife.

       4 June: Birth of Pasteur's sister Emilie.

1827   The Pasteur family moves to a tannery on the banks of the Cuisance River at Arbois. Jean-Joseph buys this property in 1833 and lives there until his death.

1829   Emilie contracts encephalitis, which leaves her permanently impaired.

1831   September: Pasteur enters the primary school of Arbois.

       October: Pasteur witnesses the treatment of several victims of bites by rabid animals; the epidemic causes sixteen deaths in the region, four of them in the immediate vicinity of Arbois.

1838   Late October: Pasteur leaves for Paris in the company of his friend Jules Vercel; he is to continue his schooling as a boarding student at the Institution Barbet (3, impasse des Feuillantines). Suffering from extreme homesickness, he returns to Arbois by mid-November. On his return he executes his first pastels.

1839   Pasteur becomes a student in the final class of the collège royal of Besançon. Here he meets Jules Marcou and Charles Chappuis, son of the notary of Saint-Vit.

1840   29 August: Pasteur receives his bachelor of arts degree at Besançon.

       28 December: He is appointed teaching assistant at the Besançon collège. At the same time, he pursues his studies of special mathematics.

1841   August: Pasteur fails the examination for the bachelor of science degree. He enrolls for a second year of special mathematics at Besançon.

1842   13 August: Pasteur receives his bachelor of science degree at Dijon, despite a mediocre grade in chemistry. 26 August: He qualifies for the competitive entrance examination to the Ecole normale supérieure, but since he is ranked fifteenth of twenty-two candidates, he prefers not to take the oral examination and to put off a second attempt until the following year.

October: He pursues his studies in Paris, residing once again at the Institution Barbet. At the same time he takes courses at the collège Saint-Louis and attends the lectures of Jean-Baptiste Dumas at the Sorbonne.

1843   Summer: Pasteur receives a first prize for physics at the lycée Saint-Louis and is ranked fourth in the entrance competition for the Ecole normale supérieure.

1844   Pasteur enters Normale.

14 October: Biot presents to the Académie des sciences Mitscherlich's note on the sodium and ammonia paratartrates and tartrates that would be the starting point of Pasteur's career.

1845   Pasteur obtains the degree of licencié ès sciences (M.S.). He attracts the attention of Balard, who has just been appointed professor at the Ecole normale supérieure.

1846   Pasteur places third in the *agrégation* in physical science. He is appointed professor of physics at the collège of Tournon (Ardèche), but Balard chooses him as his graduate assistant (*préparateur*) for his chemistry courses. Pasteur will work in his laboratory until 1848.

28 August: Pasteur meets Auguste Laurent, with whom he begins to study crystallography.

1847   28 August: Pasteur defends his theses in chemistry and physics before the Faculty of Sciences in Paris. His physics thesis: *A study of the phenomena related to the rotational polarization of liquids. Application of the rotational polarization of liquids to the resolution of several problems in chemistry*. His chemistry thesis: *Research on the saturation capacity of arsenious acid. Studies on the potassium-, sodium-, and ammonia-arsenites*.

1848   15 May: Pasteur's first report to the Académie des sciences (presented by Balard): *On the relation that can exist between crystalline form and chemical composition, and on the cause of rotational polarization*.

21 May: Pasteur's mother dies at Arbois.

16 September: Pasteur is appointed professor of physics at the lycée of Dijon.

29 December: He is appointed acting professor of chemistry at Strasbourg University.

1849   20 January: Pasteur arrives in Strasbourg.

9 April: Second report on molecular dissymmetry to the Académie des sciences.

29 May: Pasteur marries Marie Laurent, daughter of the rector of Strasbourg University.

1850   2 April: Birth of Pasteur's first child, Jeanne.

30 September: Note to the Académie des sciences on his work of 1849 (the composition of racemic acid).

15 November: Death (at age 25) of Pasteur's sister Joséphine.

1851   The Pharmaceutical Society announces a competition and offers a prize to the person who will establish whether tartaric acids containing fully formed racemic acid exist and who will provide a precise account of the modalities by which tartaric acid changes into racemic acid.

25 August: Communication of a paper on aspartic and malic acids to the Académie des sciences.

8 November: birth of Pasteur's son Jean-Baptiste.

1852   21 August: Pasteur meets Mitscherlich and Rose in Biot's laboratory at the Collège de France.

September: Journey to Germany. Pasteur visits Mr. Fikentscher's factory at Zwickau (Saxony).

25 September: Pasteur arrives in Vienna and visits several tartaric acid factories in Austria.

1 October: Arrival in Prague.

6 October: Return to Strasbourg.

7 November: Pasteur is appointed to the chair of chemistry in the Faculty of Science of Strasbourg University.

1853   3 January: Note to the Académie des sciences on the origin of racemic acid.

17 March: Death of Pasteur's sister Emilie at the Ursuline convent in Voiteur (Jura).

1 June: Pasteur informs Biot that he is able to transform tartaric acid into racemic acid.

6 June: Biot communicates this discovery to the Académie des sciences.

12 August: Pasteur is made a chevalier in the imperial order of the Légion d'honneur.

1 October: Birth of Pasteur's daughter Cécile.

9 November: Pasteur receives the prize of the Pharmaceutical Society for the synthesis of racemic acid (1,500 francs).

1854   3 July: Report to the Académie des sciences on dimorphism in optically active substances.

2 December: Pasteur takes up his functions as professor of chemistry and dean of the new Faculty of Science at Lille.

1855   20 August: Communication of a paper on amyl alcohol to the Académie des sciences.

1856   30 June: Publication on the isomorphism of active and inactive substances exposed to polarized light.

November: First meeting with E. Bigo, an industrialist of Lille, who asks Pasteur's advice concerning the production of beet root alcohol. Beginning of the work on fermentation.

20 November: Pasteur becomes a candidate for the Académie des sciences, where a vacancy has occurred in the section of mineralogy and geology.

1 December: Pasteur receives the Rumford Medal of the London Royal Society for his work in crystallography.

1857   March: Pasteur loses the election to the Académie des sciences to Gabriel Delafosse.

3 August: Publication of a paper on the so-called lactic fermentation by the Society for the Advancement of Science of Lille. This publication can be considered the birth certificate of microbiology.

22 October: Pasteur is appointed administrator and director of scientific studies at the Ecole normale supérieure in Paris.

30 November: Pasteur presents a paper on lactic fermentation to the Académie des sciences.

21 December: Presentation of a second paper (on alcoholic fermentation) to the Académie des sciences.

1858   Pasteur installs a laboratory in the attic of the Ecole normale in Paris.

29 March: Paper on the fermentation of tartaric acid.

19 July: Birth of Pasteur's daughter Marie-Louise.

August: Pasteur, vacationing at Arbois, examines diseased wines and observes the presence of germs analogous to those found in lactic fermentation.

20 December: Félix-Archimède Pouchet, director of the Muséum d'histoire naturelle of Rouen, publishes his *Note on the plant and animal proto-organisms born spontaneously in artificial air or oxygen gas*, which becomes the starting point for Pasteur's work on spontaneous generation.

1859   28 February: Pasteur's letter to Pouchet concerning spontaneous generation.

10 September: Pasteur's daughter Jeanne (age 9) dies at Arbois.

Pasteur receives the prize for experimental physiology of the Académie des sciences.

1860   Publication of his paper on alcoholic fermentation (43 pages) in the *Annales de chimie et de physique*.

20 January: Pasteur presents the first detailed report on his research on molecular dissymmetry to the Chemical Society of Paris.

Pasteur obtains the prize in experimental physiology for the year 1859 (prix Montyon) given by the Académie des sciences following the report submmitted by Claude Bernard.

3 February: Second report on molecular dissymmetry to the Chemical Society of Paris.

August: Air samples collected at Arbois for the study of spontaneous generation.

20 September: Air samples collected at Chamonix.

1861   February: Discovery of anaerobic (oxygen-deprived) life.

19 May: Lecture on the doctrine of spontaneous generation to the Chemical Society of Paris.

3 July: Lecture to the Chemical Society of Paris on the organized corpuscles existing in the atmosphere; examination of the doctrine of spontaneous generation.

26 July: Publication in the bulletin of the Chemical Society of the complete set of Pasteur's findings on acetic fermentation (vinegar).

23 December: Pasteur receives the Jecker Prize of the Académie des sciences for his work on spontaneous generation (report submitted by E. Chevreuil).

1862   2 February: Death of Biot.

10 February: Paper on the mycodermas presented to the Académie des sciences. Role of these organisms in acetic fermentation explained.

7 July: Paper on an industrial process for vinegar production presented to the Académie des sciences.

1 December: Pasteur receives the Prix Alhumbert for his research on spontaneous generation (report submitted by C. Bernard).

8 December: Pasteur is elected to the Académie des sciences (mineralogy section).

1863   19 January: Jules Raulin is appointed Pasteur's graduate assistant (*agrégé-préparateur*) in the laboratory of the rue d'Ulm.

March: Napoleon III asks Pasteur to study wine and its diseases.

20 April: Paper to the Académie des sciences on the destruction of plant and animal matter after death.

29 June: Paper to the Académie des sciences on putrefaction.

24 July: Birth of Pasteur's daughter Camille.

18 November: Pasteur is appointed professor of geology, physics, and

chemistry at the Ecole des Beaux-Arts of Paris (a position he will keep until 1867).

7 December: Paper on the role of atmospheric oxygen in vinification.

1864   2 February: First class at the Ecole des Beaux-Arts.

7 April: Lecture given in the series of "Scientific Evenings at the Sorbonne" on spontaneous generation. Continued debate with Pouchet, Joly, and Musset.

18 June: Paper on the diseases of wine delivered to the Académie des sciences.

July: Pasteur sets up a laboratory for the study of wine at Arbois.

1865   1 May: Paper on a practical process for preserving and improving wine presented to the Académie des sciences. Beginning of pasteurization.

6 June: Pasteur leaves for Alès, where he is to study the diseases of the silkworm.

15 June: Jean-Joseph Pasteur dies at Arbois at the age of 74.

14 August: Paper on the heating of wine.

11 September: Camille Pasteur dies at the age of two.

25 September: First observations on the diseases of silkworms.

October–November: Outbreak of a cholera epidemic in Paris. Pasteur is appointed to a commission charged with investigating the disease.

29 November–6 December: Pasteur is invited to the Palace of Compiègne; has personal contact with Napoleon III and Eugénie.

1866   23 May: Cécile Pasteur dies at Chambéry at age twelve.

23 July: Paper on the diseases of the silkworm.

22 September: Publication of the *Etudes sur le vin*. Controversy over the priority of the techniques for heating wine.

7 November: Pasteur publishes an article on the scientific achievements of Claude Bernard in the *Moniteur*.

26 November: A second paper on the silkworm.

1867   16 March: Lister reports on chemical asepsis.

May: Pasteur studies *flacherie* at Alès.

1 July: Pasteur receives one of the Grand Prizes of the World's Fair for his work on vinous fermentation.

3 July: Pasteur expells the student Lallier from the Ecole normale (in connection with the "Affaire Sainte-Beuve").

July–August: Unrest at the Ecole normale. Closing of the school. Resignation of the directors, including Pasteur.

5 September: Pasteur requests and obtains the creation of a laboratory for

physiological chemistry at the Ecole normale. At the same time, he succeeds Balard in the chair of organic chemistry at the Sorbonne.

11 November: lecture on vinegar making given at Orleans to the manufacturers of the region.

Pasteur resigns his position at the Ecole des Beaux-Arts.

1868   Pasteur receives a degree of Doctor of Medicine *honoris causa* from Bonn University.

1 February: Publication of the pamphlet *Le Budget de la science*.

17 February: Publication of the *Etudes sur le vinaigre*.

April: Continuation of the work on the diseases of the silkworm at Alès.

15 August: Pasteur is promoted to commander of the Légion d'honneur.

19 October: Pasteur suffers a first stroke.

1869   February: Research on the diseases of the silkworm is resumed at Alès and Saint-Hippolyte-du-Fort.

October: Debate on the heating of wine with Vergnette-Lamotte.

December: Beginning of Pasteur's stay at Villa Vicentina near Trieste, an imperial estate where he is to run an experimental silkworm farm at the request of Napoleon III.

Report on the teaching of science in France.

1870   March: Publication of the *Etudes sur la maladie des vers à soie*.

Early July: Pasteur leaves the Tyrol.

8–10 July: Returns to France via Vienna, Munich (interview with Liebig), and Stuttgart.

12–14 July: Stay at Strasbourg.

27 July: Pasteur is appointed senator of the Empire, but the decree is never promulgated because of the war.

5 September: Pasteur leaves Paris for Arbois.

1871   18 January: After learning of the bombardment of the Muséum of Paris by the Prussians, Pasteur, who is living in Arbois, returns his diploma of *doctor honoris causa* to Bonn University.

24 January: Pasteur travels from Arbois to Pontarlier in search of his son Jean-Baptiste, a corporal in Bourbaki's army.

March–April: Stays at Geneva, then Lyon.

May–August: Stay at Clermont-Ferrand with Duclaux. Visit to the Kuhn brewery at Chamalières. First research about beer.

24 June: Pasteur takes out a patent for a special beer-making process.

Mid-September: Pasteur travels to London to pursue his study of beer at the great English breweries. He meets Tyndall, who mentions Lister to him.

1872   Dispute with Frémy on the origin of the ferments.

June: renewed dispute with Vergnette-Lamotte.

25 October: Pasteur applies for early retirement as professor at the Sorbonne.

1873   25 March: Pasteur is elected to the Académie de médecine.

31 March: Pasteur is made a commander in the Brazilian Order of the Rose.

17 November: Having spent time at the Tourtel Brewery at Tantonville (Meurthe-et-Moselle), Pasteur makes public a new process that prevents the alteration of beer.

1874   February: First letters between Lister and Pasteur.

12 July: The National Assembly votes to grant Pasteur a National Reward (committee headed by Paul Bert).

8 August: Pasteur speaks at the award ceremony at the lycée of Arbois.

15 October: Jean-Baptiste Pasteur marries Jeanne Boutroux.

November: The Royal Society of London awards its Copley medal to Pasteur for his work on fermentation.

1875   June–October: Installation of a laboratory for the study of fermentation at Arbois.

December: Charles Chamberland joins the laboratory in the rue d'Ulm.

1876   January: Pasteur loses the election for a senate seat.

February: Correspondence with Tyndall about spontaneous generation.

June: Publication of the *Etudes sur la bière*.

3 July: Paper about the fermentation of urine.

Mid-September: Pasteur represents France at the International Silk-Growers' Convention in Milan.

1877   8 January: Dispute with Bastian about the alteration of urine.

30 April: Paper on anthrax presented to the Académie des sciences.

16–17 July: Paper on septicemia.

August: Dispute with Colin over the virulence of anthrax blood.

1878   10 February: Death of Claude Bernard.

March: Debates with Colin on the etiology of anthrax.

May: Journey to Italy (Milan, Lago Maggiore, Lugano).

8 July: Note on chicken cholera.

September: Experiments with anthrax conducted at Jules Maunory's farm near Chartres with the help of the veterinarian Vinsot.

October: Pasteur is promoted to grand officer of the Légion d'honneur.

November: Pasteur refutes a posthumous article of Claude Bernard on alcoholic fermentation.

Dispute with Berthelot.

December: Emile Roux joins Pasteur's laboratory.

1879　4 March: Debate on the plague in the Near East.

11 March: Paper on puerperal septicemia.

September: Discovery of a vaccine obtained from attenuated cultures.

October–November: Debates with Colin on the etiology of anthrax.

4 November: Marie-Louise Pasteur marries René Vallery-Radot.

1880　22 January: Pasteur is appointed to the Central Society of Veterinary Medicine.

9 February: Paper on virulent diseases, in which the principle of virus vaccines is enunciated for the first time.

26 April: Paper on chicken cholera.

3 May: Paper in which the germ theory is extended to the etiology of boils, osteomyelitis, and puerperal fever.

12 July: Paper on the role of earthworms in the propagation of anthrax.

30 July: Death of Pasteur's sister Virginie Vichot.

3 August: Pasteur threatens to resign from the Académie de médecine.

17 September: Birth of Pasteur's granddaughter Camille Vallery-Radot.

5 October: At the Académie de médecine, Jules Guérin challenges Pasteur to a duel.

December: Pasteur begins to work on rabies.

1881　28 February: Paper on the attenuation of viruses followed by renewed virulence.

March: Roux inoculates rabies into dogs by means of trepanation.

May: Experiments with anthrax vaccination at the farm of Pouilly-le-Fort near Melun.

2 June: Death of Emile Littré.

13–14 June: Reports on the experiments of Pouilly-le-Fort to the Académie des sciences and the Académie de médecine.

27 June: Pasteur announces his candidacy for the seat at the Académie française left vacant by the death of Littré.

4 July: Death of Sainte-Claire Deville.

July: Pasteur receives the Grand-Croix of the Légion d'honneur.

8 August: Paper on the virus vaccines (for chicken cholera and anthrax) presented at the International Congress of Medicine in London.

Late September: Pasteur journeys to Bordeaux to study yellow fever.

8 December: Pasteur is elected to Littré's seat at the Académie française.

1882　Pasteur is received at the Académie française; speeches by Ernest Renan and Pasteur.

May: Antianthrax vaccinations at Nîmes.

15 September: Paper on the attenuation of viruses to the Public Health Congress at Geneva. Dispute with Robert Koch.

30 October: Pasteur's nephew Adrien Loir goes to work in the laboratory of the rue d'Ulm as assistant préparateur.

11 November: Paper on cattle pleuropneumonia.

15 November–4 December: Pasteur studies swine erysipelas at Bollène.

25 December: Open letter to Robert Koch (published in January 1883).

December: First dog made refractory to rabies.

1883   15 March: Discovery of the agent of swine erysipelas.

April: Disputes with Michel Peter.

21 May: Response to the criticism of anthrax vaccination voiced by the committee of the Veterinary School of Turin.

13 July: Second National Reward granted to Pasteur (committee headed by P. Bert and J. Méline).

14 July: Pasteur speaks at Dole on the occasion of the placing of a commemorative plaque on the house of his birth.

August: Pasteur dispatches Roux, Straus, Nocard, and Thuillier to Egypt for the purpose of studying cholera.

18 September: Louis Thuillier dies in Alexandria.

26 November: First vaccination against swine erysipelas with an attenuated virus (developed in collaboration with L. Thuillier).

1884   28 January: Publication of the first biography of Pasteur: *Histoire d'un savant par un ignorant*, written (but not signed) by René Vallery-Radot.

February: Paper on rabies. Experimentation with the two-necked flask.

11 April: death of Jean-Baptiste Dumas.

18–20 April: Pasteur travels to Edinburgh for the tricentennial of the University.

19 May: Additional paper on rabies.

June: Experiments with refractory dogs for the rabies commission.

July: Straus and Roux dispatched to Toulon to conduct research on an outbreak of cholera, whose vibrio has been discovered by Koch.

10 August: Paper to the International Congress of Medicine in Copenhagen on the general principle of vaccination and preventive methods against rabies in humans.

August: A laboratory for the study of rabies, complete with animal pens, is set up at Villeneuve-l'Etang in the park of Saint-Cloud (village of Marnes-la-Coquette).

22 September: Letter of Pasteur to Pedro II of Brazil concerning the experimental use of the antirabies vaccine on humans.

1885   May–June: First trials of the antirabies vaccine in humans.

6 July: Antirabies vaccination of Joseph Meister.

12 July: Joseph Meister is considered out of danger.

20 October: Antirabies vaccination of Jean-Baptiste Jupille.

26–27 October: Papers on the methods of preventing rabies after bites have occurred presented to the Académie des sciences and subsequently to the Académie de médecine.

9 November: Antirabies treatment administered to Louise Pelletier.

6 December: Death of Louise Pelletier.

10 December: Joseph Bertrand delivers his inaugural address at the Académie française (eulogy of Jean-Baptiste Dumas).

December: Antirabies treatment administered to four American children sent from New York.

1886   January: First donations destined for the creation of an antirabies institute.

1 March: Official opening of a subscription for the founding of an antirabies vaccination institute.

March: Antirabies vaccination of nineteen Russians who have come from Smolensk.

11 May: Fundraising gala for the Institut Pasteur at the Trocadéro Palace organized by the "Scientia" Society.

13 May: Birth of Pasteur's grandson Louis Pasteur Vallery-Radot.

June: Gamaleïa and Metchnikoff set up an antirabies laboratory in Odessa.

July: Alexandre Yersin becomes Roux's assistant.

October: Antirabies treatment of Jules Rouyer.

26 November: Jules Rouyer dies. The child's father sues; the case is dismissed by the court in January 1887.

1 December: Pasteur leaves Paris for a long stay at Villa Bischoffsheim in Bordighera (Italy).

1887   January: First issue of the *Annales de l'Institut Pasteur*, a new review edited by Duclaux. Lively controversies over rabies carried out by Peter and certain foreign journalists.

23 February: Earthquake on the French and Italian Riviera. Pasteur leaves Bordighera and goes to Arbois, where he stays until April.

March: Grancher founds the *Bulletin médical*. Purchase of a building site in the rue Dutot (20th arrondissement of Paris), where the Institut Pasteur is to be built.

May: Official statutes of the Institut Pasteur.

18 July: Pasteur elected perpetual secretary of the Académie des sciences.

23 October: Second attack of paralysis, followed by recovery.

27 November: Pasteur's letter to the editor of *Le Temps* describing a procedure for the massive destruction of rabbits in response to an inquiry of the government of New South Wales.

Late December: Successful completion of the experimental infection of rabbits with the chicken cholera microbe on the property of the widow Pommery at Reims.

1888   January: Loir is sent to Australia to head a program of rabbit eradication.

14 November: Inauguration of the Institut Pasteur.

1889   October: Pasteur speaks at the unveiling of a statue of Jean-Baptiste Dumas at Alès; one of that town's streets is named for Pasteur.

1890   July: Albert Calmette becomes Roux's assistant at the Institut Pasteur.

1891   January: Founding of an Institut Pasteur at Saigon under the direction of Albert Calmette.

1892   September: A village in Algeria (near Constantine) is given Pasteur's name.

27 December: Pasteur's jubilee celebrated at the Sorbonne.

1893   Creation of an Institut Pasteur at Tunis.

1894   29 May: Address given at Lille on the occasion of a special meeting of the Society for the Advancement of Science.

June: Yersin identifies the plague bacillus at Hong Kong.

July–October: Pasteur's last stay at Arbois.

1 November: Pasteur suffers another stroke, which weakens him greatly.

1895   Reception for the students of the Ecole normale at the Institut Pasteur to celebrate the hundredth anniversary of the school.

12 May: Pasteur refuses to accept the Prussian Order of Merit.

13 June: Pasteur leaves the Institut and is driven to Villeneuve-l'Etang.

28 September: Pasteur dies at 4:20 P.M.

5 October: State funeral; mass at Notre-Dame of Paris.

NOTES

Chapter 1   My Father Is a Hero

1. Louis Pasteur, *Correspondance générale*, collected and annotated by Louis Pasteur Vallery-Radot (Paris: Flammarion, 1951), 1:21.
2. Ibid., 1:32.
3. *Hernani*, a play written by Victor Hugo in September 1829 and first performed on 25 February 1830. It was the occasion of a furious "battle" between "Classics" and "Romantics." During a popular uprising in response to the government's antirepublican measures, the police forces massacred a family in the rue Transnonain in Paris. This event inspired one of Daumier's most famous lithographs.
4. René Pasteur Vallery-Radot, *La Vie de Pasteur* (Paris: Hachette, 1900), 13.
5. Ibid., 10.
6. A very popular game of chance in 1830.
7. Robert Brichet, *Pasteur et le Sénat* (Besançon: Cètre, 1989), 19.
8. *Oeuvres de Pasteur*, edited by Louis Pasteur Vallery-Radot (Paris: Masson, 1922–39), vol. 7, "Mélanges scientifiques et littéraires," 360.
9. Ibid., 361.
10. Pasteur Vallery-Radot, *Vie de Pasteur*, 17.
11. Durand Gréville in the *Intermédiaire des chercheurs et des curieux* of 10 September 1888, cited by René Vallery-Radot, *Pasteur dessinateur et pastelliste* (Paris: Emile-Paul, 1912).
12. Pasteur, *Correspondance générale*, 1:20.
13. Ibid., 49, 34.
14. Ibid., 37.
15. Ibid., 58.
16. Immanuel Kant, *Prolégomènes à toute métaphysique qui pourra se présenter comme science* (1783). Original: *Prolegomena zu einer jeglichen Metaphysik die als Wissenschaft wird auftreten können* (Leipzig: Voss, 1878).
17. Pasteur, *Correspondance générale*, 1:26, 41, 37.
18. Ibid., 43.
19. Ibid., 54, 46.
20. Ibid., 48, 66.

21. Pasteur Vallery-Radot, *Vie de Pasteur*, 25.
22. Pasteur, *Correspondance générale*, 1:71.
23. Ibid., 110.
24. Ibid., 81.
25. Ibid., 77.
26. Ibid., 81.
27. Ibid., 82.
28. Ibid., 94.
29. Ibid., 99, 96.
30. Ibid., 98, 106, 123.

Chapter 2   Crystals: A New Law

1. Pasteur, *Correspondance générale*, 1:141.
2. Ibid., 148.
3. Jean Jacques and Claire Salomon-Bayet, eds., *Sur la dissymétrie moléculaire* (Paris: Bourgeois, 1986), 31.
4. *Correspondance générale*, 1:148.
5. Pasteur Vallery-Radot, *Vie de Pasteur*, 1:38.
6. "Dissymétrie moléculaire," in *Oeuvres de Pasteur*, 1:320.
7. Ibid., 410.
8. Société Philomatique de Paris, *Bulletin des Sciences* (December 1815): 176–83.
9. Cited in "Dissymétrie moléculaire," 310.
10. *Correspondance générale*, 1:155.
11. Ibid., 157.
12. "Thèse de Physique," in *Oeuvres de Pasteur*, 1:20.
13. Pasteur Vallery-Radot, *Vie de Pasteur*, 41.
14. *Correspondance générale*, 1:161.
15. Ibid., 164, 167.
16. Ibid., 172.
17. Ibid., 140.
18. Ibid., 172.
19. *Comptes rendus de l'Académie des Sciences* 19(1844):719–20.
20. "Dissymétrie moléculaire," 323.
21. Pasteur Vallery-Radot, *Vie de Pasteur*, 46.
22. "Dissymétrie moléculaire," 325.
23. *Comptes rendus de l'Académie des Sciences* (23 October 1848).
24. Ibid., 387.

Chapter 3   The Philosopher's Stone

1. *Correspondance générale*, 1:181.
2. Ibid., 101.
3. Ibid., 109.
4. "Dissymétrie moléculaire," 81–82, 196.
5. Pasteur Vallery-Radot, *Vie de Pasteur*, 57.
6. *Correspondance générale*, 1:228.
7. René Dubos, *Louis Pasteur, franc-tireur de la science,* French translation (Paris: PUF, 1955), 309. Original title: *Louis Pasteur, Free Lance of Science* (Boston: Little Brown, 1950).
8. *Correspondance générale*, 1:324.
9. Ibid., 323.
10. Ibid., 230.
11. Ibid., 234.
12. "Mélanges scientifiques et littéraires," 98.
13. *Correspondance générale*, 1:234.
14. Ibid., 248.
15. Ibid., 246.
16. Pasteur Vallery-Radot, *Vie de Pasteur*, 66.
17. *Correspondance générale*, 1:255.
18. Ibid., 256.
19. Ibid., 258.
20. Ibid., 293.
21. Ibid., 260.
22. Ibid., 275.
23. Ibid., 268.
24. Ibid., 273.
25. Ibid., 284.
26. Ibid., 284.
27. Ibid., 285.
28. Ibid., 297.
29. Ibid., 290.
30. Ibid., 295.

Chapter 4   The World of the Liquefied Forests

1. *Correspondance générale*, 1:325.
2. Pasteur Vallery-Radot, *Vie de Pasteur*, 79.

3. *Correspondance générale*, 1:315.

4. Ibid., 313.

5. Ibid.

6. Ibid., 315.

7. Ibid., 319.

8. Ibid., 320.

9. Ibid., 296.

10. "Dissymétrie moléculaire," 329.

11. Ibid., 241.

12. Pasteur Vallery-Radot, *Vie de Pasteur*, 79.

13. "Dissymétrie moléculaire," 362.

14. Ibid., 398.

15. *Correspondance générale*, 1:326.

16. Ibid., 329.

17. "Dissymétrie moléculaire," in *Oeuvres de Pasteur*, 1:375, 176.

Chapter 5   Fermentation and Life

1. Inaugural Lecture, University of Lille, 1 December 1854, in "Mélanges scientifiques et littéraires," 131.

2. "Dissymétrie moléculaire," 394.

3. "Mélanges scientifiques et littéraires," 215.

4. Ibid., 130.

5. *Correspondance générale*, 1:351.

6. Ibid., 372.

7. Ibid., 375.

8. Ibid., 363.

9. Ibid., 384.

10. *Papiers de Louis Pasteur*, Bibliothèque Nationale de France, n.a.fr. 10087–18095.

11. *Correspondance générale*, 1:412.

12. "Fermentations et générations dites spontanées," in *Oeuvres de Pasteur*, 2:81.

13. Quoted in Emile Duclaux, *Pasteur, histoire d'un esprit* (Sceaux: Charaire, 1896), 76. English translation by I. Smith and Florence Hedger, *Pasteur: The History of a Mind* (Philadelphia: W. B. Saunders, 1920), 57.

14. Antoine Laurent Lavoisier, "Mémoire sur la fermentation spiriteuse," in *Oeuvres de Lavoisier* (Paris, 1865), 3:780.

15. "Fermentations et générations dites spontanées," 52.

16. Justus von Liebig, *Traité de chimie organique* [French translation by P. Gerharot] (Paris: Fortin et Masson, 1840–41), 27.

17. Ibid., 27.

18. Chaptal, Jean Antoine Claude de, Comte, *Chimie appliquée aux arts* (Paris: chez Deterville, 1807), 2:538.

19. Eilhardt Mitscherlich, "Sur l'affinité chimique," *Annales de Physique et de Chimie*, 3rd series (1842): 34.

20. Jacques Nicolle, *Pasteur, sa vie, sa méthode, ses découvertes* (Paris: Marabout Université, 1969), 85.

21. "Etudes sur le vinaigre et le vin," in *Oeuvres de Pasteur*, 3:13.

22. "Dissymétrie moléculaire," 277.

23. "Etudes sur le vinaigre et le vin," 20.

24. Ibid., 16.

25. Ibid., 13.

26. *Correspondance générale*, 1:403, 414.

27. Ibid., 416.

28. Pasteur Vallery-Radot, *Vie de Pasteur*, 91, 422.

29. Ibid., 423.

30. "Fermentations et générations dites spontanées," 51.

31. *Papiers de Louis Pasteur*, Registres de laboratoire, Bibliothèque Nationale de France (Département des manuscrits), n.a.fr. 17923–18028.

32. "Fermentations et générations dites spontanées," 35.

33. Louis Pasteur Vallery-Radot, *Les Plus belles pages de Pasteur* (Paris: Flammarion, 1943), 71.

34. *Correspondance générale*, 2:103.

35. "Fermentations et générations dites spontanées," 47.

36. Ibid., 74.

37. Pasteur Vallery-Radot, *Vie de Pasteur*, 97.

38. "Fermentations et générations dites spontanées," 70.

39. Ibid., 414.

40. Ibid., 481.

Chapter 6   Rue d'Ulm Revisited

1. Pasteur Vallery-Radot, *Vie de Pasteur*, 93.

2. Ibid., 94.

3. Archives de l'Ecole Normale Supérieure.

4. *Correspondance générale*, 2:139.

5. Archives de l'Ecole Normale Supérieure.

6. Pierre Larousse, *Grand Dictionnaire universel du XIXᵉ siècle*, 200.

7. Archives de l'Ecole Normale Supérieure.

8. Ibid.

9. Ibid.

10. Nicolle, *Pasteur, sa vie*, 100.

11. *Correspondance générale*, 2:270.

12. "Mélanges scientifiques et littéraires," 163.

13. Pasteur Vallery-Radot, *Vie de Pasteur*, 124.

14. "Mélanges scientifiques et littéraires," 172.

15. Ibid., 177.

16. *Correspondance générale*, 2:58.

17. Ibid., 207.

18. Ibid., 197.

19. Ibid., 253.

20. Pasteur Vallery-Radot, *Vie de Pasteur*, 115.

21. Ibid., 117.

22. "Mélanges scientifiques et littéraires," 247.

23. J. Thomas, *Sainte-Beuve et l'Ecole Normale* (Paris: Les Belles Lettres, 1936), 151.

24. Ibid., 153.

25. Text published in the *Avenir National* of 2 July 1867.

26. Thomas, *Sainte-Beuve et l'Ecole Normale*, 167.

27. Pasteur, Unpublished letters (among them this draft), Archives de l'Ecole Normale Supérieure.

28. *Journal de Paris*, 8 July 1867.

29. Pasteur Vallery-Radot, *Les Plus belles pages*, 302.

30. *Correspondance générale*, 2:27.

31. Nicolle, *Pasteur, sa vie*, 77.

32. Duclaux, *Pasteur, histoire d'un esprit*, 43.

33. Adrien Loir, *A l'Ombre de Pasteur* (Paris: Le Mouvement Sanitaire, 1938), 157.

34. Ibid., 28.

35. Ibid., 157.

36. Pasteur Vallery-Radot, *Vie de Pasteur*, 191.

37. Ibid., 193.

38. Ibid., 199.

39. Ibid., 200.

40. Ibid., 202.

41. Ibid., 203.

42. *Correspondance générale*, 2:156.

43. Pasteur Vallery-Radot, *Vie de Pasteur*, 206.

Chapter 7   The So-called Spontaneous Generations

1. "Fermentations et générations dites spontanées," 191.

2. Ibid., 224.

3. Pasteur Vallery-Radot, *Vie de Pasteur*, 97.

4. "Fermentations et générations dites spontanées," 629.

5. Ibid., 294.

6. Ibid., 329.

7. Ibid., 330.

8. Ibid., 214.

9. Ibid., 288.

10. F. A. Pouchet, *Hétérogénie ou traité de la génération spontanée* (Paris: Baillière, 1859), 1.

11. Ibid., 2.

12. "Fermentation et générations dites spontanées," 335.

13. Ibid., 334.

14. F. Dagognet, *Méthodes et doctrine dans l'oeuvre de Pasteur* (Paris: PUF, 1967), 156.

15. Pasteur Vallery-Radot, *Vie de Pasteur*, 109.

16. Ibid., 110.

17. Ibid.

18. *Comptes rendus de l'Académie des sciences* 57(1863): 902.

19. "Fermentation et générations dites spontanées," 326.

20. Ibid., 333.

21. Ibid., 342.

22. *Correspondance générale*, 2:127.

23. Pasteur Vallery-Radot, *Vie de Pasteur*, 127.

24. Ibid., 128.

25. J. Rostand, *La Genèse de la vie. Histoire des idées sur la génération spontanée* (Paris: Hachette, 1943), 139.

26. "Fermentation et générations dites spontanées," 375.

27. Ibid., 398.

28. "Maladies virulentes, virus-vaccins et prophylaxie de la rage," in *Oeuvres de Pasteur*, 6:42.

29. Pasteur Vallery-Radot, *Vie de Pasteur*, 127.
30. "Mélanges scientifiques et littéraires," 30.

Chapter 8 The Microscope and the Silkworm

1. *Correspondance générale*, 2:193.
2. Ibid., 193.
3. Ibid.
4. "Etude sur la maladie des vers à soie," in *Oeuvres de Pasteur*, 4:308.
5. Loir, *A l'Ombre de Pasteur*, 58.
6. "Etude sur la maladie des vers à soie," 39.
7. Ibid., 48.
8. Loir, *A l'Ombre de Pasteur*, 57.
9. Duclaux, *Pasteur*, 117.
10. Pasteur Vallery-Radot, *Vie de Pasteur*, 154.
11. *Correspondance générale*, 2:10.
12. Duclaux, *Pasteur*, 183.
13. Pasteur Vallery-Radot, *Vie de Pasteur*, 154.
14. Duclaux, *Hommage à Pasteur*, 50.
15. *Correspondance générale*, 2:251.
16. Ibid., 256.
17. "Etude sur la maladie des vers à soie," 181.
18. Ibid., 393.
19. Ibid., 350.
20. *Correspondance générale*, 2:249.
21. Duclaux, *Pasteur*, 209.
22. Ibid., 210.
23. "Etude sur la maladie des vers à soie," 472.
24. Ibid., 99.
25. Duclaux, *Pasteur*, 215.
26. Ibid., 218.
27. Ibid., 220.
28. "Etude sur la maladie des vers à soie," 206.
29. Ibid., 238.
30. Ibid., 236.
31. *Correspondance générale*, 2:394.
32. Ibid., 399.
33. Pasteur Vallery-Radot, *Vie de Pasteur*, 206.

34. *Correspondance générale*, 2:410.
35. Pasteur Vallery-Radot, *Vie de Pasteur*, 210.
36. *Correspondance générale*, 2:416.
37. Ibid., 217.
38. Ibid.
39. Ibid., 425.
40. Pasteur Vallery-Radot, *Vie de Pasteur*, 220, 221.
41. *Correspondance générale*, 2:225.
42. Pasteur Vallery-Radot, *Vie de Pasteur*, 225.
43. *Correspondance générale*, 2:434.
44. Ibid., 459.
45. Ibid., 464.
46. Pasteur Vallery-Radot, *Vie de Pasteur*, 229.
47. *Oeuvres de Pasteur*, 4:3.

Chapter 9   Industrial Pasteurization

1. *Correspondance générale*, 2:129.
2. Ibid., 125.
3. Ibid., 216.
4. Ibid., 217.
5. Ibid., 219.
6. Ibid., 230.
7. Ibid., 226.
8. Ibid., 232.
9. Ibid., 234.
10. "Etude sur le vinaigre et le vin," 145.
11. Ibid., 146.
12. Nicolle, *Pasteur, sa vie*, 141.
13. Ibid., 152.
14. *Papiers de Louis Pasteur*, "Registres de laboratoire," 17923–18028, Bibliothèque Nationale de France (Département des manuscrits), n.a.fr. 17923–18028.
15. "Etude sur le vinaigre et le vin," 172.
16. Ibid., 191.
17. Ibid., 191.
18. Ibid., 218.
19. *Correspondance générale*, 2:278.

20. "Etude sur le vinaigre et le vin," 377.

21. Ibid., 259.

22. Ibid., 69.

23. Pasteur Vallery-Radot, *Vie de Pasteur*, 232.

24. *Papiers de Louis Pasteur*, "Coupures de presse et correspondances diverses, 1870–1871," Bibliothèque Nationale de France (Département des manuscrits), n.a.fr. 17983–84.

25. *Correspondance générale*, 2:492.

26. Ibid., 513, 514.

27. Ibid., 490.

28. Ibid., 504.

29. *Papiers de Louis Pasteur*, "Coupures de presse et correspondances diverses, 1870–1871," Bibliothèque Nationale de France (Département des manuscrits), n.a.fr. 17983–84.

30. Loir, *A l'Ombre de Pasteur*, 8.

31. "Mélanges scientifiques et littéraires," 216.

32. "Etude sur la maladie des vers à soie," 742–46.

33. *Correspondance générale*, 2:237.

34. Pasteur Vallery-Radot, *Les Plus belles pages*, 161.

35. *Correspondance générale*, 2:536.

36. Ibid., 544.

37. Pasteur Vallery-Radot, *Vie de Pasteur*, 278.

38. *Correspondance générale*, 2:549.

39. Pasteur Vallery-Radot, *Vie de Pasteur*, 283.

40. Duclaux, *Pasteur*, 238.

41. Pasteur Vallery-Radot, *Les Plus belles pages*, 168.

Chapter 10   A Chemist among the Doctors

1. "Maladies virulentes, virus-vaccins et prophylaxie de la rage," 448, 444.

2. Duclaux, *Pasteur*, 279.

3. "Maladies virulentes, virus-vaccins et prophylaxie de la rage," 161.

4. Ibid., 425.

5. *Cours de chimie* (Paris: Editions chez Delepine, 1713), 668.

6. "Mélanges scientifiques et littéraires," 8.

7. "Maladies virulentes, virus-vaccins et prophylaxie de la rage," 8.

8. *Correspondance générale*, 2:572.

9. "Maladies virulentes, virus-vaccins et prophylaxie de la rage," 167.

10. Pasteur Vallery-Radot, *Vie de Pasteur*, 306.

11. Ibid., 305.

12. Ibid., 312.

13. Ibid., 314.

14. Sherwin B. Nuland, *Héros de la médecine,* French translation (Paris: Presse de la Renaissance), 240. Original title: *Doctors: The Biography of Medicine* (New York: Knopf, 1988).

15. Ibid., 334.

16. Ibid., 339.

17. Joseph Lister, "On a New Method of Treating Compound Fractures . . . ," *Lancet* (1867): 364ff.

18. Nuland, *Héros,* 240.

19. René Dubos, *Pasteur and Modern Science,* ed. Thomas D. Brock (1960; new illus. ed., Madison, Wisc.: Science Tech, 1988), 90.

20. *Correspondance générale*, 2:578.

21. Ibid., 565.

22. Ibid., 565.

23. Pasteur Vallery-Radot, *Vie de Pasteur*, 299.

24. Ibid., 302, 123.

25. "Maladies virulentes, virus-vaccins et prophylaxie de la rage," 123.

26. *Papiers de Louis Pasteur*, "Registres de laboratoire," Bibliothèque Nationale de France (Département des manuscrits), n.a.fr., 17923–18028.

27. Pasteur Vallery-Radot, *Vie de Pasteur*, 362.

28. Ibid., 362.

29. Ibid., 322.

30. "Maladies virulentes, virus-vaccins et prophylaxie de la rage," 101.

31. Loir, *A l'Ombre de Pasteur*, 141.

32. Nicolle, *Pasteur*, 204.

33. R. Bichet, *Pasteur et le Sénat* (Lyon: Cêtre, 1989), 73.

34. Ibid., 73.

35. *Correspondance générale*, 2:627.

36. Ibid., 629.

Chapter 11   Anthrax in Sheep and Chickens

1. "Maladies virulentes, virus-vaccins et prophylaxie de la rage," 223.

2. Pasteur Vallery-Radot, *Vie de Pasteur*, 345.

3. "Maladies virulentes, virus-vaccins et prophylaxie de la rage," 184.

4. Ibid., 185.
5. Ibid., 185.
6. Ibid., 190.
7. *Correspondance générale*, 2:31.
8. "Maladies virulentes, virus-vaccins et prophylaxie de la rage," 193.
9. Ibid., 213.
10. Pasteur Vallery-Radot, *Vie de Pasteur*, 368.
11. Ibid., 366.
12. "Maladies virulentes, virus-vaccins et prophylaxie de la rage," 22.
13. Pasteur Vallery-Radot, *Vie de Pasteur*, 373.
14. Ibid., 399.
15. Loir, *A l'Ombre de Pasteur*, 52.
16. *Correspondance générale*, 2:599.
17. Ibid., 601, 582.
18. Ibid., 603.
19. *Correspondance générale*, 3:46.
20. Ibid., 103.
21. Ibid., 87.
22. Ibid., 119.

Chapter 12   Culture Broths

1. René Dubos, *Louis Pasteur, franc-tireur de la science* (Paris: PUF, 1955), 280.
2. "Etudes sur la bière," in *Oeuvres de Pasteur*, 5:245.
3. Loir, *A l'Ombre de Pasteur*, 14.
4. Ibid., 15.
5. Duclaux, *Pasteur*, 329.
6. Loir, *A l'Ombre de Pasteur*, 16.
7. Ibid., 41.
8. Ibid., 29.
9. Ibid., 23.
10. Ibid., 28, 27.
11. "Maladies virulentes, virus-vaccins et prophylaxie de la rage," 130, 133.
12. Ibid., 138.
13. Duclaux, *Pasteur*, 335.
14. "Maladies virulentes, virus-vaccins et prophylaxie de la rage," 493.
15. "Etude sur le vinaigre et le vin," 378.
16. "Maladies virulentes, virus-vaccins et prophylaxie de la rage," 294.

17. Loir, *A l'Ombre de Pasteur*, 41.

18. Ibid., 21.

19. "Maladies virulentes, virus-vaccins et prophylaxie de la rage," 608.

Chapter 13   *From One Academy to Another*

1. Claude Bernard, *Introduction à l'étude de la médecine expérimentale* (Paris: Baillière, 1865), 75.

2. Henri Mondor, *Grands médecins presque tous* (Paris: Corréa, 1943), 308.

3. Ibid., 310.

4. Ibid., 313.

5. Claude Bernard, *Lettres à Madame . . .* (Lyon: Fondation Mérieux, 1974), 30.

6. Pasteur Vallery-Radot, *Vie de Pasteur*, 161, 166.

7. L. Velluz, *Vie de Berthelot* (Paris: Plon, 1964), 224.

8. Loir, *A l'Ombre de Pasteur*, 63.

9. Pasteur Vallery-Radot, *Vie de Pasteur*, 371.

10. Ibid., 372.

11. "Etude sur le vinaigre et le vin," 567.

12. Dubos, *Louis Pasteur, franc-tireur*, 16.

13. Michel Méridiens, ed., *La Nécessité de Claude Bernard* (Paris: Klincksiek, 1891), 276.

14. L. Delhomme and Claude Bernard, *Principes de médecine expérimentale* (Paris: PUF, 1947), 25–26.

15. *Correspondance générale*, 3:211.

16. D. Hamburger, *Monsieur Littré* (Paris: Flammarion, 1988), 245.

17. Pasteur Vallery-Radot, *Vie de Pasteur*, 352.

18. "Maladies virulentes, virus-vaccins et prophylaxie de la rage," 333.

19. Hamburger, *Littré*, 89.

20. "Mélanges scientifiques et littéraires," 334, 336.

21. Ibid., 337.

22. Ibid., 293.

23. Pierre Larousse, *Grand Dictionnaire*, 1894 edition.

24. *Correspondance générale*, 3:198.

25. Ibid., 200.

26. Ibid., 211.

27. Ibid., 205.

28. *Papiers de Louis Pasteur*, "Notes de lectures," Bibliothèque Nationale de France (Département des manuscrits), n.a.fr. 18006–74.

29. "Mélanges scientifiques et littéraires," 326.
30. Ibid., 337.
31. Ibid., 347.
32. Ibid., 351.

Chapter 14   *The Wager of Pouilly-le-Fort*

1. Pasteur Vallery-Radot, *Vie de Pasteur*, 213.
2. *Correspondance générale*, 3:150.
3. "Maladies virulentes, virus-vaccins et prophylaxie de la rage," 378.
4. Jean-Louis Moreau, *Traité historique et pratique de la vaccine* (Paris: Bernard, An IX [1801]).
5. Edward Jenner, "An Inquiry into the Causes and Effects of Variolae Vaccinae . . ." (London: Sampson Low, 1798).
6. *Oeuvre complète du Dr. Jeunez*, trans. Jean-Joseph De Laroque (Ardèche: Privas, 1800), 9–12.
7. Mosseley, *Discussion historique et critique de la vaccine*, French translation (Paris, 1807), 203.
8. Loir, *A l'Ombre de Pasteur*, 102.
9. Auzias Turenne, *La Syphilisation* (Paris: Baillière, 1878), 712.
10. Ibid., 710.
11. Pasteur Vallery-Radot, *Vie de Pasteur*, 220.
12. Ibid., 221, 223.
13. *Correspondance générale*, 3:159.
14. Ibid., 169.
15. Pasteur Vallery-Radot, *Vie de Pasteur*, 227.
16. Ibid., 224.
17. "Maladies virulentes, virus-vaccins et prophylaxie de la rage," 329.
18. *Correspondance générale*, 3:143.
19. Pasteur Vallery-Radot, *Vie de Pasteur*, 261, 404.
20. Ibid., 405.
21. *Correspondance générale*, 3:179.
22. Ibid., 191.
23. Pasteur Vallery-Radot, *Vie de Pasteur*, 415.
24. *Correspondance générale*, 3:194.
25. Dubos, *Pasteur, franc-tireur*, 343.
26. *Correspondance générale*, 3:197.
27. Dubos, *Pasteur, franc-tireur*, 343.

28. *Correspondance générale*, 3:199.

29. Ibid., 224.

30. Ibid., 229.

31. Pasteur Vallery-Radot, *Vie de Pasteur*, 459.

32. Ibid., 424.

33. The experiments conducted in Hungary are reported in Charles Chamberland's book, *Le Charbon et la vaccination charbonneuse d'après les travaux de Monsieur Pasteur* (Paris: Bernard Tignol, 1883).

34. "Maladies virulentes, virus-vaccins et prophylaxie de la rage," 403.

35. Ibid., 404, 426.

36. Ibid., 425.

37. Ibid., 447.

38. Pasteur Vallery-Radot, *Vie de Pasteur*, 477.

39. Ibid., 482.

40. Ibid., 474.

41. Ibid., 475.

42. "Maladies virulentes, virus-vaccins et prophylaxie de la rage," 448.

43. Ibid., 450.

44. Pasteur Vallery-Radot, *Vie de Pasteur*, 459.

45. *Correspondance générale*, 3:319.

46. Ibid., 332.

Chapter 15   Rabies Must Be Defeated!

1. "Mélanges scientifiques et littéraires," 34.

2. Emile Roux, "L'Oeuvre médicale de Pasteur," *Agenda du chimiste* (1896): 76.

3. Claire Salomon-Bayet, *Pasteur et la révolution pastorienne* (Paris: Payot, 1986), 48.

4. "Maladies virulentes, virus-vaccins et prophylaxie de la rage," 555.

5. Ibid., 563.

6. Ibid., 572.

7. Ibid., 564.

8. *Papiers de Louis Pasteur*, "Registres de laboratoire," Cahier de l'été 1883, pp. 145 and 144bis, Bibliothèque Nationale de France (Département des manuscrits), n.a.fr., 18017.

9. "Maladies virulentes, virus-vaccins et prophylaxie de la rage," 598.

10. Pasteur Vallery-Radot, *Vie de Pasteur*, 520.

11. "Maladies virulentes, virus-vaccins et prophylaxie de la rage," 602.

12. Loir, *A l'Ombre de Pasteur*, 66.

13. "Maladies virulentes, virus-vaccins et prophylaxie de la rage," 604.

14. F. Rosset, *Pasteur et la rage* (Paris: Fondation Mérieux, 1985), 55.

15. Ibid., 83.

16. Pasteur Vallery-Radot, *Vie de Pasteur*, 429.

17. Ibid., 433.

18. Ibid., 434.

19. Rosset, *Pasteur et la rage*, 56.

20. Ibid., 76.

21. Ibid., 96.

22. *Correspondance générale*, 3:438.

23. *Papiers de Louis Pasteur*, "Registres de laboratoire," Bibliothèque Nationale de France (Département des manuscrits) n.a.fr.,17923–18028.

24. Ibid.

25. Ibid.

26. Pasteur Vallery-Radot, *Vie de Pasteur*, 292.

27. *Correspondance générale*, 4:26.

28. Pasteur, unpublished letters, Archives de l'Ecole Normale Supérieure.

29. Pasteur Vallery-Radot, *Vie de Pasteur*, 553.

30. *Correspondance générale*, 4:43.

31. Pasteur Vallery-Radot, *Vie de Pasteur*, 562.

32. *Correspondance générale*, 4:52.

33. Ibid., 51.

34. Pasteur Vallery-Radot, *Vie de Pasteur*, 585.

Chapter 16   The Capstone

1. "Maladies virulentes, virus-vaccins et prophylaxie de la rage," 627.

2. Dubos, *Pasteur, franc-tireur*, 353.

3. "Maladies virulentes, virus-vaccins et prophylaxie de la rage," 822.

4. *Correspondance générale*, 4:160.

5. "Maladies virulentes, virus-vaccins et prophylaxie de la rage," 810, 811.

6. Ibid., 824.

7. Dubos, *Pasteur, franc-tireur*, 350.

8. Loir, *A l'Ombre de Pasteur*, 77.

9. Ibid., 77.

10. Ibid., 79.

11. Salomon-Bayet, *Pasteur et la révolution pastorienne*, 61.

12. *Correspondance générale*, 6:169.

13. Loir, *A l'Ombre de Pasteur*, 87.

14. "Maladies virulentes, virus-vaccins et prophylaxie de la rage," 657.

15. Ibid., 657, 658.

16. *Correspondance générale*, 4:19.

17. "Maladies virulentes, virus-vaccins et prophylaxie de la rage," 619.

18. Pasteur Vallery-Radot, *Vie de Pasteur*, 575.

19. *Correspondance générale*, 4:65.

20. Ibid., 68.

21. Pasteur Vallery-Radot, *Vie de Pasteur*, 579.

22. "Maladies virulentes, virus-vaccins et prophylaxie de la rage," 637.

23. "Mélanges scientifiques et littéraires," 419.

Chapter 17   The Last Life Sciences

1. *Correspondance générale*, 4:222.

2. Loir, *A l'Ombre de Pasteur*, 106.

3. Ibid., 131.

4. "Maladies virulentes, virus-vaccins et prophylaxie de la rage," 301.

5. *Correspondance générale*, 4:255.

6. Ibid., 354.

7. A. M. Moulin, *Le Dernier langage de la médecine. Histoire de l'immunologie de Pasteur au SIDA* (Paris: PUF, 1991), 54.

8. Louis Pasteur Vallery-Radot, *Pasteur, son esprit, son caractère. Un chapitre des héros de l'esprit français* (Paris: Amyot-Dumont, 1952), 32.

9. H. Mollaret, *Alexandre Yersin* (Paris: Fayard, 1985), 50.

10. Nicolle, *Introduction à la médecine expérimentale* (Paris: Alcan, 1932), 19.

11. "Mélanges scientifiques et littéraires," 427.

12. *Correspondance générale*, 4:337.

13. Ibid., 348.

14. Ibid., 365.

15. *Correspondance générale*, 2:103.

BIBLIOGRAPHY

Pasteur's Works

These works appear in chronological order and in the original French, with translations of titles supplied.

Thèses de physique et de chimie présentées à la Faculté des sciences de Paris [Theses on Physics and Chemistry Presented to the Faculty of Sciences at Paris]. Bachelier, 1847.

"Mémoire sur la relation qui peut exister entre la forme cristalline et la composition chimique et sur la cause de la polarisation rotative" [Note on the Relation Which Can Exist between the Form of a Crystal and Its Chemical Composition and on the Causes of Rotary Polarization]. *Comptes rendus de l'Académie des sciences* 26 (1848).

"Etude sur les modes d'accroissement des cristaux" [Study on the Modes of Growth in Crystals]. *Annales de chimie et de physique* 49 (1856).

"Mémoire sur la fermentation appelée lactique" [Note on "Lactic" Fermentation]. *Comptes rendus de l'Académie des sciences* 45 (1857).

"Mémoire sur la fermentation de l'acide tartrique" [Note on the Fermentation of Tartaric Acid]. *Comptes rendus de l'Académie des sciences* 46 (1858).

"Expériences relatives aux générations dites spontanées" [Experiments Relative to Spontaneous Generations]. *Comptes rendus de l'Académie des sciences* 50 (1860).

"Des générations spontanées" [On Spontaneous Generations]. *Bulletin de la Société de chimie* 2 (1860).

"Animalcules infusoires vivant sans gaz oxygène libre et déterminant des fermentations" [Infusorial Microorganisms Living without Free Oxygen and Causing Fermentation]. *Comptes rendus de l'Académie des sciences* 51 (1861).

"Sur les fermentations acétique et butyrique" [On Acetic and Butyric Fermentations]. *Bulletin de la Société de chimie* 3 (1862).

"Mémoire sur les corpuscules organisés qui existent en suspension dans l'atmosphère. Examen de la doctrine des générations spontanées" [Note on Organized Corpuscles Which Exist in Suspension in the Atmosphere.

An Examination of the Doctrine of Spontaneous Generations]. *Comptes rendus de l'Académie des sciences* 54 (1862).

"Etude sur les vins. De l'influence de l'oxygène de l'air dans la vinification" [Study on Wine. On the Influence of Oxygen in the Air on Winemaking]. *Comptes rendus de l'Académie des sciences* 57 (1863).

"Des générations spontanées. Soirées scientifiques de la Sorbonne" [On Spontaneous Generations. Scientific Evenings at the Sorbonne]. *Revue scientifique* 1 (1864).

"Lavoisier." *Le Moniteur universel*, 4 September 1865.

"Sur l'emploi de la chaleur comme moyen de conservation du vin" [On the Use of Heat as a Means of Preserving Wine]. *Comptes rendus de l'Académie des sciences* 61 (1865).

"Claude Bernard." *Le Moniteur universel*, 7 November 1866.

Etudes sur le vin, ses maladies, causes qui les provoquent, procédés nouveaux pour le conserver et le vieillir [Studies on Wine, Its Disorders, Causes Which Provoke Them, New Procedures for Its Preservation and Aging]. Imprimerie impériale V. Masson, 1866.

"Le Budget de la science" [The Budget for Science]. *Revue des cours scientifiques*, 1 February 1868.

Etudes sur le vinaigre, sa fabrication, ses maladies [Studies on Vinegar, Its Manufacture, Its Disorders]. Gauthier-Villars, 1868.

Etudes sur la maladie des vers à soie [Studies on Silkworm Diseases]. Gauthier-Villars, 1870.

"Des forces dissymétriques naturelles" [Natural Dissymmetrical Forces]. *Comptes rendus de l'Académie des sciences* 78 (1874).

"Observations sur la méthode de traitement des amputés de M. A. Guérin" [Observations on the Method of Treating M. A. Guerin's Amputees]. *Comptes rendus de l'Académie des sciences* 80 (1875).

Etudes sur la bière, ses maladies, causes qui les provoquent, procédés pour la rendre inaltérable avec une théorie nouvelle de la fermentation [Studies on Beer, Its Disorders, What Causes Them, Procedures for Keeping Beer from Deteriorating, with a New Theory of Fermentation]. Gauthier-Villars, 1876.

"La Théorie des germes et ses applications à la médecine et à la chirurgie" [The Theory of Germs and Its Application to Medicine and Surgery]. *Comptes rendus de l'Académie des sciences* 86 (1878); en collaboration avec Joubert et Chamberland.

"Sur la théorie de la fermentation, à l'occasion d'un article publié dans la *Revue scientifique* sous le titre de 'La fermentation alcoolique. Dernières expériences de Claude Bernard'" [On the Theory of Fermentation, in Reply to an Article Published in the *Revue scientifique* Entitled "Alcoholic Fermentation: Recent Experiments of Claude Bernard"]. *Comptes rendus de l'Académie des sciences* 87 (1878).

Examen critique d'un écrit posthume de Claude Bernard sur la fermentation [A Critical Examination of a Posthumous Work of Claude Bernard on Fermentation]. Gauthier-Villars, 1879.

"De la septicémie puerpérale" [On Puerperal Sepsis]. *Bulletin de l'Académie de médecine* 8 (1879).

"De l'atténuation du virus du choléra des poules" [On the Mitigation of the "Hen Cholera Virus"]. *Comptes rendus de l'Académie des sciences* 91 (1880).

Sur les maladies virulents et en particulier sur la maladie appelée vulgairement choléra des poules [On Virulent Diseases and in Particular on the Disease Popularly Called Hen Cholera]. Masson, 1880.

"De la possibilité de rendre les moutons réfractaires au charbon par la méthode des inoculations préventives" [On the Possibility of Rendering Sheep Resistant to Anthrax Using Preventive Inoculations] et "Compte rendu sommaire des expériences faites à Pouilly-le-Fort sur la vaccination charbonneuse" [Summary Account of Experiments at Pouilly-le-Fort on Anthrax Vaccinations]. *Comptes rendus de l'Académie des sciences* 92 (1881); en collaboration avec Roux et Chamberland.

"Sur la rage" [On Rabies]. *Comptes rendus de l'Académie des sciences* 92 (1881); en collaboration avec Chamberland, Roux et Thuillier.

"Des virus-vaccins" [Virus Vaccines]. *Revue scientifique* 4 (1881).

Discours de réception de M. Louis Pasteur à l'Académie française. Réponse de M. Ernest Renan [Acceptance Discourse by M. Louis Pasteur to the Académie Française. Response by M. Ernest Renan]. Calmann-Lévy, 1882.

"La Vaccination charbonneuse. Réponse à un Mémoire de M. Koch" [The Anthrax Vaccine: Response to a Note of M. Koch]. *Revue scientifique* 5 (1883).

Réponse au discours de réception de M. Joseph Bertrand à l'Académie française [Response to the Acceptance Discourse of M. Joseph Bertrand to the Académie Française]. Didot, 1885.

"Méthode pour prévenir la rage après morsure" [Method for Preventing Rabies after Animal Bites]. *Comptes rendus de l'Académie des sciences* 101 (1885).

Le Traitement de la rage [The Treatment of Rabies]. Marpon et Flammarion, 1886.

"Lettre à M. Duclaux sur la rage" [Letter to Mr. Duclaux on Rabies]. *Annales de l'Institut Pasteur* 1 (1887).

"Sur la destruction des lapins en Australie et dans la Nouvelle-Zélande" [On the Destruction of Rabbits in Australia and New Zealand]. *Annales de l'Institut Pasteur* 2 (1888).

"Discours prononcé dans le grand amphithéâtre de la Sorbonne le 27 décembre 1892" [Lecture Given in the Grand Amphitheater of the Sorbonne on 27 December 1892]. *Jubilé de M. Pasteur*, Gauthier-Villars, 1892.

"Note remise à Son Excellence le ministre de l'Instruction publique et des Cultes sur sa demande par M. Pasteur, directeur des Etudes Scientifiques de l'Ecole normale supérieure (avril 1862)" [Note Submitted, at His Request, to His Excellency the Minister of Public Education and Religion by M. Pasteur, Director of Scientific Studies at the Ecole Normale Supérieure (April 1862)]. Note inédite découverte en 1923 par M. Dupuy, secrétaire de l'Ecole normale supérieure [unpublished note discovered in 1923 by M. Dupuy, secretary of the Ecole normale supérieure]. Archives nationales, 1923.

"Trois plis cachetés de Louis Pasteur" [Three Sealed Letters of Louis Pasteur]. *Comptes rendues de l'Académie des sciences*, 1964.

Sur la dissymétrie moléculaire [On Molecular Dissymmetry]. Recueil de textes de L. Pasteur, J.-H. van t'Hoof et A. Werner, présentés par J. Jacques et C. Salomon-Bayet. Christian Bourgois, 1986.

For a complete rendering of Pasteur's writings, see:

Oeuvres de Pasteur, réunies par Louis Pasteur Vallery-Radot, Masson, 1922–1939. Vol. 1, *Dissymétrie moléculaire* [Molecular Dissymmetry]; vol. 2, *Fermentations et générations dites spontanées* [Fermentation and So-called Spontaneous Generations]; vol. 3, *Etudes sur le vinaigre et sur le vin* [Studies on Vinegar and on Wine]; vol. 4, *Etudes sur la maladie des vers à soie* [Studies on Silkworm Diseases]; vol. 5, *Etudes sur la bière* [Studies on Beer]; vol. 6, *Maladies virulents, virus-vaccins et prophylaxie de la rage* [Virulent Diseases, Virus Vaccines, and Rabies Prophylaxis]; vol. 7, *Mélanges scientifiques et littéraires* [Miscellanies, Scientific and Literary].

Correspondance générale, réunie et annotée par Louis Pasteur Vallery-Radot. 4 vols. Flammarion, 1951.

Works on Pasteur

Published in Paris unless otherwise specified.

Bichet, R. *Pasteur et le Sénat*. Lyon: Cêtre, 1989.

Bournand, F. *Un bienfaiteur de l'humanité. Pasteur, sa vie, son oeuvre*. Tolra, 1896.

Boutet, J.-F. *Pasteur et ses élèves*. Garnier, 1898.

Blaringhem, L. *Pasteur et le transformisme*. Masson, 1923.

Caddeder, A. "Pasteur et le choléra des poules. Révision critique d'un récit historique." *History and Philosophy of the Life Sciences* 7 (1985).

Calmette, A. "Pasteur et les Instituts Pasteur." *Revue d'hygiène* 45 (1923).

Cochlin, D. *Quatre Français*. Hachette, 1912.

Cuny, H. *Louis Pasteur et le mystère de la vie*. Seghers, 1963.

Dagognet, F. *Méthodes et doctrine dans l'oeuvre de Pasteur*. PUF, 1967.

———. *Pasteur sans la légende*. Synthélabo, 1994.

Delaunay, A. *L'Institut Pasteur des origines à aujourd'hui*. France-Empire, 1962.

———. *Pasteur et la microbiologie*. PUF, 1962.

———. *Présence de Pasteur*. Fayard, 1973.

Diara, A. *Un débat français vu par la presse (1858–1869): L. Pasteur, F.-A. Pouchet et la génération spontanée*. Actes du Muséum de Rouen, 1984.

Dubos, R. *La Leçon de Pasteur*. Albin Michel, 1988.

———. *Louis Pasteur franc-tireur de la science*. PUF, 1955.

Duclaux, E. "Le Laboratoire de Monsieur Pasteur." *Le Centenaire de l'Ecole normale 1795–1895*. Hachette, 1895.

———. "Le Laboratoire de Monsieur Pasteur." *Centième anniversaire de la naissance de Pasteur*. Institut Pasteur, Hachette, 1922.

———. *Pasteur. Histoire d'un esprit*. Sceaux: Charaire, 1896.

Epardant, E. *Pasteur glorifié à l'écran. Le Film du centenaire*. Tedesco, 1922.

Farley, J., and G. Geison. "Science, Politics, and Spontaneous Generation in 19th Century France. The Pasteur-Pouchet Debate." *Bulletin of the History of Medicine* 48 (1974); French translation in B. Latour et M. Callon, *La Science telle qu'elle se fait*. La Découverte, 1991.

Fleury, M. de. *Pasteur et les Pastoriens*. Rueff, 1895.

Gascar, P. *Du côté de chez Monsieur Pasteur*. Odile Jacob, 1986.

Geison, G. "Pasteur." In Gillespie (Ch. C.), *Dictionary of Scientific Biography*. Vol. 10. New York: Scribner, 1974.

Jacques, J. *Pasteur et son double*. Hachette, 1992.

Kopaczewski, W. *Pasteur et la bactériologie*. Rabat, 1945.

Latour, B. *Pasteur, Bataille contre les microbes*. Fernand Nathan, 1985.

———. "Pasteur et Pouchet, hétérogenèse de l'histoire des sciences." In Michelle Serres, *Eléments d'histoire des sciences*. Bordas, 1989.

Loir, A. *A l'Ombre de Pasteur*. Le Mouvement sanitaire, 1938.

Metchnikov, E. *Trois fondateurs de la médecine moderne: Pasteur, Lister, Koch*. Alcan, 1933.

Mondor, H. *Pasteur*. Corréa, 1945.

Nicol, L. *L'Epopée pastorienne et la médecine vétérinaire*. Chez l'auteur, 1974.

Nicolle, J. *Un maître de l'enquête scientifique. Louis Pasteur*. La Colombe, 1953.

——. *Pasteur, sa vie, sa méthode, ses découvertes*. Marabout Université, 1969.

Parker, S. *Louis Pasteur et les microbes*. Sorbier, 1993.

Pasteur: Documents sur sa vie et son oeuvre. Donation de Louis Pasteur Vallery-Radot. Bibliothèque Nationale, 1964.

Pasteur Vallery-Radot, L., R. Legroux, and M. Schoer. *Microbiologie: l'Oeuvre de Pasteur et ses conséquences*. Palais de la découverte: Masson, 1937.

Pasteur Vallery-Radot, L. *Images de la vie et de l'oeuvre de Pasteur*. Documents photographiques, Flammarion, 1956.

——. "Les Instituts Pasteur d'outre-mer." *La Presse Médicale* 21 (1939).

——. *Louis Pasteur inconnu*. Flammarion, 1954.

——. *Madame Pasteur*. Flammarion, 1941.

——. *Pasteur, son esprit, son caractère. Un chapitre des héros de l'esprit français*. Amyot-Dumont, 1952.

——. *Les Plus belles pages de Pasteur*. Flammarion, 1941.

Pennetier, G. *Pouchet et Pasteur*. Actes du Muséum d'histoire naturelle de Rouen, Giriend, 1907.

Perreux, G. *Pasteur au pays d'Arbois. Histoire d'un Arboisien par un Arboisien*. Dole: Presses jurassiennes, 1962.

Préville, H. de. *Pasteur, un médecin sans diplôme*. Tolra, 1923.

Ramon, G. "Ce que Pasteur doit aux vétérinaires et ce que le médecin vétérinaire doit à Pasteur." *Revue médicale et vétérinaire* 112 (1936).

Rosset, R. *Pasteur et la rage*. Fondation Mérieux, 1985.

Rostand, J. *Hommes de vérité*. Stock, 1942.

Roux, E. "L'Oeuvre agricole de Pasteur." *Société nationale d'agriculture, compte rendu de la séance solenelle du 2 mars 1911*.

——. "L'Oeuvre médicale de Pasteur." *Agenda du chimiste*, 1896.

Salomon-Bayet, C. *Pasteur et la révolution pastorienne*. Payot, 1986.

Sédillot, C. "De l'influence des découvertes de Pasteur sur les progrès de la chirurgie." *Comptes rendus de l'Académie des sciences*, 1876.

Vallery-Radot, M. *Pasteur, un génie au service de l'homme*. Fauré, 1985.

Vallery-Radot, R. *La Vie de Pasteur*. Hachette, 1900.

[Vallery-Radot, R.] *M. Pasteur. Histoire d'un savant par un ignorant*. Hetzel, 1883.

——. *Pasteur dessinateur et pastelliste*. Emile-Paul, 1912.

Vallette, J. *Quelques remarques sur l'attitude religieuse de Louis Pasteur. Thèse présentée à la Faculté de Théologie protestante de Montauban*. Angoulême: Coquemard, 1904.

Wrotnowska, D. "Le Vaccin anticharbonneux. Pasteur et Toussaint d'après des documents inédits." *Histoire des sciences médicales* 12 (1978).

On the recent controversy over Pasteur and his methods, which Gerald Geison first unleashed at the meeting of the American Association for the Advancement of Science, Boston, January 1993, see:

Anderson, C. "Pasteur Notebooks Reveal Deception." *Science* 19 (February 1993).

Bynum, W. F. "The Scientist as Anti-Hero." *Nature* 4 (May 1995).

"L'Affaire Pasteur." *Le Nouvel Observateur*, 4 March 1993.

Makle, R. "Louis Pasteur, Genius, Pioneer—and a Cheat." *The Observer*, 14 February 1993.

Mallaval, C. "Louis Pasteur traité de tricheur." *Libération*, 3 March 1993.

Perutz, M. F. "The Pioneer Defended." *New York Review of Books*, 21 December 1995.

Porter, Roy. "Lion of the Laboratory." *Times Literary Supplement*, 16 June 1995.

Russell, C. "Louis Pasteur and Questions of Fraud." *Washington Post* (1 February 1993).

INDEX

Library of Congress Cataloging-in-Publication Data

Debré, P. (Patrice)
    Louis Pasteur / Patrice Debré : translated by Elborg Forster.
        p.   cm.
    Includes bibliographical references and index.
    ISBN 0-8018-5808-9 (alk. paper)
    1. Pasteur, Louis, 1822–95.   2. Science — History — 19th century.
3. Scientists — France — Biography.   I. Title.
Q143.P2D33 1998
509'.2 — dc21
    [B]   97-43686
                                                                    CIP